Mastering Windows Security and Hardening

Second Edition

Secure and protect your Windows environment from cyber threats using zero-trust security principles

Mark Dunkerley

Matt Tumbarello

BIRMINGHAM—MUMBAI

Mastering Windows Security and Hardening
Second Edition

Group Product Manager: Vijin Boricha
Publishing Product Manager: Prachi Sawant
Senior Content Development Editor: Sayali Pingale
Technical Editor: Nithik Cheruvakodan
Copy Editor: Safis Editing
Project Manager: Neil Dmello
Proofreader: Safis Editing
Indexer: Rekha Nair
Production Designer: Jyoti Chauhan
Senior Marketing Coordinator: Hemangi Lotlikar

First published: July 2020
Second edition: July 2022

Production reference: 1120722

Published by Packt Publishing Ltd.
Livery Place
35 Livery Street
Birmingham
B3 2PB, UK.

978-1-80323-654-4

www.packt.com

Contributors

About the authors

Mark Dunkerley is a cybersecurity and technology leader with over 20 years of experience working in higher education, healthcare, and Fortune 100 companies. Mark has extensive knowledge in IT architecture and cybersecurity through delivering secure technology solutions and services. He has experience in cloud technologies, vulnerability management, vendor risk management, identity and access management, security operations, security testing, awareness and training, application and data security, incident and response management, regulatory and compliance, and more. Mark holds a master's degree in business administration and has received certifications through (ISC)², AirWatch, Microsoft, CompTIA, VMware, AXELOS, Cisco, and EMC. Mark has spoken at multiple events, is a published author, sits on customer advisory boards, has published several case studies, and is featured as one of Security magazine's 2022 Top Cybersecurity Leaders.

> *Thank you to my wife, Robin, and children, Tyne, Isley, and Cambridge, for all your continued support. To my parents for shaping me into the person I am today. My brother for his ongoing service in the British Army. My co-author, Matt, for his dedication to the success of this book. To anyone I missed, thank you! Without you all, this book would have not been possible.*

Matt Tumbarello is a senior solutions architect. He has extensive experience working with the Microsoft security stack, Azure, Microsoft 365, Intune, Configuration Manager, and virtualization technologies. He also has a background working directly with Fortune 500 executives in a technical enablement role. Matt has published reviews for Azure security products, privileged access management vendors, and mobile threat defense solutions. He also holds several Microsoft certifications.

> *I would like to acknowledge all my peers, business partners, and coworkers, who continuously help me learn and grow. Without your support, this book wouldn't be possible.*

About the reviewer

Premnath Sambasivam is a server engineer with 10 years experience in Windows, Azure, VMware, and SCCM administration. He is an MCSE Cloud Platform, and Infrastructure certified professional and Microsoft certified Azure Architect. He has developed and deployed Microsoft System Center Configuration Manager solutions to manage more than 6,000 assets in his client's environment and various VMware solutions. Premnath is a technology enthusiast who loves learning and exploring new technologies. He is currently working as a Senior Cloud Engineer for one of the major retail brands in the USA. He reviewed the books *Mastering Windows Server 2019* and *Windows Server 2022 Fundamentals*, which is also published by *Packt Publishing*.

I want to thank my wife and son for encouraging me to spend time learning. Reviewing books also refreshes our memory and lets us learn about the latest technologies and software improvements. Special thanks to my mom and dad for always being supportive.

Table of Contents

3

Hardware and Virtualization

4

Networking Fundamentals for Hardening Windows

5
Identity and Access Management

Part 2: Applying Security and Hardening

6
Administration and Policy Management

7

Deploying Windows Securely

8

Keeping Your Windows Client Secure

9

Advanced Hardening for Windows Clients

10
Mitigating Common Attack Vectors

11
Server Infrastructure Management

12

Keeping Your Windows Server Secure

Part 3: Protecting, Detecting, and Responding for Windows Environments

13

Security Monitoring and Reporting

14

Security Operations

15
Testing and Auditing

16
Top 10 Recommendations and the Future

Index

Other Books You May Enjoy

Preface

Throughout this book, you will be provided with the knowledge needed to protect your Windows environment and the users that access it. The book will cover a variety of topics that go beyond the hardening of just the operating system to include the management of devices, baselining, hardware, virtualization, networking, identity management, security operations, monitoring, auditing, and testing. The goal is to ensure that you understand the foundation of, and multiple layers involved in, providing improved protection for your Windows systems.

Since this book focuses on security, it's important to understand the core principles that form an information security model. These principles are known as the **CIA** triad, which stands for **confidentiality**, **integrity**, and **availability**. If you have pursued a security certification, such as the CISSP or Security+ certifications, you will be very familiar with this model. If not, it is recommended that you familiarize yourself with them as a security professional. This book will not go into detail about the CIA triad, but the concepts provided in this book will support the foundation of ensuring the confidentiality, integrity, and availability of information on the Windows systems you manage. At a high level, CIA means the following:

- **Confidentiality** involves ensuring that no one other than those who are authorized to can access information.

- **Integrity** involves ensuring that the information being protected is original and has not been modified without the correct authorization.

- **Availability** involves ensuring that information is always available when access is needed.

This book is split into three sections to help guide you and provide the understanding and knowledge that's needed to implement a solid Windows security foundation within your organization. The first section will cover getting started and foundations for Windows security. The second section will focus on applying security and hardening with the third section providing information to protect, detect, and respond for Windows environments.

Who this book is for

This book is intended to educate the technical and security community, which includes the following roles:

- Microsoft security, cloud, and technical roles, such as engineers, analysts, architects, and administrators

- Anyone involved in the management of a Windows environment

- All technical-related security roles

- Technical/security managers and directors

What this book covers

Chapter 1, Fundamentals of Windows Security, introduces the security world within IT and the enterprise. It covers how security is transforming the way we manage technology and discuss threats and breaches. We will look at the challenges organizations currently face and discuss a concept known as zero trust.

Chapter 2, Building a Baseline, provides an overview of baselining and the importance of building a standard that's approved by leadership and adopted by everyone. We will cover what frameworks are and provide an overview of the more common frameworks used in securing and hardening an environment. We will then look at operational best practices within enterprises and cover the importance of change management to ensure that anything that falls outside the scope of policy receives the correct approvals.

Chapter 3, Hardware and Virtualization, provides an overview of physical servers and virtualization. The chapter will cover hardware certification, enhancements in hardware security, and virtualization-based security concepts to secure and harden devices, including overviews of BIOS, UEFI, TPM 2.0, and Secure Boot.

Chapter 4, Networking Fundamentals for Hardening Windows, provides an overview of networking components and their role in hardening and securing your Windows environment. You will learn about the software-based Windows Defender Firewall and how to configure it on Windows devices. Additionally, you will be provided with knowledge of network security technology from Microsoft as it relates to Windows VMs running in Azure.

Chapter 5, Identity and Access Management, provides a comprehensive overview of identity management and the importance it plays in securing Windows systems. Identity has become the foundation of securing users – this chapter will cover everything you need to do within the identity and access management area. We will provide details on account and access management, authentication, MFA, passwordless authentication, conditional-based access controls, and identity protection.

Chapter 6, Administration and Policy Management, provides details about different methods for the administration and modern management of Windows endpoints. You will be provided with the knowledge needed to ensure best practices are applied, looking at topics around enforcing policies and security baselines with Configuration Manager and Intune.

Chapter 7, Deploying Windows Securely, provides an overview of the end user computing landscape. We will discuss device provisioning, upgrading Windows, and building hardening images. You will learn about modern methods used to deploy Windows using Intune and Windows Autopilot and deploying images in virtualized Windows environments.

Chapter 8, Keeping Your Windows Client Secure, covers Windows clients and the concepts used to keep them secure and updated. You will learn how to stay updated with Windows Updates for Business, protect data with BitLocker encryption, enable passwordless sign-in with Windows Hello for Business, and how to enforce policies, configurations, and security baselines.

Chapter 9, Advanced Hardening for Windows Clients, provides a comprehensive review of advanced hardening configurations that are applied to Windows clients to protect enterprise browsers, secure Microsoft 365 apps, and apply zero-trust security principals to reduce the attack surface. You will learn advanced techniques for applying policies to third-party products using Intune, how to enable advanced features of Microsoft Defender to protect against unwanted apps and ransomware, and how to enable hardware-based virtualized isolation for Microsoft Edge and Office. You will also learn how to enable a removable storage access control policy to protect against data loss with removable media.

Chapter 10, Mitigating Common Attack Vectors, covers common attack techniques used by attackers to intercept communications and try to move laterally throughout the network. You will learn about different types of adversary-in-the-middle attacks and how to prevent them, as well as ways to protect against lateral movement and privilege escalation through Kerberos tickets. You will also learn about using Windows privacy settings to safeguard users' privacy from apps and services that run on Windows clients.

Chapter 11, Server Infrastructure Management, provides an overview of the data center and cloud models that are used today. We will then go into detail on each of the current models as they pertain to the cloud and review secure access management to Windows Server. We will also provide an overview of Windows Server management tools, as well as Azure services for managing Windows servers.

Chapter 12, Keeping Your Windows Server Secure, looks at the Windows Server OS and introduces server roles and the security-related features of Windows Server 2022. You will learn about the techniques used to keep your Windows Server secure by implementing **Windows Server Update Services** (**WSUS**), Azure Update Management, onboarding machines to Microsoft Defender for Endpoint, and enforcing a security baseline. You will also learn how to implement application control policies and PowerShell security.

Chapter 13, Security Monitoring and Reporting, talks about the different tools available to collect telemetry data, as well as insights and recommendations for securing your environment. This chapter will inform you about the ways in which to act on these recommendations. The technologies covered include Microsoft Defender for Endpoint, Azure Log Analytics, Azure Monitor, and Microsoft Defender for Cloud.

Chapter 14, Security Operations, talks about the **security operations center** (**SOC**) in an organization and discusses the various tools used to ingest and analyze data to detect, protect, and alert you to incidents. The technologies covered include **Extended Detection and Response** (**XDR**), the Microsoft 365 Defender Portal, Microsoft Defender for Cloud Apps, Defender for Cloud, Microsoft Sentinel, and Microsoft Defender Security Center. This chapter also talks about data protection with Microsoft 365 and the importance of ensuring that up-to-date business continuity and disaster recovery plans are in place.

Chapter 15, Testing and Auditing, discusses validating that controls are in place and enforced. You will learn about the importance of continual vulnerability scanning and the importance of penetration testing to ensure that the environment is assessed in terms of protecting against the latest threats.

Chapter 16, Top 10 Recommendations and the Future, provides recommendations and actions to take away after reading this book. It also provides some insight into the direction that device security and management is headed, as well as insights into our thoughts on the importance of security in the future.

To get the most out of this book

We will primarily focus on the most current versions of Windows available today, including Windows Server 2022, Windows 11, and the resources available within Microsoft Azure. We understand migrating to the latest Windows OS and shifting workloads from on-premises to the cloud is not an overnight task and may take years. In general, the concepts we provide throughout this book can be used within most configurations of Windows but could vary slightly depending on the build or version. Upgrading to the latest supported versions of Windows is critical to provide for the effective hardening of your systems and should be a driving factor to push your migrations forward. It is strongly encouraged to upgrade as soon as possible as Microsoft will no longer release security patches or offer support for deprecated versions.

To get the most out of this book, the following items will be needed to follow along with any provided examples. Thanks to cloud technology, you will be able to quickly enable an environment to build the infrastructure and foundation needed to support your journey throughout this book.

It is recommended that you set up an Office 365 subscription (add your own custom domain), which will in turn create an **Azure Active Directory** (**AAD**) tenant. Once the AAD tenant has been set up, this will allow you to add an Azure subscription to begin consuming Azure resources tied to your Office 365 subscription and your custom domains.

Office 365 E5 30-day free trial: `https://go.microsoft.com/fwlink/p/?LinkID=698279&culture=en-US&country=US`

Azure Account with $200 credit for 30 days: `https://azure.microsoft.com/en-us/free/`

Cloud subscriptions required:

- An Azure subscription
- Microsoft Enterprise E5 (M365 E5 includes Intune licensing, Microsoft Defender for Endpoint, and Windows Enterprise)
- An Intune subscription and license
- Windows 10 E3 or E5
- Enterprise Mobility + Security E3 or E5 (includes Azure AD Premium P2)

Permissions:

- Global administrator rights to your Office 365 subscription
- Owner role or appropriate RBAC to your Azure subscription to deploy resources
- Domain admin rights on your domain controller or equivalent rights to modify Group Policy

Azure resources:

- Azure VMs (Windows 11 and Windows Server 2022 Core and Desktop versions from Marketplace)
- A virtual network, subnet, network security group, and resource group
- AAD
- Defender for Cloud
- Microsoft Sentinel
- Azure Bastion
- Microsoft Defender for Cloud Apps
- Azure Log Analytics workspace
- Azure Automation account
- Azure Update Management
- Azure Privileged Identity Management

Applications, tools, and services:

- PowerShell (version 5.1 recommended) with the AAD module and the Azure PowerShell Az module
- Text viewer to edit and open JSON files
- Windows Assessment and Deployment Kit
- Windows Deployment Services (Windows Server roles and features)
- Microsoft Deployment Toolkit
- Microsoft Endpoint Manager (Configuration Manager) hierarchy
- Windows 2016 Active Directory and domain functional level
- Microsoft Security Compliance Toolkit

- **Windows Server Update Services (WSUS)**

- Windows 10+ Pro/Enterprise, Windows Server 2016+ Core/Datacenter

All licensing and pricing is subject to change by Microsoft. Additionally, many of the products that are mentioned are covered under a license bundle, or available à la carte if you only want to enable a small subset of features.

For information about licensing Microsoft 365, visit this link:

`https://www.microsoft.com/en-us/microsoft-365/compare-microsoft-365-enterprise-plans`

To compare the different products available in the Microsoft 365 plans, visit this link:

`https://www.microsoft.com/en-us/microsoft-365/compare-microsoft-365-enterprise-plans`

For AAD pricing and features, visit this link:

`https://azure.microsoft.com/en-us/pricing/details/active-directory/`

If you are using the digital version of this book, we advise you to type the code yourself. Doing so will help you avoid any potential errors related to the copying and pasting of code.

Download the color images

We also provide a PDF file that has color images of the screenshots/diagrams used in this book. You can download it here: `https://packt.link/jSZjR`

Conventions used

There are a number of text conventions used throughout this book.

`Code in text`: Indicates code words in text, database table names, folder names, filenames, file extensions, pathnames, dummy URLs, user input, and Twitter handles. Here is an example: "Open the registry editor, go to the `HKEY_LOCAL_MACHINE\System\CurrentControlSet\Control\Terminal Server\WinStations\RDP-Tcp` registry subkey, and look for the `DWORD` port number."

A block of code is set as follows:

```
If (!($NBTNS.NetbiosOptions -eq "2")){ $NBTNSCompliance = "No"
} Else { $NBTNSCompliance ="Yes" }
```

Code output or a command-line entry is set as follows:

```
Get-CimInstance -ClassName Win32_DeviceGuard -Namespace root\
Microsoft\Windows\DeviceGuard
```

Bold: Indicates a new term, an important word, or words that you see onscreen. For example, words in menus or dialog boxes appear in the text like this. Here is an example: "Select **Port** as the rule type to create and click **Next**. Select **TCP** and enter 65001 in the box to specify **Specific local ports** and click **Next**."

> **Tips or Important Notes**
> Appear like this.

Get in touch

Feedback from our readers is always welcome.

General feedback: If you have questions about any aspect of this book, mention the book title in the subject of your message and email us at customercare@packtpub.com.

Errata: Although we have taken every care to ensure the accuracy of our content, mistakes do happen. If you have found a mistake in this book, we would be grateful if you would report this to us. Please visit www.packtpub.com/support/errata, click Submit Errata, and fill in the form.

Piracy: If you come across any illegal copies of our works in any form on the Internet, we would be grateful if you would provide us with the location address or website name. Please contact us at copyright@packt.com with a link to the material.

If you are interested in becoming an author: If there is a topic that you have expertise in and you are interested in either writing or contributing to a book, please visit authors.packtpub.com.

Share Your Thoughts

Once you've read *Mastering Windows Security and Hardening*, we'd love to hear your thoughts! Scan the QR code below to go straight to the Amazon review page for this book and share your feedback.

https://packt.link/r/180323654X

Your review is important to us and the tech community and will help us make sure we're delivering excellent quality content.

Part 1: Getting Started and Fundamentals

In this section, you will learn about the fundamentals of protecting your Windows environment. It is important that you understand the basics and the different layers involved to better protect your Windows endpoints. First, you will be provided with an overview of the security fundamentals, followed by looking at the importance of building a baseline and why you need to adopt a framework to build a successful security program. Then, we will cover hardware-based and virtualization-based security features, as well as network hardening fundamentals, to ensure your Windows endpoints have the correct protections in place. Finally, we will complete the section with a focus on one of the most important aspects of your security program, the protection of your identities and access management.

This part of the book comprises the following chapters:

- *Chapter 1, Fundamentals of Windows Security*
- *Chapter 2, Building a Baseline*
- *Chapter 3, Hardware and Virtualization*
- *Chapter 4, Networking Fundamentals for Hardening Windows*
- *Chapter 5, Identity and Access Management*

1
Fundamentals of Windows Security

In recent times, cybersecurity has become a hot topic throughout the world, and even more so with leadership teams and board members of major organizations asking the question: *Are we secure?* The short answer is *no*: no one is secure in today's digital world, and the time has never been more critical to ensure that you are doing everything within your power to protect your organization and its users.

As we continue to receive daily news of breaches throughout the world, it is clear how severe the issue of cybercrime has become. To put it bluntly, we simply need to do a better job of protecting the data that we collect and manage within our organizations today. This isn't an easy task, especially with the advancement of organized cyber and state-sponsored groups with budgets most likely far exceeding that of most organizations. As security professionals, we need to do our due diligence and ensure we identify all risks within the organization. Once identified, they will need to be addressed or accepted as a risk by leadership.

As a consumer, it is most likely that your data has already been breached, and there's a chance your account information and passwords are sitting on the dark web somewhere. We need to work with the assumption that our personal data has already been breached and build better barriers around our data and account information. For example, in the US, purchasing identity protection as a service to monitor your identity can serve as an insurance policy if you incur any damages. In addition to this, the ability to place your credit reports on hold to prevent bad actors from opening accounts under your name is another example of a defensive approach that you can take to protect your personal identity.

> **Important Note**
> There are many identity protection plans available today; a couple of notable ones include Norton LifeLock (`https://www.lifelock.com/`) and Aura Identity Guard (`https://www.identityguard.com/`). For those in the US, you can lock your credit for free online at each of the Credit Bureau's websites: Experian, Equifax, and TransUnion.

As the cybersecurity workforce continues to evolve and strengthen with more and more talented individuals, we want to help contribute to the importance of securing our data, and we hope this book will provide you with the necessary knowledge to do the right thing for your organization. As you read this book, you will not only learn the technical aspects of securing Windows, but you will also learn what else is necessary to ensure the protection of Windows and those who use it. Protecting Windows has become a lot more than making a few simple configuration changes and installing an **antivirus** (**AV**) tool. There is an entire ecosystem of controls, tools, and technologies to help protect your Windows systems and users. This means adopting a layered approach to protecting your devices, taking into consideration the network, applications, infrastructure, hardware, identity, monitoring, auditing, and much more.

As you read through this chapter, you will learn about the broader fundamentals of security and the principles behind the foundation that is needed to protect your Windows environment. Specifically, you will learn about the following topics:

- Understanding the security transformation
- Living in today's digital world
- Today's threats
- Ransomware preparedness
- Identifying vulnerabilities
- Recognizing breaches

- Current security challenges
- Focusing on zero trust

Understanding the security transformation

Over the years, security has evolved from being a shared role or a role that was non-existent within a business. Today, well-defined teams and organizational structures exist or are being created to focus solely on security. Not only are these teams maturing constantly, but the **Chief Information Security Officer** (**CISO**) has become a person of significant importance and may report directly to a **Chief Executive Officer** (**CEO**) and not the **Chief Information Officer** (**CIO**).

Over the years, many roles that never existed before have begun to appear within the security world, and new skill sets are always in demand. As an overview, here are some of the more common security roles that you can expect to see within a security program: CISO/CSO, **Information Technology** (**IT**) Security Director, IT Security Manager, Security Architect/Engineer, Security Analyst, Security/Compliance Officer, Security Administrator, Security Engineer, Software/Application Security Developer, Software/Application Security Engineer, Cryptographer/Cryptologist, Security Consultant/Specialist, Network Security Engineer, and Cloud Security Architect.

As an example, the following screenshot shows what a security organization may look like through an organization chart. Every organization is different, but this will provide you with a basis of what can be expected:

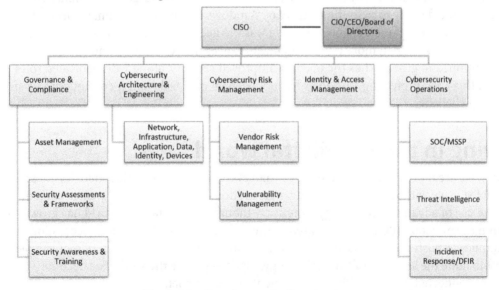

Figure 1.1 – Sample organization structure

One thing to point out regarding these roles is the shortage of a cybersecurity workforce throughout the world. Although an ongoing concern, the great news is that since the original version of this publication, there has been a significant increase in cybersecurity professionals worldwide, according to the cybersecurity workforce study by the **International Information System Security Certification Consortium ((ISC)²)**. The updated *(ISC)² 2021 Cybersecurity Workforce Study* shows that a worldwide growth of 65% is still needed to meet the demand for cyber experts, which is significantly down from 2019 when 145% growth was needed. The study estimates that there are approximately 4.19 million cybersecurity professionals globally. This is an increase of more than 700,000 from 2020. The ongoing challenge continues to be with the growth of new positions that are continuously being created as cybersecurity programs continue to enhance. This makes it difficult to find well-seasoned talent and may require you to think outside the box as you look to onboard those new to the field or looking to shift careers. You can read more about the *(ISC)² 2021 Cybersecurity Workforce Study here*: `https://www.isc2.org/Research/Workforce-Study`.

One of the primary factors for a growing need for security experts correlates to the advancement of the **personal computer** (or **PC**) and its evolution throughout the years. The PC has changed the way we connect, and with this evolution comes the supporting infrastructure, which has evolved into the many data centers seen throughout the world.

As we are all aware, Windows has been the victim of numerous vulnerabilities over the years and continues to be a victim even today. The initial idea behind the Windows **Operating System (OS)** was a strong focus on usability and productivity. As a result of its success and adoption across the globe, it became a common target for exploits. This, in turn, created many gaps in the security of Windows that have traditionally been filled by many other companies. A good example was the need for third-party AV software. As the world has turned more toward digitization over the years and Windows usage has continued to grow, so has the need for improved security, along with dedicated roles within this area. Protecting Windows has not been an easy task, and it continues to be an ongoing challenge.

Living in today's digital world

Today, we are more reliant on technology than ever and live in a world where businesses cannot survive without it. As our younger generations grow up, there is greater demand for the use of advanced technology. One scary thought is how fast the world has grown within the previous 100 years compared to the overall history of mankind. Technology continues to push the boundaries of innovation, and a significant portion of that change must include the securing of this technology, especially since the world has become a more connected place with the advancement of the internet.

To give you a rough idea of technology usage today, let's take a look at the current desktop usage throughout the world. For these statistics, we will reference an online service called *StatCounter GlobalStats*: `https://gs.statcounter.com/`. This dataset is not all-inclusive, but there is a very large sampling of data used to give us a good idea of worldwide usage. *StatCounter GlobalStats* collects its data through web analytics via a tracking code on over 2 million websites globally. The aggregation of this data equates to more than 10 billion page views per month. The following screenshot shows the OS market share that is in use worldwide. More information from *StatCounter* can be viewed at `https://gs.statcounter.com/os-market-share/desktop/worldwide`:

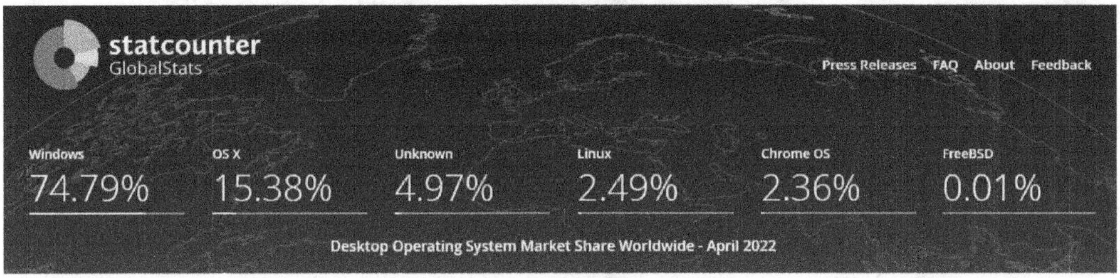

Figure 1.2 – StatCounter desktop OS market share worldwide

As you can see, the Windows desktop market is more widely adopted than any other desktop OS available today. Seemingly, Windows has always had negative connotations because of its ongoing vulnerabilities in comparison to other OSs. Part of this is due to how widely used Windows is—a hacker isn't going to waste their time on an OS that isn't widely adopted. We can assume there would be a direct correlation between OS adoption rates and available security vulnerabilities. Additionally, the Windows OS is supported across many types of hardware, which opens opportunities for exploits to be developed. One reason why we see significantly fewer macOS vulnerabilities is due to the hardware control with which Apple allows its software to run on. As the platform has grown, though, we have seen an increase of vulnerabilities within its OS too. The point I'm making is that we tend to focus our efforts on areas where it makes sense, and Windows has continued to be a leader in the desktop space, making it a very attractive source to be attacked. This, in turn, has created an ecosystem of vendors and products over the years, all aimed at helping to protect and secure Windows' systems.

Let's look at the current adoption of the different Windows OSs in use. The following screenshot from *StatCounter* shows the current Windows desktop version usage around the world today. To view these statistics, visit `https://gs.statcounter.com/os-version-market-share/windows/desktop/worldwide`:

Win10	Win7	Win11	Win8.1	Win8	WinXP
73.24%	12.62%	8.89%	3.24%	1.22%	0.44%

Desktop Windows Version Market Share Worldwide - April 2022

Figure 1.3 – Desktop Windows version market share worldwide

As you can see, Windows 10 is the most adopted OS at 73%. In addition, Microsoft has recently released Windows 11, accounting for almost 9% of the desktop market share already. Microsoft continues to push more users and organizations to the latest version of Windows, and this is where it spends the majority of its development resources. There are also major changes to Windows 11 compared with older versions, which is why it is critical to migrate from older versions, especially for security-specific reasons. Microsoft ended its support (including security updates) for Windows XP in April 2014 and Windows 7 in January 2020. It has also announced the retirement of Windows 10 support for October 14, 2025.

A recent buzz term you have most likely heard in recent years is that of **digital transformation**. This refers to the shift from a legacy on-premises infrastructure to a modernized cloud-first strategy to support the evolving need for big data, machine learning, **Artificial Intelligence** (**AI**), and more. A significant part of this shift also falls within Windows systems and management. In *Chapter 11*, *Server Infrastructure Management*, we will look at the differences between a data center and a cloud model, including where the responsibilities fall for maintaining and securing underlying systems. Prior to digital transformation, we relied heavily on the four walls of the corporation and its network to protect a data center and its systems. This included a requirement for client devices to be physically on the corporate network in order to access data and services. With this model, our devices were a little easier to manage and lock down, as they never left the corporate office. Today, the dynamics have changed. Referencing back to *StatCounter*, in the following screenshot, you can see a significant shift from traditional desktop usage to a more mobile experience. The **Mobile** percentage reflects an increase of over 2% since the initial release of this publication 2 years ago. To view the source of this screenshot, visit `https://gs.statcounter.com/platform-market-share/desktop-mobile-tablet/worldwide/#monthly-200901-202110`:

Figure 1.4 – StatCounter platform comparison (January 2009 - May 2022)

Focusing on Windows security, the traditional model of an organization would have typically included the following security tools as part of its baseline:

- AV software
- Windows firewall
- Internet proxy service
- Windows updates

Depending on your organization or industry, there may have been additional tools. However, for the most part, I'd imagine the preceding list was the extent of most organizations' security tools on Windows client devices. The same would have most likely applied to the Windows servers in the traditional model. As the digital transformation has brought change, the traditional method of Windows management has become legacy. There is an expectation that we can work and access data from anywhere, at any time. With the rapid increase in remote working during 2020 and 2021, this model and expectation have been fast-tracked. We live in an internet-connected world, and when we plug our device in, we expect to access our data with ease. With this shift, there is a major change in the security of the systems we manage and—specifically—the Windows server and client. As we shift our infrastructure to the cloud and enable our users to become less restricted, the focus of security revolves not only around the device itself but that of the user's identity and, more importantly, the data. Today, the items we listed earlier will not suffice in the enterprise. The following tools are those that would be needed to better protect your Windows devices:

- **Advanced Threat Protection** (ATP): AV and threat protection, **Endpoint Detection and Response** (EDR), advanced analytics and behavioral monitoring, network protection, exploit protection, and more

- **Data Loss Prevention** (DLP) and information protection

- **Identity protection**: Biometric technology, **Multi-Factor authentication** (MFA), and more

- Application control

- Machine learning and advanced AI security services

Today's threats

The threat landscape within the cyber world is extremely diverse and is continually becoming more complex. The task of protecting users, data, and systems is becoming more difficult and requires the advancement of even more intelligent tools to keep bad actors out. Today, criminals are more sophisticated, and large groups have formed with significant financial backing to support the wrongdoings of these groups. The following are common cyber threats: national governments, nation-states, terrorists, spies, organized crime groups, hacktivists, hackers, business competitors, and insiders/internal employees.

> **Tip**
>
> To learn more about these cyber-threat sources, the **Department of Homeland Security (DHS)** has a great reference here: `https://us-cert.cisa.gov/ics/content/cyber-threat-source-descriptions`.

To shed some light on real-world examples of data-breach sources today, *Verizon* releases an annual report, *Data Breach Investigations Report*. You can view their latest report here: `https://enterprise.verizon.com/resources/reports/dbir/`. The report is built on a set of real-world data and contains some eye-opening data on data breaches, such as the following revelations highlighted in the 2021 report:

- 85% of breaches involved a human element.

- 61% of breaches involved credentials.

- 3% of breaches involved vulnerability exploitation.

- Action variants in breaches: phishing 36%, up by 25% from 2020; use of stolen credentials 25%; ransomware 10%, which more than doubled from 2020.

- Credentials remain one of the most wanted data types.

- The most common motivation for attacks continues to be financial.

- The number-one threat actor is currently organized crime.

The full 2021 report can be found here: `https://www.verizon.com/business/resources/reports/dbir/2021/masters-guide/`.

There are many types of cyberattacks in the world today, and this creates a diverse set of challenges for organizations. While not all threats are Windows-specific, there's a chance that Windows is the median or attack vector in which an attacker gains access by exploiting a vulnerability. An example of this could be an unpatched OS or an out-of-date application. Next, we list many types of threats that could cause damage directly using a vulnerability within the Windows OS or by using the Windows OS as an attack vector.

Malware is software or code designed with malicious intent that exploits vulnerabilities found within the system. The following types of threats are considered malware: adware, spyware, virus (polymorphic, multipartite, macro, or boot sector), worm, Trojan, rootkit, bots/botnets, ransomware, and logic bombs.

In addition to malware, the following types of attack techniques can be used to exploit vulnerabilities:

- Keylogger
- Phishing (email phishing, spear phishing, whale phishing, vishing, smishing, or pharming)
- Social engineering
- **Business Email Compromise (BEC)**
- **Structured Query Language (SQL)** injection attack
- **Cross-Site Scripting (XSS)**
- **Denial of Service (DoS)** and **Distributed Denial of Service (DDoS)**
- Session hijacking
- **Man-in-the-Middle (MITM)** attacks
- Password attacks (brute-force, dictionary, or birthday attacks)
- Credential stuffing or reuse
- Identity theft
- **Advanced Persistent Threats (APTs)**
- Intellectual property theft
- Shoulder surfing
- Golden Ticket: Kerberos attacks
- **Domain Name System (DNS)** tunneling and dangling DNS
- Zero-day

> **Tip**
>
> To learn more about the threats listed earlier, the **National Institute of Standards and Technology (NIST)** has a glossary that provides more information on most, if not all, of the threats in the preceding list: `https://csrc.nist.gov/glossary`.

Now that we've just reviewed today's threats, let's take a look at an extremely important topic that has everyone's attention. Ransomware preparedness is on everyone's security priority list.

Ransomware preparedness

In the previous section, we introduced many of the threats and cyberattacks that continue to challenge us as cybersecurity professionals. One specific type of malware we want to cover in more detail is ransomware. Since the original release of this publication, ransomware incidents have grown exponentially, and ransomware is currently one of the biggest threats to organizations today. In short, a ransomware attack refers to an intruder encrypting data belonging to a user or organization, making it inaccessible. For the user or organization to gain access back to their data, they are held to a ransom in exchange for the decryption keys. The intruders will use many tactics to try to force payment, including threats to leak the data and list the data for sale on the dark web, to the extent of erasing backups, to name a few.

Ransomware has been around for a long time, and the first documented incident occurred in 1989, known as *PC Cyborg* or the *AIDS Trojan*. Since then, ransomware has evolved substantially into a business with high payoffs for attackers. There is even a **ransomware-as-a-service (RaaS)** model that allows hackers to subscribe and use the service to commit their own attacks. A report released by the **Federal Bureau of Investigation's (FBI's) Internet Crime Complaint Center (IC3)** (`https://www.ic3.gov/`) received 2,084 ransomware incidents between January and July 2021 alone. The cost of complaints from the 2,084 incidents totaled over **US dollars (USD)** $16.8 million. This is a 62% increase in reported incidents and a 20% increase in losses from the same period in 2020. These numbers alone show a significant increase in ransomware. You can read more on the report here: `https://www.ic3.gov/Media/News/2021/210831.pdf`.

When it comes to ransomware (or any other threat), the first action should be to protect your environment as best as possible. Although there is a lot that can be done to prevent ransomware from occurring, there is no way to make your environment 100% resilient from such an attack. Because of this, the second action you need to take is being prepared to respond. Ransomware can impact anyone at any time, and the better prepared you are, the better you will be able to handle the situation and the quicker you will be able to recover your environment. Time is of the essence in these situations as you may be losing millions of dollars, customers, and a reputation that has taken years to build.

There are many great resources available for ransomware preparedness and response. Our review of ransomware preparedness and response will be referencing the following two excellent resources for recommendations and information:

- **Cybersecurity and Infrastructure Security Agency (CISA)**: `https://www.cisa.gov/stopransomware`

- NIST: `https://csrc.nist.gov/projects/ransomware-protection-and-response`

First, let's review some best practices for protecting your environment from a ransomware attack. A lot of the following recommendations should be part of your standard security best practices, but it's best to review and validate any gaps you may have in your infrastructure:

- Enforce MFA, use least privileges or just-enough privilege, and implement **Privileged Access Management (PAM)** and **Privileged Identity Management (PIM)**.

- Patch and update all software and OSs (including network devices) to the latest supported versions.

- Ensure you are using the latest protection solutions including EDR or **Extended Detection and Response (XDR)**.

- Implement next-generation network protection: firewalls, **Intrusion and Detection Prevention (IDP)**, **Intrusion Prevention Systems (IPSs)**, and so on.

- Implement network segmentation.

- Restrict the use of scripting to approved users.

- Secure your **Domain Controllers (DCs)**.

- Block access to malicious sites.

- Only allow trusted devices on your network.

- Disable the use of macros.

- Only allow approved software to be used by your users.

- Remove local admin permissions.

- Enable advanced filtering for email.

- Use **Sender Policy Framework (SPF)**, **DomainKeys Identified Mail (DKIM)**, and **Domain-based Message Authentication, Reporting, and Conformance (DMARC)**.

- Block the **Server Message Block (SMB)** outbound protocol and remove outdated versions.

- Follow best practices to harden your end-user and infrastructure devices.

- Protect your cloud environment with best practices, especially public file shares.

- Review your remote strategy and ensure outside connections into your environment are secure. If **Remote Desktop Protocol (RDP)** is needed, ensure best practices are deployed.

- For backups, maintain an offline backup or air gap, encrypt all backups, and validate recovery by testing regularly.

- Implement and focus attention on a well-defined **Vulnerability Management Program (VMP)**.

- Implement a good cybersecurity and awareness program. Train users not to click on links or open attachments unless they are confident they are legitimate.

- Build a mature **Vendor Risk Management (VRM)** program.

This is a very high-level summary and may not be inclusive of everything that you may need to account for when protecting your network from a ransomware attack. This is a great starting point, but a lot of effort and time will be needed to best protect against malicious attackers.

> **Additional Information**
>
> The preceding list has been derived from the *CISA Ransomware Guide* (`https://www.cisa.gov/sites/default/files/publications/CISA_MS-ISAC_Ransomware Guide_S508C_.pdf`) and the *NIST Tips and Tactics | Preparing Your Organization for Ransomware Attacks* document (`https://csrc.nist.gov/CSRC/media/Projects/ransomware-protection-and-response/documents/NIST_Tips_for_Preparing_for_Ransomware_Attacks.pdf`), which contain a lot more detail for review.

As already stated, there is no way of completely preventing a ransomware attack in any environment—it could happen to anyone. No matter how much you protect your environment, there will always be a way to circumvent it. Because of this, the next best action you can take is responsiveness. Being prepared and ready to respond to a ransomware attack will allow you to handle the situation much more efficiently and get your environment up and running much faster. Here are some critical items to help with your response:

- Have an up-to-date **Incident Response Plan (IRP)**. This should include all critical information needed to respond to a ransomware attack, including contact information (local law enforcement), responsibilities, communications, and so on.

- Ensure you have a ransomware playbook as part of your IRP.

- Ensure you have a well-documented **Disaster Recovery (DR)** plan and **Business Continuity Plan (BCP)** that is up to date and tested.

- Ensure you have a mature **Security Operations Center (SOC)** or **Managed Security Service Provider (MSSP)**.

- Conduct a tabletop exercise using ransomware as your theme. Ensure executives are involved in these exercises as they will ultimately need to make some of the important final decisions.

- Carry cybersecurity insurance and understand what options are available with ransomware payment from your cyber insurance policy if this becomes a decision point. Check if you need to obtain a cryptocurrency account or if this is part of the service they provide. Make sure you are not breaking any laws if payment will proceed.

- At a minimum, contract with a couple of respected **Incident Response (IR)** vendors with whom you have **Service Level Agreements (SLAs)** for engagement. Ensure that they also provide **Digital Forensics Incident Response (DFIR)** and are approved for use by your cybersecurity insurance policy. A couple to review are *Secureworks* (`https://www.secureworks.com/services/incident-response/incident-management-retainer`) and *KPMG* (`https://advisory.kpmg.us/services/cyber-response-services.html`).

> **Information**
> We will cover the IRP in more detail in *Chapter 14, Security Operations*.

Everything we have covered for ransomware response should be part of the overall hardening of your environment as part of your security program. Implementing these recommendations will help protect you against many threats. In addition, having an IRP is intended for any security incident within your environment, and this should be a requirement for your security program in general. As we take you through this book, you will learn how to harden your Windows environment to best protect yourself from a ransomware attack.

Identifying vulnerabilities

Now that we know more about the kinds of threats you may face, it's even more important for you to know where to access information about these vulnerabilities. You also need to be aware of any resources that are available so that you can educate yourself on what's required to remediate any vulnerabilities. As you are already aware, Windows is renowned for its ongoing vulnerabilities, and patching/updating these systems has morphed into a full-time and very specialized role over the years. The following website is the authoritative source regarding Microsoft security updates: `https://msrc.microsoft.com/update-guide/en-us`.

Useful Information

Here is a link to the **Microsoft Security Response Center (MSRC)**:
`https://www.microsoft.com/en-us/msrc?rtc=1`.

As shown in the following screenshot, you will be provided with a list of all identified vulnerabilities from Microsoft within a selected time range, with additional filtering options. To give you an idea of the risk profile for Windows, the following filter is scoped to Windows 11 for x64-based systems over November 2021, which returned 47 uniquely addressed vulnerabilities:

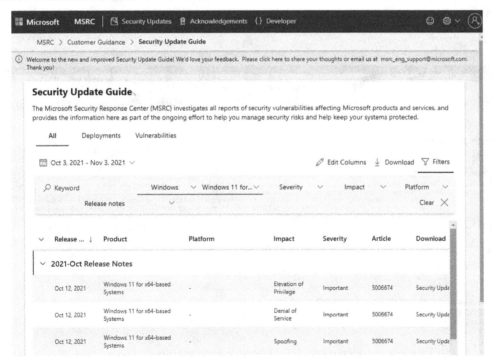

Figure 1.5 – Microsoft Security Update Guide

One term you may have heard as part of vulnerability management with Microsoft is the famous **Patch Tuesday** (also referred to as **Update Tuesday**). Patch Tuesday occurs on the second Tuesday of every month and is the day that Microsoft will release its monthly patches for Windows and other Microsoft products. There are many references on the internet for Patch Tuesday, in addition to the MSRC. One example of a common resource used to track Patch Tuesday releases is the Patch Tuesday dashboard: `https://patchtuesdaydashboard.com/`.

As you review the updates needed for your Windows systems, you will notice that each of them has a **unique identifier** (**UID**) to reference the update, beginning with *CVE*. **CVE** stands for **Common Vulnerabilities and Exposures** and is the standard for vulnerability management, allowing one source to catalog and uniquely identify vulnerabilities. CVE is not a database of vulnerabilities but a dictionary providing definitions for vulnerabilities and exposures that have been publicly disclosed. The US DHS and CISA sponsor the CVE.

> **Tip**
> Visit this website to learn more about CVE: `https://www.cve.org/About/Overview`.

The following screenshot shows an overview of what the CVE provides and can be found at `https://cve.mitre.org/cgi-bin/cvekey.cgi?keyword=Windows`:

Figure 1.6 – CVE Windows search results

In addition to the CVE is the NVD. **NVD** is the **National Vulnerability Database,** which is an additional resource for vulnerability management provided by NIST. The NVD is synced with the CVE to ensure the latest updates appear within its repository. NVD provides additional analysis of the vulnerabilities listed in the CVE dictionary by using the following:

- **Common Vulnerability Scoring System (CVSS)** for impact analysis
- **Common Weakness Enumeration (CWE)** for vulnerability types
- **Common Platform Enumeration (CPE)** for structured naming standards

The following screenshot shows an overview of what the NVD provides and can be found via `https://nvd.nist.gov/vuln/search/results?form_type=Basic&results_type=overview&query=Windows&search_type=all&isCpeNameSearch=false`:

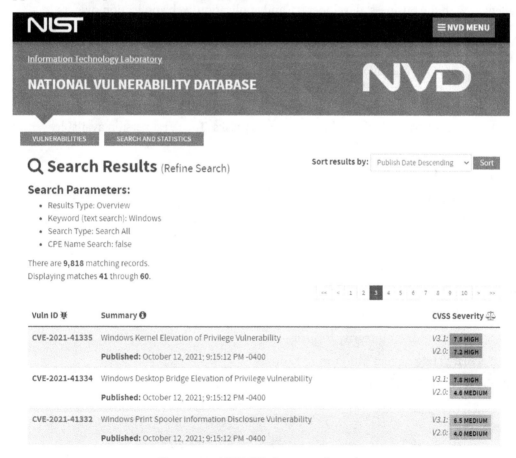

Figure 1.7 – NVD Windows search results

> **Information**
> CISA has released a known exploited vulnerabilities catalog that will be continuously maintained. This should be incorporated into your VMP for review. You can find the catalog here: `https://www.cisa.gov/known-exploited-vulnerabilities-catalog`.

One additional resource that we should mention is the **Open Web Application Security Project (OWASP)**. OWASP is a nonprofit organization that helps improve the security of software for individuals and enterprises. It provides a tremendous amount of resources such as tools, documentation, and a community of professionals, all looking to continually enhance software security. Although OWASP is focused on application and web application security, there is a high possibility that an application or a web application will be running on your Windows servers and Windows clients. Because of this, it is critical, as a security professional, for you to be able to intelligently discuss these concerns and challenges with the business, developer(s), and application/web application owners.

> **Tip**
> You can find more information about OWASP at `https://owasp.org/`.

One of the more common projects that OWASP provides is the *OWASP Top 10*, which provides the most critical web application security risks. The latest version available was recently published in 2021 and is presented here:

1. Broken Access Control
2. Cryptographic Failures
3. Injection
4. Insecure Design
5. Security Misconfiguration
6. Vulnerable and Outdated Components
7. Identification and Authentication Failures
8. Software and Data Integrity Failures
9. Security Logging and Monitoring Failures
10. **Server-Side Request Forgery (SSRF)**

> **Tip**
> View *OWASP Top Ten Web Application Security Risks* for 2021 here: `https://owasp.org/www-project-top-ten/`.

Recognizing breaches

If you follow the news, you are probably aware that there is no shortage of breaches nowadays. They are happening so frequently that it is not uncommon for several breaches to occur weekly or even daily. What is even scarier is that these are just the ones that we hear about. To give you an idea of how serious the issue has become, the following list has some of the more notable breaches that are documented on Wikipedia's *List of data breaches* page. There are many sources on the internet identifying top breaches, but Wikipedia has the most comprehensive information we have found with references to each of the listings:

Company	Year	# breached records	Reason
Yahoo	2013	3,000,000,000	Hacked
First American Financial Corp.	2019	885,000,000	Poor security
Facebook	2019	540,000,000	Poor security
Marriott International	2018	500,000,000	Hacked
Adobe Systems	2013	152,000,000	Hacked
Under Armour	2018	150,000,000	Hacked
eBay	2014	145,000,000	Hacked
Equifax	2017	143,000,000	Poor security
Target Corporation	2013	110,000,000	Hacked
Capital One	2019	106,000,000	Hacked
Solarwinds	2020	Source Code Compromised	Hacked

Figure 1.8 – Wikipedia list of data breaches

You can find the source of the preceding screenshot at `https://en.wikipedia.org/wiki/List_of_data_breaches` and a list of security incidents here: `https://en.wikipedia.org/wiki/List_of_security_hacking_incidents`.

As you review the breaches and understand how they occurred, you will see a common trend where, for the most part, the breach occurred from hacking or poor security practices. You might also notice that other common methods of breaches include lost or stolen equipment. These statistics are alarming, and they indicate how critical it is to secure and harden our systems as best as possible.

It is also important to point out that the tactics of some malicious actors are not to breach records but to hold a company at ransom for a large payout. One of the more notable ransomware attacks recently was against *Colonial Pipeline*, which is one of the largest fuel pipelines in the US. This ransomware was so impactful that it forced the company to shut down its fuel distribution operations, causing gas shortages for consumers throughout the east coast. Another attack becoming more common is that of the supply chain, where hackers look to compromise a vendor that can then in turn compromise all its downstream customers. One of the most infamous such attacks was the SolarWinds cyberattack, where hackers implanted malicious code into their software, which was received by thousands of customers. Once installed, hackers were provided the ability to infiltrate customers' networks.

To give you an idea of the importance of securing and hardening your environment, the **International Business Machines Corporation** (**IBM**) data breach report of 2021 provides some data points that are not to be taken lightly. In 2021, the average cost of a data breach was $4.24 million, which is the highest average cost since the report began. The most common initial attack vector was compromised credentials. In addition, the report shows that the average cost of a user record from a data breach is $161 per record. A quick calculation of this multiplied by 100,000 customers calculates a potential loss estimated at $16.1 million. When you look at the number of breached records shown in *Figure 1.8*, you will understand how this could be extremely damaging to a business's value and reputation.

You can download and view more details on the *IBM Cost of a Data Breach Report* here: `https://www.ibm.com/security/data-breach`.

> **Tip**
>
> An interesting site for reference is *Have I Been Pwned*. This site will show you whether any of your accounts that use your email address have ever been breached and, if so, where the breach was: `https://haveibeenpwned.com/`. You can also sign up for notifications for any breaches using your email address or submit a specific domain to be notified on.

There are many sources available where you can view security news and follow the latest trends and best practices. Here are some recommended resources to help keep you up to date with the latest happenings in the security world today:

- *DarkReading*: `https://www.darkreading.com/`
- *Cyware*: `https://cyware.com/cyber-security-news-articles` (recommended phone app)

- *Cybersecurity Insiders*: `https://www.cybersecurity-insiders.com/`
- *CSO*: `https://www.csoonline.com/`
- *Krebs on Security*: `https://krebsonsecurity.com/`
- *The Hacker News*: `https://thehackernews.com/`
- *Darknet Diaries* podcast: `https://darknetdiaries.com/`
- *Risky Business* podcast: `https://risky.biz/`

Next, we will discuss the security challenges we face in today's world and within the enterprise.

Current security challenges

By the time you have finished reading through the chapter, you will have hopefully been provided with a sense of how important security has become today and the challenges that come with it. We are continually becoming more reliant on technology than ever before, with no signs of slowing down. We have an expectancy of everything being digitized, and, as the IoT is taking off, everything around us will be connected to the internet, thus creating even more challenges to ensure security is efficient.

As we briefly covered earlier, attacks are becoming more and more sophisticated every day. There is an ever-growing army of bad actors working around the clock, trying to breach any data they can get their hands on because the cost of private data is very expensive. There is also a shift in the way bad actors are threatening organizations by looking for weakness in the supply chain and holding companies to ransom. With the advancement of cloud technology, supercomputers, and the reality of quantum computing coming to light, hackers and organized groups now have access to much more powerful systems and are easily able to crack passwords and their hashes much more easily, making them obsolete as the only factor of authentication. No one should be using just passwords anymore; however, the reality is, most still are. The same applies to encryption. The advancement of computers is making algorithms insecure, with the ongoing need for stronger encryption. These are just some of the ongoing challenges we are faced with when protecting our assets.

Keeping up with vulnerabilities today is a full-time role. It's critical that we keep on top of what they are and which Windows systems need to be updated. We will discuss the management of Windows updates later in the book, but having a program in place to manage the overwhelming amount of Windows updates is critical. Additionally, third-party applications will need to be carefully monitored and updated accordingly. An example of a commonly used application is Adobe Acrobat Reader DC to view **Portable Document Format** documents (**PDFs**). The following screenshot is a vulnerability report from *Microsoft Defender Security Center*. It provides a software inventory of all machines with the application installed and lists the number of vulnerabilities detected across all machines in your organization:

Figure 1.9 – Acrobat Reader DC identified vulnerabilities

As you can see, out-of-date applications have critical known vulnerabilities that are used by attackers.

Most organizations are reluctant to release the latest Windows updates to their servers straight away because of the risk that a patch could break a production system. The downside to this is that your system will have a known vulnerability, which opens up an opportunity for it to be exploited between the time of the patch release and the system being patched. Another challenge we are faced with is zero-day vulnerabilities. A zero-day vulnerability is one that has been identified but currently has no remediation or mitigation available from the vendor. Because of these challenges, it is critical we build a layered defense strategy into our Windows clients and servers. For example, never make your database server accessible via the internet, encrypt the traffic to your web servers, and only open the ports needed to communicate, such as allowing port 443 for secure (**HyperText Transfer Protocol Secure (HTTPS)**) traffic only.

As we focus on securing Windows devices within our environments, we can't turn a blind eye to the fundamentals, including the overarching ecosystem that also needs to be considered when protecting your Windows devices. This book will cover a lot of detail on the specifics of securing and hardening your Windows systems and devices, but we also want to ensure the bigger picture is covered—for example, simple concepts of **identity and access management (IAM)**. A user whose account has been compromised to allow an intruder on your Windows system has just made all the securing and hardening of that system irrelevant. The concept of weak physical access controls and policies could allow someone to simply walk into a server room and gain physical access to your systems. Other examples are allowing a developer to install an insecure web application with vulnerabilities on it, or a business that develops a process without security best practices in mind. All the controls you put in place with Windows become irrelevant, as an educated hacker could use the web application or exploit a process as an attack vector to gain access to your system. These examples show the criticality of not only being familiar with how to secure and harden the Windows OS but also ensuring all the other factors that fall within a mature security program work together to ensure your environment is as secure as possible. This, of course, doesn't come easily, and it is critical you stay current and continue to learn and learn and learn!

Managing and securing your Windows systems is not a simple task, especially if you are working toward securing them correctly. There is a lot involved, and to efficiently and effectively secure your Windows systems, you need well-defined policies, procedures, and standards in place, along with a rigorous change-control process to ensure anything that falls outside of the standards receives the appropriate approval to minimize risk. Full-time roles exist today to manage and secure your Windows systems, along with specialized roles that are necessary to manage your Windows environments. Examples include Windows desktop engineers, Windows server engineers, Windows update administrators, Windows security administrators, Windows **Mobile Device Management (MDM)** engineers, and more. As part of these roles, it is critical that staff are continuously educated and trained to provide the best security for Windows. The landscape is changing daily, and if your staff aren't dynamic or don't stay educated, mistakes and gaps will occur with your security posture.

Another task to think about that must be addressed with your Windows devices is inventory management. It is important to ensure you know where all your devices are and who has access to them. Even more important is ensuring that devices are collected upon any terminations, especially those pertaining to disgruntled employees. Enforcing policies on your Windows devices is also another challenge; for instance, how do you ensure all your devices have the latest policies, and how can you ensure accurate reporting on non-compliant devices? Remote management can also be a challenge—that is, to make sure that not just anyone can remotely access your devices, including the auditing of support staff for anything that they shouldn't be doing. Running legacy applications on your Windows devices creates an instant security concern, and making sure they are patched to the latest supported version is critical. This list goes on, and we will be diving into much greater detail within the following chapters to help provide the information you need to protect your Windows environment.

Before we move on to the next topic, one additional challenge that needs mentioning is shadow IT. In short, shadow IT is the setup and use of servers and infrastructure without IT or the security team's approval or knowledge—for example, a business function. This instantly creates a significant security concern as the Windows systems will most likely be used with no standards or hardening in place. In addition, hackers are known to target application-managed identities to gain access to other systems due to their privileged permissions. This can be a challenge to manage, but it is something that needs to be understood and prevented within any business.

Focusing on zero trust

Within only a couple of years since the first edition of the book was released, zero trust has gained tremendous momentum and has become a *buzz* and marketing term for most security vendors. As a reminder, the zero-trust architecture model was created by John Kindervag while he was at *Forrester Research Inc.* back in 2009. If you are not clear on what exactly zero trust is, essentially, it is a model where we trust no one until we can validate who they are, who they are meant to be, and whether they are authorized to have access to the system or information. In simple terms, zero trust is well known for the concept of *never trust, always verify*.

There are a lot of zero trust references from many vendors, and depending on which vendor you work with, there will be slight differences in their approach to zero trust. No matter which vendor you work with or which approach you take, the core of a zero-trust model will always fall back on the principle of *never trust, always verify*. Effectively implementing a zero-trust model requires a multilayered approach with your security strategy, along with the use of the most current and modern technology available. The method of allowing a user to access the environment with only a username and password is outdated and insecure.

Since we are focusing our security on Microsoft products, we will review zero trust as Microsoft has approached it. With Microsoft's approach to zero trust, they have created their strategy for customers around six different pillars, as follows:

1. **Identities** are the new perimeter of zero trust. An identity is something (typically a user) that needs to access an app, data, or some other form of resource. It is critical that identities have multiple layers of protection to prevent unauthorized access.

2. **Device and endpoint** protection is an essential component of zero trust. Whether a mobile device, laptop, server, IoT, and so on, we need to ensure recommended baselines are deployed and that devices stay compliant and are constantly being scanned for vulnerabilities.

3. **Data** is at the core of the zero-trust model. It is ultimately data that the intruders are looking to exfiltrate from your environment. This is the true asset that needs to be protected. Because of this, it is critical that you know where all your data lives, who has access to it; whether it's classified correctly and encrypted, and that you have the correct controls to prevent it from being removed from your environment.

4. **Applications and application programming interfaces (APIs)** are gateways to your data. They need to be governed and deployed with best practices to prevent unauthorized access to data, whether intentionally or unintentionally. Ensuring the business is following enterprise standards is critical for preventing shadow IT.

5. **Infrastructure** pertains to everything within your environment that provides the means to store your data and/or run applications such as servers, VMs, appliances, **Infrastructure as a Service (IaaS)**, **Platform as a Service (PaaS)**, and **Software as a Service (SaaS)**. Preventing unwanted access to your infrastructure is crucial by ensuring you have best practices and baselines in place, along with effective monitoring.

6. A **network** is a medium where your data travels. Once considered the perimeter for defense, this pillar still holds a critical role as part of zero trust to ensure all data is encrypted during transport, next-generation protection is deployed, micro-segmentation is in place, and ongoing monitoring is taking place to detect unauthorized access to your data in transit.

> **Information**
>
> You can read more about the zero-trust Microsoft model here: `https://www.microsoft.com/en-us/security/business/zero-trust`.

The following screenshot represents the six pillars, with some examples of the technologies and solutions that should be implemented to support your broader zero-trust strategy:

Figure 1.10 – Microsoft's six pillars of zero trust

Now that you have an idea of what zero trust is and the technologies involved, you will need to have a plan in place to ensure success. Zero trust is not going to happen overnight or within a few weeks. Zero trust is a strategy and a foundation that needs to be planned over months or even years, depending on your organization's maturity. With this, you will need to be able to measure your success, along with your current maturity. Microsoft has a very high-level and simplified maturity model that can help you better understand your current state and what will be needed to obtain zero trust. The following screenshot provides an example of how the Microsoft maturity model is presented:

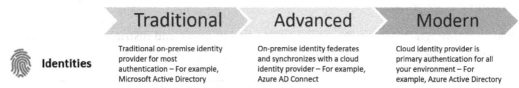

Figure 1.11 – An example based on Microsoft's zero-trust maturity model

For more details on the maturity model example, browse to `https://download.microsoft.com/download/f/9/2/f92129bc-0d6e-4b8e-a47b-288432bae68e/Zero_Trust_Vision_Paper_Final%2010.28.pdf` to view Microsoft's maturity model with more details across all six pillars.

Once you have a good understanding of the technologies involved in zero trust and how the maturity model works, the next step is to build your strategy around accomplishing this transition. Remember—this transformation isn't going to happen overnight, so you'll need to build a roadmap based on where you are today with the technologies you have in place versus where you would like to be. Accomplishing this journey is not going to be easy, but the better you document your strategy and vision, the easier your journey will become. Here is a high-level idea of how you could build out your roadmap:

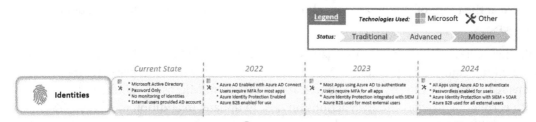

Figure 1.12 – Example of a zero-trust roadmap

There are many variations and vendors providing guidance on how to adopt a zero-trust approach. A couple of additional notable models to review as part of your planning include the NIST zero-trust architecture and the CISA zero-trust maturity model, as listed here:

- NIST zero trust: `https://www.nist.gov/publications/zero-trust-architecture`

- CISA zero trust: `https://www.cisa.gov/publication/zero-trust-maturity-model`

As you read through this book, you will find that the guidance and instructions provided will ultimately lead to a zero-trust model.

> **Tip**
> Microsoft also provides deployment guides to each of its pillars. The following link provides specific guidance to securing endpoints with zero trust. The same **Uniform Resource Locator** (**URL**) also has links to the other deployment guidelines for each of the pillars: `https://docs.microsoft.com/en-us/security/zero-trust/deploy/endpoints`.

Summary

In this chapter, we covered—at a high level—what you can expect to read throughout this book. We provided an overview of security in an enterprise and covered the different roles that you can expect to see within security departments. Next, we looked at how security relates to the digital world and its relevance as the world becomes more digital. We also looked at the usage of Windows throughout the world to better understand its adoption by users.

We then reviewed the current threat landscape and the types of cyber threats and covered ransomware preparedness. After reviewing threats in the enterprise, we then provided details on where you can go to learn about recent Microsoft vulnerabilities, with correlating patches and instructions on how to update. In addition to Microsoft's vulnerability resources, we provided insight into where patches get their naming standards via CVE, along with NVD (*NIST*). Next, we looked at some of the biggest breaches that have occurred to date and provided some popular sources to keep you up to date with the latest cyber news. We finished the chapter with an insight into some of today's general security challenges and—more specifically—those with Windows systems, before closing with an overview of zero-trust security and what it entails.

In the next chapter, we will review building a baseline. This chapter will review what a baseline is and then go into detail as to why a baseline must be formed. As part of the baseline, you need to ensure your policies, standards, and procedures are in place and are well defined and signed off by the leadership team and all the stakeholders who are liable for protecting the data. Having these documented is important for security reasons, as well as for compliance and auditing purposes. Following this, we will briefly cover change management and its importance as it relates to baselining. We will then review frameworks and what they entail before moving on to some common frameworks that should be referenced when building your baseline. We will finish the chapter with a review of baseline controls and how to implement them.

2
Building a Baseline

In this chapter, we are going to cover the importance of building a baseline for your Windows systems. As you deploy tens to hundreds to thousands of Windows devices, you need to ensure you are deploying your devices pre-hardened and secured. In addition to this, you need to ensure consistency with every system that is deployed. This is where building a foundation is critical, and that must be followed. In addition to the baseline requirements, very stringent policies, standards, and procedures must be followed. Anyone falling outside of these boundaries will create additional risk for the organization, so it is critical a well-defined program is put in place early with the backing and support of leadership.

The following will be covered in this chapter:

- Overview of baselining
- Introduction to policies, standards, procedures, and guidelines
- Incorporating change management
- Implementing a security framework
- Building baseline controls
- Incorporating best practices

Overview of baselining

Security baselining is the practice of implementing a minimum set of standards and configuration within your environment, more specifically, capturing a minimum configuration for your Windows devices. Building a baseline provides a minimum defined standard that will help ensure a more secure environment as you deploy systems and devices within the enterprise. Depending on the size of your organization, baselines could be in the form of checklists or spreadsheets that someone follows to ensure the predefined security controls are in place. A more advanced method includes capturing a snapshot or image that is already configured with predefined security controls. In addition to the starting baseline, there are additional management tools to layer and enforce baseline configurations. A couple of examples include **Group Policy Objects (GPOs)** and **Mobile Device Management (MDM)**.

Unless you are a small business with under 100 employees, it will be impractical to deploy any type of system or device and individually configure it every time a new one is built, especially if your user count and servers start reaching hundreds to thousands with an extremely high volume of device deployments on a day-to-day basis. This could also be very error-prone. Because of this, it will be extremely important that a well-defined program is put in place to minimize potential error-prone steps involved with deploying systems, and devices receive their baseline and hardening configurations systematically.

Another important factor to consider with baselining is that your organization may be required to follow strict compliance regulations that will force the need to ensure specific security requirements are adhered to. Baselines will help when being audited, or when the need to provide evidence arises. Some regulatory compliance examples include the following:

- **Sarbanes-Oxley Act (SOX)**
- **Health Insurance Portability and Accountability Act (HIPAA)**
- **Federal Information Security Management Act (FISMA)**
- **Payment Card Industry Data Security Standard (PCI DSS)**
- **General Data Protection Regulation (GDPR)**
- **California Consumer Privacy Act (CCPA)**
- **Gramm Leach Bliley Act (GLBA)**

We can expect this to continue to grow as privacy continues to become a big discussion point and challenge. It is important to have a minimum understanding of what regulatory compliances are and how they directly relate to your organization's sector. They will play a big part in planning your overall security baselining.

As you begin to define and deploy your baselines, you will find that one will not fit all. You are going to need to document and build them for different use cases. These are some examples of where unique baselines may need to be defined:

- Network devices (switches, routers, firewalls, and so on)
- Windows systems: servers and clients
- Linux/Unix systems
- Storage/file servers
- Database servers
- Web servers
- Application servers
- **Operational Technology (OT)**
- **Internet of Things (IoT)**

As we look more specifically at the Windows environment, you may end up with baselines for different architectures:

- For Windows Server, you have the following: Windows **Domain Controller (DC)** Server, Windows Server **Internet Information Services (IIS)**, Windows SQL Database Server, Windows DNS Server, Windows Remote Desktop Services, and others.
- For a Windows client, you have the following: Windows standard client (user workstation), **Privileged Access Workstation (PAW)**, Windows virtual client, OT and IoT clients, and others.

Now that we've provided an overview of what baselines are, the next few sections will cover details around the foundation and overall strategy that supports the ability to build well-defined baselines and ensure consistency. Deploying baselines without well-defined policies, processes, and a framework will not be successful in the long term and can leave your organization vulnerable. In addition, having these foundations in place provides a platform to ensure leadership engagement and sign-off, which drives a consistent message to the organization and the important role each associate plays in its success.

Introduction to policies, standards, procedures, and guidelines

A follow-on to the baselining overview section is policies, standards, procedures, and guidelines. This section works hand in hand with baselining and holds extreme importance within an organization. It is critical as part of your security program that well-defined policies, standards, and procedures are in place and being followed by everyone. In addition, it is important that the policies are signed off and enforced by leadership. Without this support, it becomes more difficult to enforce and collectively get behind security at an organizational level.

Start by defining and creating your company policies. As a result, your standards can then be built to form the foundation of your baselines. Once these baselines are created, procedures and guidelines can be built to implement the baselines and help accomplish the end goal. Keeping this strategy in mind will drive compliance with your company policies.

The following provides a brief overview, with recommendations of policies, standards, procedures, and guidelines.

Defining policies

A security policy is the first level of formalized documentation for your organization's security program and is mandatory. Policies are a critical component of your overall security program, which requires sign-off and support from the leadership team to ensure success. Policies should be very broad and general with no direct tie to the technology or solutions within the organization. In general, they should not change often, but periodic review is critical. Some examples of policies may include acceptable use policy, change management policy, disaster recovery policy, privacy policy, information security program policy, and so on.

If you don't have any policies in place, it is highly recommended you begin with some basics, as it relates to your Windows security today. The following, at a minimum, should be included as part of securing your devices and referenced in a policy:

- Security updates
- Encryption
- Firewall
- Password policy, **Multi-Factor Authentication (MFA)**, and biometrics
- Local administrative access strategy

- Security protection tools and antivirus
- Compliance and protection policies
- Data loss prevention and information protection

An example of a documented policy may include *All systems must be kept up to date with the most recent security updates.*

Next, let's look at setting standards to follow the defined policies.

Setting standards

Standards follow policies in that they define the specifics of the policy and are mandatory. These provide the direction needed to support the policies. Standards help enforce consistency throughout the organization and will provide specifics on the technology to be deployed.

Referring to the recommended items listed in the policy, the following are examples of standards for policies:

- All Windows workstations will be configured using Windows Update for Business and Windows servers using WSUS or Azure Update Management. Update schedules will be defined and documented by the business use case.
- All Windows servers and end user workstations will be encrypted using BitLocker and/or Azure Disk Encryption.
- A Windows firewall will be enabled and configured on all Windows end-user devices and servers. Connection rules will be documented.
- A PIN and biometrics with *Windows Hello* must be used, and accounts will be required to use a password with a minimum of 12 characters. Passwords must contain a lowercase, uppercase, numerical, and special character and are required to be changed annually.
- MFA will be required for all users accessing the corporate environment and resources.
- There will be no standard user accounts assigned with local admin access on any Windows device.
- All Windows end-user devices and servers will be enabled with Microsoft Defender for Endpoint.

- Compliance policies for conditions such as device risk and minimum OS version will be assigned and enforced with Conditional Access on Windows devices.

- Unified labeling with data loss prevention and information protection will be deployed to all Windows end-user devices.

Next, let's look at building procedures to define a set of instructions used to accomplish tasks.

Creating procedures

Procedures are the step-by-step instructions used to accomplish a repeatable task or process. The instructions are intended to achieve a specific goal and assist with implementing the defined policies, standards, and any guidelines that may apply. Procedures may change frequently depending on software version changes, hardware replacement, and so on. For better organization of procedures, you may want to look at third-party tools for help. One example is a tool known as Nintex Promapp that helps document and share your organization's processes: `https://www.nintex.com/process-automation/process-mapping/`.

Here is an example of a procedure:

1. Deploy a new device with Windows.

2. Ensure the device is connected to the internet.

3. Verify that the device configurations and applications have been installed.

4. Check whether the device is compliant.

Finally, let's look at creating guidelines to act as recommended best practices.

Recommending guidelines

Guidelines provide recommendations or best practices and are not mandatory requirements. They can be complementary controls in addition to standards, or even provide guidance where a standard may not apply.

An example of a guideline might be: *Ensure you save and close all documents and programs before rebooting after receiving the latest Windows updates.*

Although not mandatory, guidelines provide a lot of value to users to help them to be more productive with technology. When building guidelines, it's important to think about how to efficiently provide the users with visibility and access to the guidelines. An effective communication plan is critical to ensure the users read and use the guidelines. The following are five ideas that will help with communicating your guidelines:

- Build a theme around your guideline communications, for example, Smart Tech Guidelines.

- Insert a section on guidelines in the company newsletters and/or communications.

- Link your guidelines back to a central repository for users to come back and access it.

- Keep your guidelines short and to the point.

- Make your guidelines relevant to both professional and personal usage.

The following diagram illustrates the hierarchy of policies, standards, procedures, and guidelines with the addition of where baselines fall within the model:

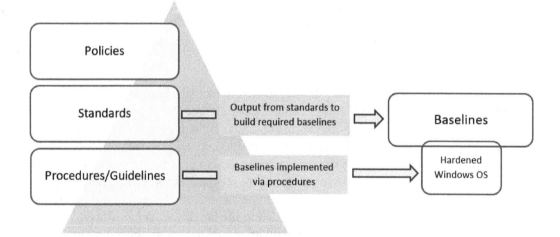

Figure 2.1 – Policies, standards, procedures, guidelines, and baselines

In the next section, we will provide an overview of the change management process. It's important to follow a change control process whenever you're implementing change in an environment.

Incorporating change management

It is critical that you understand the importance of change management and its place as part of the overall security program. Your organization most likely has some form of change control process in place today. If not, it is highly recommended that one be enabled to provide a more structured and reliable environment.

The following diagram provides an example of a change flow process that you should implement if one doesn't exist within your environment today:

Figure 2.2 – Change management flow process

Change management is typically part of a larger program, more specifically around service management. One of the more common frameworks to help with change management is **Information Technology Infrastructure Library (ITIL)**.

> **Tip**
> Visit this website to learn more on ITIL: `https://www.axelos.com/certifications/itil-service-management/what-is-itil`.

As part of your security program, you will want to ensure that you have all your baselines signed off from management and that they are well documented. More importantly, you will need to ensure that the baselines are being implemented with every deployment. If any exceptions or deviations from the baselines are needed, it will be extremely important that the requests are pushed through the change management process and audited. They will need to be reviewed and approved by the appropriate teams, which will most likely include sign-off from someone in the security team who will be part of the change process. The same will apply to any changes needed to the baselines. As hardware, software, and operating systems change, there will be a need to modify the baseline to meet the changes. These changes should also go through a change control process to ensure everyone agrees to and approves the changes.

> **Important Note**
> If a security incident occurs on a system where a baseline wasn't correctly applied and approvals were not received for that exception, you could be putting the company and your job at risk.

Next, let's look at security frameworks and widely adopted frameworks that can be incorporated into your own security program.

Implementing a security framework

There's a possibility that your organization may have an information security framework in place today. If not, it's highly recommended you begin that journey straight away to help lay the foundation of your security program and strategy. There are many different frameworks available for implementation, and the direction you take may depend on multiple factors according to your business type, industry requirements, and regulations.

An information security framework is designed to build a well-defined basis for your organization's security program. One of the primary reasons to implement an information security framework is to help reduce risk as much as possible. It will help cover the foundation of everything you need to be aware of within your security program and help to identify any gaps within the organization.

Implementing an information security framework isn't easy and can be extremely complex, requiring a major time investment. Implementing a framework won't just happen overnight; it will take a lot of planning and many months, and even years, to implement correctly. It will be important to think of the framework as more of a journey that continues to evolve and improve over time.

A significant benefit of implementing a framework is the ability to provide a well-constructed overview of your security program and strategy to executive management and leadership. The framework will help provide the executive team with a comprehensive view of what security controls are in place and a roadmap of work to be completed. This will also allow them to provide feedback, prioritize needs, provide valuable input, and provide justification for budget allocation. The ability to make the security program and strategy transparent to leadership is a significant advantage.

The following are some of the more common and widely adopted frameworks in the world today:

- **Control Objectives for Information and Related Technology (COBIT):** `https://www.isaca.org/resources/cobit`

- **International Standards Organization 27000 Family (ISO):** `https://www.iso.org/isoiec-27001-information-security.html`

- **National Institute of Standards and Technology (NIST)** Framework for Improving Critical Infrastructure Cybersecurity, also known as the NIST Cybersecurity Framework: `https://www.nist.gov/cyberframework`

- NIST SP 800-53: Security and Privacy Controls for Federal Information Systems and Organizations: `https://csrc.nist.gov/publications/detail/sp/800-53/rev-5/final`

- NIST SP 800-171: Protecting Controlled Unclassified Information in Nonfederal Systems and Organizations: `https://csrc.nist.gov/publications/detail/sp/800-171/rev-2/final`

- **Health Information Trust Alliance Common Security Framework (HITRUST CSF)**: `https://hitrustalliance.net/hitrust-csf/`

Your industry and location may dictate which framework is to be used, but in general, they can all be used throughout any industry as a foundation. As an example, healthcare will most likely adopt the HITRUST framework. ISO 27000 and COBIT most likely will have a more global presence over NIST, which is leveraged by the US government primarily.

To help with your implementation, let's take a closer look at the NIST framework for improving critical infrastructure cybersecurity. Although the framework was initially created for critical infrastructure, it can be used by any organization of any industry and size. This framework has gained a lot of popularity and is being adopted by many. The NIST Cybersecurity Framework is built around five core functions, as shown here:

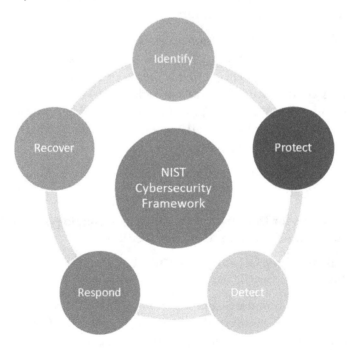

Figure 2.3 – NIST Cybersecurity Framework core functions

More information about the five functions in the NIST framework can be found at this link: `https://www.nist.gov/cyberframework/online-learning/five-functions`

Within these functions are categories that break down the focus to a subcategory, with a set of references on how to manage the risk within that given subcategory. To take this a step further, let's review the specific category that relates to the baseline configuration that you will follow as part of your overall implementation. The following figure breaks down the **Protect** function of the NIST framework:

Function	Category	Subcategory	Informative Reference
Protect (PR)	**Information Protection Processes and Procedures (PR.IP):** Security policies (that address purpose, scope, roles, responsibilities, management commitment, and coordination among organizational entities), processes, and procedures are maintained and used to manage the protection of information systems and assets.	**PR.IP-1:** A baseline configuration of information technology/industrial control systems is created and maintained incorporating security principles for example, the concept of least functionality)	**CIS CSC** 3, 9, 11 **COBIT 5** BAI10.01, BAI10.02, BAI10.03, BAI10.05 **ISA 62443-2-1:2009** 4.3.4.3.2, 4.3.4.3.3 **ISA 62443-3-3:2013** SR 7.6 **ISO/IEC 27001:2013** A.12.1.2, A.12.5.1, A.12.6.2, A.14.2.2, A.14.2.3, A.14.2.4 **NIST SP 800-53 Rev. 4** CM-2, CM-3, CM-4, CM-5, CM-6, CM-7, CM-9, SA-10

Figure 2.4 – Example from the NIST Cybersecurity Framework

> **Tip**
>
> The NIST framework for improving critical infrastructure cybersecurity web page that contains the preceding example can be found at `https://nvlpubs.nist.gov/nistpubs/CSWP/NIST.CSWP.04162018.pdf`.

As you can see from the figure, the NIST Cybersecurity Framework provides the needed guidance and resources that can be used to meet the controls. Ensuring a framework is adopted will build a solid foundation to ensure the required baseline controls to harden your systems are put in place. Frameworks represent the overall controls at a higher level and help ensure there are no gaps in your security program, including any gaps in your Windows infrastructure.

Next, let's look at baseline controls. Baseline controls are set to define a standard set of configurations for your devices.

Building baseline controls

Following on from the previous section, we will cover more specifics of the baseline controls that can be used for your Windows devices. Here we will cover the following:

- **Center for Internet Security (CIS)**
- Windows security baselines
- Intune's security baselines

Let's look at each of these options in detail.

CIS

First, we will look at CIS. You may already be familiar with CIS, and you will see CIS listed on a lot of most-popular-framework lists, although it's not a fully comprehensive framework like others listed previously. Instead, CIS is more of a tactical compilation of controls and guidelines that allow organizations to meet the requirements of a chosen framework. The following screenshot is of the current CIS home page, which can be visited at this link: `https://www.cisecurity.org/`.

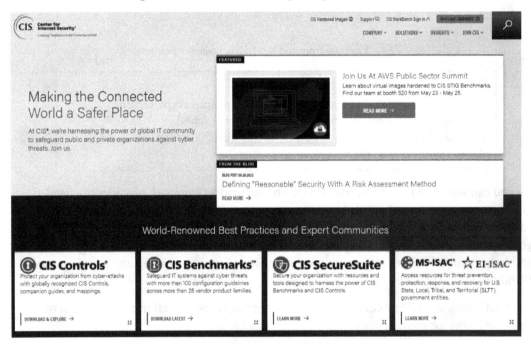

Figure 2.5 – CIS home page

CIS is a non-profit organization comprising a global community to provide protection against the ongoing cybersecurity threat landscape. More specifically, the CIS vision is *"Leading the global community to secure our ever-changing connected world."* Their mission is *"... to make the connected world a safer place by developing, validating, and promoting timely best practice solutions that help people, businesses, and governments protect themselves against pervasive cyber threats."*

> **Tip**
>
> You can learn more about CIS at `https://www.cisecurity.org/about-us/`, and sign up for the MS-ISAC advisories, newsletters, and webinars at `https://learn.cisecurity.org/ms-isac-subscription`.

CIS has an overwhelming number of tools and resources available, with many being free of charge. More specifically, CIS provides two sets of best practices widely adopted throughout the world: CIS Controls and CIS Benchmarks. As of version 8, the CIS Controls are a broader set of 18 foundational and advanced controls that provide a more comprehensive approach to overall security protection for your organization, whereas the CIS Benchmarks are focused more on the specific hardening of your systems, software, networks, and more.

> **Tip**
>
> CIS cybersecurity best practices can be found at `https://www.cisecurity.org/cybersecurity-best-practices/`.

If you opt to move forward with the CIS Benchmarks, you will need to download the checklist and customize it for your specific needs. The list of available benchmarks is extensive and includes categories for the following:

- Desktop and web browsers
- Mobile devices
- Network devices
- Security metrics
- Server operating systems
- Server application and roles
- Virtualization platforms
- Cloud and Microsoft apps

CIS also has the option of purchasing hardened images to provide for easier deployment moving forward.

To download the latest CIS Benchmarks, follow these steps:

1. Open a browser and navigate to `https://www.cisecurity.org/cis-benchmarks/`.

2. Click on **Access all Benchmarks**.

3. Enter the required information, agree to the terms, and then click **Get Free Benchmarks Now**.

CIS Benchmarks
Download Our Free Benchmark PDFs

The CIS Benchmarks are distributed free of charge in PDF format to propagate their worldwide use and adoption as user-originated, de facto standards. CIS Benchmarks are the only consensus-based, best-practice security configuration guides both developed and accepted by government, business, industry, and academia.

View Our Extensive Benchmark List:

Desktops & Web Browsers:

- Apple Desktop OSX
- Apple Safari Browser
- Google Chrome
- Microsoft Internet Explorer
- Microsoft Windows Desktop XP/NT
- Mozilla Firefox Browser
- Opera Browser

Mobile Devices

- Apple Mobile Platform iOS
- Google Mobile Platform

Network Devices

- Agnostic Print Devices
- Checkpoint Firewall
- Cisco Firewall Devices

FREE BENCHMARKS

Complete the form below and get access to ALL of our benchmark PDFs for free.

First Name *

Last Name *

Organization *

Sector *

Role *

Email *

Country *

☐ I have read and agree to the Terms of Use*.
Commercial use is prohibited without a CIS SecureSuite Membership permitting such use. I may receive emails from CIS related to submitting this form and marketing emails if outside the EU unless I opt out. *

Get Free Benchmarks Now

* Terms of Use.
* By submitting the form, I have reviewed the CIS Privacy Policy.

Figure 2.6 – CIS Benchmarks download page

4. Go to your mailbox and look for an email from CIS (check the junk email folder also).

5. Open the email and click **Access PDFs**.

6. You will be provided with a list of all the available CIS Benchmarks in PDF format. **Microsoft Windows Desktop** will be at the top of the list:

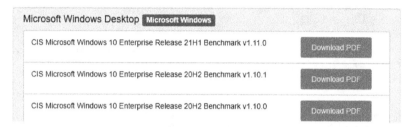

Figure 2.7 – CIS Benchmark Microsoft Windows Desktop PDFs

7. Scroll down and you will see **Windows Server** CIS Benchmarks.

8. Keep scrolling down and you will also see **Microsoft Azure** CIS Benchmarks:

Figure 2.8 – CIS Benchmark Microsoft Azure PDFs

9. In addition, there are many more Windows-specific CIS Benchmarks for specific roles, such as IIS, SQL, and Exchange.

10. Once you download the PDFs, follow and implement the recommendations to harden your systems.

> **Tip**
> Visit here to access CIS Hardened Images that map back to the CIS Benchmarks: `https://www.cisecurity.org/cis-hardened-images`.

If you are using Azure as your data center, there are also CIS Hardened Images available in Azure Marketplace for simpler, hardened deployment for your servers. Log in to https://portal.azure.com, navigate to **Marketplace**, and search for CIS Windows.

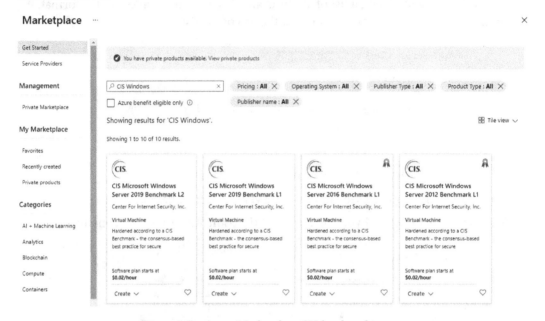

Figure 2.9 – Azure Marketplace CIS hardened images

Next, let's look at security baselines specifically for Windows and the Microsoft Security and Compliance toolkit baselines for Windows.

Windows security baselines

As part of their services, Microsoft offers Windows security baselines from the Microsoft **Security Compliance Toolkit** (**SCT**), which provides recommended configurations to harden your Windows systems. The following options are available for selection from the Microsoft download site:

- Windows 11 security baselines
- Windows 10 security baselines
- Windows Server security baselines
- Microsoft Office security baselines
- Microsoft Edge security baselines
- Policy Analyzer and **Local Group Policy Object** (**LGPO**) tools

To give an idea of the complexity of securing Windows, the **Group Policy Object (GPO)** consists of thousands of configurable settings. This clearly shows the need to leverage pre-defined baselines to help with the hardening of your Windows devices. The more common Microsoft tools used to implement these baselines will consist of the following:

- Microsoft Intune

- Group Policy Objects

- Microsoft Endpoint Configuration Manager

> **Tip**
>
> Go here to view additional information on Windows security baselines:
> `https://docs.microsoft.com/en-us/windows/security/`
> `threat-protection/windows-security-configuration-`
> `framework/windows-security-baselines.`

If you go in the Microsoft direction with Windows security baselines, the available resources can be downloaded for free to allow the implementation of a baseline. To download the resources, follow these steps:

1. Go to `https://www.microsoft.com/en-us/download/details.aspx?id=55319`.

2. Click **Download**.

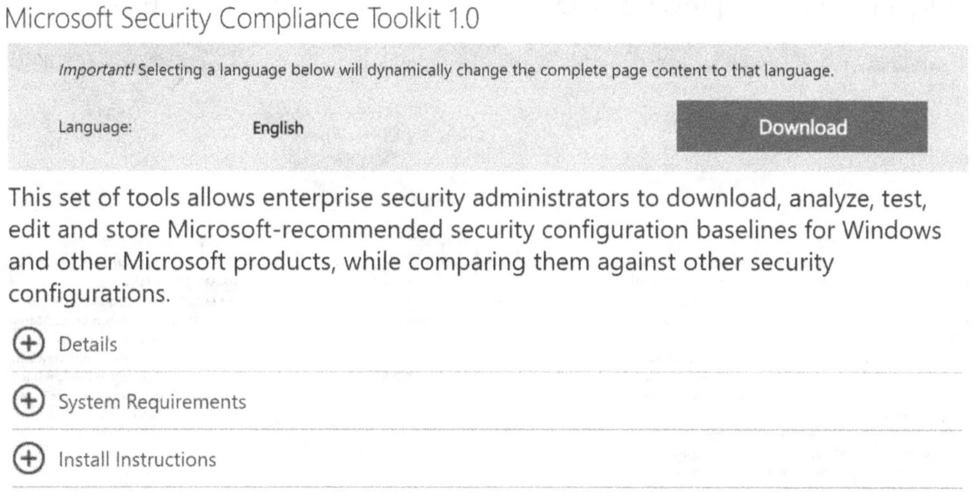

Microsoft Security Compliance Toolkit 1.0

Important! Selecting a language below will dynamically change the complete page content to that language.

Language: English   Download

This set of tools allows enterprise security administrators to download, analyze, test, edit and store Microsoft-recommended security configuration baselines for Windows and other Microsoft products, while comparing them against other security configurations.

⊕ Details

⊕ System Requirements

⊕ Install Instructions

Figure 2.10 – Microsoft SCT 1.0 download

3. Select the desired versions, or click the box next to the filename to select all.

4. Click **Next**.

5. You will receive all the toolkits in `.zip` format.

Downloading the toolkit referenced previously will provide you with everything needed to deploy the recommended baselines from Microsoft. The following describes what's included in the security baselines ZIP file(s):

- The `documentation` folder includes an `Announcement` file that summarizes the recommendations and new settings. It also contains the full baseline in XLSX format and a `New Settings` spreadsheet. There are `PolicyRules` files that are useful when viewing comparisons with the Policy Analyzer tool.

- The `GP Reports` folder lists the outputs of each GPO in HTML format.

- The `GPOs` folder contains the **Globally Unique Identifiers (GUIDs)** for each GPO setting.

- The `Scripts` folder contains helpful scripts that can be used to map the GPO GUIDs to friendly names or to import baselines into Active Directory.

- The `Templates` folder includes `ADMX` and `ADML` files for Group Policies that are referenced in the new baselines and might not be included in the latest available download of Administrative Templates.

The following figure shows a quick snapshot of the baseline settings in Excel format that can be deployed using the provided GPOs within the toolkit for Windows 11:

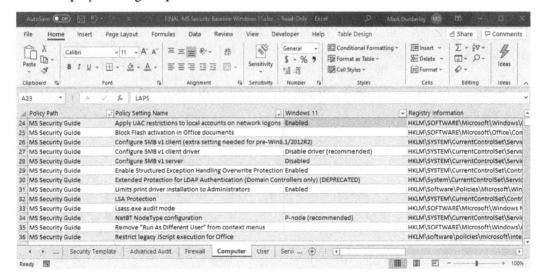

Figure 2.11 – FINAL-MS Security Baseline Windows 11.xlsx

The following screenshot shows the download in Excel format for Windows Server 2022:

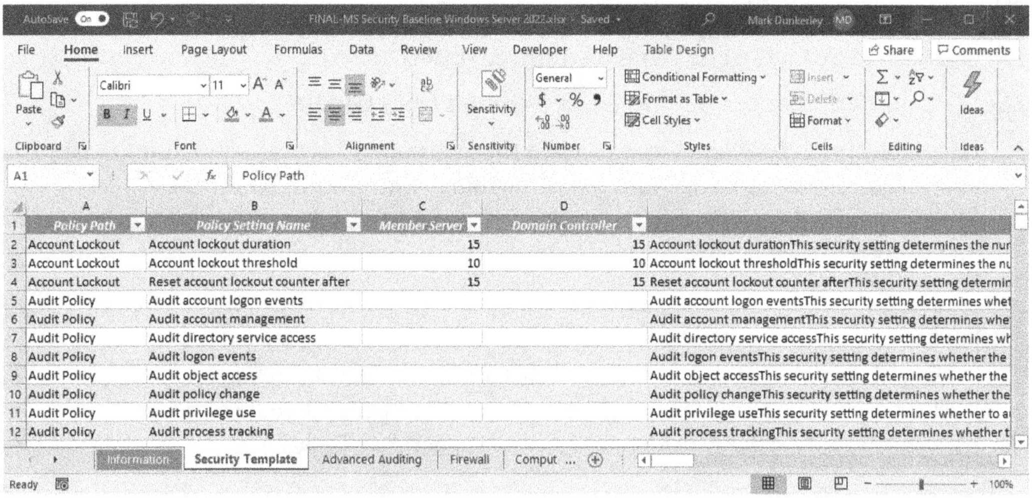

Figure 2.12 – FINAL-MS Security Baseline Windows Server 2022.xlsx

Notice that Microsoft provides separate settings for a member server versus a **Domain Controller (DC),** providing additional settings specifically for your DCs as part of the Windows Server baseline. Also, if you look at the bottom of the spreadsheet, you will see the different categories the hardening is being applied to.

Comparing policies with Policy Analyzer

The Policy Analyzer tool is useful for comparing GPOs against other GPOs, the local policy settings on the computer, and the local registry. If you have downloaded the Microsoft SCT already, ensure you have selected `PolicyAnalyzer.zip` and `Windows 11 Security Baseline.zip` at a minimum. Extract them if you wish to follow along.

Let's compare the out-of-the-box Windows 11 settings to the Windows 11 security baseline by following these steps:

1. Open `PolicyAnalyzer.exe` as **Administrator**.

2. Click on **Add**, choose **File**, and select **Add files from GPO(s)…**.

3. Open the directory where you extracted the Windows 11 security baselines, navigate to the `GPOs` folder, and select the folder with the GUID for the **MSFT Windows 11 – Computer policy** (`9FE25A81-CB6B-4F76-B9D2-147E9BED9A06`).

> **Tip**
>
> There are many GUIDs inside the GPO folder of the extracted baseline. Clicking on the `gpreport.xml` file will display the policy friendly name, or use the `MapGuidsToGpoNames.ps1` file in **Scripts | Tools** for help.

4. In **Policy File Importer**, click on **Import**. Give it a friendly name such as `MSFT Windows 11 – Computer`, then click on **Save**.

5. This will bring you back to the main menu of Policy Analyzer. Ensure the policy you imported is selected, then click on **View/Compare** to bring up **Policy Viewer**.

The column headings are straightforward and provide an overview of the policy group or registry key as well as the policy setting. If you click on a row, detailed information will be presented about the policy in the details pane below the policies. Here are some notes to keep in mind about the menu options:

- Selecting **View | Show only Differences** will display all of the differences between the recommended baseline and what's currently set in the local registry.

- Selecting **View | Show only Conflicts** will display settings that differ from the recommended baseline to the current setting. These will be highlighted in yellow.

- Selecting **Export | Export table to Excel** will export only the table with differences or conflicts.

- Selecting **Export | Export all data to Excel** will include the details pane explanation.

The following screenshot shows the **Policy Viewer** output after clicking on **View/Compare**. Anything highlighted in yellow is a conflict between the GPO and current setting in the local registry:

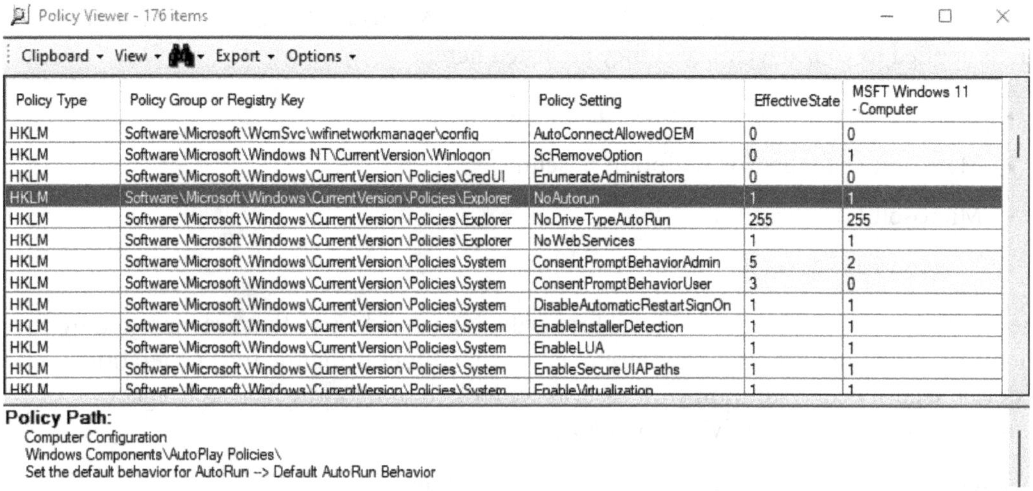

Policy Type	Policy Group or Registry Key	Policy Setting	EffectiveState	MSFT Windows 11 - Computer
HKLM	Software\Microsoft\WcmSvc\wifinetworkmanager\config	AutoConnectAllowedOEM	0	0
HKLM	Software\Microsoft\Windows NT\CurrentVersion\Winlogon	ScRemoveOption	0	1
HKLM	Software\Microsoft\Windows\CurrentVersion\Policies\CredUI	EnumerateAdministrators	0	0
HKLM	Software\Microsoft\Windows\CurrentVersion\Policies\Explorer	NoAutorun	1	1
HKLM	Software\Microsoft\Windows\CurrentVersion\Policies\Explorer	NoDriveTypeAutoRun	255	255
HKLM	Software\Microsoft\Windows\CurrentVersion\Policies\Explorer	NoWebServices	1	1
HKLM	Software\Microsoft\Windows\CurrentVersion\Policies\System	ConsentPromptBehaviorAdmin	5	2
HKLM	Software\Microsoft\Windows\CurrentVersion\Policies\System	ConsentPromptBehaviorUser	3	0
HKLM	Software\Microsoft\Windows\CurrentVersion\Policies\System	DisableAutomaticRestartSignOn	1	1
HKLM	Software\Microsoft\Windows\CurrentVersion\Policies\System	EnableInstallerDetection	1	1
HKLM	Software\Microsoft\Windows\CurrentVersion\Policies\System	EnableLUA	1	1
HKLM	Software\Microsoft\Windows\CurrentVersion\Policies\System	EnableSecureUIAPaths	1	1
HKLM	Software\Microsoft\Windows\CurrentVersion\Policies\System	EnableVirtualization	1	1

Policy Path:
Computer Configuration
Windows Components\AutoPlay Policies\
Set the default behavior for AutoRun --> Default AutoRun Behavior

Figure 2.13 – Policy Viewer in Policy Analyzer

We will provide more details on the implementation of security controls in *Section 2, Applying Security and Hardening.*

> **Tip**
> The Microsoft **Security Compliance Toolkit (SCT)** also provides additional details on the available tools to manage your Windows baselines: `https://docs.microsoft.com/en-us/windows/security/threat-protection/security-compliance-toolkit-10`.

Next, let's look at Intune's security baselines to configure Windows devices managed by Intune, to secure and enhance protection for your users and devices.

Intune's security baselines

If you are using Microsoft Endpoint Manager to manage your end-user Windows devices, you can take advantage of the built-in Intune security baselines to secure and harden your devices for your users. The baselines can be used pre-configured as is, or they can be customized to better meet your requirements. To use Intune's security baselines, your device needs to be running the following at a minimum:

- Windows 10 version 1809 and later
- Windows 11

There are currently four security baselines available in Microsoft Endpoint Manager that can be applied to your devices, and they are listed here:

- Security Baseline for Windows 10 and later
- Microsoft Defender for Endpoint Baseline
- Microsoft Edge Baseline
- Windows 365 Security Baseline

These baselines are extremely powerful for hardening your devices that are managed by Microsoft Endpoint Manager. The Intune security baselines bring the next generation of device security configuration with ease, as we look to move away from the legacy management of using and relying on GPOs.

To access the Intune security baselines, do the following:

1.   Log in to `https://endpoint.microsoft.com/`.
2.   Click on **Endpoint Security** on the left menu.
3.   Click on **Security baselines**.
4.   Here you will see the available security baselines for use:

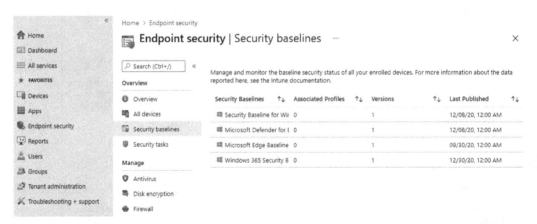

Figure 2.14 – Intune security baselines

As already mentioned, the baselines all come pre-configured with default values based on Microsoft's recommendations for a baseline. Microsoft has a very comprehensive overview of the configurations and what the defaults are for each of their baselines. They also maintain a history of the released versions for review. You can link to each of the baseline settings here: `https://docs.microsoft.com/en-us/mem/intune/protect/security-baselines#available-security-baselines`.

> **Tip**
> We will cover Intune security baselines in more detail in *Section 2, Applying Security and Hardening*.

No matter which method you use to implement your baselines, it is highly recommended when making any configuration changes from newly released baselines to ease them into production and thoroughly test them first. In addition, ensure any changes go through your change control process for tracking and offer transparency to the business.

Next, let's recap the chapter by providing a checklist of best practices that will help when building a security framework and implementing your baselines.

Incorporating best practices

To finish off the chapter, we wanted to provide a checklist of the most important items that will help enforce your security baselines. The following list is ranked in order of importance as you look to build and enforce your baselines:

1. Select and deploy a framework to build a foundation; the NIST Cybersecurity Framework is a great place to start.

2. Select a baseline foundation; we covered CIS, Windows security baselines, and Intune's security baselines.

3. For your Windows devices, use Policy Analyzer from the Microsoft SCT to review your baselines, or use the CIS-CAT Pro tools to review your system configurations against the CIS Benchmarks.

4. Create or use (and re-use) a *Golden Image* template for each use case and always keep up with the latest updates. CIS has pre-defined hardened images that can be used.

5. Build well-documented and easy-to-follow procedures that others can use and follow.

6. Use automation of controls and tools to reinforce the baseline; MDM with Intune or Active Directory Group Policy as an example.

7. Use compliance policies to validate controls are in place. This will also help with auditing devices that are non-compliant.

8. Implement a quarantine or risk access policy with non-compliant devices.

9. Implement efficient monitoring and reporting for device compliance. Microsoft Power BI is a great way to visually provide reports.

10. Always keep up with the latest Windows versions and the technology used to manage the devices. The modern world is very dynamic and moving at an extremely fast pace.

It's important to note that while creating a security framework and enforcing controls with full compliance is desirable, exceptions will need to be accounted for. It is recommended that your organization also includes a risk register that clearly documents the systems and applications that cannot comply with the defined policies and standards. The register should identify all risks as well as rate the implications or severity each risk has for potential impacts on the organization. These implications should not only be focused on from the security lens, but also identify potential legal liabilities and cost implications if the risks were exploited. Leadership should be made aware of these risks and sign off on their acceptance. Furthermore, a stakeholder should be named to act as the accountable party and the register should be reviewed frequently to identify any possible solutions to mitigate the risks.

Summary

Throughout this chapter, we provided an overview of baselining to help you understand its importance and role within the overall security program. Next, you learned about policies, standards, procedures, and guidelines and the importance of them as part of your overall security strategy. We also looked at how these policies, standards, procedures, and guidelines interact and build upon each other to structure the baseline model. We then covered the change management process with respect to baseline management.

Finally, we reviewed frameworks and their role within the security function of your organization, discussing the more widely adopted frameworks implemented. Following this section was an overview of baseline controls that are available for Windows. These option. included CIS, Windows security baselines, and Intune's security baselines, and we gave you an idea of where to retrieve pre-defined templates, configurations, and images before providing best practices as part of baselining.

In the next chapter, we will cover hardware and virtualization. The chapter will review the importance of ensuring both your hardware and virtualized machines adhere to the same level of security that you implement on the Windows OS.

3
Hardware and Virtualization

In this chapter, we will cover the importance of hardware and virtualization as it relates to security. These items can be easily overlooked but are critical components of the overall strategy for securing your Windows systems. As you purchase hardware, you need to consider the process that exists for the supply chain. Who is manufacturing the hardware and how do we trust that the components building the final product are clean and free from vulnerabilities? How do we validate that no additional components have been added that could compromise our security and privacy? We also need to take into consideration the existence of hardware vulnerabilities that become extremely difficult to manage. Ensuring your hardware is the latest available version and secure is just as important as protecting your **operating system (OS)**. Vulnerabilities such as Meltdown and Spectre are prime examples of this. You can learn more about Meltdown and Spectre by reading this article: https://www.us-cert.gov/ncas/alerts/TA18-004A.

In addition to the underlying physical hardware is the hardening and securing of virtualized infrastructure that's used to deploy your data center and virtual workstations. Chances are, most of you have some form of virtualization deployment within your environment. Moving from a decentralized management model to a centralized model for your systems could allow a **single point of compromise (SPOC)** to morph into a major compromise of many systems within your virtual infrastructure. For this reason, it is critical that you take time to understand the hardware being used within your environment and ensure it is protected correctly and is secured. A weakness within your hardware is a vulnerability to the running OS, and all the investment you've spent on hardening the OS becomes obsolete.

Throughout this chapter, we will provide the awareness needed to ensure you know the best hardware-based security features for protecting your Windows OS. As you read through this chapter, you will learn more about the following topics:

- Physical servers and virtualization
- Introduction to hardware certification
- The firmware interface, **Trusted Platform Module (TPM)**, and Secure Boot
- Isolated protection with **virtualization-based security (VBS)**
- Protecting data from lost or stolen devices
- Hardware security recommendations and best practices

Technical requirements

For the *Physical servers and virtualization* section, you will need a Windows OS running with the supported hardware listed throughout to set up Hyper-V. You will also need access to an Azure subscription to set up a **virtual machine (VM)** within Azure. To do this, go to https://portal.azure.com/.

For the remainder of this chapter, each section will have baseline requirements needed to turn on specific hardware-based security features and will include links for further reference.

Physical servers and virtualization

Today, your organization likely has physical hardware for both your data center and end users. In both scenarios, the Windows OS will be running on top of the physical hardware layer. This adds an extra layer of concern as it relates to security. Within the physical device, your OS requires interaction with the hardware and your data will interact with hardware components such as the **central processing unit (CPU)** and **random-access memory (RAM)**. The same will apply to hard drives, which contain the OS and any personal data stored locally at rest. If no action is taken regarding your storage devices, your data will be in cleartext and easily readable. Understanding the physical layer of your devices and what can be done to better protect them is a critical step in protecting data and the Windows OS.

In addition to running a single OS on a physical device comes the concept of virtualization. Virtualization, in its simplest form, allows you to take a physical server and install multiple isolated VMs and OSs on top of shared underlying physical hardware. This allows for greater efficiency and workloads with your current hardware and resources. Prior to cloud and virtualization technology, you were required to deploy a physical server for each app/service you wanted to deploy (not accounting for **High Availability (HA)** configurations). This was a manageable process in the early days of server compute data centers, but as more demand came along with the extremely fast pace of technology and the need for more apps, the ability to deploy physical servers quickly became very challenging and expensive.

Fortunately, the advancement of data center virtualization became available, and the ability to deploy multiple OSs on a single piece of server hardware has been a game-changer within the enterprise. In the following diagram, the physical server deployment underlying hardware only has one OS, while the virtualized deployment has many sharing the same hardware host:

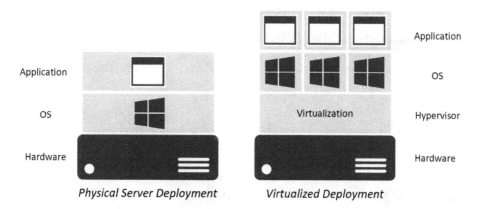

Figure 3.1 – Physical server deployment versus virtualized deployment

Within the end user world, virtualization has also been widely adopted. Virtualization of the desktop has allowed companies to overcome many challenges as it relates to quickly deploying desktops to contractors/vendors and offshore employees. Virtualization also provides access to legacy applications that may not be supported on the latest OS and provides additional desktops to users for development and testing. It is also a great scenario for part-time workers or those who only need limited access intermittently. Virtualization provides ample opportunity with great flexibility, but we must remember that it still runs on physical hardware with similar security concerns when it comes to protecting Windows.

Microsoft virtualization

Let's review the virtualization technologies Microsoft provides for both the Windows Server environment and end-user desktop.

Hyper-V

For traditional on-premises deployments, Microsoft has Hyper-V technology, which is a hypervisor-based virtualization platform. As with other enterprise-grade platforms, Hyper-V allows you to manage, deploy, and run multiple VMs on a single piece of hardware, thus allowing better use of your hardware resources. The following requirements are needed to run VMs in Hyper-V:

- Windows hypervisor
- The Hyper-V Virtual Machine Management service
- The virtualization **Windows Management Instrumentation** (**WMI**) provider
- The **Virtual Machine Bus** (**VMBus**)
- **Virtualization service provider** (**VSP**)
- **Virtual infrastructure driver** (**VID**)

The following tools can be used for managing the Hyper-V environment:

- Hyper-V Manager
- Hyper-V module for Windows PowerShell
- VM connection (sometimes called VMConnect)
- Windows PowerShell Direct
- System Center **Virtual Machine Manager** (**VMM**)

For further information about Hyper-V technology, visit this link:

`https://docs.microsoft.com/en-us/windows-server/virtualization/hyper-v/hyper-v-technology-overview`

Hyper-V can be run on both Windows Server and Windows desktop, with the latest versions being Windows Server 2022 and Windows 11. Hyper-V on Windows Server is designed for more of an enterprise-grade deployment, thus allowing the migration of VMs between hosts and access to enterprise-grade storage for improved VM performance. It also supports advanced hardware protection features using TPM attestation with Guarded Fabric and Shielded VMs. Hyper-V for Windows desktops allows users to spin up multiple VMs for testing/development purposes and can run multiple OSs. Hyper-V also supports multiple OSs aside from Windows.

> **Tip**
> To view a list of supported OSs that can run on Hyper-V, visit this link:
>
> `https://docs.microsoft.com/en-us/windows-server/virtualization/hyper-v/supported-windows-guest-operating-systems-for-hyper-v-on-windows`

To get started with Hyper-V on Windows Server, you will need the following minimum requirements:

- 64-bit processor with **second-level address translation (SLAT)**.

- VM monitor mode extensions.

- At least 4 **gigabytes (GB)** of RAM. This will increase the more VMs you would like to run concurrently.

- Virtualization support turned on in the **Basic Input/Output System (BIOS)** or **Unified Extensible Firmware Interface (UEFI)** with the following:

  - Hardware-assisted virtualization

  - Hardware-enforced **Data Execution Prevention (DEP)** > Intel: **execute disable bit (XD bit)**; **Advanced Micro Devices (AMD)**: **no execute bit (NX bit)**

In addition to the basic requirements for Hyper-V, you will want to ensure you enable Guarded Fabric and Shielded VMs for the best protection and security. Guarded Fabric is the infrastructure component used to enable and protect Shielded VMs from being compromised. Shielding a VM allows it to only be run on an approved host and prevents unauthorized access within the environment, offline or outside of the protected environment. Shielded VMs first became available in Hyper-V 2016 and require the following features:

- UEFI 2.3.1 (Ensure boot is configured to use UEFI)
- TPM v2.0
- **Input-output memory management unit** (**IOMMU**) and SLAT
- Secure Boot enabled

You will need to be using **generation 2** (**Gen 2**) VMs and will require a minimum Windows Server 2012 OS to enable the Shielded VM feature. More information on Guarded Fabric and Shielded VMs can be found at this link: `https://docs.microsoft.com/en-us/windows-server/security/guarded-fabric-shielded-vm/guarded-fabric-and-shielded-vms`.

> **Tip**
> The following link is for the installation documentation for Hyper-V on Windows Server. This deployment is more involved and will require multiple physical devices in order to be set up and secured correctly: `https://docs.microsoft.com/en-us/windows-server/virtualization/hyper-v/get-started/install-the-hyper-v-role-on-windows-server`.

Getting started with Hyper-V on Windows desktop is a painless task. The feature is free to enable, and you will only need licenses for the respective OS you would like to run.

The minimum requirements needed to enable Hyper-V on Windows desktop are listed here:

- Latest Windows OS
- 64-bit processor with SLAT
- CPU support for VM Monitor Mode Extension (**Virtualization Technology** (**VT**)-c on Intel CPUs)
- Minimum of 4 GB memory
- Enabled in BIOS: VT and hardware-enforced DEP

Before you enable this feature, verify the hardware is compatible by opening PowerShell or Command Prompt, type `systeminfo`, and press *Enter*. Scroll down to the Hyper-V requirements and verify `Yes` is listed next to all items.

To view the additional Windows requirements to run Hyper-V, visit the following informational link: `https://docs.microsoft.com/en-us/virtualization/hyper-v-on-windows/reference/hyper-v-requirements`.

To install Hyper-V through the Windows **graphical user interface** (**GUI**), follow these instructions:

1. On your supported Windows device, go to **Search** and type `Turn Windows features on or off`.

2. Select **Turn Windows features on or off** to open the **Windows Features** window.

3. Search for `Hyper-V` and select it. Ensure all sub-options are selected.

4. Click **OK** in the **Windows Features** window, as shown in the following screenshot, and click **Restart Now** when prompted to reboot your device:

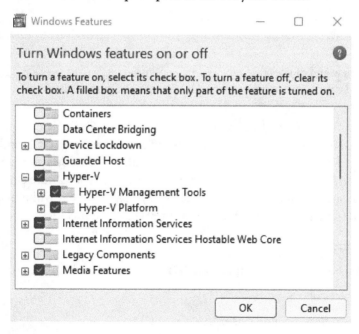

Figure 3.2 – Hyper-V in Windows Features

> **Tip**
>
> For additional methods to enable Hyper-V, visit the following link:
>
> `https://docs.microsoft.com/en-us/virtualization/`
> `hyper-v-on-windows/quick-start/enable-hyper-v`

Now that Hyper-V is enabled, you can start creating VMs and set up your own lab to implement and test the recommendations being provided in this book.

For more information on setting up your first VM, follow the instructions provided here: `https://docs.microsoft.com/en-us/virtualization/hyper-v-on-windows/quick-start/quick-create-virtual-machine`. You will need a license to run any Windows VMs that you have set up.

Azure Stack HCI

To date, Microsoft has announced that Hyper-V Server 2019 will be the last version of the product, but will continue to be available on newer versions of Windows and supported through January 2029.

> **Tip**
>
> You can view **end of life** (**EOL**) for all Microsoft products at `https://docs.microsoft.com/en-us/lifecycle/products/`.

Microsoft's next generation of a hypervisor platform is Azure Stack HCI. Azure Stack HCI is an Azure service that provides a **hyper-converged infrastructure** (**HCI**) and acts as a virtualized host. The solution allows you to run virtualized Windows and Linux similar to Hyper-V and bridges your on-premises infrastructure with your Azure cloud services. Azure Stack HCI can be managed with current tools you are familiar with, including Windows Admin Center and PowerShell. An Azure Stack HCI includes the following:

- An OS (Azure Stack HCI)
- An **original equipment manufacturer** (**OEM**) partner's validated hardware
- Hybrid services on Azure
- The Windows Admin Center **user interface** (**UI**)
- Compute resources within Hyper-V
- Software-defined storage using Storage Spaces Direct
- **Software-defined networking** (**SDN**) with an optional network controller

You can learn more about the Azure Stack HCI solution, including how to deploy, requirements, how to manage, and more here:

`https://docs.microsoft.com/en-us/azure-stack/hci/overview`

Azure VMs

Microsoft's Azure cloud **Infrastructure-as-a-Service (IaaS)** offering allows us to set up and consume VMs on-demand with no underlying infrastructure required from the consumer. To create a VM within Azure, follow the next steps. As a reminder, you will need an active Azure subscription to create VMs:

1. Log in to the Azure portal at `https://portal.azure.com`.
2. Click on the **Portal** menu at the top left (if the menu is hidden).
3. Select **All services** and choose **Compute**.

    Here, you will see all the compute services, including VMs. You can either click on the **Virtual machines** option to be directed to the VM portal or you can hover over **Virtual machines** to view additional information and options, as illustrated in the following screenshot:

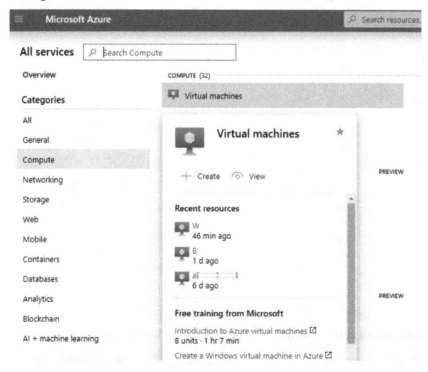

Figure 3.3 – VM tooltip in Microsoft Azure

In addition to the VM management portal, the Azure Marketplace has predefined images provided by Microsoft available to deploy. To view the available images within the Marketplace, follow these steps:

1. Search for `Marketplace` within the top search menu and choose **Marketplace**.

2. Select **Compute** in the **Categories** blade.

3. Search for `Windows` to view all the available Windows images, as illustrated in the following screenshot:

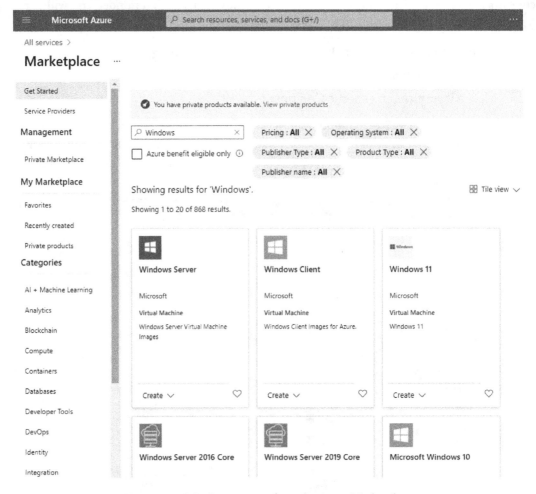

Figure 3.4 – Windows images from the Azure Marketplace

For more information about Windows VMs running in Azure, visit this link:

```
https://docs.microsoft.com/en-us/azure/virtual-machines/
windows/overview
```

We recommend that you use Azure to deploy and test security configurations for Windows Server and desktop for topics that are covered in this book.

Azure Virtual Desktop

Another option available for virtualization in Azure is **Azure Virtual Desktop** (**AVD**). This **Platform-as-a-Service** (**PaaS**) offering from Microsoft provides both desktop and app virtualization within the Azure cloud. The following features are provided with the AVD service:

- Multi-session Windows 11 hosts for non-persistent and persistent VMs

- Desktop and application virtualization with **Microsoft Software Installer Package** (**MSIX**) support and app attach for app streaming

- FSLogix profile containers using Azure NetApp files or Azure files for highly available streaming profiles

- Office 365 ProPlus and OneDrive are fully compatible

- Unified management portal to deploy hosts, manage user sessions, publish applications, manage scaling, and monitor performance

Visit the following link to learn more about the AVD service running in Azure:

```
https://docs.microsoft.com/en-us/azure/virtual-desktop/
overview
```

Windows 365 Cloud PC

Microsoft's latest Windows virtualization service is Windows 365 Cloud PC. Windows 365 Cloud PC helps simplify the ability to consume Windows VMs. With the recent adoption and ongoing need for a hybrid work model, Microsoft has built a service that allows for a secure Windows PC to be personalized and easily accessed from any device, including personal devices. Windows 365 Cloud PC is considered a **Software-as-a-Service** (**SaaS**) in which Microsoft manages the platform for you, taking away the need for any virtualization **subject-matter experts** (**SMEs**). The model is designed for ease of use with the ability for anyone to manage and consume, whereas AVD is a **virtual desktop infrastructure** (**VDI**) platform that requires some experienced expertise to architect, deploy, and manage, but provides greater flexibility over the management plane.

Some of the benefits of using Windows 365 Cloud PC are outlined here:

- Simple to set up, manage, and consume.

- Security is built into the model with a focus on Zero Trust.

- Access your Windows Cloud PC from any device.

- Latest OS with Windows 10 or Windows 11.

- Cloud PC can be personalized to meet your needs.

- Two editions are available: Windows 365 Business and Windows 365 Enterprise.

- Ideal for contractors, developers, and interns.

To learn more on Windows 365 Cloud PC, visit `https://docs.microsoft.com/en-us/windows-365/`.

Next, we'll look at the security concerns that affect the underlying hardware of our systems.

Hardware security concerns

Protecting the hardware of your systems is a critical task and one that may not have been at the forefront of your security priorities in the past. Over the years, we have invested heavily in the software side of security but recently, the impact of hardware vulnerabilities has shown the criticality of ensuring your hardware is protected from exploits, thus requiring more attention in this area to address risks.

The following list outlines some risks with hardware that must be addressed:

- Hardware vulnerabilities, including the following:

  - Rootkits embedded in the BIOS and UEFI

  - Side-channel attacks toward CPUs

  - Kernel-level exploits

  - Firmware attacks

  - Memory vulnerabilities

- Insider threats such as those with physical access to hardware and privileged access to your environment.

- Referencing back to the **National Institute of Standards and Technology (NIST)** Cybersecurity Framework, Cybersecurity Supply Chain Risk Management was added to the latest version (*1.1*) in 2018. NIST references the following cybersecurity supply chain risks to be aware of:

  - Any insertion of counterfeit items as part of the supply-chain process

  - The production of items that have not been approved

- Tampering with any items within the process

- Theft of any items

- The insertion of malicious software and hardware at any time during the process

- Manufacturing and development practices not maintaining expected standards during the supply-chain process

You can view additional information about the NIST *Cybersecurity Supply Chain Risk Management* project and the risks at this link:

```
https://csrc.nist.gov/Projects/Supply-Chain-Risk-Management
```

In 2021, MITRE and the **Cybersecurity and Infrastructure Security Agency (CISA)** released the first version of the **Common Weakness Enumeration (CWE)** *Most Important Hardware Weaknesses* list. You can view the list here: `https://cwe.mitre.org/scoring/lists/2021_CWE_MIHW.html`.

Now that we've covered various security concerns and common scenarios that can cause exposure to risk, let's look at concerns for the virtualized environment.

Virtualization security concerns

Now that it's commonplace to centralize hundreds and thousands of standalone workloads requiring significantly less hardware as a result of virtualization, it is just as critical that we ensure our virtualized infrastructure is protected correctly. A breach of a virtualized host could mean a compromise to hundreds of servers over a single physical server in past models.

Here are some risks with virtualization that must be addressed:

- Everything listed in the *Hardware security concerns* section we just covered. Your virtualized infrastructure will be running on the same hardware.

- Hypervisor threats.

- VM escape or the ability to interact with the physical host OS or hypervisor directly from a VM.

- Non-segregation of resources, network, and data.

- VM sprawl.

- Non-encrypted storage, physical drives, virtual disk files, and network traffic.

Although there are security tools and configurations that help ensure the security of your VMs and the data within them, you may want to consider the physical separation of specific functions within your virtualized infrastructure. For example, the management plane, the production environment, the **demilitarized zone** (**DMZ**), and highly confidential applications and databases should be separated. Ensure the separation of these functions includes the physical host, network, and storage—this will help safeguard against the risks. The following diagram shows a representation of using virtualization and isolation to separate different functions within an infrastructure deployment:

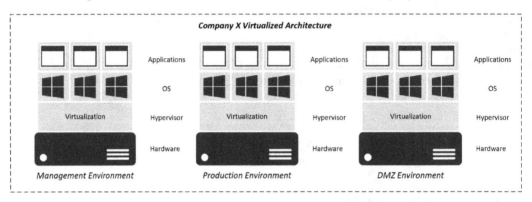

Figure 3.5 – Architecture separating core functions on its own hardware

> **Tip**
>
> *Microsoft Docs* has an article on Hyper-V security in Windows Server that can be found at this link:
>
> ```
> https://docs.microsoft.com/en-us/windows-server/
> virtualization/hyper-v/plan/plan-hyper-v-security-
> in-windows-server
> ```

Next, let's briefly review some concerns relating to cloud hardware and virtualization.

Cloud hardware and virtualization

As you move workloads to the cloud, you will, at a minimum, subscribe to an IaaS. Because of the dynamic changes with the cloud offering, you will no longer have any responsibilities regarding the hardware layer and the virtualized hypervisor layer. This will shift your efforts from needing to implement any best practices being recommended in this chapter to validating that the provider has these best practices implemented and in place. With Microsoft Azure, Microsoft is only responsible for implementing the security requirements for both the hardware and virtualized hypervisor layer of the services being provided to you. Any additional security hardening on the software layer—including the OS—is the customer's responsibility. You will need to work with Microsoft (or your cloud provider) to ensure evidence is provided to you, including responses to surveys and/or questionnaires, that ensures they have implemented the best security practices with their cloud hardware and virtualization offerings. This could also include audits, penetration tests, and security assessments.

Microsoft currently invests more than **United States dollars** (**USD**) $1 billion on cybersecurity research and development annually. With this investment, it continues to release and evolve new security features within Azure to provide the best protection against hardware and virtualization services. A couple of solutions that are worth becoming familiar with (as they relate to hardware and virtualization) include Azure Confidential Computing (`https://azure.microsoft.com/en-us/solutions/confidential-compute/`) and Microsoft Azure Attestation (`https://azure.microsoft.com/en-us/services/azure-attestation`), which provide an additional layer of security by applying encryption to data in memory, along with validations that the hardware and software being used is approved before allowing access.

As we go through the remainder of this chapter, you will be provided with an overview of the hardware security components and recommendations to ensure your systems are as secure as possible. In the next section, we will review hardware certification.

Introduction to hardware certification

Ensuring your hardware is certified is a critical process of the overall security program. As you purchase new servers, PCs, storage, and peripherals, it is critical you validate that the hardware is compatible with your deployed systems. Using non-compliant hardware could make your hardware vulnerable to a compromise, or the additional hardware components could even have a compromise already embedded in them.

An example would be allowing the use of **Universal Serial Bus** (**USB**) drives on your devices. Users receiving a free USB drive don't realize that the drive itself could be infected and that, once inserted into a company device, it could compromise the entire organization. Because of this, it is critical you only allow pre-certified USB drives that are encrypted and provided by the organization to be used by employees. Any data that is copied from a USB drive to a company device must require encryption. Another concern, as mentioned previously, is the supply-chain process. Ensuring the vendor has certified the hardware for Windows significantly reduces the risk the hardware could be pre-infected with vulnerabilities. This doesn't necessarily mean it will be 100% guaranteed, but your risk is reduced significantly.

Purchasing and procurement teams are a critical component in validating that the hardware you purchase has been through an appropriate certification process. These teams will help during the supply-chain process to ensure vendors are compliant with your requirements and contracts are maintained as promised, and to maintain vendor relationships. You will need to work closely with procurement as they work through the **request for proposal** (**RFP**) and requirements for hardware-related requests. To ensure the procurement process is optimized, guidelines from technical and security experts should be clearly defined. One important point to call out is that you should be careful with going cheap on your hardware purchases. There is always a drive to bring costs down, but opting for hardware that is cheaper than certified hardware could be a costly mistake. It may prove more cost-effective to get a contract with a vendor, standardize on hardware, and purchase a warranty program. There's a good saying in life: you get what you pay for!

> Tip
> Auditing, including vendor management, will be covered in more detail in *Chapter 15, Testing and Auditing*. This chapter will cover vendor management as it relates to compliance with your suppliers.

In addition to verifying the supply-chain process is clean, you will want to ensure you are using hardware that has been certified and approved by Microsoft to run the Windows OS. Microsoft has a very well-defined Windows Hardware Compatibility Program for vendors to follow to ensure they are maintaining the highest standards of security and compatibility for running Windows. Using any hardware outside of this compatibility list could render your Windows OS unstable and, even more importantly, create security gaps within your systems.

> **Tip**
>
> Information about the Windows Hardware Compatibility Program can be found at this link:
>
> `https://docs.microsoft.com/en-us/windows-hardware/design/compatibility/`

To view the Windows Compatible Products List, browse to `https://partner.microsoft.com/en-us/dashboard/hardware/search/cpl`, type in a product name, and click **Search**. You will be provided with a list of compatible hardware based on your search. The following example shows the first couple of items returned with **Laptop** in the product name:

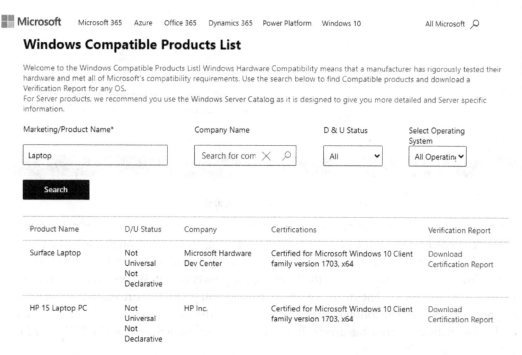

Figure 3.6 – Windows Compatible Products List for hardware compatibility

There is an additional portal where you can view certified products specific to Windows Server within the Windows Server Catalog. To view supported hardware for Windows Server, browse to `https://www.windowsservercatalog.com/`. Within the landing page, go to the **Hardware** section and click on the version of Server or the specific product category that you would like to view. As you browse through the items, you will see which specific version is supported by Microsoft, along with the badges that were awarded for certification. As you can see, a Windows Server 2022 badge has already been created and applied to identify hardware that supports Windows Server 2022.

Figure 3.7 – Windows Server Catalog badges

As you read through this section, you will have probably realized there is a lot more due diligence needed before you go out and purchase any hardware. Although there is an incredible ecosystem of hardware that supports Windows, you will want to ensure that the hardware you are purchasing is the latest and supports the latest versions of Windows Server and Windows 11. The most current hardware will provide more enhanced features over older hardware. The following sections will cover the hardware security features that need to be enabled on your hardware. In the next section, we will review the BIOS, UEFI, Secure Boot, and TPM components.

The firmware interface, TPM, and Secure Boot

A firmware interface is typically low-level software that acts as the medium between the OS and hardware to provide a basic UI for configuring device features and providing instructions for the boot procedure. BIOS and UEFI are the standard firmware interfaces used in these operations. Just as with the OS, firmware is also at risk to vulnerabilities and will need to be updated to remain secure. Next, we will review the different interfaces and some of their security features.

Protecting the BIOS

The BIOS is loaded directly onto a PC motherboard. Its purpose is to initialize the physical hardware, go through a series of processes, and eventually boot into Windows. Just as with the OS or PC software, the BIOS in your systems can become outdated and vulnerable to unauthorized modification. Furthermore, the BIOS initializes privileged hardware processes with greater rights than the OS itself. As a result, malware developers not only target the OS but other mechanisms in the boot process, including the boot loader and hypervisor used for virtualization, to gain access to these privileges. To mitigate these vulnerabilities, it's important to have a system of authorized update mechanisms for updating the BIOS and ensure it's only configured and signed by an authentic source such as the device manufacturer. To help maintain the integrity of the BIOS and protect against malware such as bootkits, digital signature verification should be used for updates and include a manual rollback and recovery process.

> **Tip**
> A bootkit is a manipulation of the **Master Boot Record** (**MBR**), which allows malicious software to load prior to the OS and remain active after the OS loads.

According to the NIST *BIOS Protection Guidelines*, organizations should have an authenticated BIOS update mechanism using the **Root of Trust for Update** (**RTU**) measurement with approved digital signature algorithms for verification, as specified in *NIST FIPS 186-3, Digital Signature Standard*. The updated standard (*186-4*) can be found at this link: `https://nvlpubs.nist.gov/nistpubs/FIPS/NIST.FIPS.186-4.pdf`.

The NIST security guidelines for protecting BIOS are specified in four system BIOS functionalities and state the following:

- Use an authenticated BIOS update mechanism with digital signatures to validate the integrity of updates.

- Secure the **local** update process with system passwords and physical locks, or only allow BIOS updates through a local update process with a physical **information technology** (**IT**) presence.

- Use integrity protection features to prevent modifications to BIOS.

- Implement non-bypassability features to ensure only the authenticated update mechanism is used.

To read more about the NIST security guidelines for protecting BIOS, visit this link:

```
https://nvlpubs.nist.gov/nistpubs/Legacy/SP/
nistspecialpublication800-147.pdf
```

Understanding UEFI

UEFI is the successor to BIOS and is now standard on most manufactured hardware. Both BIOS and UEFI have a similar boot process regarding the initializing flow, but with some differences. UEFI does not rely on a boot sector to copy an MBR and uses what's known as a boot manager to determine what to boot. The traditional BIOS runs 16-bit code and leverages only the MBR, which presents limitations such as support for drives larger than 2 **terabytes** (**TB**). UEFI uses the **GUID Partition Table** (**GPT**) and supports 32-bit or 64-bit code processes in a *protected mode* before transferring control over to the OS during the **runtime** (**RT**) processes. The higher bit support has more space that allows for friendlier UIs and faster boot times. UEFI incorporates security technologies such as Secure Boot and supports traditional BIOS methods, such as booting over the network or from flash memory.

As with the BIOS, the first phase in the UEFI boot process is known as the **Security Phase** (**SEC**). This acts as a core root of trust (or boot block in BIOS) to validate the integrity of the code and other firmware components before moving to the rest of the boot process. One of the biggest security advantages of using UEFI over BIOS is the ability to validate boot loaders by checking their digital signature using Secure Boot. If a boot loader has been replaced by malicious code, it won't be allowed to execute based on an invalid or revoked digital signature. This check starts the entire **trusted boot** process, which secures the boot chain all the way until Windows loads.

There are many types of security features built into the UEFI setup, and they will vary by vendor. Some of the more common security features found in the UEFI setup are outlined here:

- Password settings that include setting a supervisor password and lock settings, preventing users from making changes without entering the supervisor password. You can also require a password at unattended boot, at restart, and even at the boot device list.

- Fingerprint settings to use biometrics during pre-desktop authentication as an alternative or in conjunction with entering a supervisor password.

- Security chip settings for the TPM.

- UEFI BIOS update settings to add protection regarding BIOS updates, including rollbacks.

- Memory protection for execution prevention against virus and worm attacks that create memory buffer overflows.

- Virtualization settings to enable or disable virtualized hardware support such as Intel VT-x or AMD-V.

- I/O port access to enable or disable the use of devices such as wireless, Bluetooth, USBs, cameras, and microphones.

- Internal device access tamper detection of the physical covers of storage devices.

- Anti-theft to enable a *lo-jack* for your PC using a third-party provider.

- Secure Boot settings.

- **Trusted execution environments** (**TEEs**) such as Intel **Software Guard Extensions** (**SGX**) for hardware-based isolation of application code in memory.

- Device Guard, which is a feature set that consists of **Configurable Code Integrity** (**CCI**), **Virtual Secure Mode** (**VSM**) Protected Code Integrity, and Secure Boot. Device Guard features set the foundation for VBS, which we will discuss in more detail later.

Now you better understand what UEFI is, we will next cover UEFI Secure Boot and how to enable it.

UEFI Secure Boot

Secure Boot is a hardware-based security feature available in the UEFI environment that ensures only trusted software and firmware can execute in the boot chain. Each software, driver firmware, and OS boot loader (Windows Boot Manager) has a digital signature or hash that is validated by referencing signature keys stored in the Secure Boot database. Secure Boot consists of a **Platform Key** (**PK**) created by the OEM that creates the trust to the **Key Exchange Key** (**KEK**) with a public/private key pair. The KEK has public signing keys and is used to modify or add signatures to the safelist **database** (or **db**) and the **revoked signature database** (or **dbx**), to accommodate for new releases of Windows and known bad signatures for revocation purposes. These *allow* or *deny* databases are used for validation against the certificates, keys, or image hashes of boot loaders, firmware, and drivers. For example, if malware such as a bootkit invalidates the boot loader, then UEFI checks the Secure Boot database and won't allow the OS to boot when the signature doesn't exist or is blacklisted. This is important to note because rootkits are typically low-level software and hide from the OS, making them undetectable by most antivirus software.

The following diagram provides an example of Secure Boot validating the signature of Windows Boot Manager:

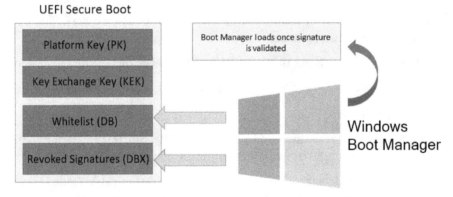

Figure 3.8 – Sample flow of Secure Boot validating the signature of Windows Boot Manager

Secure Boot is typically easily configured in UEFI by following these similar steps, depending on the manufacturer and BIOS:

1. Boot into the UEFI setup. Typically, pressing *F12* or *F10* during startup will load a boot device list where you can choose this setup.

> **Tip**
>
> If you're logged in to Windows, hold down *Shift* and choose **Restart** to be presented with an **Advanced Options** feature to boot into UEFI.

2. Secure Boot settings are typically found under the **Security** tab.
3. Change **Secure Boot** to **On**, as shown in the following screenshot:

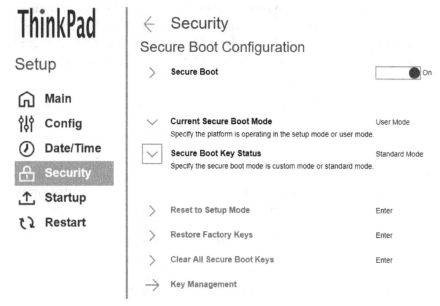

Figure 3.9 – Secure Boot configuration on a Lenovo ThinkPad workstation

Next, we'll look at the features of the TPM security chip that is embedded in the hardware.

TPK (TPM 2.0)

A TPM provides hardware-based security, typically in the form of a tamper-resistant chip built directly onto a motherboard. A TPM helps by providing an additional hardware layer separated from the memory and OS. Its main purpose is to perform cryptographic functions in isolation and is often considered the hardware *root of trust*. TPMs primarily deal with the encryption and decryption of security keys and run passively, so they do not rely on the OS to process instructions. Each TPM chip has its own unique **Rivest-Shamir-Adleman (RSA)** private key that's imprinted directly on the chip itself. This private key is never exposed to another external process, therefore only allowing the decryption of TPM-encrypted keys to be handled by the same TPM chip.

The TPM also has built-in protection against dictionary attacks, which are used to guess the authorization value for gaining access to protected keys. This is known as a TPM lockout, and in Windows, a maximum count threshold of 32 is set with a 10-minute healing time to protect against this type of tampering. TPM lockout settings can be managed using Group Policy and are located under `Computer Configuration\ Administrative Templates\System\Trusted Platform Module Services`.

For more information about managing TPM lockout in Windows 11 and Windows Server 2016 and later, visit this link:

`https://docs.microsoft.com/en-us/windows/security/information-protection/tpm/manage-tpm-lockout`

Windows computers containing a TPM provide enhanced hardware security features that can be used across a variety of functions, such as the following:

- Encryption, decryption, and other cryptographic functions
- Key storage and generation
- Integrity validation (security such as Secure Boot for firmware and boot loaders)
- Strong user and device authentication technologies (Windows Hello and **virtual smart cards (VSCs)**)
- Antimalware boot measurements for start state integrity checks
- Virtualized-based security features such as Windows Device Guard and Credential Guard
- BitLocker drive encryption
- Health attestation services for both local and remote attestation

In some use cases, such as BitLocker drive encryption, a USB drive can act as a TPM alternative and must be present for the computer to start Windows. This is known as a TPM start up key and allows the use of BitLocker without a physically compatible TPM chip. In conjunction with a TPM chip or PIN, this enables a form of BitLocker **two-factor authentication (2FA)**.

Unlike in earlier versions of Windows, Windows handles most of the provisioning of the TPM and reduces the need for manual configuration. For more information on how Windows uses the TPM, visit this link: `https://docs.microsoft.com/en-us/ windows/security/information-protection/tpm/how-windows-uses- the-tpm`.

TPM 2.0 and Windows support a feature known as the Health Attestation service. This allows **mobile device management (MDM)** services such as Intune to collect telemetry data from Windows-managed devices in your organization for remote device health attestation reports. The state of your device's health attestation can be used with risk-based conditional access and block access to resources if certain compliance conditions aren't met. The additional level of analysis provides RT protection of hardware after boot. The following screenshot shows a Windows Health Attestation telemetry report from Microsoft Endpoint Manager:

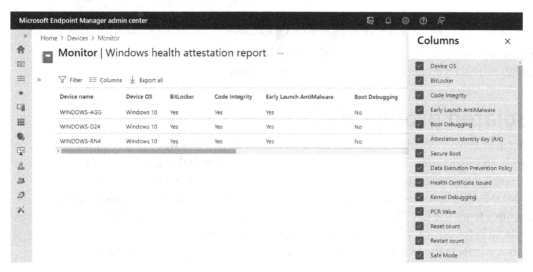

Figure 3.10 – Windows Health Attestation report from Microsoft Endpoint Manager

The TPM security chip is typically enabled in the **Security** tab of the UEFI setup, as illustrated in the following screenshot:

Figure 3.11 – TPM 2.0 enabled on a Lenovo ThinkPad workstation

In the next section, we are going to talk about how VBS can protect your system with isolation and by enabling VBS-based security features.

Isolated protection with VBS

First available in Windows 10 and Windows Server 2016, VBS was not a requirement but highly recommended. In Windows 11 Enterprise, many available features of VBS will be enabled by default on supported hardware and can be managed with Group Policy or MDM. VBS leverages hardware virtualization and the Windows hypervisor to isolate memory from the OS. This separation is known as virtual secure mode, which provides protection for critical system processes to help prevent exploitation. For example, if malware infects the OS, it will remain contained to the OS and be inaccessible to VSM.

For a system to be considered VBS-capable, it needs to meet the following minimum hardware requirements:

- TPM 2.0

- 64-bit processor running Intel VT-x or AMD-v virtualization extensions

- IOMMU or **system memory management unit** (**SMMU**) (Intel VT-D, AMD-Vi, ARM64 SMMUs) IOMMU

- SLAT for **virtual address translation** (**VAT**)

- UEFI memory reporting

- Compatible drivers with **Hypervisor-Protected Code Integrity** (**HVCI**)

- Secure **Memory Overwrite Request** (**MOR**) v2

- **System Management Mode** (**SMM**) protection

For more detailed information on VBS and the hardware requirements for VBS, visit this link:

```
https://docs.microsoft.com/en-us/windows-hardware/design/
device-experiences/oem-vbs
```

The following diagram shows the level of separation that is accomplished through VBS security. Normal Windows operating processes occur on the left and are blocked from accessing secure mode on the right.

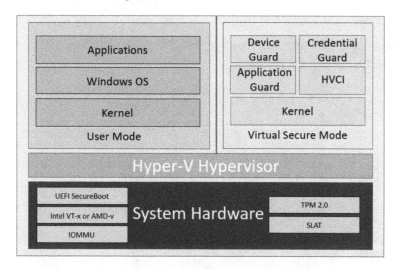

Figure 3.12 – A hypervisor isolates normal user and kernel modes

The following security features leverage VBS or are enhanced with VBS enablement:

- Windows Defender Credential Guard
- HVCI
- Microsoft Defender Application Guard
- System Guard
- Kernel DMA Protection

The following screenshot is from **System Information** on Windows and shows that VBS is running:

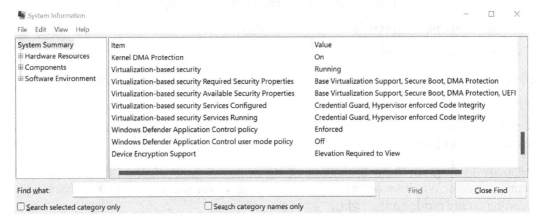

Figure 3.13 – MSinfo32.exe displaying VBS features on this system

Let's look at some of the features that leverage VBS. First, we will cover Microsoft Defender Credential Guard.

Windows Defender Credential Guard

Credential Guard helps protect user authentication and access tokens in the **Local Security Authority Subsystem Service (LSASS)** or `Lsass.exe` file from being stolen. Without Credential Guard enabled, derived credentials such as Kerberos tickets and password hashes are stored in memory without the secure isolated protection of a VBS hypervisor and are vulnerable to password-stealing malware. With Credential Guard enabled, credentials are stored in a protected isolated process called `LsaIso.exe`. **LSA Isolated (LSAISO)** is only accessible by LSASS using secured **Remote Procedure Calls (RPCs)**, and credentials stored in LSAISO are not exposed to processes outside of this protected container. This mitigates tools from stealing **New Technology LAN Manager (NTLM)** password hashes and Kerberos **ticket-granting tickets (TGTs)** stored in the memory thanks to VSM's isolation.

The following screenshot shows an LSAISO process running in Task Manager:

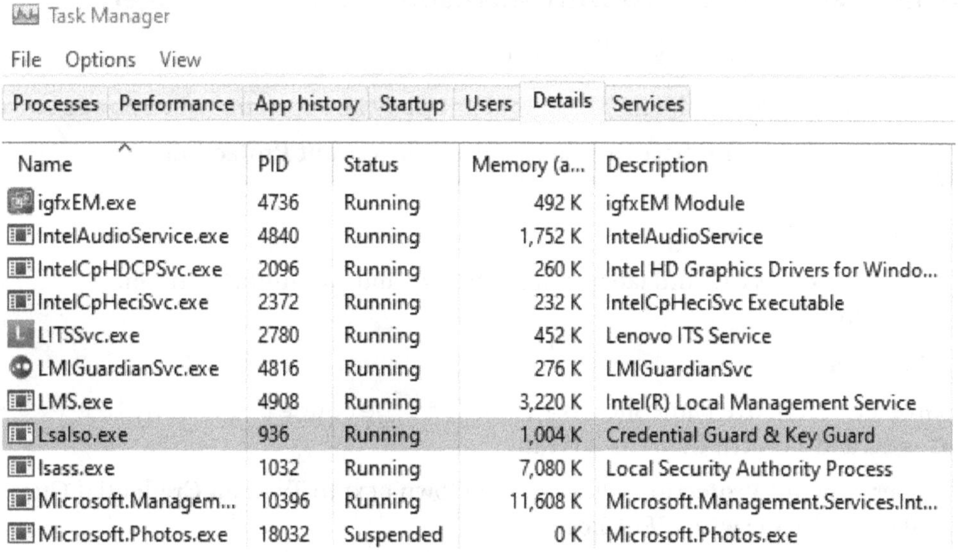

Figure 3.14 – LsaIso.exe process running in Task Manager

Here are some types of attacks that Credential Guard helps protect us against:

- **Pass-the-Hash (PtH)** is where an attacker can bypass entering a user's credentials by passing a captured password hash as an alternative authentication method.

- **Pass-the-Ticket (PtT)** is like PtH but where an attacker authenticates by using a Kerberos ticket instead of the user's password.

> **Tip**
>
> With Credential Guard enabled, **single sign-on (SSO)** does not work with
> NTLMv1, MS-CHAPv2, Digest, and CredSSP authentication methods, making
> applications prompt for credentials.

If you're considering enabling Credential Guard, it is recommended to read about the
important considerations mentioned in the preceding information tip before turning it
on. More information can be found at this link: `https://docs.microsoft.com/`
`en-us/windows/security/identity-protection/credential-guard/`
`credential-guard-considerations`.

Next, we'll look at how to enable Credential Guard using Endpoint Security in Microsoft
Endpoint Manager.

Enabling Credential Guard with Microsoft Endpoint Manager

To enable Credential Guard, follow these steps:

1. Log in to Microsoft Endpoint Manager at `https://endpoint.microsoft.com`.
2. Choose the **Endpoint Security** blade and select **Account Protection**
    under **Manage**.
3. Click **Create Policy**.
4. Choose **Windows 10 and later** under **Platform** and **Account Protection**
    under **Profile**.
5. Click **Create**.
6. Give it a descriptive name such as `Microsoft Defender Credential`
    `Guard` and a description. Click **Next**.
7. Under **Account Protection**, click the dropdown next to **Turn on Credential Guard**
    and select **Enable with UEFI Lock**.
8. Click **Next**.
9. Set a scope tag if required or click **Next**.
10. Under **Assignments**, select a security group to assign the policy to and click **Next**.
11. Review the summary and click **Create**. Your policy should look like this:

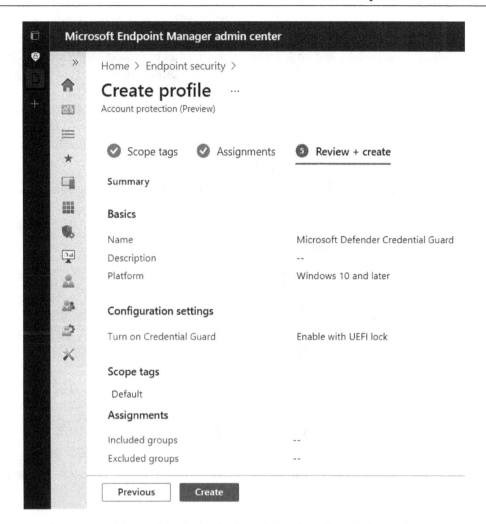

Figure 3.15 – Microsoft Defender Credential Guard profile in Endpoint Security

> **Tip**
> **Enable with UEFI Lock** (recommended) ensures that Credential Guard cannot be remotely disabled. If you need to disable it remotely, choose **Enable without UEFI lock**.

For more information about how Windows Defender Credential Guard works, visit this link:

```
https://docs.microsoft.com/en-us/windows/security/identity-
protection/credential-guard/credential-guard-how-it-works
```

HVCI

HVCI combines virtualization and code integrity verification to protect the Windows OS from malicious code. HVCI is the key component of the **Memory integrity** functionality of the **Core isolation** security features in Windows system security. This helps to ensure that the code integrity service used to validate the signatures of drivers and kernel-mode processes are contained with hypervisor isolation, thus making them protected from tampering. HVCI is not only applicable to a single system running on physical hardware, but can also protect Gen 2 VMs with a Hyper-V host in Server 2016 or later.

In the following screenshot, you can see the **Memory integrity** feature being enabled in **Core isolation**:

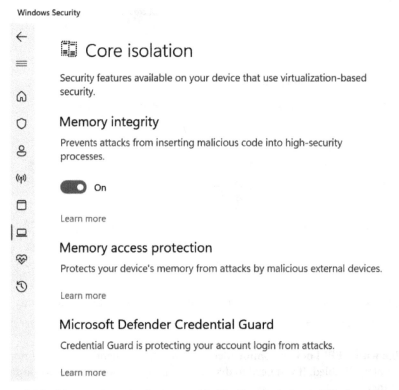

Figure 3.16 – Memory integrity feature enabled in Core isolation in Windows Security

HVCI can also be combined with **Windows Defender Application Control (WDAC)** code integrity policies, designed for whitelisting what applications, drivers, and file paths can run in Windows. Hypervisor-protected code integrity enhances protection by adding an extra layer on top of code integrity policies and expanding them into the VBS features of hypervisor isolation. This helps to ensure that WDAC policies remain resilient against tampering and modification. WDAC working together with VBS used to be branded as key features known in Device Guard.

> **Tip**
>
> While most modern systems have the hardware and drivers to fully support HVCI, it is important to point out that **all** drivers must be HVCI-compatible; otherwise, it can result in blue screens or system failures.

To view a list of baseline protections that are required for both VBS and HVCI, visit this link:

```
https://docs.microsoft.com/en-us/windows/security/threat-
protection/device-guard/requirements-and-deployment-planning-
guidelines-for-virtualization-based-protection-of-code-
integrity
```

Let's review how to enable HVCI using Group Policy.

Enabling HVCI

To enable HVCI, follow these steps:

1. Open **GPMC** and create a new **Group Policy Object (GPO)** or modify an existing one.
2. Go to **Computer Configuration | Policies | Administrative Templates | System | Device Guard**.
3. Open **Turn on Virtualization Based Security** and choose **Enabled** (radio button).

4. Ensure the dropdown under **Virtualization Based Protection of Code Integrity** is set to **Enabled** with or without a UEFI lock, as follows:

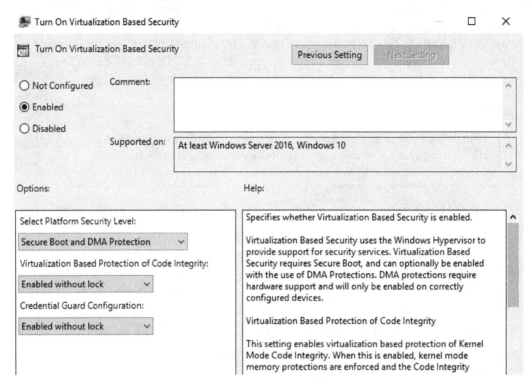

Figure 3.17 – Group Policy for HVCI and VBS features

HVCI and other hardware-based security features can be validated using **Windows Management Instrumentation (WMI) class Win32_DeviceGuard** or in **System Information**. The WMI class is only available on enterprise versions of Windows. To check hardware-based security features, open an administrative PowerShell window and run the following command:

```
Get-CimInstance -ClassName Win32_DeviceGuard -Namespace root\
Microsoft\Windows\DeviceGuard
```

The following output shows the hardware-based security features on your device:

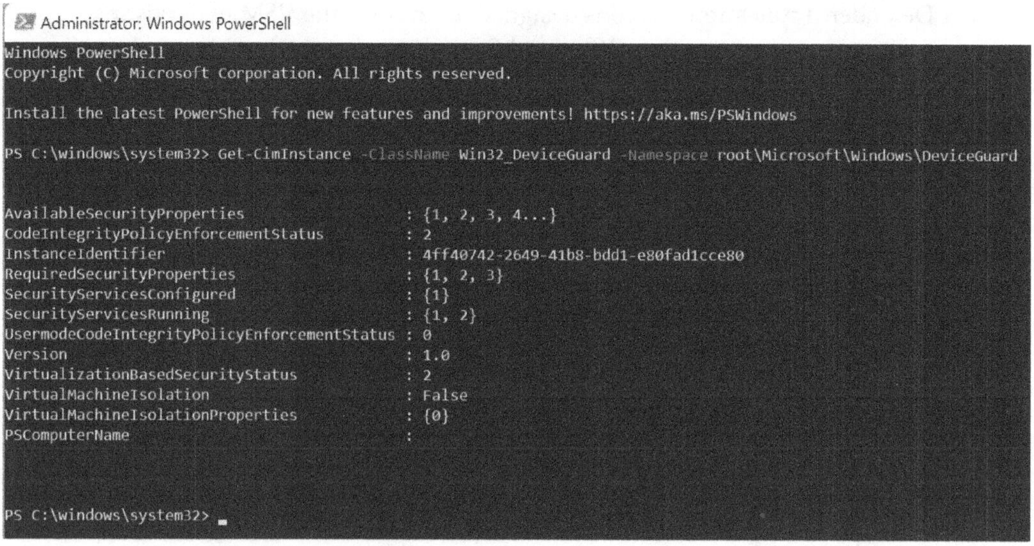

```
Administrator: Windows PowerShell

Windows PowerShell
Copyright (C) Microsoft Corporation. All rights reserved.

Install the latest PowerShell for new features and improvements! https://aka.ms/PSWindows

PS C:\windows\system32> Get-CimInstance -ClassName Win32_DeviceGuard -Namespace root\Microsoft\Windows\DeviceGuard


AvailableSecurityProperties                     : {1, 2, 3, 4...}
CodeIntegrityPolicyEnforcementStatus            : 2
InstanceIdentifier                              : 4ff40742-2649-41b8-bdd1-e80fad1cce80
RequiredSecurityProperties                      : {1, 2, 3}
SecurityServicesConfigured                      : {1}
SecurityServicesRunning                         : {1, 2}
UsermodeCodeIntegrityPolicyEnforcementStatus    : 0
Version                                         : 1.0
VirtualizationBasedSecurityStatus               : 2
VirtualMachineIsolation                         : False
VirtualMachineIsolationProperties               : {0}
PSComputerName                                  :


PS C:\windows\system32> _
```

Figure 3.18 – Win32_DeviceGuard class from WMI

For additional details on the available hardware-based security features, visit this link:

```
https://docs.microsoft.com/en-us/windows/security/threat-
protection/device-guard/enable-virtualization-based-
protection-of-code-integrity
```

Next, we'll have a look at Microsoft Defender Application Guard.

Microsoft Defender Application Guard

Microsoft Defender Application Guard is designed to leverage the VSM hypervisor isolation of VBS to protect Microsoft Office and Edge through containerization. What's great about this protection is that it helps protect against novel types of attacks due to the nature of the isolation of the application from the host OS. Microsoft Office and Edge are constant targets for malware developers due to their wide usage. Application Guard has built-in support for Microsoft Office and Edge browser but also extends these features to Google Chrome and Firefox through browser extensions. This allows untrusted sites that are opened in Chrome or Firefox to be redirected to a protected Edge browsing session. In the following screenshot, you can see how to obtain the extension from the Google Chrome web store:

Figure 3.19 – Application Guard Extension available in the Google Chrome web store

> **Tip**
> When enabling the extension or deploying it through Intune, a Win32 component will need to be installed to activate Application Guard. A restart will be required.

Application Guard for web browsers protects users from sites that aren't defined as trusted in a network isolation policy configuration. Whenever a site is opened that's not in this policy, a new containerized browsing session is opened in Microsoft Edge, leveraging the VSM of VBS and isolating this session from user and kernel-mode attacks on the underlying system.

A few examples of attacks that Application Guard protects against are given here:

- Zero-day and unpatched vulnerabilities exploited from a website
- Drive-by attacks where malicious code is injected into a website
- **Cross-site scripting** (**XSS**) attacks and other web-based malware

There are specific hardware and software requirements needed to support enabling Application Guard, as outlined here:

- 64-bit CPU processor
- SLAT and Intel VT-x or AMD-v virtualization features
- 8 GB RAM
- Windows 10 or later (Enterprise, Professional, or Education)

Application Guard has two working modes, as follows:

- **Standalone mode** allows the user to control when to enable or disable Application Guard and does not require the endpoint to be managed by policy.
- **Enterprise-managed mode** is the policy-managed version that can be configured and controlled by MDM or Group Policy.

Application Guard can be enabled through Endpoint Security in **Microsoft Endpoint Manager**, **Group Policy**, or **Configuration Manager**. The following screenshot shows an Application Guard policy in Endpoint Security in **Microsoft Endpoint Manager**:

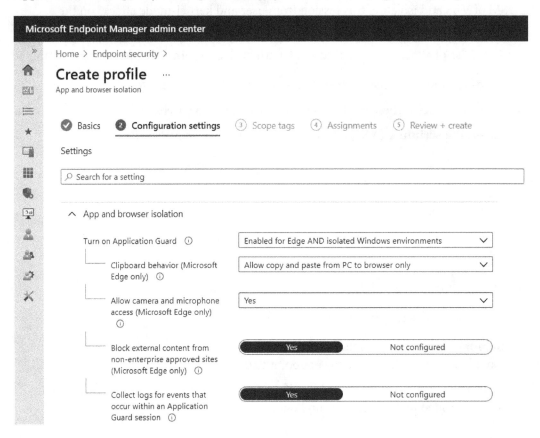

Figure 3.20 – App and browser isolation profile in Microsoft Endpoint Manager Endpoint security

The following screenshot shows how to configure network boundaries using the Windows network isolation policy settings:

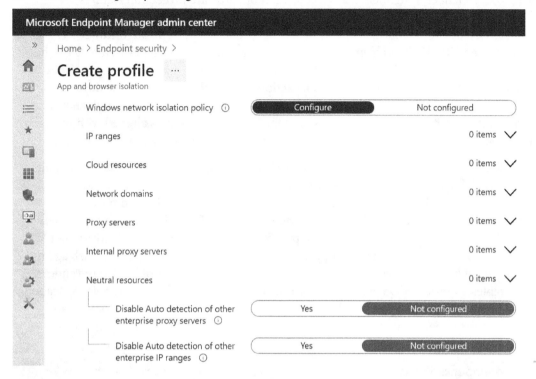

Figure 3.21 – Windows network isolation policy configuration in Endpoint Security

A network boundary policy can also be configured in Intune using device configuration templates.

When turning on Application Guard, you can select to enable it for Edge and/or isolated Windows environments. Isolated Windows environments allow other applications that support Application Guard (including Office) to be enabled. The policy is flexible and allows administrators to determine which features can interact with the host OS, such as copying and pasting or printing.

Next, we'll look at Windows Defender System Guard and how it uses a set of integrity checks to keep your system safe.

Windows Defender System Guard

Windows Defender System Guard is a collection of hardware protection features (including Secure Boot) that ensure the integrity of your system is protected during boot, as well as after Windows is running. Using local attestation through **Root of Trust Management** (**RTM**) measurements and remote attestation, RTM measurements can be sent for analysis to an MDM solution such as Intune or Configuration Manager. If a PC is determined to be *untrustworthy*, remote action can be taken to isolate the device or prevent access to resources. This helps to protect the system after boot through analysis and monitoring of these measurements. System Guard, by design, is aligned with the Zero Trust security model and assumes that all processes could be compromised until verified.

For added flexibility, System Guard supports **Dynamic Root of Trust for Measurement** (**DRTM**) with a process known as Secure Launch to dynamically protect and add management flexibility over the **Static RTM** (**SRTM**) that's typically used for local attestation of boot processes. As we discussed earlier, the SEC in UEFI is loaded at the start of boot and begins the Trusted Boot process. To achieve compliance with an SRTM, each piece of code in the entire boot process must be validated from start to finish against a whitelist of known trusted measurements. Due to a large number of hardware vendors, executables, libraries, BIOS versions, code updates, and length of code, support for scaling this whitelist is difficult. Also, SRTM does not protect the RT after load operations and requires trust to be maintained through the entire boot process. DRTM and the Secure Launch technology of System Guard help overcome these challenges.

DRTM simplifies SRTM management by allowing for the boot process (sometimes unvalidated) to be executed and immediately controlled by being forced into a secure *launch code* container. This launch code is known-good code that's used to continue the boot process and is independent of specific hardware configurations. This dynamic measurement allows for more flexibility, and code that's not added to SRTM measurements can boot the system before Secure Launch takes control. The system can maintain a **Trusted Computing Base** (**TCB**) standard without you having to reset the entire TPM and validate the entire boot chain **end to end** (**E2E**). The **Trusted Computing Group** (**TCG**) has defined a TCB as something that protects the system and enforces computer security policy.

For more information on the DRTM specification, visit this link:

```
https://trustedcomputinggroup.org/wp-content/uploads/DRTM-
Specification-Overview_June2013.pdf
```

> **Tip**
> Only TPM 2.0 supports System Guard Secure Launch and remote attestation.
> To enable System Guard Secure Launch, all requirements for Device Guard,
> Credential Guard, and VBS must be met.

We have just covered many of the hardware-based security features that are used to protect your systems all the way through the boot process. Now, we'll review an additional hardware security feature known as Kernel DMA Protection.

Kernel DMA Protection

Kernel DMA Protection helps protect PCs from external devices that can be plugged into exposed hot-plugged **Peripheral Component Interconnect Express (PCIe)** ports such as a Thunderbolt connection. PCI hotplug ports are **direct memory access (DMA)**-capable, which makes them a vulnerable target for attackers due to their privileged access in the system. A malicious actor can quickly insert a USB such as a peripheral in an unattended device and execute code or steal data without any interaction or having to log in to the computer. A DMA-capable device has direct access to system memory by design to allow it to support high-powered peripherals such as graphics cards.

By leveraging the IOMMU hardware feature, enabling DMA protection will keep these peripherals isolated from memory until the user logs back in to the computer. DMA protection also works in conjunction with DMA remapping-compatible driver standards. A driver that meets these standards can automatically be started without requiring a user to log in.

Kernel DMA Protection can be enabled in Endpoint Security in Intune or with Group Policy. The following screenshot is from the Device Control profile in an attack surface reduction policy in Endpoint Security:

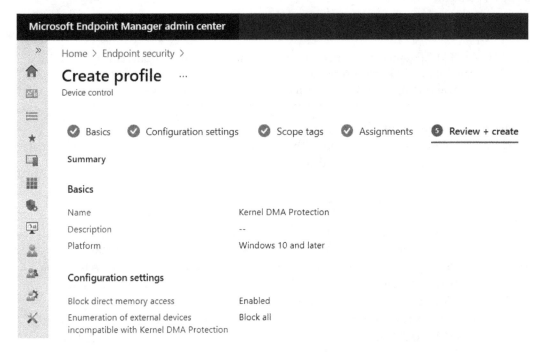

Figure 3.22 – DMA Protection settings in the Device control profile in Endpoint Security

> **Tip**
> Setting **Block all** for the enumeration of external devices incompatible with Kernel DMA Protection will block devices not compatible with DMA remapping

VBS security is not required to enable Kernel DMA protection settings.

Next, let's look at hardware security features available on the market to help protect against data stored in memory from lost or stolen devices.

Protecting data from lost or stolen devices

Unfortunately, physical devices such as laptops can be prone to being lost or stolen. Certain protections need to be in place to help protect the data on the system from being exfiltrated, assuming the device will not be recovered. Fortunately, for devices under corporate management, MDM remote actions can be triggered to wipe a device the next time it comes online. Additionally, BitLocker encryption will prevent data loss from the hard drive, but what kinds of protections are available for data such as credentials or encryption keys stored in memory? Recently, both Intel and AMD have made significant hardware security advances to cover these types of scenarios and protect data stored in memory with encryption.

Secure Memory Encryption (AMD)

Secure Memory Encryption (**SME**) is a feature specific to AMD-based systems. Pages in the memory are encrypted through page tables using a 128-bit ephemeral **Advanced Encryption Standard** (**AES**) key that is generated at random during boot time and is not accessible by external software. Additionally, AMD extends this protection to VMs with **Secure Encrypted Virtualization** (**SEV**), which allows VMs to run fully encrypted and can only be decrypted by the underlying VM itself. SME can be enabled in the BIOS of AMD systems.

Total Memory Encryption (Intel TME)

Total Memory Encryption (**Intel TME**) is a similar feature to SME developed by Intel. As with SME, Intel TME encrypts memory using AES *XTS* with a randomly generated encryption key that is not accessible to software. Intel has recently expanded on this technology to add support for multiple encryption keys called **Multi-Key Total Memory Encryption** (**MKTME**).

Recently, Microsoft has also announced a collaboration effort with multiple CPU vendors to build and release the Microsoft Pluton processor. The Pluton processor was built to offer enhanced protection around vulnerabilities in the communications bus that transfers data from the TPM cryptography chip to the CPU and helps mitigate the risks of an attacker stealing data with physical access to the device through that communication channel. Visit this link to read more: `https://aka.ms/pluton`.

Purchasing devices that support these hardware-based security features will help ensure data will remain protected if a device is lost or stolen. Next, let's wrap up the chapter by recapping hardware security recommendations and best practices.

Hardware security recommendations and best practices

When looking at the security of hardware, it's important to keep these considerations in mind:

- Only purchase hardware that has been through a proper hardware certification program. The *Windows Hardware Compatibility Program* certification process is a great resource to help ensure the hardware is reliable and compatible with Windows.

- Keep your hardware up to date. Just as with software, hardware continues to evolve to become more secure.

- Have an effective and secure system for upgrading firmware/BIOS and ensure the proper protections are enabled to ensure only approved sources can update them.

- Purchase physical hardware that supports BitLocker (TPM 2.0), DRTM, SMM, Secure Boot, DMA Protection, Memory Encryption (AMD/Intel), and hardware-based isolation of application code in memory (TEE with Intel SGX). This will allow you to enable software features that support hardware-based security.

- Turn on VBS as soon as possible and enable Credential Guard, HVCI, Application Guard, Windows Defender Application Control, and Kernel DMA Protection to put the power of your hardware into action.

- Ensure your data is also protected if lost or stolen by protecting your data stored in memory with encryption. SME (AMD) and Intel TME features provide this level of security for your data.

- Ensure you are aware of the latest risks and hardware weaknesses with resources available to you, such as the NIST *Cybersecurity Supply Chain Risk Management* project and CWE's *Most Important Hardware Weaknesses* list.

- Subscribe to receive notifications as vulnerabilities are identified. A couple of examples include the CISA alerts (`https://us-cert.cisa.gov/ncas/alerts`) and advisories from the **Center for Internet Security Multi-State Information Sharing and Analysis Center (CIS MS-ISAC)** (`https://learn.cisecurity.org/ms-isac-subscription`).

Summary

In this chapter, we provided an overview of the hardware-based security features used to protect Windows from the boot chain, the OS layer, and for virtualization of the OS. We covered hardware concerns in terms of vulnerabilities such as rootkits and bootkits and the importance of the supply chain to ensure your organization purchases hardware that has been properly certified. Next, we covered BIOS, Secure Boot, and TPM and how these hardware components are the framework for hardware-backed VBS. We talked about the latest advanced protection features leveraging VBS, such as Credential Guard and Microsoft Defender Application Guard, as well as how to enable them using MDM or through Group Policy.

Finally, we finished by discussing how System Guard uses dynamic root-of-trust measurements and remote attestation to help protect your systems and hardware-based security features to secure data from lost or stolen devices, along with an overview on hardware security recommendations and best practices.

In the next chapter, we will discuss networking and the fundamentals that play a large role in securing your Windows systems. We will discuss physical hardware components used in the network infrastructure and the Windows Defender Firewall software, including configurations with Intune, Group Policy, and Configuration Manager. Finally, we will provide an overview of Azure network solutions that are used to secure Windows systems inside an Azure **virtual network** (**VNet**).

4

Networking Fundamentals for Hardening Windows

In this chapter, we will cover the importance of Windows networking components for the overall security and hardening of your Windows systems. Network security has traditionally been at the heart of security, but recently this has shifted. It hasn't become any less important, but with the adoption of cloud computing and the rapid acceleration of remote work, security strategies need to shift from a strong focus on the network perimeter to the application, device, and identity levels of focus. Devices are no longer sitting within corporate offices and users require the flexibility to access company resources from anywhere.

Even though the workforce has become more decentralized, there is still a need to maintain stringent network security controls within your offices and any on-premises data centers. In addition, you will need to implement strong controls at the desktop level and adopt a strategy for securing **Virtual Networks** (**VNETs**) and cloud resources. With this comes new complexity, and it's important to have the right tools and skill sets that support the strategic vision, deployment, and maintenance of your network tools and solutions.

Throughout this chapter, we will cover security fundamentals to raise awareness of the supporting network infrastructure of a Windows environment. We will review some of the network security tools available, including the software-based Windows Defender Firewall. Finally, we will cover network security solutions that protect and allow access to your Windows workloads running in Azure. This chapter includes the following topics:

- Network security fundamentals
- Understanding Windows network security
- Windows Defender Firewall and Advanced Security
- Web protection features in Microsoft Defender for Endpoint
- Introducing Azure network security

Technical requirements

We recommend having the following to follow along with the examples in the chapter:

- An **Active Directory** (**AD**) domain with Group Policy and an updated central store.
- A Windows 10/11 Pro or Enterprise workstation domain joined and/or Intune enrolled.
- A Microsoft Intune license
- A Microsoft Defender for Endpoint license
- An Azure tenant

Network security fundamentals

Networking can be a very challenging task for technology teams. Networks can be very sensitive and commonly take the blame for most outages, without people even knowing the true root cause of an issue. This is simply because most of our data traverses over a network, so it's critical that it performs optimally. If it doesn't, it can bring a business to its knees because of how dependent we have become on the network. In addition to the already challenging task of network operations, we are also concerned with network security. Ensuring that the data we transmit is secure, no perpetrators are accessing our network who shouldn't be, preventing traffic that isn't welcome, and ensuring confidential data is isolated are some of the challenges faced in this space.

As we mentioned previously, this shift in security is mainly due to the evolution of device access and cloud technologies that have forced us to change our strategies. Although this has shifted the core from a network security perimeter-focused strategy, network security has never been more important than it is today. Additional advanced security features are now required to secure and harden both the device and the cloud technologies in use.

Before we review some of the core network security technologies, it's important that we review and cover the **Open Systems Interconnection** (**OSI**) and **Internet Protocol Suite** (**TCP/IP**) models. These models have been built to allow for an open standard/framework to be referenced. The OSI model is a framework that is used as more of a guideline that provides a standard for network communications. It provides a great reference that allows us to understand the flow of network traffic from one endpoint to the other and serves as a great troubleshooting tool for us to understand where any breakdowns or failures may be occurring. The TCP/IP model is comprised of open standard protocols for network communication and has become more widely adopted over the OSI model.

The following diagram provides a comparison of the OSI model and TCP/IP model, including examples of what falls within each layer. Understanding where communication is failing within the network will significantly help a security expert with any investigative and/or troubleshooting tasks:

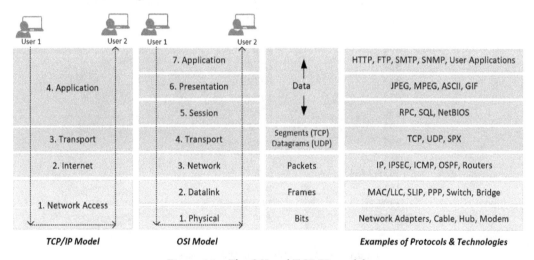

Figure 4.1 – The OSI and TCP/IP models

In addition to being familiar with the OSI and TCP/IP models, understanding common ports used in communications is somewhat an expectation of any network or security professional. As you build out and architect your solutions and integrate your technologies, knowing which ports are used by which protocols allows for more intelligent decisions so that you can provide better controls.

> **Tip**
>
> When building **Network Security Groups (NSGs)** or firewall rules, a best practice is to limit communications from known sources using only the required ports needed to create the connections.

Being able to quickly identify the type of traffic and which ports are being used may speed up your ability to mitigate an attack. The following is a list of some of the more common ports and the protocols/services that use them:

Port #	Service
20-21	FTP
22	SSH
23	Telnet
25	SMTP
53	DNS
69	TFTP
80	HTTP
88	Kerberos
110	POP3
115	SFTP
123	NTP
137-139	NETBIOS
143	IMAP
161-162	SNMP
179	BGP
389	LDAP
443	HTTPS
464	Kerberos
636	LDAPS
993	IMAPS
995	POP3S
989-990	FTPS
1433-1434	SQL
1512	WINS
3306	MYSQL
3389	RDP

Figure 4.2 – Common ports and services

The **Internet Assigned Numbers Authority (IANA)** website provides a list of all registered service names and port numbers at the following link:

```
https://www.iana.org/assignments/service-names-port-numbers/
service-names-port-numbers.xhtml
```

There are many components involved within a network architecture and the topology can be extremely complex. The following technologies are considered more critical for your enterprise deployment as they relate to your network security and should be implemented to protect your Windows environment:

- Routers and switches using VLANs

- Next-generation-type firewalls

- A **Virtual Private Network (VPN)** to encrypt connections

- **Intrusion Detection Systems (IDSs)/Intrusion Prevention Systems (IPSs)** to proactively detect and prevent threats

- Wi-Fi with a minimum of WPA2-Enterprise security

- **Network Access Control (NAC)** to better manage endpoint access to your network.

- Proxy/web content filters to prevent malicious websites

- Next-generation antivirus and anti-malware tools for more intelligent protection

- **Data Loss Prevention (DLP)** to prevent the loss of sensitive data

- Email/spam filtering to protect users against spamming, phishing, and so on

- **Security Information and Event Management (SIEM)** to help you detect abnormal activity

- DNSSEC to protect your DNS services

- **Public Key Infrastructure (PKI)** to provide digital certificates for encryption

From a network device management perspective, the following are important:

- Ensure you keep the software of your network devices current.

- Enable auditing on devices.

- Integrate authentication using LDAP.

- Use a **Privileged Access Management (PAM)** solution.

- Disable or prevent local account access and change default usernames and passwords.

- Ensure the management of devices is encrypted (SSH).

- Isolate the management network.

- Don't allow the management of network devices from the internet.

For a more detailed review on securing network infrastructure devices, the Department of Homeland Security has a security tip reference available at this link:

```
https://www.us-cert.gov/ncas/tips/ST18-001
```

These technologies are very complex and, in most cases, require specialized skill sets to implement and manage daily. Some of these technologies are both hardware- and software-based. Hardware-based technologies are typically the rack-mounted gear in your **Main Distribution Frame** (**MDF**) and primarily protect your facilities and data centers. Software-based technologies protect the OS, end users, and VNETs. An example would be your computer's software firewall. Ensure you deploy the latest next-generation hardware or virtual-based firewalls at your data center locations (including the cloud) and physical offices and enable the software-based firewalls on your Windows devices for additional protection. Software-based technologies are becoming more critical for end user devices due to the shift from centralized offices to a dispersed and remote workforce.

The Microsoft technology stack offers many solutions that can be compared to other networking vendor offerings either as an alternative or as a complement. As a security professional, it is important you are aware of and understand each of the technologies referenced earlier to provide the best protection within your organization. Throughout the remainder of this chapter, we will review the Microsoft-specific network technologies that provide the best protection for Windows devices.

Understanding Windows network security

In this section, we will review the core networking components of Windows clients and servers. Having familiarity with these components is a must for any security professional when managing and troubleshooting Windows devices. It's also recommended that you apply the baseline recommendations for network-related settings to ensure your system is hardened correctly.

Network baselining

Referencing back to *Chapter 2*, *Building a Baseline*, you will want to ensure that your network-specific components have been configured based on the recommendations. There are many network-related settings for Windows and implementing the baseline recommendations is the more practical approach compared to building your own standards from scratch. As an example, referencing back to the latest Microsoft security baseline for Windows clients, simply filtering for the `network` keyword in the `Security Template` worksheet provides 40 settings.

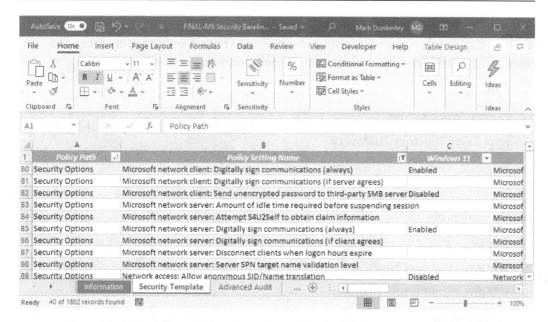

Figure 4.3 – Network-specific configurations within the Windows 11 security baselines

The preceding settings will need to be configured using Group Policy or **Mobile Device Management (MDM)** to ensure enforcement and consistency. Unless you only have a few devices or servers to manage, individually configuring these settings is not realistic.

The CIS security benchmarks make a great complement to the Microsoft security baselines when building out your security controls. The CIS benchmarks are typically more recognized as a standard for hardening your systems and although there will be a lot of overlap between the Microsoft and CIS baselines, you will find that CIS tends to include tighter security controls. If using the Microsoft security baselines and CIS security benchmarks, we recommend focusing on the following areas regarding network communications between Windows clients and servers:

- **Microsoft network client** settings focus on how clients communicate with servers and how those communications are handled.

- **Microsoft network server** settings focus on how servers accept communications from clients and handle those sessions.

- **Network access** settings control how clients and users connect and interact remotely with each other and to file shares.

- **Network security** settings focus on hardening the protocols and authentication methods used to authorize communication between clients and servers.

In addition to the recommended local policies, there are additional recommendations within the CIS Benchmarks document you will want to review and configure related to network settings. Some of these recommended settings include the following:

- DNS client

- LanmanWorkstation

- Link Layer Topology Discovery

- Peer Name Resolution Protocol

- Network connections

- Network provider

- TCPIP

- Windows Connect Now

- Windows Connection Manager

- WLAN services

> **Tip**
> As a reminder, be extremely cautious when enabling any new settings and, more specifically, network settings on any devices or servers. They can be very disruptive to production if they're not tested correctly.

Next, we will review the network and internet management console on Windows clients.

Windows clients

To access the network and internet settings, open **Windows Settings** and click on **Network & internet**. You will be presented with the **Network & internet** management console, as shown in the following screenshot taken on Windows 11:

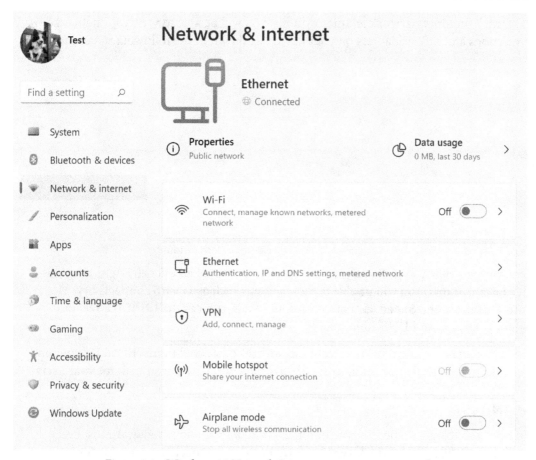

Figure 4.4 – Windows 11 Network & internet management console

Here, you will have access to all the networking-related components on your Windows device. You can view your current network status, adapter-specific settings (**Wi-Fi**, **VPN**, **Ethernet**, and so on), **Mobile hotspot**, **Airplane mode**, **Proxy**, and **Dial-up** settings. Within **Advanced network settings**, you will have the ability to disable your network adapters, view data usage, view hardware and connection properties, network reset your adapters, and link out to **More network adapter options** and **Windows Firewall** settings.

If you click on **More network adapter options**, you will be able to view your network connections and active adapters on your device, as shown in the following screenshot:

Figure 4.5 – Windows 11 Network Connections

Here, you can view your network settings for any connected adapters by right-clicking and clicking **Status**. You will be able to view settings such as **IPv4/6 connectivity**, **Media State**, **Duration**, and **Speed**, and more specific details, such as the **IP**, **MAC**, **Default Gateway**, **DHCP**, and **DNS** addresses.

The following network connection technologies are considered more critical for your end user devices and need to be set up correctly to ensure a safer environment for your users.

WLAN/Wi-Fi

Wireless technology is a necessity within the world of technology today. Almost every laptop and mobile device will have some form of Wi-Fi connectivity available for use. With all technologies, there are threats, and this applies to Wi-Fi. Unfortunately, Wi-Fi is much more susceptible to vulnerabilities than **Local Area Network** (**LAN**) technologies due to the information being transmitted over the air and not through a cable, which is much more difficult to breach. There are many threats when it comes to Wi-Fi, and some of the more known ones include rogue access points or networks, man-in-the-middle attacks, and unauthorized access to insecure **Wireless Local Area Network** (**WLAN**) systems.

Securing your corporate Wi-Fi is not a small task and will require very skilled network engineers to architect and implement correctly, especially with implementing enterprise-grade security. Here are a few important tips for Wi-Fi security and your Windows clients:

- Do *not* use **Wired Equivalent Privacy** (**WEP**).

- Enable enterprise-grade authentication – WPA2-Enterprise with EAP-TLS: `https://docs.microsoft.com/en-us/windows/win32/nativewifi/wpa2-enterprise-with-tls-profile-sample`.

- Ensure any guest networks, **Operational Technology (OT)** networks, and **Internet of Things (IoT)** networks are isolated from more sensitive production networks and ensure all guest access is protected. If using a guest SSID with a password, implement a process to rotate the password regularly.

- Ensure the WLAN infrastructure is kept current and up to date.

For a more comprehensive list regarding how to secure your enterprise-grade wireless infrastructure, the Department of Homeland Security has a guide available for reference: `https://www.us-cert.gov/ncas/tips/ST18-247`.

As part of your baselines and the management of your devices, you will want to ensure you enforce the required security settings for your users. This can be accomplished using device administration tools such as Group Policy or Intune MDM. The following is an example of configuring Wi-Fi settings using the Intune settings catalog:

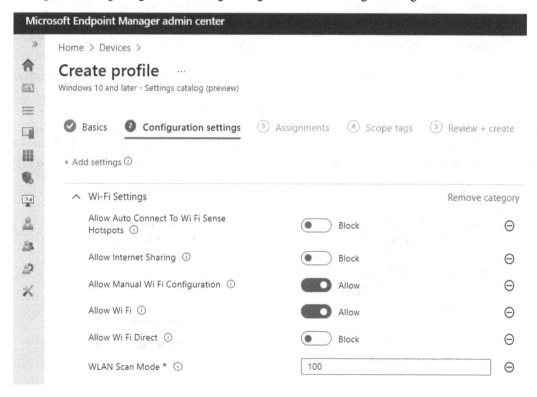

Figure 4.6 – Wi-Fi settings in Intune

We will cover creating device configuration profiles using Intune in more detail in *Chapter 6, Administration and Policy Management*.

We recommend configuring the following security policies for Wi-Fi/WLAN settings on your corporate devices:

- Block connections to suggested open hotspots and networks shared by contacts or paid services. These features are known as Wi-Fi Sense.

- Block internet sharing.

- Block Wi-Fi Direct.

- Minimize the number of simultaneous connections to the internet or a Windows domain. Set *preventing Wi-Fi* when on Ethernet.

- Block connections to a non-domain network when connected to an authenticated domain network.

- Block access to the **Windows Connect Now (WCN)** wizards.

- Block the configuration of wireless settings using WCN.

- Remove any unused Wi-Fi networks if possible.

Always be cautious when using open Wi-Fi in public places because of the ongoing threats from attackers. We don't know how well other Wi-Fi networks have been configured, and vulnerabilities may exist within their networks. When traveling, use cellular data to connect to the internet if it's an option or ensure you connect to a VPN when connecting to any public Wi-Fi. It's important to provide security awareness to your users and advise them about the risks of public Wi-Fi and what they should be doing to protect themselves.

> **Tip**
>
> Your home is just as vulnerable to threats regarding Wi-Fi and even more so as we connect more devices (IoT) within our home to it. Educate your users and provide them with the awareness needed to protect themselves from the ongoing threat landscape. The Department of Homeland Security provides home network security tips here: `https://www.us-cert.gov/ncas/tips/ST15-002`.

Next, let's look at Bluetooth technology and discuss a few recommendations to ensure secure Bluetooth connections.

Bluetooth

Your Windows device most likely has Bluetooth as a connectivity option since many peripherals, such as keyboards, mice, and audio speakers, use it as a standard connection type. Although an extremely convenient technology, it also comes with many flaws. There are many Bluetooth threats today that you should be familiar with. Some well-known threat types include **bluejacking**, **bluesnarfing**, and **bluebugging**. The best protection against Bluetooth is to disable it and prevent your users from using it. Unfortunately, this may not be a reality for most, so it is important to understand the technology and the risks associated with it. To help you understand Bluetooth and the risk it entails, NIST has published a *Guide to Bluetooth Security*, also known as *NIST Special Publication 800-121 Revision 2*: `https://nvlpubs.nist.gov/nistpubs/SpecialPublications/ NIST.SP.800-121r2-upd1.pdf`.

The *Guide to Bluetooth Security* is an extremely comprehensive document that provides you with all the knowledge needed to secure your Windows devices using Bluetooth, including a security recommendation checklist. In addition, the following recommendations are provided in the guide to help you improve your Bluetooth security:

- Ensure the strongest Bluetooth security mode is enforced for all users where Bluetooth is enabled. Bluetooth operates in two security modes with four levels of differing variance that can be mixed if supported by the device.

- For the strictest security, use security mode 2 with level 4 and require communications that are paired, encrypted, and signed. For the least restrictive, you would use level 1, which will not require any security at all.

- Ensure Bluetooth is listed and referenced in the company security policies and that the device settings have been modified to reflect these policies.

- Ensure any users enabled to use Bluetooth are fully aware of security issues with Bluetooth and their responsibilities while using it.

- Delete unused Bluetooth pairings or configure allowed Bluetooth services and policies using MDM.

Bluetooth security should also be part of your defined baseline and you will want to configure and deploy specific security settings for your users. The following screenshot shows some of the available Bluetooth settings from the Intune settings catalog:

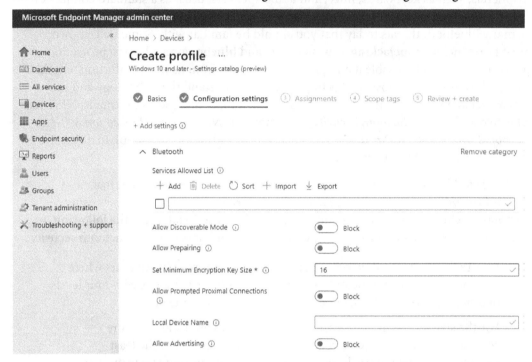

Figure 4.7 – Bluetooth settings in the Intune settings catalog

Bluetooth settings can also be configured in the **Attack Surface Reduction** rules within **Endpoint security**.

By configuring the Bluetooth allowed services policy, it will only allow pairings with explicitly defined Bluetooth profiles and services using **Universally Unique Identifiers (UUIDs)**. For more information about configuring the UUID allowed services list for Bluetooth using MDM, visit this link:

```
https://docs.microsoft.com/en-us/windows/client-management/
mdm/policy-csp-bluetooth#servicesallowedlist-usage-guide
```

Unfortunately, many of the inherent flaws and vulnerabilities in Bluetooth are typically not fixed simply by updating firmware or applying updates and require the hardware to be physically updated. Due to how these chips share physical resources with other system components, a successful exploit can allow attackers to invoke code execution and try to read data on your systems. If you're interested in reading more about notable Bluetooth vulnerabilities, visit this link:

```
https://awesomeopensource.com/project/engn33r/awesome-
bluetooth-security
```

Now that we have covered Wi-Fi and Bluetooth connections, let's discuss VPN and its use within an organization to connect to internal resources.

VPN

The more remote the workforce has become, the more we rely on VPN connectivity. VPN is essentially a technology that allows users to connect to their corporate network over an encrypted secure connection over the internet. VPN allows a user to be anywhere at any time and access corporate data securely. As part of your policies and remote strategy, it is critical that you ensure users are connecting to a VPN when working remotely. The risk of connecting your work device (or any device) to open Wi-Fi connections in public places creates significant risk. When connecting to any network outside of your corporate office, a VPN should be connected to ensure a secure working session. VPN technology has been around for a long time and is a tool that users are most likely familiar with when working remotely.

One primary challenge with VPNs is that they require user interaction to connect once logged into the device. For the most secure working environment, the VPN should automatically connect once connected to an internet connection. Microsoft has a technology known as **Always On VPN**, which is a configurable Windows VPN profile that can automatically connect when off your corporate network. This is a great technology and works very well! To use the Always On VPN technology with Windows 11, you need several components for the infrastructure to support it. The following documentation provides more details on the Always On VPN configuration for Windows clients: `https://docs.microsoft.com/en-us/windows-server/remote/remote-access/vpn/vpn-device-tunnel-config`. The following screenshot shows what a Windows Always On VPN connection looks like in Windows 11:

Figure 4.8 – Windows 11 VPN connection

To manually configure or modify a VPN profile in Windows, open **Windows Settings**, click on **Network & internet**, and click on **VPN**.

To administer a VPN profile using device management, the Windows Always On VPN technology uses the VPNv2 configuration service provider in MDM. There are quite a few settings to configure manually, but all the profile properties can be configured by using a `ProfileXML` schema file instead of configuring each node individually. The `ProfileXML` file can be enforced on clients using PowerShell, Configuration Manager, or Intune through MDM or the WMI-to-CSP Bridge component. The following screenshot shows how to configure the Windows auto VPN using the Intune Windows 10 and later custom profile type and a `ProfileXML` file:

Figure 4.9 – Always On VPN in Intune

Additional *always-on* VPN solutions for Windows are available from Palo Alto Networks through their GlobalProtect client and Cisco's AnyConnect. There is also an option within Windows 11 to use the built-in VPN client to connect to other third-party services or providers. To learn more about setting up a VPN connection from your Windows client, visit this link:

```
https://support.microsoft.com/en-us/windows/connect-to-a-vpn-
in-windows-3d29aeb1-f497-f6b7-7633-115722c1009c#WindowsVersion
=Windows_11
```

> **Tip**
> For personal use, you may want to consider using a consumer-grade VPN service if working from a café or public Wi-Fi hotspot.

Next, let's look at the network security components in Windows Server and the roles and features that can be enabled as components of a network infrastructure.

Windows Server

For Windows Server, the networking-based components are also available in the **Network & Internet** settings. To open them, search for **Network**, click on **Settings**, and then select **Network & Internet**.

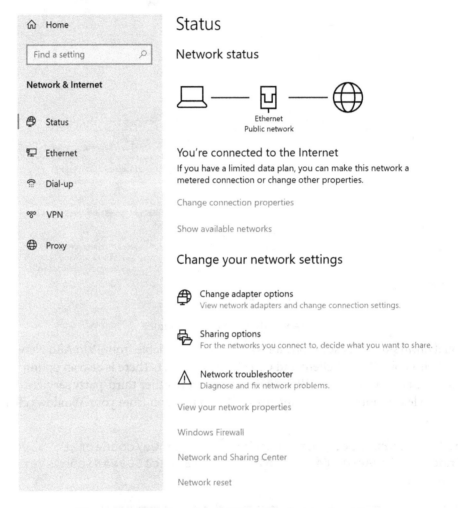

Figure 4.10 – Windows Server 2022 network status in settings

Just like on client devices, the **Network & Internet** management console for Windows Server will provide access to all the Windows network components, such as the current network status, Windows Firewall, the Network and Sharing Center, adapter-specific settings for VPN and Ethernet, proxy settings, and network reset functions.

LAN/Ethernet

Windows server implementations should only be connected using Ethernet for security purposes and for any needed internet access. Ethernet is much more secure than Wi-Fi and provides greater reliability and performance. In addition to using Ethernet, servers should be on a separate network segment from your client devices if possible. Separation should go as far as containing segments for different data classification types, including a **Demilitarized Zone (DMZ)**, and segments for traffic that flows to apps and databases. To accomplish this, you will need to implement VLANs on your LAN and organize the segmentation of the network based on the classification level of the information stored on the servers.

Server roles and features

In addition to the base OS for Windows Server, there are server roles and features. There are many available roles and features that will provide separate services for your enterprise. A few of the more common ones include **AD Domain Services**, **Web Server (IIS)**, and **SMTP Server** for email relay.

To access these roles and features, search for Server Manager within the search area and click on it. Once **Server Manager** is open, ensure you are within the dashboard and click **Add Roles and Features**. Here, you can add your desired network-related server roles and features, as seen in the following screenshot:

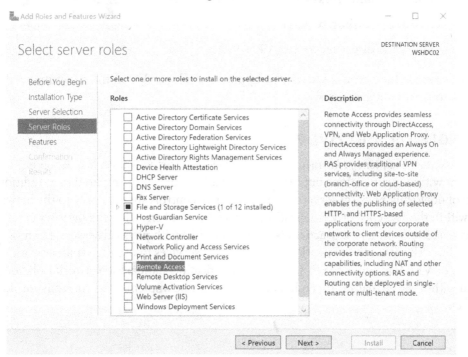

Figure 4.11 – Windows Server roles and features

There are many roles and features available that support network-specific functions. Some of them include the following:

- DNS server
- DHCP Server
- Active Directory Certificate Services
- Network Policy and Access Services
- Remote Access
- Network Load Balancing
- SMTP Server
- SNMP service

To implement the Microsoft's Always On VPN we discussed earlier, the following needs to be deployed within your server environment:

- AD Domain Services
- DNS server
- Network policy and access services (NPS-RADIUS)
- AD **Certificate Authority** (**CA**)
- Remote access (direct access and VPN-RAS)

Next, let's look at the network security components in Hyper-V and the recommended network security configurations.

Networking and Hyper-V

As discussed in *Chapter 3*, *Hardware and Virtualization*, virtualization has become prevalent within organizations but brings a lot of additional risk compared to a traditional model of deploying physical servers. Similar security controls should apply to the network layer within the Hyper-V or virtualization architecture and ensuring the network is set up correctly and following best practices is a must. Just like with physical servers, the most concerning risk within the network layer is allowing services to use the same network segment or VLAN as servers with strict data classification standards. Network isolation is critical within the virtualization architecture and must be implemented correctly for the best security.

The following documentation provides an overview of networking within Hyper-V:

```
https://docs.microsoft.com/en-us/windows-server/virtualization/
hyper-v/plan/plan-hyper-v-networking-in-windows-server
```

There is also this reference for Hyper-V security for Windows Server that should be reviewed as part of your baseline:

```
https://docs.microsoft.com/en-us/windows-server/virtualization/
hyper-v/plan/plan-hyper-v-security-in-windows-server
```

Within this reference, the network-specific security items include the following:

- Use a secure network for both host management and **Virtual Machines** (**VMs**).
- Use separate networks and dedicated physical adapters for the physical hosts.
- Use a separate secure network to access virtual hard disk files and VM configurations.
- Use a separate secure network for any VM migrations and ensure encryption is enabled.
- For VMs, ensure the virtual **network interface cards** (**NIC**) are connected to the correct virtual switch and are configured with the correct security settings.

Now that we have reviewed the network components of Windows Server and tips for securing virtualized networking, let's review some of the tools that are helpful when troubleshooting network-related issues.

Network troubleshooting

As a security professional, you are going to need to be familiar with troubleshooting and investigative work especially when trying to build a timeline during an incident. Chances are that the network layer will be a key component in the troubleshooting and investigative process. Microsoft initially had its own tool that was able to capture and analyze network traffic, known as Microsoft Network Monitor, which was replaced with **Microsoft Message Analyzer** (**MMA**). Unfortunately, MMA has been retired and Microsoft has no plans to replace it. Fortunately, there are many alternatives, and these tools are widely used today by network and security professionals alike. A couple of the more widely adopted freeware tools that allow for the inspection and analysis of network traffic include the following:

- Wireshark: `https://www.wireshark.org/`
- Telerik Fiddler: `https://www.telerik.com/fiddler`

Before using any of these tools, ensure you have the authority or have been provided with permission to run them on a network you do not own.

In this section, we reviewed some of the basic Windows networking components found in Windows clients, servers, and Hyper-V. We resurfaced the topic of baselining for network-specific hardening within your environment. We also provided recommendations for configuring Wi-Fi and Bluetooth settings using Intune. In the next section, we will look at network security features within Windows, including Windows Firewall and enabling web protection features with Microsoft Defender for Endpoint.

Windows Defender Firewall and Advanced Security

Windows Defender Firewall is a software-based firewall that's enabled out of the box and used to control network connections to your PC. To view the basic firewall settings, including the status of each profile, open **Windows Security** from the **Settings** app and select **Firewall & network protection**. There are local security settings you can change from here, including configurations specific to each network profile, such as blocking incoming connections, allowing an app through the firewall, and restoring the default firewall settings.

The three network profile types in Windows Firewall are **domain**, **private**, and **guest/ public**, as follows:

- **Domain profile** settings are defined by the domain profile and are set systemically using Group Policy or from network devices located on the corporate network. Local policy settings are typically overwritten if they're managed systemically.

- **Private profile** is used for home networks or **Small Office Home Networks (SOHOs)** where a domain controller may not be present. A private profile can be configured by a local security policy and by default, incoming connections to apps are blocked if they are not on the list of allowed apps.

> **Tip**
> If an app is blocked, you can view the event log or firewall log for more details. `EventID 5031` in **Windows Logs/Security** will show you if the firewall blocked an incoming connection from an application.

- **Public profile** is used for guest or public networks. Network discovery is turned off by default in Windows for this profile, which blocks file and print sharing. Incoming connections are set identical to the private profile where apps are blocked that are not on the list of allowed apps.

Clicking on **Advanced Settings** and elevating user account control with an administrative account will open **Windows Defender Firewall with Advanced Security**. With Advanced Security, you have fine-grained control over inbound, outbound, and connection security rules for both ingress and egress flows. The inbound/outbound rules can specify specific ports or programs or use custom settings that may include a combination of ports, programs, and physical network adapters. Windows Defender Firewall already comes configured with a set of predefined rules and these rules cannot be directly modified, but they can be enabled or disabled.

> **Tip**
>
> A green checkbox next to the rule name means that it is enforced. If many modifications are needed in addition to the predefined rules, consider building them into a prehardened image or use MDM to configure them.

Connection security rules are used to secure the communications between the source and destination using IPsec and define the authentication conditions needed to establish the connections. Typically, advanced connection security rules will be managed through third-party devices or software. An example of a connection security rule would be to specify a required method of authentication needed to establish a secure connection. If the source and destination systems do not meet the conditions, then the connection is denied. Connection security rules can be defined to require and/or request authentication, both inbound and outbound, and include settings for common authentication methods, such as using certificates or Kerberos and NTLMv2 for computer and user authentication.

> **Important Note**
>
> An important difference between firewall rules and connection security rules is that, simply put, firewall rules are used to allow or deny traffic.

Advanced Security also includes a monitoring section that shows a high-level overview of the status of each network profile. It has helpful links to view active firewall rules, active connection security rules, security associations, and logging settings.

Typically, Windows does a good job of allowing applications through the firewall that are known and trusted. In fact, when enabling certain Windows features or installing software, rules are automatically enabled. In some scenarios, a rule may need to be added to allow inbound/outbound access to an application. This can be accomplished using Group Policy in a Windows Server AD environment or through MDM on Windows clients. Let's look at how to use Group Policy to configure a custom line-of-business app through Windows Firewall.

Configuring a firewall rule with Group Policy

For this example, we need to allow an app named `BusinessApp.exe` whose binaries are located under the `C:\Program Files (x86)\MyLOBApp` path. We want to allow both inbound/outbound connections over the domain profile only. This will ensure that connectivity will work only while connected to the corporate network. Follow these steps to allow the app through the firewall with Group Policy:

1. Open the **Group Policy Management** snap-in from your management workstation and create a new GPO linked to an **organizational unit (OU)** that contains the computer systems you wish to target.

2. Give it a friendly name, such as `Windows Defender Firewall - Inbound Allow Rules`, right-click it, and choose **Edit**.

3. Navigate to **Computer Configuration | Policies | Windows Settings | Security Settings | Windows Firewall with Advanced Security**, expand it, and then expand it again.

4. Right-click **Inbound Rules** and choose **New Rule** to open **New Inbound Rule Wizard**.

5. Select **Program**, click **Next**, and select the radio button next to **This program path**.

6. Enter the path of the executable file for the custom line-of-business app from the install directory, for example, `%ProgramFiles% (x86)\MyLOBApp\ BusinessApp.exe`. Click **Next**.

7. Select the **Allow the connection on Action** menu.

8. Select the domain profile only and click **Next**.

9. Give it a friendly name, such as `Allow BusinessApp`, and click **Finish**.

10. Repeat the same process, but for **Outbound Rules**.

The following screenshot shows the applications allowed in the inbound rules of Windows Defender Firewall after a Group Policy refresh:

Figure 4.12 – Inbound rules in Windows Defender Firewall with Advanced Security

> **Tip**
>
> Configuring firewall rules through Group Policy does not support environmental variables used to resolve the context of the current user. This can cause some challenges if non-administrative users are being prompted to allow an app through the firewall that is running from the `%APPDATA%` or `%USERPROFILE%` locations.

To configure a Windows Defender Firewall rule using Intune, use the **Endpoint protection** template in a device configuration profile for Windows 10 and later. In this template, you can also manage other firewall settings, including the ability to create a rule for a specific network profile.

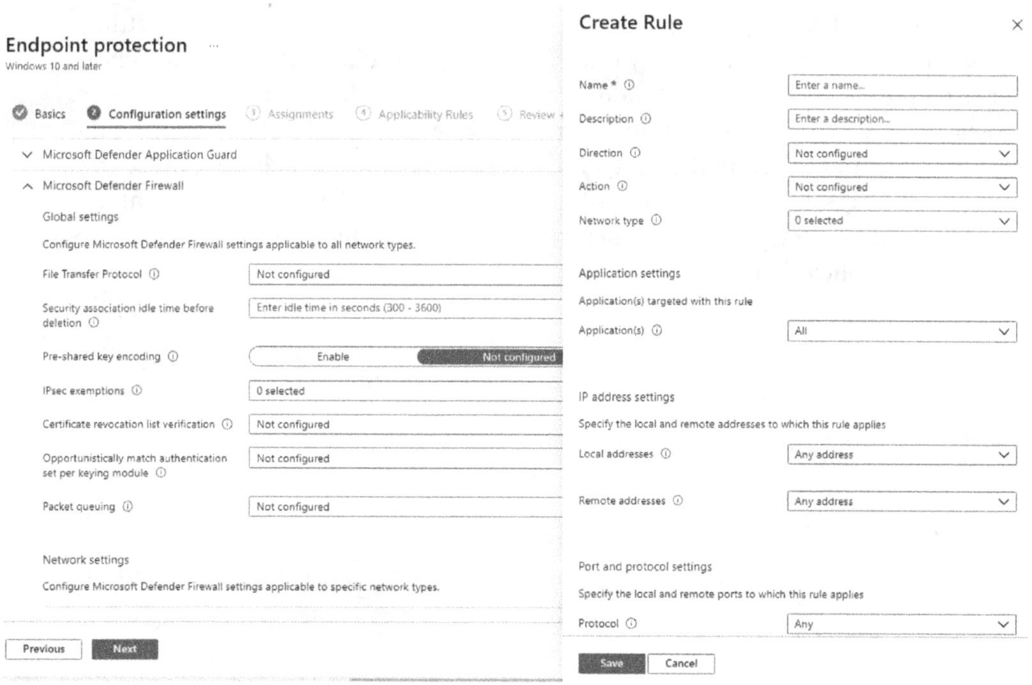

Figure 4.13 – Microsoft Defender Firewall template in Intune

Next, let's look at configuring a security baseline for Windows Firewall.

Configuring a Windows Defender Firewall security baseline

As mentioned earlier, Windows is already configured with a great set of default settings for Defender Firewall. In fact, Microsoft recommends keeping the default settings and not changing any of the defaults for the most secure configuration. However, it is recommended to enforce the default policies to avoid end users from making changes. Intune has a recommended security baseline available in the **Intune Security Baselines** feature located in the **Firewall** section of the Microsoft Defender for Endpoint baselines. The preconfigured values will help get you started with enforcement.

Two additional configuration policies worth noting, known as rule merging, can be set independently in each network profile. **Ignore all local firewall rules** and **Ignore connection security rules** control whether rules created locally by administrators can be merged with rules from a policy-managed firewall.

> **Tip**
> If disabling the merge of local firewall and security rules, set the **Display a notification** policy to **No**, so end users are not prompted when a firewall rule blocks a connection.

Windows Firewall has another built-in feature known as **Shielded mode**. This setting can help isolate a client from the network by blocking all traffic, including connections for apps on the allow list. **Shielded mode** can be set independently on each profile type by going to the **Windows Security** app | **Firewall & network protection**, choosing the network profile, and selecting **Blocks all incoming connections, including those in the list of allowed apps**. Later, we will also discuss a feature in Defender for Endpoint that allows you to isolate devices from the network if they are deemed compromised without having to push policies directly to a device.

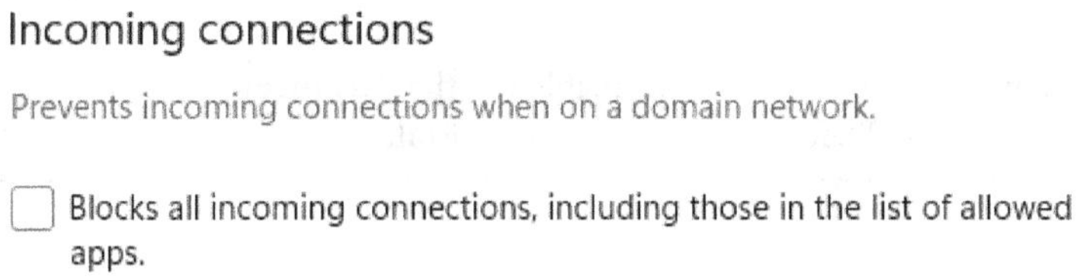

Incoming connections

Prevents incoming connections when on a domain network.

☐ Blocks all incoming connections, including those in the list of allowed apps.

Figure 4.14 – Shield mode in Firewall & network protection in Windows Security

Next, we will look at the web protection features available in Microsoft Defender for Endpoint.

Web protection features in Microsoft Defender for Endpoint

Microsoft Defender for Endpoint has a list of features known as next-generation protection that are available for customers who have purchased a Microsoft Defender Endpoint Plan 1 or 2 license type. Included in these features are a set of network and web protection solutions that can help control network connections to your clients. These features, known as web protection and network protection, can complement or possibly replace a traditional network proxy service for web traffic control. Let's look at some of these features. To view the endpoint settings of Defender for Endpoint, log in to the Microsoft 365 Defender portal at `https://security.microsoft.com`, choose **Settings**, and then select **Endpoints**.

Using custom indicators

Custom **Indicators of Compromise** (**IOCs**) allow security teams to create blocklists of file hashes, IP addresses, URLs/domains, and certificates. When an entry is added to the custom indicators list, a response action can be set to allow, audit, warn, or block. If using Defender for Endpoint operationally for SecOps, you can also generate alerts and set the severity and category for each item. The indicators list can be found at this link in the Microsoft 365 Defender security portal for endpoints and will look as in the following screenshot:

```
https://security.microsoft.com/preferences2/custom_ti_
indicators/files
```

> **Tip**
>
> Custom network indicators must be enabled in the **Advanced Features** section of the Microsoft 365 Defender for Endpoints portal.

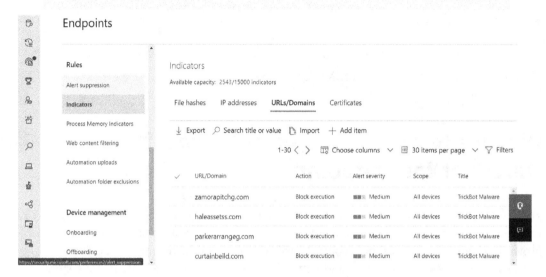

Figure 4.15 – Custom indicators in Microsoft 365 Defender

Custom indicators are a great way to block not only IOCs but also any other specific websites or connections you don't want your end users to visit. For example, you can add a streaming service such as Netflix and when a user tries to visit it in a web browser, they will receive a Windows Security toast notification informing them the content is blocked. To enable the blocking feature on the client, it will require a Defender SmartScreen (for Edge) or network protection (other browsers) policy to be enabled.

Figure 4.16 – Windows Security toast notification for a blocked network connection

The flexibility to manually add URLs and IPs to block is a great added feature available in Microsoft Defender for Endpoint. If you don't have the bandwidth to manage these lists manually, use Microsoft Defender's web content filtering policy to configure policies based on specific content categories. Let's take a look at the web content filtering capabilities in Defender for Endpoint.

Web content filtering

By creating a web content filtering policy, you can rely on Microsoft's security stack to help filter websites based on content category. This solution works in all major browsers and can be set up to enforce block mode if using the network protection policies. When configuring a web content filtering policy, there are five parent categories, and each has several child categories to further refine the scope. The following content categories are available:

- **Adult content** contains child categories for cults, gambling, nudity, pornography or sexually explicit content, sex education, tasteleess images, and violence.

- **High bandwith** allows you to block sites that are categorized as download sites, image sharing, peer-to-peer, or streaming media and downloads.

- **Legal liability** contains categories for child abuse, criminal activity, hacking, hate and intolerance, illegal drugs, illegal software, school cheating, self-harm, and weapons.

- **Leisure** allows you to block chat and gaming sites, instant messaging, professional networking sites, social networking, and web-based email.

- **Uncategorized** lets you block websites that are newly registered or parked domains.

Just like creating a custom indicator, you can scope a web content filtering policy to all devices or a subset of devices in your organization. If the content categories are blocking a website needed for business purposes, you can allow it by adding it to the URL list in the custom indicators and setting the action to allow. Microsoft 365 Defender includes a **Web Threat Protection** report so you can get an idea about the types of websites that would be blocked under a content policy. To view these reports, log in to the Microsoft 365 Defender portal at `https://security.microsoft.com`, choose **Reports**, and select **Web Protection**. Click on the details under the web activity summary, and you will gain better insight into the clients and domains that were visited under each content category. This will be helpful when planning your filtering policies.

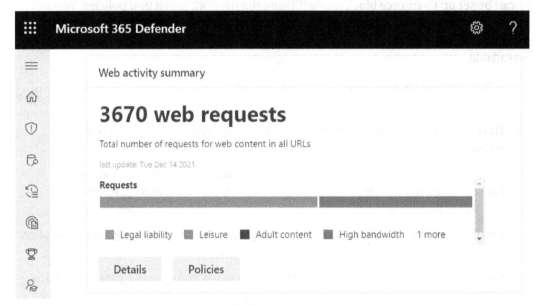

Figure 4.17 – Microsoft 365 Defender report for web content filtering

Next, let's look at how to turn on network protection to configure the blocking of network connections on the client.

Blocking connections with network protection

Network protection is a security feature that can be enabled through the Exploit Guard features of Microsoft Defender. Network protection helps to reduce attacks from low-reputation IP and URL sources known for phishing, social engineering, and malicious browser redirects based on analytics from Microsoft's threat intelligence service. Its protection covers all major browsers, including Google Chrome and Mozilla Firefox. If you're licensed for Microsoft 365 Defender for Endpoint, activity logs from devices that trigger the policy are sent into the portal for the advanced hunting of queries, alerting, and reporting. Network protection is a great feature for small- or medium-sized businesses or those looking to get away from third-party proxy services and the administration required to maintain them. It supports the following three modes:

- **Not Configured** keeps the feature disabled, which is the default Windows setting.

- **Enabled** will block connections and send activity logs to Microsoft 365 Defender.

- **Audit only** will allow connections but will also send activity logs to Microsoft 365 Defender.

> **Tip**
> If network protection is alerting false positives, Microsoft recommends opening a support case so that their threat team can investigate.

When a connection is blocked, a notification informs the user of the blocked connection through Windows Security notifications. You can also customize the details in the action buttons on the notification, such as setting a support email or web URL to the IT support portal. The notification from Windows Security looks like the one previously shown in the custom indicators policy, *Figure 4.16*.

In the following screenshot, **Kusto Query Language** (**KQL**) is being used to query the Microsoft 365 Defender logs for Exploit Guard network protection events in **Advanced hunting**:

Figure 4.18 – KQL query in Advanced hunting in Microsoft 365 Defender

Network protection can be enabled through PowerShell, Group Policy, or MDM. For information about managing a Group Policy central store and the latest downloads for administrative templates, visit the following link:

`https://docs.microsoft.com/en-US/troubleshoot/windows-client/ group-policy/create-and-manage-central-store`

After importing the ADMX templates, the feature can be found under **Windows components** | **Microsoft Defender Antivirus** | **Windows Defender Exploit Guard** | **Network Protection**.

To enable network protection using Group Policy, follow these steps:

1. Open the **Group Policy Management** snap-in console from a management workstation and create a new **Group Policy Object (GPO)** linked to an OU that contains the computer systems you wish to target.

2. Give it a friendly name, such as `Microsoft Defender Exploit Guard - Network Protection`, right-click it, and choose **Edit**.

3. Navigate to **Computer Configuration | Policies | Administrative Templates | Windows Components | Microsoft Defender Antivirus | Windows Defender Exploit Guard | Network Protection**.

4. Open the **Prevent users and apps from accessing dangerous websites** policy setting and set it to **Enabled**.

5. Choose the enforcement mode of **Block** or **Audit Mode**.

> **Tip**
>
> If a user turns off notifications, they may not receive any notice that a connection has been blocked.

If using Intune, network protection can be configured using templates or through Endpoint Security by going to the Microsoft Defender Antivirus profile and choosing the Windows 10, Windows 11, or Windows Server platform, and then set **Enable Network Protection** to Enabled (block mode) as in the following screenshot:

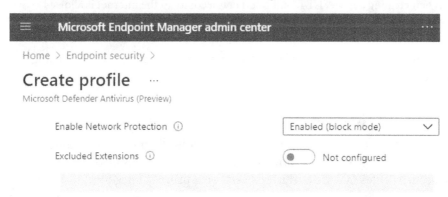

Figure 4.19 – Enable network protection in Endpoint security

In the previous sections, we covered Windows Defender Firewall with Advanced Security, including how to create custom rules using Group Policy. We discussed enforcing a security baseline and recommended a few specific policies to enforce. Next, we covered web protection features of Microsoft Defender, including custom indicators, web content filtering, and enabling network protection to block connections to suspicious hosts. In the next few sections, we'll shift focus and discuss Azure cloud solutions that can be used to provide network security to your Windows-based endpoints in Azure.

Introducing Azure network security

At its foundation, the Azure networking plane in regard to private addressing consists of a VNET containing a defined address space. Just like traditional networking concepts, the VNET can then be further segmented into subnets, where resources are assigned to a designated space. Resources inside the same VNET are typically allowed to communicate with each other as well as with other Azure services using the underlying networking fabric or service endpoints. Depending on the networking topography in your environment, if you need to enable communications into other VNETs, a feature known as VNET peering will allow cross-VNET communication. When protecting your Windows resources in Azure, there are a few features available for controlling traffic flow inbound and outbound to your endpoints. Using a combination of **User-Defined Routing** (**UDR**), NSGs, Azure Firewall, and **Network Virtual Appliances** (**NVAs**) will help ensure communications are locked down to only allow the appropriate traffic to reach your resources.

> Tip
> When creating a new VNET, outbound connectivity to the internet is allowed by default.

In the next section, we are going to review network security access controls using a feature known as NSGs.

Controlling traffic with NSGs

An NSG is an Azure resource that acts as a stateful firewall for evaluating inbound and outbound traffic. It is used to allow or deny communications through a set of weighted security rules that are evaluated based on a priority integer value using five-tuple information. NSG resources can be associated with subnets or VNET interfaces. As a best practice, it is recommended that NSGs are applied at the subnet level over a direct assignment to a network interface. This helps minimize the amount of NSGs for simplification purposes. A security rule inside an NSG has the following five-tuple properties:

- **Name**, which is used to identify it; for example, `AllowRDP`.
- **Priority**, between 100 and 4,096, which is used as the weight during evaluation.
- **Source or Destination**. The options can include **ANY**, an individual IP, a range specified in **CIDR notation**, or a service tag.
- **Protocol** support for **TCP, UDP, ICMP**, or **Any**.

- **Direction** (inbound or outbound).

- **Port range**.

- **Action**, such as allow or deny.

> **Tip**
>
> When assigning a priority, it is recommended to assign them at intervals of 50 or 100 to ensure there is plenty of space to insert rules in the future.

To help simplify the rule definitions in your NSG, a service tag can be used to help minimize this complexity for connectivity to Azure services.

Using service tags for Azure services

The destination value for a security rule in an NSG can also be set to an Azure service tag. Service tags are available to help simplify the creation and maintenance of security rules. Instead of manually specifying IP ranges for common connection points or managing connections that have changing details, Microsoft maintains the connection information for these services inside of a service tag. Examples of service tags in Azure include values for VNETs, storage, SQL, or the internet. The full list of available VNET service tags can be found at this link:

```
https://docs.microsoft.com/en-us/azure/virtual-network/
service-tags-overview
```

> **Tip**
>
> Security rules can also be augmented security rules that contain a comma-separated list of IP ranges in CIDR notation instead of you having to create separate rules for each IP block.

The following screenshot shows the NSG inbound security rules. You can see the service tags depicted as **Internet** and **VirtualNetwork** in the **Source** and **Destination** sections.

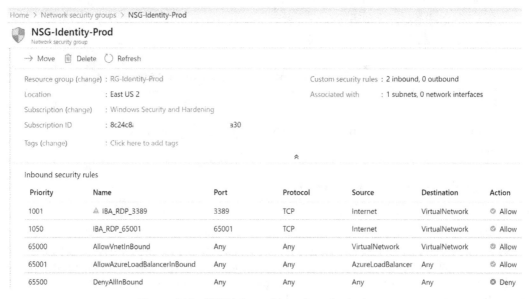

Home > Network security groups > NSG-Identity-Prod

NSG-Identity-Prod
Network security group

→ Move   🗑 Delete   ↺ Refresh

Resource group (change) : RG-Identity-Prod		Custom security rules : 2 inbound, 0 outbound	
Location : East US 2		Associated with : 1 subnets, 0 network interfaces	
Subscription (change) : Windows Security and Hardening			
Subscription ID : 8c24c8· a30			
Tags (change) : Click here to add tags			

Inbound security rules

Priority	Name	Port	Protocol	Source	Destination	Action
1001	⚠ IBA_RDP_3389	3389	TCP	Internet	VirtualNetwork	⊘ Allow
1050	IBA_RDP_65001	65001	TCP	Internet	VirtualNetwork	⊘ Allow
65000	AllowVnetInBound	Any	Any	VirtualNetwork	VirtualNetwork	⊘ Allow
65001	AllowAzureLoadBalancerInBound	Any	Any	AzureLoadBalancer	Any	⊘ Allow
65500	DenyAllInBound	Any	Any	Any	Any	⊗ Deny

Figure 4.20 – NSG inbound security rules in Azure

Next, we will review grouping resources with **Application Security Groups** (ASGs).

Grouping resources with ASGs

ASGs are an additional enhancement for simplifying NSG rules as they allow you to group together multiple components that may power the backend of an application. You can use this custom group and use it inside a security rule, such as a service tag. For example, let's say you have a few shared backend app servers and database servers. As the business requirements evolve, there may be a need to use these resources to service other business functions that are housed in different subnets. By grouping these resources together and creating an ASG, you can specify the source as this grouping in the NSG rule and don't have to granularly define each component moving forward.

Creating an NSG in Azure

It's common when building new infrastructure in Azure to deploy a jump box server. This is helpful to allow connectivity from the internet to access resources inside your private network without needing a VPN or **virtual desktop infrastructure (VDI)** environment. Although we strongly recommend only using a jump box as needed for emergency use and shutting it down when not in use, we want to demonstrate how to create an NSG rule that allows RDP traffic over the internet to a Windows server. To do this, we will set a non-common port to help deter malicious actors actively looking for connections listening on port 3389 (RDP) over the internet.

This demo assumes the following resources have already been configured in Azure:

- A resource group, Azure VNET, and a subnet
- Windows VM with a public IP address assigned

The new NSG will contain two inbound security rules. One will use 3389 to allow connections using the standard default RDP port, and the other set to port TCP 65001, which will be the new, modified port for RDP on the underlying VM. Finally, we will create an inbound firewall rule in Windows Firewall and set the listening port to 65001 for RDP through the registry on the VM. Let's get started:

1. Log in to the Azure portal at https://portal.azure.com.
2. Search for Network Security Groups and select it.
3. Click **Add** or **Create network security group** if none have been created.
4. Choose the subscription and then select the resource group that contains your Windows VM. Give it a friendly name, such as NSG-Identity-Prod. In our example, we are creating an NSG for a subnet that contains domain controllers.
5. Select the region that your subnet resides in and choose **Review + Create**. Then, select **Create** after the validation passes. Go to the resource after the deployment completes.

   NSGs have predefined security rules out of the box and cannot be deleted. Their priority is specifically set high to allow for plenty of space for custom security rules. The maximum integer for a custom-defined security rule is 4096.
6. Click on **Inbound security rules** under **Settings**.

7. Click **Add** to create a new inbound rule with the following settings:

    ▪ **Source**: Service Tag

    ▪ **Source service tag**: Internet

    ▪ **Source port ranges**: *

    ▪ **Destination**: VirtualNetwork

    ▪ **Destination port ranges**: 3389

    ▪ **Protocol**: TCP

    ▪ **Action**: Allow

    ▪ **Priority**: 1001

    ▪ **Name**: IBA_RDP_3389

8. Click **Add** to create the rule.

9. Repeat these steps to create the inbound allow rule for the custom TCP port 65001, but with the following changes:

    ▪ **Destination port ranges**: 65001.

    ▪ **Priority**: 1050.

    ▪ **Name**: IBA_RDP_65001. This is short for inbound allow, Remote Desktop Protocol, and port number 65001.

10. Click **Add** to create the second rule.

    Notice the warning symbol in the following screenshot next to the inbound allow rule for 3389. Azure does a good job of warning you that a common port is open to the internet:

Inbound security rules

Priority	Name	Port	Protocol	Source	Destination	Action
1001	⚠ IBA_RDP_3389	3389	TCP	Internet	VirtualNetwork	⊘ Allow
1050	IBA_RDP_65001	65001	TCP	Internet	VirtualNetwork	⊘ Allow
65000	AllowVnetInBound	Any	Any	VirtualNetwork	VirtualNetwork	⊘ Allow
65001	AllowAzureLoadBalancerInBound	Any	Any	AzureLoadBalancer	Any	⊘ Allow
65500	DenyAllInBound	Any	Any	Any	Any	⊗ Deny

Figure 4.21 – Inbound security rules inside an NSG in Azure

Now, let's associate the NSG with a subnet.

11. Click on **Subnets** in **Settings**.

12. Choose **Associate**, select a VNET from the dropdown, and select the subnet that contains your Windows VMs. Click **OK**.

    Now that we have created the NSG rules and associated them with a subnet, we can remote into the Windows VM and modify the settings to change the RDP listening ports to 65001. Follow the next steps to do so.

13. Find the VM by searching for it by name in the Azure portal. On the **Overview** tab, click **Connect** to download the RDP file and connect over a public IP with the default RDP port of 3389.

14. Log in to the VM using the administrative account you used when creating the VM to load the desktop.

15. Once at the desktop, search for Windows Defender Firewall with Advanced Security to open the advanced menu. Click on **Inbound Rules**.

16. Click **New Rule** and create a new inbound rule.

17. Select **Port** as the rule type to create and click **Next**. Select **TCP** and enter 65001 in the box to specify **Specific local ports** and click **Next**.

18. Keep the default of **Allow the connection** selected and click **Next**.

19. Keep all three profiles selected and click **Next**.

20. Give the rule a friendly name, such as Remote Desktop - IBA 65001, and click **Finish**.

21. Open the **registry editor** (**regedit**), go to the HKEY_LOCAL_MACHINE\System\ CurrentControlSet\Control\Terminal Server\WinStations\ RDP-Tcp registry subkey, and look for the DWORD port number.

22. Open the DWORD port number, choose **Decimal**, and modify the port number so that it states 65001. Click **OK**.

23. Restart the VM.

24. Validate the port number change worked by connecting to the VM with RDP by specifying the IP and port, for example, 40.70.223.4:65001.

If using the Azure portal, the following screenshot shows the **Connect to virtual machine** menu that appears after clicking **Connect**:

Figure 4.22 – Connect to virtual machine menu in Azure VM overview

To recap, we just created a new NSG with two inbound rules to allow TCP ports 3389 and 65001 to accommodate RDP traffic. Then, we modified the default RDP listening port on the VM host to TCP 65001 and created a new inbound allow firewall rule in Windows Defender Firewall to allow for TCP 65001. Next, let's review a few security recommendations when working with Azure networking.

Security recommendations for Azure Networking

While working with Azure networking can be highly complex, here are a few recommended security best practices to consider whenever creating new resources, in particular, Windows VMs:

- Ensure RDP access is not allowed directly from the internet. Use a VPN deployment or VDI environment for connectivity internally.

- Disable SSH access from the internet using an NSG.

- Disable UDP services from the internet using an NSG.

- Only allow the necessary ports, protocols, and services to connect to your resources through NSGs.

Connecting privately and securely to Azure services

Earlier, we mentioned how, by default, internal resources connect to Azure services using service endpoints over the Azure networking fabric. This communication doesn't occur over your own private and secure network and doesn't prevent access to other resources inside of Azure. To accommodate this, Azure has a service called **private endpoints**. Private endpoints allow internal resources to connect directly to PaaS services privately over your own VNET, allowing for greater control and privacy. When creating a private link resource, an internal private IP is assigned to it and, therefore, you will need to use DNS lookups for name resolution and connecting using an alias. To date, there are many services that support private links. One example is using Azure Files to host company file shares in the cloud over a standard on-premises file server. Without private endpoints, client devices would connect to file shares directly over the internet. If using private endpoints on the Azure file share, clients would then only be able to connect through a VPN connection or from a VM already in your private IP space. You could also connect Azure App Service to a private link if you wanted to host traditional intranet sites in Azure and not have to deploy IIS on a traditional physical server.

Protecting Windows workloads in Azure

Two additional network security solutions in Azure worth noting for Windows deployments are Azure **Web Application Firewall** (**WAF**) and **Azure DDoS Protection**. Both services can be deployed to sit in front of your Windows IaaS and PaaS workloads and help protect against common attacks from over the internet.

WAF is a service that is typically used to complement the frontend entry point of a web app and is added to Azure Front Door or Application Gateway. The WAF uses OWASP rules to help protect your web app from common generic attack techniques, such as cross-site scripting, local file inclusion, and SQL injection. By deploying a WAF in front of your web application, it helps to reduce the need to prevent these attacks directly in the application code itself. To read more about Azure WAF, visit the following link:

```
https://docs.microsoft.com/en-us/azure/web-application-
firewall/overview
```

Azure DDoS Protection is a service that helps protect against **Distributed Denial of Service** (**DDoS**) attacks. DDoS attacks are typically used to target resources with the purpose of taking them offline or making them unusable by consuming all available resources. If you're deploying services in Azure, there is a basic level of DDoS protection from Microsoft available by default through Azure's global infrastructure. For other Windows workloads deployed into your own VNET, Microsoft offers an enhanced version called Azure DDoS Protection Standard. This will help to directly protect your specific resources that are available over the public internet. To read more about Azure DDoS, visit this link:

```
https://docs.microsoft.com/en-us/azure/ddos-protection/ddos-
protection-overview
```

Summary

In this chapter, we started by looking at network security fundamentals and covered the OSI and TCP/IP models, reviewed common ports and protocols, and looked at the important technologies needed to help with network security. After that, we covered Windows network security, which started with network baselining for your Windows devices. This section then covered the Windows 11 network management pane, before moving on to securing WLAN/Wi-Fi, Bluetooth, and VPN, including Microsoft's Always On VPN. The following section covered the Windows Server network management pane, including LAN and Ethernet best practices for your servers, and provided an overview of the server roles and features. The final sections overviewed Hyper-V networking and security before finishing off with network troubleshooting tools.

Following Windows network security, we covered Windows Defender Firewall and Advanced Security, which provided steps you can follow to configure firewall rules and enforce policies. Then, we discussed web protection features that are available in Microsoft Defender for Endpoint, including Web Threat Protection, a web content filtering policy, and how to enable network protection in block mode. The final topic of this chapter discussed Azure network security features, where we covered NSGs, connecting privately to Azure services, and additional features for protecting Windows workloads running in Azure.

In the next chapter, we will review identity and access management and its importance in Windows management and zero-trust security. We will review components involved in the life cycle of accounts and the types of access needed. Then, we will review different authentication technologies and provide recommendations on how you should be using these protocols in today's world. We will finish with a review of implementing Azure Conditional Access and Identity Protection to help protect from identity-based attacks.

5
Identity and Access Management

In this chapter, we will be reviewing identity and access management in depth, as well as their importance within enterprises today. In a zero-trust model, identity has become extremely critical, and the need to protect them has never been more important. Through third-party SaaS providers and the data center shift to the cloud, users can now access their corporate information from anywhere over the internet. A simple breach of just their identity can allow an intruder to log in and access their information and a greater variety of data within the environment. Because of this, we need to revisit the traditional authentication methods and add enhanced protection to our identity and access model.

In today's world, the use of passwords is becoming obsolete. Due to major breaches and the advancement of technology, it's not implausible to assume our passwords have been breached and are sitting on the dark web for sale. In addition, bad password management leads to the reuse of passwords on other accounts or with slight variations that are easy to crack. What does this mean? Essentially, someone with your password will be able to easily start accessing all your accounts, especially if no additional security protections are in place. It's commonplace to see attackers use the same user accounts and passwords within services that haven't been breached to try and gain additional access. This type of attack is known as credential stuffing, and unfortunately, it is working!

In this chapter, we will provide an overview of identity and access management and many of the components required to build a robust program. We will cover account and access management, followed by authentication and **Multi-Factor Authentication (MFA)**, and discuss moving away from passwords with passwordless technologies. We will finish off the chapter with powerful Microsoft cloud technologies that can better protect your identities – Conditional Access and Identity Protection.

This chapter will include the following topics:

- Identity and access management overview
- Implementing account and access management
- Understanding authentication, MFA, and going passwordless
- Using Conditional Access and Identity Protection

Technical requirements

In this chapter, we will refer to identity management solutions, such as **Microsoft Identity Manager (MIM)**. We will also provide overviews of different access and identity management services within Azure and the varying levels of licenses and requirements. It is encouraged that you research the specific licensing requirements for each solution independently if they fit your needs. At a minimum, you can follow along with the referenced solutions using the following:

- An Azure tenant with subscription owner and Global Administrator rights
- Azure AD Premium P2

Now, let's take an overview of identity and access management.

Identity and access management overview

Identity and access management has never been as important as it is today. Identity can be somewhat considered as the foundation for security within your organization. Although there are other methods of compromising data, simply gaining access to a user or administrative account can be destructive. If an intruder compromises an account, they now acquire the same account level of access across all systems and data. All this can take place without anyone being alerted. It is very important that you are rigid with your identity and access policies. The role of **least-privilege** is a must! This is a role where no access is added to your account until needed, based on your job function. We will cover this in more detail in the *Authorization* section of this chapter that follows. Essentially, if you don't need access, you don't get it. In addition, ensuring that you use separate user accounts for administrative accounts is critical. There should never be elevated or privileged permissions added to your regular user accounts where general day-to-day tasks are performed. As already stated, passwords have become obsolete and do not provide the level of security needed today. You need to start implementing multi-factor-level authentication as soon as possible, along with more modern technologies, to protect your users' and your organization's data.

Let's look at the foundation or framework of what comprises a solid identity and access management program. If you have studied for your **Certified Information Systems Security Professional (CISSP)** certification or some other security-related exam, you will be familiar with the term **Identification, Authentication, Authorization, and Accountability (IAAA)** or just **Authentication, Authorization, and Accountability (AAA)**. Here, we will use IAAA to emphasize identity as the first component of the process because, without an identity, the proceeding components cannot exist.

Identity

The identity portion of your access relates to something that identifies who you are. Simply put, it is a unique identifier that you enter to let a system know that it is you who is trying to access it. A simple example could be a username, an email address, or an employee ID number. Traditionally, within a windows environment, you may be familiar with sAMAccountName, which was essentially a username that only worked within your corporate network. Using this outside your corporate network was not possible, as there was no unique identifying factor that accompanied the username on places such as the internet. Today, you will need to adopt the **User Principal Name (UPN)** method for your identity to best support internet and cloud-based technologies. The UPN appends a domain that you own to the end of your username to provide a unique or immutable identity no matter where you are and how you access it. An example of a UPN identity could be user@windowssecurityandhardening.com

Authentication

After you have entered your identity, you will typically be presented with some form of authentication to validate that you are the person who should be gaining access to the system. The most common form of this is entering a password. As part of authentication, there are four different methods to authenticate your identity that are worth mentioning:

- **Type 1** is something you know. This is currently the most common and widely adopted method of authentication. Some examples include a password, PIN, or passphrase.

> **Tip**
>
> An example of a password is `W!nd0ws11`, while a similar passphrase is `W!nd0ws $ecure 11`. Typically, passphrases can contain spaces, are longer, and are easier to remember than a password.

- **Type 2** is something you have. This authentication method consists of something you have with you to confirm it is you. Some examples include a hard token, a soft token on your phone, or a smart card. With Microsoft, the Authenticator app is an example of this.

- **Type 3** is something that you are. This authentication method is commonly known as biometrics or something physical that is used to authenticate you. Some examples of this include your fingerprints, and facial or iris scans. On Microsoft, Windows Hello falls under this category of authentication.

- **Type 4** is location-based. This authentication method may not be as commonly used but can be extremely powerful with the advancement of cloud compute and AI technologies. Authenticating based on location allows you to authenticate users based on a defined geolocation. Think of a company that only conducts business within the United States. Why would they expect anyone to access their servers from outside the country? Anyone accessing them from outside the United States is expected to be an intruder and should be blocked. In Azure, Conditional Access can provide this level of authentication.

Next, let's look at authorizing your identity against the system to gain access to data.

Authorization

Once you have been authenticated into your environment or system, you will need to access data or systems that you have been authorized to access. Authorization is a method in which permissions have been added to your identity to allow access to some information or system. Just because you have been authenticated doesn't mean that you are authorized to access data or a system. This is where the principle of least-privilege comes into play, and as a best practice, there should be no authorization added to a new identity by default. Authorization isn't an easy task, especially as an organization grows. Ensuring that someone only has access to what they are authorized to do can be a full-time role, especially ensuring that authorization is updated and removed as roles change. To better help with authorization and to define your users' job functions, a well-defined access model should be in place to implement **Role-Based Access Control** (**RBAC**), using identity management tools and centralized management of users and groups that use Azure AD or on-premises AD.

Now that the identity has been authenticated and authorized to your system, let's look at what accountability is in terms of actions and data accessed inside an environment.

Accountability

The last function is known as accountability, also referred to as accounting. This is the function that allows us to ensure that an identity or user is not abusing their rights or the authorization provided. Here, we need to ensure that we track all the activities of our users and ensure auditing is in place to hold anyone accountable for any misuse of their identity and access. This is also a critical process as part of any investigative and forensic work that is carried out for attempted breaches or compromised accounts. Knowing exactly where an account has been authenticated and what data it has accessed must be captured and audited for historical purposes.

As you can see, the full scope of an identity and access management program takes great effort and time to implement correctly. With the identity of your users being at the forefront and serving as the key to your data and systems, organizations must implement a well-rounded and robust identity and access management program. As we discussed earlier in this book, ensure your policies and standards have leadership support to ensure your program is effective.

> **Tip**
>
> As a start, to assist with password management practices, it is highly recommended that you provide a form of password manager for your users. It is also recommended that you educate your users about identity protection services. There are many services available, and you can view the service as insurance coverage for your identity. It is worth the investment!

To finish off this section, we wanted to remind you of the importance of physical access security. Ensuring that an intruder can't easily walk onto your site is not only critical to your data but also the safety of your employees. Nothing is more important than human safety! Referring back to the **National Institute of Standards and Technology (NIST)** cybersecurity framework, there is a specific category and subcategory to ensure you implement the minimum recommended guidelines for physical security, as shown here:

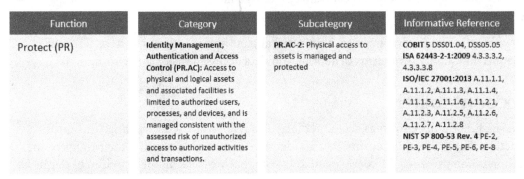

Function	Category	Subcategory	Informative Reference
Protect (PR)	**Identity Management, Authentication and Access Control (PR.AC):** Access to physical and logical assets and associated facilities is limited to authorized users, processes, and devices, and is managed consistent with the assessed risk of unauthorized access to authorized activities and transactions.	**PR.AC-2:** Physical access to assets is managed and protected	**COBIT 5** DSS01.04, DSS05.05 **ISA 62443-2-1:2009** 4.3.3.3.2, 4.3.3.3.8 **ISO/IEC 27001:2013** A.11.1.1, A.11.1.2, A.11.1.3, A.11.1.4, A.11.1.5, A.11.1.6, A.11.2.1, A.11.2.3, A.11.2.5, A.11.2.6, A.11.2.7, A.11.2.8 **NIST SP 800-53 Rev. 4** PE-2, PE-3, PE-4, PE-5, PE-6, PE-8

Figure 5.1 – The NIST cybersecurity framework's physical access to the assets guidelines

The following link gives you access to *NIST Special Publication 800-53*, referenced in the preceding figure: `https://csrc.nist.gov/publications/detail/sp/800-53/rev-5/final`.

Next, we will review account and access management and what is involved in the identity management life cycle.

Implementing account and access management

One of the most important tasks with your identity and access management program is the management and life cycle of accounts and the auditing of the access they have. The whole life cycle process in account and access management may involve multiple teams to make the process efficient and successful. Likely, there's a chance that multiple systems and tools are involved in the life cycle management, including manual human processes that increase vulnerability due to poor housekeeping and being error-prone. Account and access management is a complex process and only becomes more challenging through the ongoing expansion of the application portfolio, as well as the shift to the cloud. A typical account and access management program may involve resources from HR, identity and security teams, technical operations teams, hiring managers, and potentially others.

To ensure success with your account and management program, it must start with well-defined policies that enforce the correct standards and procedures throughout the life cycle process. Any mismanagement of this process can result in significant damage to your organization due to the critical nature of the identity in security. Some examples of this include applying the principle of least privilege to a user account, untimely terminations, or not managing and auditing your privileged accounts. One method to help with any mismanagement or even manual errors is implementing automation. The more you can automate, the fewer errors will occur. There will always be a need for manual processes, whether it be the initial input of user data, physical validation that the information is correct, or applying an end date to a user, but the more you can reduce these processes, the less error-prone they will be.

Every organization is structured differently, but for the most part, you will likely have accounts categorized as **Full-Time Employees** (FTE), contractors, vendors, guests, or service accounts and service principals. For each of these categories, you will need separate policies and procedures for their management. Your FTE accounts may be managed differently than your contractor accounts, as well as your vendor and guest accounts. Ensuring all these accounts are managed correctly will require close collaboration between HR, the identity and security teams, and the hiring or reporting managers of these accounts.

> **Tip**
> Never use shared accounts within your environment. All users who access your environment should have an account assigned that identifies them. Without an assigned identity, individual accountability is difficult.

Let's look at a typical account and access life cycle scenario that starts with HR as the source of authority for the identity.

HR and identity management

Your identity management life cycle needs to start somewhere, and that is most likely with your HR department. This is where your employees start their journey within your organization. Once an employee has accepted a position, their digital profile is created. HR software is very specialized and is typically independent of the core IT identity services. Within an HR system, you can expect to manage your personal information, time off, payroll, performance, the employees you manage, and employee training, to name a few. The big challenge with your HR system is efficiently integrating the application into your core identity services.

> **Tip**
> There are many HR platforms available on the market today. Some examples include Microsoft Dynamics 365 Human Resources, SAP SuccessFactors, Workday, and Oracle PeopleSoft.

When you look at your core identity service, it's common that it will consist of an on-premises AD deployment that's synchronized to Azure AD, creating a hybrid identity model. One of the challenges with the traditional deployment using on-premises AD is the limitation on delivering a well-rounded identity service beyond basic account management. Having the ability for employees to request contractor accounts, manage their own AD groups, update profile information, use a self-service portal, manage the account life cycle process, and build automation are not native features of AD. This is where an identity management solution sits between your HR source and AD to provide the required efficiency.

From a Microsoft standpoint, this solution is known as MIM, which is the latest edition of **Forefront Identity Manager** (**FIM**). MIM is an extremely powerful management tool and serves as a critical component of the overall identity life cycle. MIM deployments can be highly complex, depending on the level of customizations needed. To learn more about MIM and its capabilities, visit `https://docs.microsoft.com/en-us/microsoft-identity-manager/microsoft-identity-manager-2016`.

> **Tip**
> The HR system will contain **Personal Identifiable Information** (**PII**), so it is critical that you work closely with the HR team to ensure that data is correctly secured and encrypted both in transit and at rest.

As cloud adoption increases, an ideal scenario is to directly integrate your HR system with a cloud identity provider, such as Azure AD. Unfortunately, while many organizations have compliance requirements to keep identities and passwords on-premises, many applications still rely on AD for authentication, with on-premises architecture fully deployed, making it hard to shift this model.

At some point soon, you are going to need to shift to a direct provisioning model from your HR system to Azure AD. Because of this, it is recommended you start planning out this journey, as a transition this complex will take time. Microsoft already supports provisioning from some of the more common HR platforms, including Workday and SAP SuccessFactors. You can find more information here for planning purposes: `https://docs.microsoft.com/en-us/azure/active-directory/ app-provisioning/plan-cloud-hr-provision.`

Next, we will review the available Microsoft directory service technology to support your identity life cycle.

Integrating directory services

Beyond an HR system with integration into your identity management solution or MIM is **Active Directory (AD)**. AD is an on-premises hierarchical directory that stores objects, such as user accounts, passwords, user information, computer objects, and security groups. This is where the user objects will be active and enabled for accessing the IT systems within your environment.

Referring back to the overall identity life cycle, once a record has been established in MIM based on the HR feed, it will then provision an account in AD for the user. The MIM object isn't the account that the user will use but is typically set up as the authoritative source to the AD object. What this means is if the MIM object is set to be termed because the HR system sent a term instruction, the AD object will be disabled. Depending on your configuration, re-enabling the account directly within AD will eventually revert it to a disabled state, since MIM is authoritative. This is extremely powerful and exactly how the process should work. It also helps to prevent anyone with access to AD from creating and enabling accounts that they shouldn't. All requests should filter through MIM, or other identity management tools, for better control and accountability, as they serve as the centralized place for all identity and access requests.

> **Tip**
> Protecting your active identities is critical, and AD needs to be at the core of this protection. The following is Microsoft's best practices for securing Azure AD: `https://docs.microsoft.com/en-us/windows-server/identity/ad-ds/plan/security-best-practices/best-practices-for-securing-active-directory`.

For on-premises identities to work in a cloud directory, a copy will need to be synchronized to support the identity and authentication requirements of the modern world. Typically, a synchronization tool is used between your on-premises AD and the cloud provider. This will enable what's known as the hybrid identity for your organization. Microsoft's Azure AD Connect provides that synchronization for all or any selected objects, **Organizational Units** (**OUs**), and user attributes from your AD into Azure AD. With Azure AD Connect, you can provide a single identity for your user to access both on-premises resources as well as cloud-based resources.

Azure AD is Microsoft's enterprise cloud identity provider, providing the next generation of identity management and security for your users. Some examples of cloud resources that a user will log in to using their Azure AD identity are Exchange Online, OneDrive for Business, SharePoint Online, Windows (Azure AD Join), and so on. The user will not know the difference between the traditional AD account and the Azure AD account that is optimal for the user experience. Let's review a little more about hybrid identities with Azure AD Connect.

Hybrid identities with Azure AD Connect

Azure AD connect supports the synchronization of users, user attributes, security groups, and other on-premises objects. One important decision to make about Azure AD Connect is the method of authentication for your users to sign in. For hybrid identities, the two supported models are cloud authentication and federated authentication. While there are many considerations when choosing the authentication model, in cloud authentication, Azure AD handles the sign-in, whereas with the federated model, the sign-in is passed back to on-premises **AD Federation Services** (**FS**). To learn more about the considerations for each authentication model, visit this link: `https://docs.microsoft.com/en-us/azure/active-directory/hybrid/choose-ad-authn`.

For the remainder of this section, we will be referencing features of the cloud authentication model. Let's now review how passwords can be managed between on-premises and Azure AD.

Since Azure AD only syncs a copy of your identity, you need to determine how to manage your passwords within the cloud. Azure AD Connect provides the option to sync the password hash (known as password hash synchronization) to Azure AD. It also supports pass-through authentication back to on-premises so that users can use the same password both in the cloud and on-premises. If you have an on-premises deployment of AD FS, you can also use the federated integrated feature to take advantage of additional AD FS capabilities. Let's review the difference between pass-through authentication and password hash sync. If you plan to invest in Microsoft-based cloud identity security solutions, it is strongly recommended to enable password hash sync.

> **Tip**
> Microsoft does not sync the actual passwords of your user accounts. For optimal security of your user passwords, a hash of the password with a per-user salt is synced to Azure AD.

Azure AD pass-through authentication validates passwords between on-premises and the cloud against AD. If you wish to enforce on-premises security and password policies and do not want passwords to be stored with the Azure AD authentication service, choose pass-through authentication.

Azure AD password hash synchronization synchronizes the stored on-premises password hash to the Azure AD authentication service with Azure AD Connect. Authentication takes place against Azure AD, not against on-premises AD. Whenever a password is changed on-premises, the password is synced (within 2 minutes) to Azure AD. If password writeback is configured, then a user can change the password from the cloud and sync it back on-premises. Some on-premises password policies will still apply to the cloud identity if configured for password hash sync. For example, an on-premises password complexity policy will override any policies configured in the cloud. Additionally, any on-premises expiration policy will not affect the cloud identity, as the default configuration is set to never expire. If a user's password expires on-premises, cloud services will not be interrupted.

Azure AD Connect also supports two additional features to help with password management.

The `EnforceCloudPasswordPolicyForPasswordSyncedUsers` feature of the MSOnline PowerShell module allows you to enforce users to comply with a password expiration policy in Azure AD. To learn more about setting this cmdlet, visit this link:

`https://docs.microsoft.com/en-us/azure/active-directory/`
`hybrid/how-to-connect-password-hash-synchronization`.

If using on-premises tools such as **Active Directory Users and Computers (ADUC)** to reset passwords, you can set the `ForcePasswordChangeOnLogOn` feature of the Azure AD Connect PowerShell module to support the temporary password flag in Azure AD. This will allow the user to reset their password the next time they log in to Azure AD resources if it needs to be reset by the helpdesk.

Figure 5.2 – The Reset Password option in ADUC

Azure AD Connect monitoring is available to check the health status through the Azure portal. To view Azure AD Connect Health, log in to Azure AD at `https://portal.` `azure.com` and search for `Azure AD Connect Health`. You can click on your service name to view additional details about your servers and alerts.

Figure 5.3 – The Azure AD Connect Health dashboard in the Azure portal

Now that we have covered Azure AD Connect and options for managing passwords, let's review the life cycle process.

Completing the account life cycle process

Simply put, once the identity is synchronized to the cloud, this completes the life cycle of a user's identity from the system of record. While there may be variances in the products between vendors, the life cycle framework and policy provided can be applied using any identity service or vendor. To recap, the following workflow should be referenced to process your identities:

HR > identity management system > traditional directory service > identity synchronization > modern directory services

It's also critical to ensure the off-boarding process works correctly throughout this life cycle. If properly configured, when HR initiates a termination, the account should be disabled throughout all directories. Any failures in this process can allow a user to access resources after they have left the organization, which can cause a serious security risk.

> **Tip**
> Make sure you are fully aware of synchronization times within each of your identity systems. Terminating a user within HR can take time to synchronize downstream through your systems. For immediate action, you may need to manually intervene to ensure identities are correctly disabled in a timely fashion.

In addition to your identity life cycle is your group management strategy. Ensure you constantly audit group memberships and understand what services, applications, and data your users have access to. Ensure users don't have access to groups that they shouldn't, and as users change roles, it is critical they are removed from groups that they no longer need access to. Group management must be a part of your identity life cycle, and the correct processes and procedures are critical to ensure they are managed correctly.

The following diagram illustrates a high-level architecture of what your identity life cycle may look like:

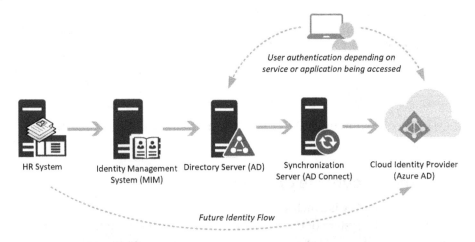

Figure 5.4 – The identity life cycle architecture example

The account life cycle can be very complex, and mismanagement of this process can easily create vulnerabilities. Some best practices with this architecture include working closely with your HR team, ensuring all traffic, feeds, and integrations are encrypted, minimizing the number of privileged identities, and enforcing MFA on all accounts. Be sure to enable auditing to hold accountability and add safeguards to ensure the identity portal is secure and not accessible over the internet if possible.

Next, let's look at managing external or guest identities in Azure AD, which are commonly referred to as B2B.

Managing Azure external user access (B2B)

Azure AD B2B is a game-changer within the enterprise identity space. When looking back at the traditional model with AD and on-premises applications, providing access to external partners was far from an easy task. For most organizations, setting up an identity within their local AD was probably a standard practice. This meant the need for local AD accounts to be managed and governed within your environment, as with contractors and employees. Many organizations may have set up AD FS or third-party federation applications to support the ability to allow an external party to use their own ID. The issue with this is the involvement, complexity, and effort to set up the federation between both parties. The more external vendors you work with, the more integrations you will need to set up and systems to manage.

Azure AD B2B greatly simplifies this process by allowing you to invite a guest user into your Azure environment and provide them with access to your applications and services. Once invited into your organization, the guest user then authenticates using their current organization's work identity to access approved applications hosted in your subscription. There is no need to provision new domain accounts or add any additional infrastructure to support this collaboration model. The best part of this process is that the user's identity is managed and maintained by their hosting organization, which is a significant improvement from an account management perspective. Although housekeeping will be required in your Azure AD to clean up stale accounts, once the external user is disabled within their hosting environment, they can no longer access your environment using that identity, which is a major advantage. Azure B2B also supports collaboration with organizations through external identity partners, such as Facebook, Google, Microsoft accounts, and other **Security Assertion Markup Language (SAML)/Web Services Federation (WS-Fed)** identity providers.

You can learn more about Azure AD B2B at `https://docs.microsoft.com/ en-us/azure/active-directory/b2b/what-is-b2b`.

> **Tip**
>
> Entitlement management is a feature of Azure AD that can help manage the identity life cycle of B2B accounts. To learn more about this feature, visit `https://docs.microsoft.com/en-us/azure/active- directory/governance/entitlement-management- overview`.

To view and access the external collaboration settings for your tenant, log in to `https://portal.azure.com`, click on the portal menu at the top left, and choose **Azure Active Directory**. Select **External Identities** and click on **External collaboration settings**. The following screenshot is of the **Guest user access** settings in the Azure portal:

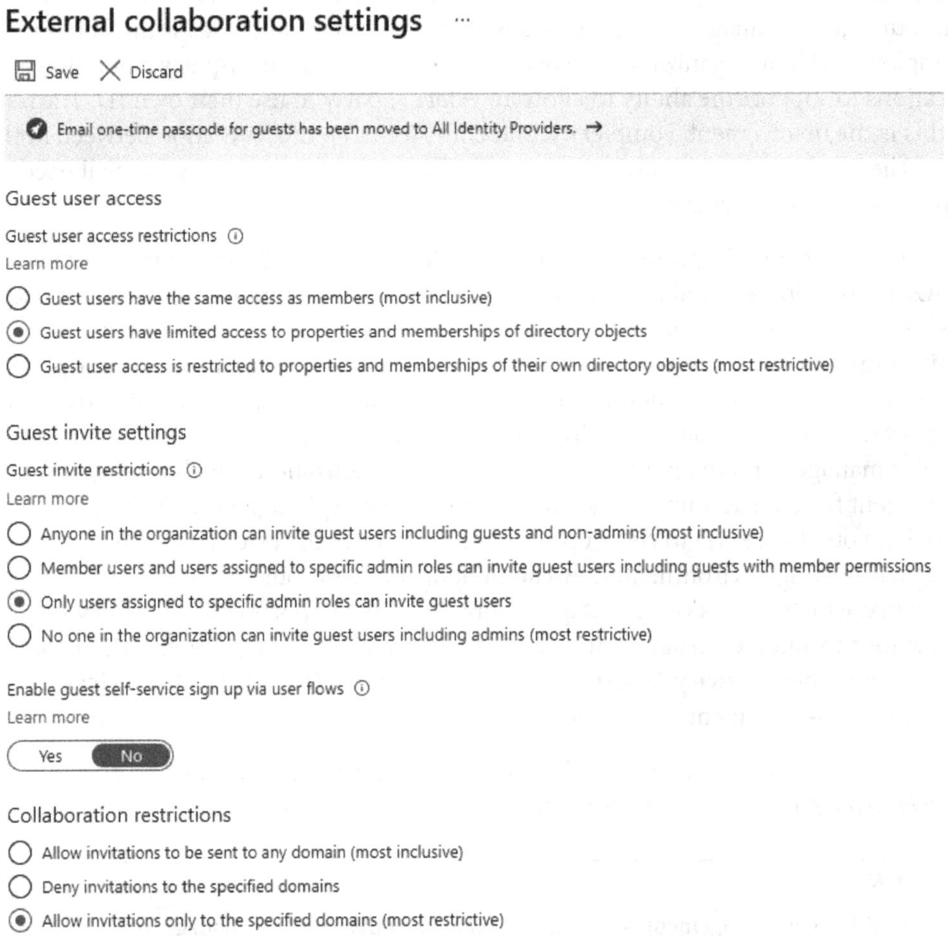

External collaboration settings ···

🖫 Save   ✕ Discard

> ⚫ Email one-time passcode for guests has been moved to All Identity Providers. →

Guest user access

Guest user access restrictions ⓘ
Learn more

○ Guest users have the same access as members (most inclusive)

◉ Guest users have limited access to properties and memberships of directory objects

○ Guest user access is restricted to properties and memberships of their own directory objects (most restrictive)

Guest invite settings

Guest invite restrictions ⓘ
Learn more

○ Anyone in the organization can invite guest users including guests and non-admins (most inclusive)

○ Member users and users assigned to specific admin roles can invite guest users including guests with member permissions

◉ Only users assigned to specific admin roles can invite guest users

○ No one in the organization can invite guest users including admins (most restrictive)

Enable guest self-service sign up via user flows ⓘ
Learn more

(Yes   **No**)

Collaboration restrictions

○ Allow invitations to be sent to any domain (most inclusive)

○ Deny invitations to the specified domains

◉ Allow invitations only to the specified domains (most restrictive)

Figure 5.5 – The Azure AD Guest user access settings

For a more secure configuration, it is recommended to select the following external collaboration settings:

- **Guest user access** is set to **Guest users have limited access to properties and memberships of directory objects**.

- **Guest invite settings** is set to **Only users assigned to specific admin roles can invite guest users**.

- **Enable guest self-service sign up via user flows** is set to **No**

- **Collaboration restrictions** is set to **Allow invitations only to the specified domains (most restrictive)**.

To invite a new guest user into your organization, go back to the Azure AD management screen and click on **Users**, and then click on **New Guest User**. You will be asked to provide information, but only an email address is required to send an invite. By default, the user will have very limited access. Once this information is entered, click on **Invite**.

The external user will receive a welcome email to accept the invitation and gain access to your tenant. The welcome email can be customized to an extent like the example in the following screenshot:

Sender: Matthew Tumbarello (mtumbarello@windowssecurityandhardening.com)
Organization: Windows Security and Hardening
Domain: windowssecurityandhardening.com

This message was provided by the sender and is not from Microsoft Corporation.

MT

**Message from
Matthew Tumbarello:**

❝❝ Hi! You now have access ❞❞

If you accept this invitation, you'll be sent to https://account.activedirectory.windowsazure.com/?
tenantid=tenantid&login

Accept invitation

Figure 5.6 – An invitation example for B2B access

If you have selected the **Allow invitations only to the specified domains (most restrictive)** collaboration restriction, make sure to whitelist the domains under **Target domains** or you will receive an error sending any guest invites.

Microsoft has both B2B and B2C as part of its guest access services. You can visit `https://docs.microsoft.com/en-us/azure/active-directory/b2b/compare-with-b2c` to view the differences, as B2C requires a separate tenant from your organization's Azure AD.

The Azure AD B2B model is extremely powerful and allows you to move away from old methods of needing to provision accounts within your environment. The model is very simple and, from a security and auditing perspective, is a great improvement on previous processes.

Next, let's discuss the different types of administrative roles available in Azure AD and using RBAC. It's important to cover these roles, as they can directly relate to your Windows systems due to their control over the management plane in your Azure tenant.

Understanding the Azure cloud administrative roles

In your Azure subscription, there are many roles for controlling access to resources and services. There is Azure RBAC, which is the authorization system that is built around **Azure Resource Manager** (**ARM**) and is used to control access to different ARM-based resources, such as resource groups or virtual machines. There are Azure AD roles that are used for managing all the different services that interact with Azure AD and Microsoft 365. An example of this is the Guest Inviter role, which can grant permission to helpdesk staff to invite B2B guests. Additionally, if you're using Microsoft Endpoint Manager for managing endpoints, Intune also has its own set of roles for better fine-grained access control over enterprise-managed devices. Let's look at the Azure AD and Microsoft 365 admin roles.

Azure AD and Microsoft 365 admin roles

Administrative role permissions are worth mentioning, as they can have a direct impact on your Windows clients and servers that rely on cloud-based services to run effectively. It is important that these permissions are only assigned as needed and managed through privileged identity management for just-in-time access. While there are many cloud-based administrative roles that serve different purposes, there will be overlap in permissions, so it's important to review them in detail before assigning them to your staff. A few different types of admin roles available across Azure AD and Microsoft 365 services include the following:

- Azure AD and Azure RBAC roles
- Endpoint Manager roles
- Security and Compliance Center admins
- Exchange Online admin toles

- Teams admin roles

- SharePoint admin

- Power BI and Power Platform admins

Many of the services in Microsoft 365 will also have additional fine-grained permissions available that can be assigned from their respective admin portals. For example, Microsoft Endpoint Manager and Exchange Online have their own subsets or roles that allow you to customize permissions for admins to perform different functions. Both examples also have the capability to create custom roles if the built-in role templates don't meet your needs. The following screenshot is from the Microsoft Endpoint Manager admin center, showing the different available Intune built-in roles for assignment:

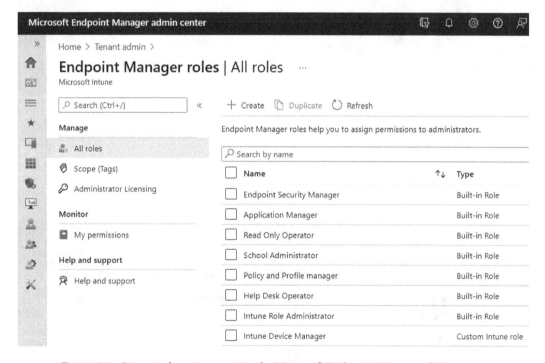

Figure 5.7 – Intune role assignments in the Microsoft Endpoint Manager admin center

The Microsoft 365 admin center also has a useful feature that lets you compare Azure AD roles. To use compare roles, log in to the Microsoft 365 Admin center at `https://admin.microsoft.com`. Click **Show All** to view all the navigation options and select **Roles | Role assignments**. Here, you can select up to three Azure AD roles and choose **Compare roles** to view their permissions side by side. The following screenshot shows output comparing two admin roles:

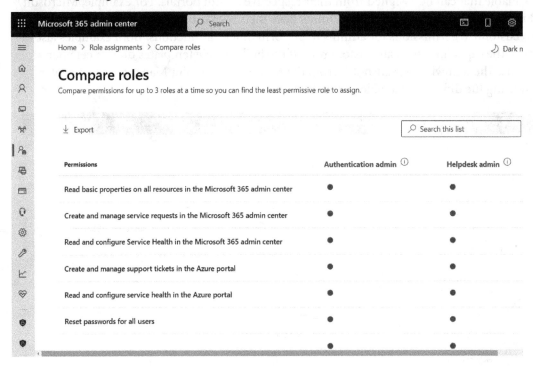

Figure 5.8 – The Microsoft 365 admin center Compare roles page

Now that we have reviewed the different types of privileged roles in Azure AD and Microsoft 365, let's discuss Azure RBAC.

Azure RBAC

RBAC in Azure is used for the authorization of access to Azure resources by using role assignments. An example of an Azure resource could be anything from a virtual machine and a virtual network interface to Azure Firewall and a load balancer, to name a few. RBAC is built around a role definition that defines the permissions based on `Actions`, `DataActions`, `NotActions`, and `NotDataActions`. They can be assigned a specific scope as well that dictates what set of resources the role applies to. RBAC roles can be assigned directly to users, groups, service principals, or managed identities. Azure already has many built-in roles for use, but custom roles can be built using the pre-defined roles as a template and customizing them to fit your needs. To read the official documentation on RBAC for Azure resources, visit `https://docs.microsoft.com/en-us/azure/role-based-access-control/`. Assigning roles to a set of resources in the Azure portal is accomplished through the **access control (IAM)** feature. You can use IAM to also view available roles and their permissions, current role assignments, and how they are inherited. To access the access control (IAM) feature, log in to `https://portal.azure.com`, search for and select **Subscriptions**, select your subscription, and then click on **Access control (IAM)**.

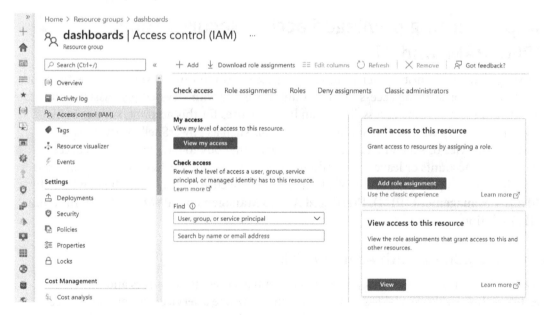

Figure 5.9 – Access control in Azure AD

Here are a few security recommendations to think about when assigning security roles in Azure AD or RBAC:

- Enable MFA for all admins and privileged users.
- Notify admins when other admins reset their passwords.
- Restrict access to the Azure AD portal for admin accounts to compliant devices or privileged access workstations.
- Limit the number of Global Administrators.
- Assign the least permissive roles for admins to perform their job.
- Use privileged identity management with approvals to grant access to roles.
- If no privileged identity management is available, use a separate account for any privileged activity.

Next, let's look at implementing PAM, PIM, and JIT, which are common acronyms for privileged access management security tools.

Implementing privileged access security tools (PIM, PAM, and JIT)

Privileged access security, in terms of securing access to Windows systems, can be thought of as the duty of securing access to an account or service with elevated permissions on a system for just-in-time access. There can be many uses for this account, such as being used interactively with remote desktop, using tools such as PowerShell, or running a Windows service or scheduled task. We don't want anybody to have unmanaged access to use these accounts or leave freestanding access to your admins, as this goes against the principle of least privilege. Let's look at three ways to secure access, using the **Privileged Identity Management (PIM)**, **Privileged Access Management (PAM)**, and **Just-In-Time (JIT)** solutions.

Enabling admins with Azure AD PIM

Azure AD PIM is a service that allows you to control access to resources and services in Azure at the identity level. Let's consider a scenario where a service desk employee needs certain administrative access to perform duties to help resolve support tickets. Using the standard model of assigning free-standing access, it's likely at some point the service desk agent has been over-provisioned and has more access than typically needed to perform normal daily tasks. Their roles may include local administrator rights to Windows devices, access to reset users' passwords and MFA methods, or the ability to perform a remote wipe on Intune-managed devices.

This service desk employee's account would be considered a privileged identity. By allowing free-standing access, not only is the principle of least privilege not being implemented, but waiting for the scheduled auditing of account permissions to flag this account as *high risk* could also be too late and leave your organization at risk if the account is compromised. Implementing Azure PIM helps balance these security risks while still allowing your employees to perform their duties safely and effectively. In fact, a key benefit of PIM is detecting which users are assigned privileged roles as part of the built-in auditing capability. PIM is built around a concept referred to as eligibility. An eligible assignment allows the user to activate their role JIT when they need to perform an administrative function. Azure PIM can manage both Azure AD and ARM-based roles.

For more information on the Azure PIM service, go to `https://docs.microsoft. com/en-us/azure/active-directory/privileged-identity- management/pim-configure`.

In the following example, we are going to set up an eligible admin in Azure AD PIM to allow a user to request the Intune Administrator Azure AD role. In the role settings, we will require a justification and a ticket number reference to submit the request. The associate will also need to specify how long they need the role. Once it's submitted, it will need to be approved to complete the activation.

> **Tip**
> The Privileged Role Administrator in Azure AD has permissions to manage the Azure AD PIM service.

To use Azure AD PIM, an Azure AD Premium P2 or an **Enterprise Mobility + Security** (**EMS**) E5 license is required: `https://docs.microsoft.com/en-us/azure/ active-directory/privileged-identity-management/pim-getting- started`.

Follow these steps to enable a workflow approval and configure an eligible admin for the Intune Administrator role:

1. Log in to the Azure portal at `https://portal.azure.com`.
2. Search for `Azure AD Privileged Identity Management` and select it.
3. Select **Azure AD Roles** from the **Manage** section.
4. Select **Settings** under **Manage**.
5. Find and select **Intune Administrator**. Click **Edit**.
6. Change the **Activation maximum duration (hours)** setting to **2**.
7. For **On activation, require**, select **Azure MFA**.

8. Check the **Require Justification on activation, Require ticket information on activation**, and **Require approval to activate** fields.

9. Scroll down and click on **Select approvers** and choose your account or an approver. Choosing at least two approvers is recommended.

10. Click **Save**.

Now, let's make a user an eligible admin for this role:

1. Under **Manage**, click on **Roles**, and select **Intune Administrator**.

2. Keep **Eligible assignments** selected and click on **Add assignments**.

3. Choose **No member selected** under **Select members** and find an eligible user, and then click on **Select**.

4. Click **Next**.

5. Leave **assignment type** as **Eligible** and keep **Permanently eligible** selected. Click on **Assign** to complete the assignment.

Once the assignment is complete, the next time the user logs in to the Azure portal, they can search for `Azure AD PIM` and request the role, requiring them to fill out the details configured in the previous steps, as shown in the following screenshot:

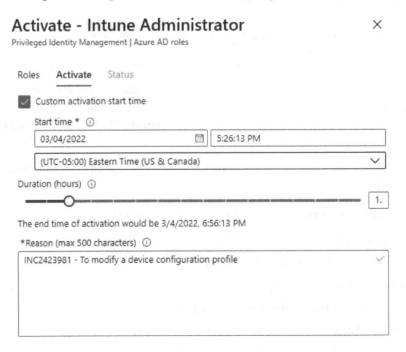

Figure 5.10 – Activating the Intune Administrator role in Azure PIM

This is a very basic example of how Azure PIM can be used for JIT access. PIM also supports privileged access groups where roles can be assigned directly to Azure AD security groups. There is also a discover and monitor feature that can find and audit existing privileged roles in your subscription.

> **Tip**
> Azure PIM also supports external users to collaborate on Azure resources using B2B.

Next, let's review what it means to grant JIT access to resources.

Granting JIT access

The JIT concept is useful for providing access to an account or assigning a role to an identity that adds the permissions needed to perform administrative actions for a set duration. JIT access provides an additional security layer that helps to mitigate risks from free-standing or always-available accounts. Typically, access is granted through an approval workflow process with a set of policies that defines who can request a defined role or account. Naturally, JIT helps enforce the principle of least privilege by eliminating permanent assignment. Many PAM solutions have a JIT solution built into their product today as standard functionality. In Azure, two solutions that provide JIT access are Azure PIM and Microsoft Defender for Cloud's (JIT) **Virtual Machine** (**VM**) access. JIT VM access works by gating the inbound traffic ports used for the remote management of your VMs by creating rules enforced through network security groups. By incorporating an approval workflow through Defender for Cloud, the locked-down ports are opened after approval for the specified duration.

> **Tip**
> *Chapter 11, Server Infrastructure Management,* includes a walk-through of how to configure JIT VM access, but further information can be found at `https://docs.microsoft.com/en-us/azure/defender-for-cloud/just-in-time-access-overview`.

Next, let's look at managing privileged accounts using PAM.

Protecting privileged accounts with PAM

A PAM solution can be critical in helping organizations secure access to systems, meet compliance, and monitor the privileged accounts used in their environment. With an ever-growing threat landscape and a seemingly unlimited amount of attack vectors, implementing a PAM solution can provide a huge benefit in strengthening your company's security posture. Free-standing accounts will always exist in the environment, whether they are actively in use or not, and make great attack targets for malicious actors. By implementing a PAM solution, coupled with a JIT request approval process and password rotation, the risk of compromise is greatly reduced. PAM solutions typically include auditing capabilities that help organizations maintain compliance and avoid penalties from access violations.

When choosing a PAM solution, look for the following core set of features:

- Account discovery for all systems, devices, and applications, which includes, but is not limited to, Windows, Linux, Unix, Cisco devices, and other networking equipment
- The ability to manage local accounts and service accounts and orchestrate restarting services with minimal impact
- Credential management capabilities, including password rotation, a workflow approval process, and auditing trails
- Expiration policies with password rotation for stale, free-standing (always-existing) privileged accounts
- The ability to provide access to systems using middleware tools, such as web password fillers and **Remote Desktop Protocol (RDP)/Secure Shell (SSH)** launchers, that never expose credentials to users
- Discovery alerts to notify the security team whether new backdoor accounts were created without approval
- Automation workflows for privileged execution using scripted tasks
- Monitoring and auditing capabilities through remote session recording, command logging, and analytics
- The capability to link to an **IT Service Management (ITSM)** tool or a change management process
- SIEM integration for leveraging **Security Operations Center (SOC)** services
- Third-party access leveraging **Single Sign-On (SSO)** authentication and MFA

SaaS solutions are becoming a popular option in the PAM space. Depending on your use case for PAM, a SaaS solution can offer a lighter footprint and eliminate the need to deploy an extensive infrastructure.

> **Tip**
>
> As part of a **Critical Incident Response Plan** (**CIRP**), it's a good idea to have dedicated accounts in PAM, only to be used by first responders. This will help ensure that audit logs are kept clean when helping identify the source of the intrusion.

In this section, we covered various topics that covered the identity and access management space. We reviewed the account life cycle process and talked about the importance of controls to ensure timely housekeeping. Throughout the rest of this section, we reviewed different admin roles in Azure AD and Microsoft 365, as well as guest B2B user accounts. Finally, we discussed managing privileged identities with PIM, PAM, and JIT.

Next, let's review the use of local administrator accounts on your Windows systems.

Securing local administrative accounts

Local administrator account access and built-in or default administrator accounts can be a challenge from both a risk and management perspective when securing your Windows devices. Two main problems exist with the local or built-in accounts in Windows (such as administrator or guest) and business users requesting administrative access to their accounts for installing applications or performing certain functions. For the latter, there may be legitimate scenarios where a business user needs these rights, but generally, this is not the case, and they should be leveraging support teams or a temporal admin elevation process that is closely audited. If they are running a process locally that requires elevated rights, that process should be revisited and transitioned to where it can be monitored under tighter controls. Allowing business users this access is an obvious vector for privilege escalation through account compromise on some of the most exposed attack surfaces, the endpoints.

Privilege escalation is a common attack tactic, considering that most applications, including malicious payloads, typically require some form of elevated rights in Windows to be installed. Many times, privilege escalation is accomplished simply by compromising a local or user account due to a vulnerability or misconfiguration. If the account has permanent admin rights, this becomes a problem. Additionally, any other stagnant local accounts left enabled on a system can be used to gain access, maintain persistence, and for dumping credentials of other accounts to move laterally. To help combat this, let's look at some security recommendations for dealing with local accounts in Windows:

- Disable both the built-in administrator account and guest account. If these accounts aren't needed, it's best to disable them. Managing passwords for local admin accounts can be a challenge, and the built-in administrator account cannot be locked out, making it a good target for an attacker. Additionally, the guest account allows users to log in without a password, opening a situation for reconnaissance activities, depending on your policies.

- While disabling these accounts is preferred, if they must exist, it is recommended to recreate them with different names that are not easily identifiable. If you just rename the built-in accounts, the **security identifiers (SID)** still exist and can be leveraged instead of the account name.

- Enable the **Do not enumerate connected users on domain-joined computers** policy and disable **Enumerate local users on domain-joined computers**. Configuring these policies can help prevent reconnaissance and discovery tactics, such as gathering information about account names that can later be targeted in other attacks, such as brute force or password spraying.

- Enable success and failure auditing for sensitive privilege use. This audit policy will log when a user account uses sensitive privileges on a system. Having these logs enabled is helpful for security forensics teams.

- For enterprise scenarios, disable the **Accounts: Block Microsoft Accounts** policy. In most cases, a user should not be able to log in to their work device with a Microsoft account. They should only be using corporate-issued identities.

Managing user requests for local admin accounts can be challenging, and some organizations have adopted processes that allow users to temporarily request admin rights for a duration of time. This can be accomplished in many ways, such as through temporary group membership in an AD, using a unified endpoint management solution, or implementing a request workflow in Azure AD PIM for Azure AD joined devices. Here are a few recommendations to better manage local admin accounts:

- For domain-joined systems, implement a system known as a **Local Admin Password Solution (LAPS)** to rotate local administrative account passwords if they must exist.

- Use a PAM solution from reviewed vendors, such as Delinea, CyberArk, or BeyondTrust, to manage the local administrative accounts on your workstations. These vendors offer robust solutions, can manage multiple account types, and have reporting and auditing functionalities built into them.

- Use Windows Autopilot or bulk enrollment methods to join devices to Azure AD. This will allow a standard user to enroll a device on Azure AD and Intune without becoming an administrator.

- For Azure AD joined devices, leverage the **Azure AD Joined Device Local Administrator** role.

> **Tip**
> Using the Azure AD role is a convenient way to add accounts as local administrators, but be mindful that this will make them an administrator on all Windows devices joined to Azure AD.

Next, let's look at the methods used for authentication and how to use them to provide users with a seamless and secure sign-in experience.

Understanding authentication, MFA, and going passwordless

In this section, we will review authentication as commonly used today. We will discuss MFA and passwordless methods used to protect users' identities. As already stated, a compromise of credentials is one of the most common methods of a breach today, and commonly used authentication models are outdated and need updating. The traditional method of entering a username and password is simply not acceptable. If you don't have a strategy in place to improve your authentication posture, add it to your top three security priorities. Investing in a zero-trust strategy means that you always assume the possibility of a breach, and account credentials are no exception.

Looking at an on-premises AD deployment, authentication methods consist of Kerberos, Integrated Windows authentication, Digest Authentication, NTLM authentication, or **Transport Layer Security (TLS)/Secure Sockets Layer (SSL)**, depending on what you are accessing from your device and how you are accessing it. Modern authentication within Azure AD typically uses an access security token with claims, OAuth 2.0, and SAML. Using a hybrid model will likely include a blend of all these protocols as part of your authentication process, depending on where the service is being provided.

> **Tip**
>
> For your on-premises deployment of AD, it is highly recommended that you enforce secure **Lightweight Directory Access Protocol (LDAP)**. By default, LDAP traffic is not encrypted and can easily be viewed by an attacker within your network.

Let's now look at password management and recommendations for how to manage and secure passwords.

Securing your passwords

Since passwords are still relevant and will be for a while, it is important to review your current policies, and it could be time to make some changes based on updated guidance. Common policies include changing them every 90 days, using 8 characters as a minimum, and ensuring complexity. Now, security research has suggested new recommendations due to the advances in cracking tools, password leaks on the dark web, predictable passwords due to frequent change, and users writing them down due to complexity. The following guidelines are now widely recommended to counter these challenges:

- Consider using passphrases over passwords.
- Use a minimum of 12 characters – the more, the better.
- Remove the periodic requirements to change passwords and only change them in the event of account compromise.
- Ban common passwords through a blocklist.
- Reduce complexity rules to allow passphrases to be easier to remember.
- Go passwordless.

In your Windows AD environment, a domain-level password policy can be configured easily if you only require one policy for all users. If there is a requirement to deploy additional password policies, then a fine-grained password policy can be configured and scoped to users and groups, allowing for multiple password policies. A use case for this would be to enforce stronger password requirements for privileged users and system accounts.

> **Tip**
> You can only apply a fine-grained password policy against global security groups or user objects. Furthermore, to use fine-grained password policies, the domain functional level must use Windows Server 2008 or newer.

Follow these steps to implement a password policy using Group Policy:

1. To change the password policy using Group Policy, open the **Group Policy Management** snap-in from your management workstation.

2. Right-click on **Default Domain Policy**, select **Edit,** and then browse to **Computer Configuration | Policies | Windows Settings | Security Settings | Account Policies | Password Policy**. In the following screenshot, you can see where to configure your domain password policy:

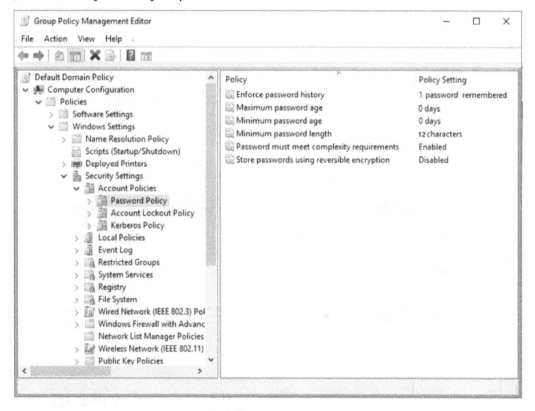

Figure 5.11 – The default AD domain password policy

Follow these steps to implement more than one password policy, using a fine-grained password policy:

1. To use fine-grained password policies, log in to your domain controller (Windows 2012 or newer) with the appropriate permissions.

2. Search for AD Administrative Center and click on **domain (local)**, double-click on **System**, and then double-click on **Password Settings Container**.

3. Right-click on the main screen and select **New**, and then click on **Password Settings**.

4. Fill out the requirements for your password policy and select the groups of users that the policy will apply to, and then click **OK**. The following screenshot shows the new policy screen:

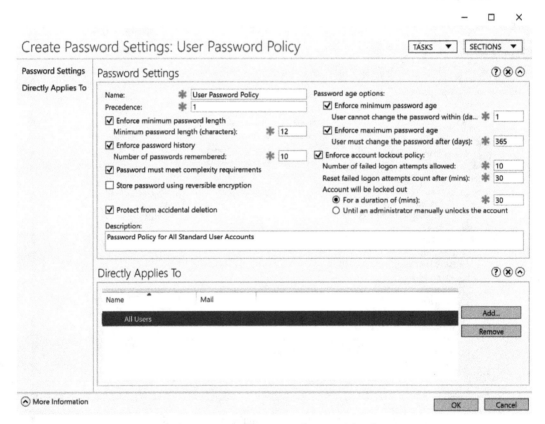

Figure 5.12 – The fine-grained password policy

If you are using cloud-managed identities in Azure AD and Microsoft 365, your options to modify the password policy are different. By default, any account created in Azure AD can use a password that is between 8 and 256 characters, requires complexity, expires every 90 days, receives a notification after 14 days of setting it, and doesn't allow the last used password to be used again when changed. Most Azure AD password policies are configured with PowerShell using the `Set-MsolPasswordPolicy` cmdlet. There are also options to configure a custom banned password list and enforce Azure AD Password Protection settings back to on-premises AD through an agent deployment. Let's look at where to set the password expiration policy in the Microsoft 365 admin center. To modify these settings, follow these steps:

1. Log in to `https://admin.microsoft.com` with the correct permissions.

2. Click on the menu at the top left. Click on **Show All** and select **Settings**. Click on **Org settings** in the sub-menu. When the page opens, choose the **Security & Privacy** tab, and select **Password expiration policy**.

3. Select **Set user passwords to expire after a number of days** to configure the two options. These policies apply to any accounts created directly in Azure AD.

> **Tip**
>
> As mentioned earlier, the Azure AD Password expiration policy does not apply to hybrid identities synced with Azure AD Connect.

Next, let's look at where to configure the **Azure AD Password Protection** options in Azure AD.

Configuring Azure AD Password Protection

The Azure AD Password Protection service helps to block weak and well-known passwords using a global banned password list managed by Microsoft, which is generated by analyzing telemetry data of commonly used and weak passwords. By default, this is enforced in Azure AD for cloud-only users; otherwise, you will require an Azure AD Premium P1 or P2 license if you synchronize users from on-premises AD or use the custom banned password list. Password Protection has a custom banned password list that can be used to blacklist organizational-specific terms, such as brand or product names, that an attacker may attempt to use for password cracking.

To modify these settings from the Azure portal, follow these steps:

1. Log in to `https://portal.azure.com` with the correct permissions.

2. Click on the menu at the top left. Click on **Azure AD** and then the **Security** blade under **Manage**. Choose **Authentication Methods** under **Manage** and select **Password Protection**.

Here, you can modify the lockout threshold and the lockout duration in seconds, enable a custom banned list, and enable password protection on-premises.

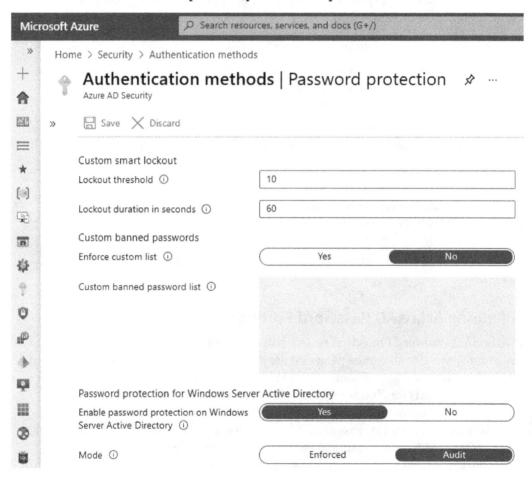

Figure 5.13 – Azure AD Password protection

To learn more about deploying Azure AD Password Protection on-premises, visit this link: `https://docs.microsoft.com/en-us/azure/active-directory/authentication/howto-password-ban-bad-on-premises-deploy`.

Next, let's look at configuring **Self-Service Password Reset (SSPR)** as a self-help tool for your users.

Enabling SSPR

SSPR is a feature of Azure AD that allows users to reset, change, or update their passwords. This is a free feature for cloud-only identities looking for password change functionality, but it requires an Azure AD Premium P1 or equivalent license if syncing from an on-premises directory. SSPR can considerably reduce support calls and save users time if they forget their password or device PINs. Depending on the SSPR configuration, at a minimum, all that is required for user self-service is an internet connection and an alternative method of authentication. Let's look at how to enable SSPR functionality for Windows clients and configure a link on the Windows 11 login screen.

To enable SSPR and set up the authentication and registration methods for hybrid identities with Azure AD Connect, follow these steps:

1. Log in to the Azure portal at `https://portal.azure.com`.

2. Go to **Azure AD**, select **Users**, and choose **Password Reset.** Choose **All** (or **Selected** to test) and then click **Save**.

    > **Tip**
    >
    > You can only choose one security group if you are using the **Selected** option to enable SSPR. Keep this in mind if you're thinking about rolling this out.

3. Choose **Authentication methods** under the **Manage** menu. Select **1** for the number of methods required to reset and select **Mobile app code** and **Mobile phone** as the methods available to users. Click **Save**.

4. Choose **Registration** under the **Manage** menu. Leave **Yes** selected to require users to register when signing in and leave the default value of **180** days to require reconfirmation of their authentication information.

    > **Tip**
    >
    > If SSPR is enabled for **All users**, then *all* user accounts (including service accounts) will be required to reconfirm their verification methods every 180 days, which can cause interruptions for automated processes.

5. Choose **Customization** under the **Manage** menu. Choose **Yes** to customize the helpdesk link and enter an email or URL for your service desk.

6. Choose **On-Premises integration** under the **Manage** menu. To configure password writeback, Azure AD Connect must be set up with the password writeback configuration. Choose **Yes**, and select **No** for the option to allow users to unlock accounts without resetting their password.

The following screenshot shows the **Write back passwords to your on-premises directory** option on the self-service password reset page:

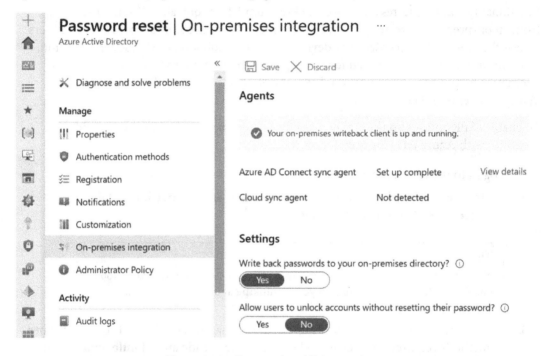

Figure 5.14 – On-premises AD integration

For more information about how to configure password writeback with Azure AD Connect, visit `https://docs.microsoft.com/en-us/azure/active-directory/authentication/howto-sspr-writeback`.

Once the self-service password reset has been configured, users can go directly to `https://aka.ms/sspr` to reset their passwords.

After they enter their user ID and enter the characters for the CAPTCHA, they will be asked to verify their identity with an alternative method, as shown here:

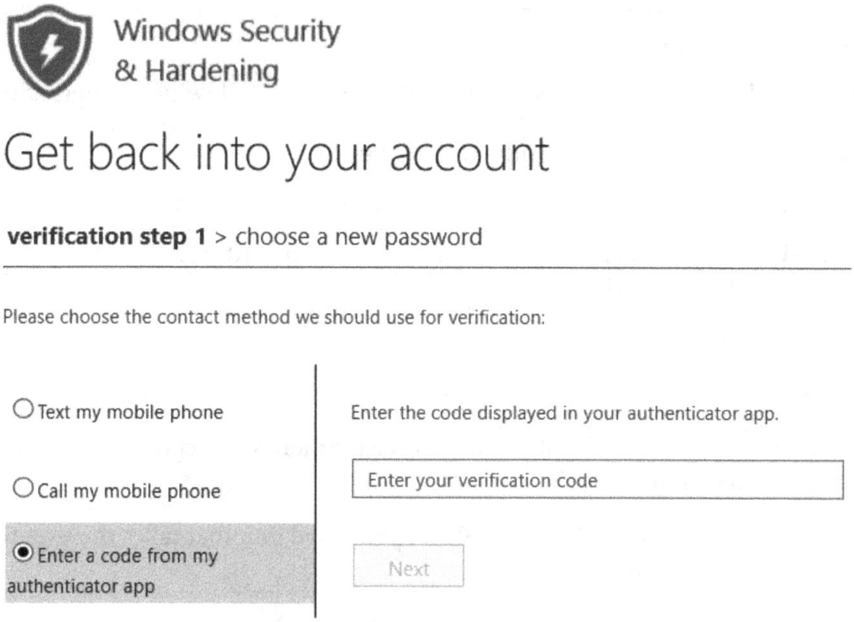

Windows Security & Hardening

Get back into your account

verification step 1 > choose a new password

Please choose the contact method we should use for verification:

○ Text my mobile phone

○ Call my mobile phone

◉ Enter a code from my authenticator app

Enter the code displayed in your authenticator app.

[Enter your verification code]

[Next]

Figure 5.15 – The self-service password reset portal asking for additional security verification

Now that we have enabled SSPR, let's look at how to implement the **Reset password** link on the Windows sign-in screen.

Enabling the self-service password link for Windows sign-in

To complete this section, Windows clients need to be Azure AD joined and Intune-managed. We will be using a custom OMA-URI setting on Intune device restriction profiles to deploy the **Configuration Service Provider (CSP)**.

For more information about the limitations and restrictions of using the **Reset password** link for Windows clients, visit `https://docs.microsoft.com/en-us/azure/active-directory/authentication/howto-sspr-windows`.

To enable the **Reset password** link on the Windows sign-in screen, follow these steps:

1. Log in to the Microsoft Endpoint Manager admin center at `https://endpoint.microsoft.com`.

2. Click on **Devices** and choose **Configuration profiles** under **Policy**. Create a new profile.

3. Select **Windows 10 and later** for **Platform** and choose **Templates** and select **Custom** under **Template name**.

4. Give the profile an identifiable name and click on **Next** to view the configuration settings.

5. Click on **Add** to add a new OMA-URI setting with the following configurations:

   - **Name**: `Self-Service Password Reset`

   - **Description**: `Windows Sign-On Reset Password link`

   - **OMA-URI**: `./Vendor/MSFT/Policy/Config/Authentication/AllowAadPasswordReset`

   - **Data Type**: `Integer`

   - **Value**: `1`

6. Click on **Save** and select **Create**. Click on **Assignments** and choose a security group or specific users or devices.

   The following screenshot shows the **Reset password** link that takes the user to the Azure AD SSPR web page:

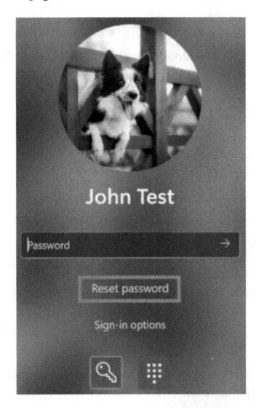

Figure 5.16 – The Self-Service Password Reset link

> **Tip**
> SSPR is also available in the **Settings** catalog under **Authentication in Intune**.

We just covered passwords and the best practices for password management. Next, let's look at one way to eliminate the need for users to enter passwords multiple times, creating a better and more secure user experience, with **Azure AD Seamless SSO**.

Authenticating with Azure AD from Windows

Full Azure AD joined Windows devices can enjoy a seamless SSO experience during sign-on and across other applications running on windows without the need to enter a password. When a user signs in to Windows, the Azure AD **Cloud Authentication Provider (CloudAP)** is the modern authentication provider used to verify the user credentials against those stored in Azure AD. This includes support for Windows Hello for Business, biometrics, a device PIN, and password-type credentials. For devices with a supported hardware, **Trusted Platform Module (TPM)**, hardware-based security is used to bound private cryptography keys to the TPM, and public keys are sent into Azure AD when the device joins Azure AD. This authentication model uses a **Primary Refresh Token (PRT)** formatted as a **JSON Web Token (JWT)** that contains the claims necessary to identify the device to Azure. This identification is also compatible with Azure AD-registered devices and other third-party applications running on Windows by using the Azure AD **Web Account Manager (WAM)** plugin to leverage the PRT token and enable SSO support. An example of this is a web browser such as Google Chrome with the Windows 10 Accounts extension. When the extension is enabled, users on Azure AD joined devices can visit websites and services hosted in your organization's Azure and Microsoft 365 subscription without the need to enter credentials.

Figure 5.17 – The Windows 10 Accounts extension for Google Chrome

The following is another example that shows the flow when using a PRT for SSO for an Azure URL within a browser:

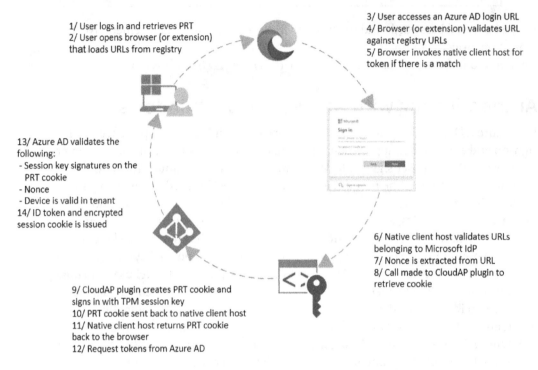

Figure 5.18 – Using PRT for SSO with the Azure URL in a browser

To learn more about Azure AD authentication and the PRT, visit this link: `https://docs.microsoft.com/en-us/azure/active-directory/devices/concept-primary-refresh-token`.

Next, let's look at how Azure AD can be used as an identity provider for other third-party applications.

Enabling SSO for apps with an Azure identity

Having SSO in your environment provides significant improvement from both a security and user experience perspective. Without SSO, users would need to manage usernames and passwords for every unique application they access. Not only is managing passwords a nightmare and susceptible to bad practices, but it can also quickly become overbearing on IT staff from an administration and auditing perspective. Using one identity to access many applications also comes with its own risks, and therefore it is important to ensure strong access and management policies are in place with your user accounts. Access requirements such as MFA, device-based compliance requirements, and location-based access conditions significantly reduce risks if an identity becomes compromised.

SSO with Azure identity isn't only limited to Azure AD joined devices. Microsoft also supports integration with other third-party SaaS providers. You can integrate directly with most SaaS, on-premises, or custom-developed apps that support standard SSO protocols, such as the SAML. Historically, many deployments of SSO used AD FS. This has served a great purpose traditionally for federation and SSO, but the infrastructure needed to support and maintain deployment can become very complex. Shifting this service to Azure requires no infrastructure to maintain, and integrating SSO has become very simple, with minimum configuration and clicks needed. In addition to the standard SSO integration, Azure AD supports the **System for Cross-Domain Identity Management** (**SCIM**) standard for provisioning users and groups directly into applications and enables advanced cloud-based security features, such as Azure AD Conditional Access or a **Cloud Access Security Broker** (**CASB**).

A simple SSO setup for a non-gallery enterprise app using SAML can be configured by following these steps:

1. Log in to `https://portal.azure.com`.

2. Search for **Enterprise Applications** and navigate to the management console.

3. Click on **All applications** on the left, and then click on **New Application**.

4. Select **Create your own application** and choose **Integrate any other application you don't find in the gallery (Non-gallery)**.

5. Enter a name for the app and click on **Create**.

6. After the app is created, click on **Single sign-on** on the left-hand-side menu, and then click on **SAML**.

Here is where you can configure a basic SAML configuration. Most third-party SaaS providers will provide you with the correct SAML information unique to your app instance or have an XML metadata file that can be imported to pre-populate SSO settings. Finally, you can export the SAML signing certificate to be used when signing requests and responses and import it into the third-party app. Most applications support the default attributes and claims that are already pre-configured in the SAML settings.

The following screenshot is of the SAML SSO configuration page:

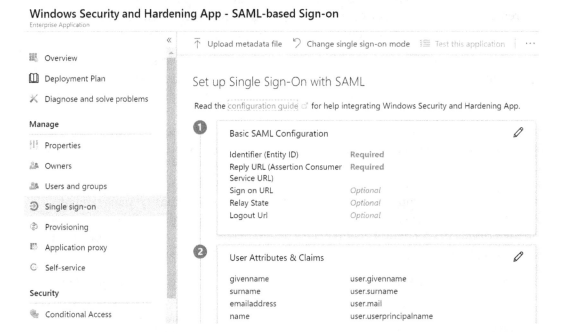

Figure 5.19 – The Azure Enterprise app with the Azure SSO SAML configuration screen

On the left-hand-side menu in the preceding screenshot, you can see additional options to configure app access with **Users and groups**, and you can apply **Conditional Access** policies, view audit logs, and much more to ensure access to your enterprise applications is hardened and secured.

Next, let's look at how to enable MFA.

Configuring MFA

One of the more important technologies in a zero-trust strategy that we recommend you deploy immediately is MFA. Passwords have become obsolete, especially when used as the only factor for authentication. If you aren't using MFA with your Azure subscription, then make doing so a priority now. Implementing MFA in Azure can be completed in a variety of ways, including Conditional Access policies, Azure AD Identity Protection, or enabling security defaults. Enabling MFA doesn't necessarily force the user to use MFA all the time, but instead, you can apply rules or risk-based conditions that will assess the need to prompt for additional verification. This provides a very powerful secure deployment and allows for minimum disruption for your users.

> **Tip**
> Azure MFA is included in the free Azure AD license, but to take advantage of the advanced conditions, you will need to upgrade to a premium license. The following link provides an overview of the different license types: `https://azure.microsoft.com/en-us/pricing/details/active-directory/`.

To access the Azure MFA settings, follow these steps:

1. Log in to `https://portal.azure.com/`.
2. Click on the portal menu at the top left and choose **Azure AD**.
3. Select **Security** and choose **MFA**. You will see the following screen:

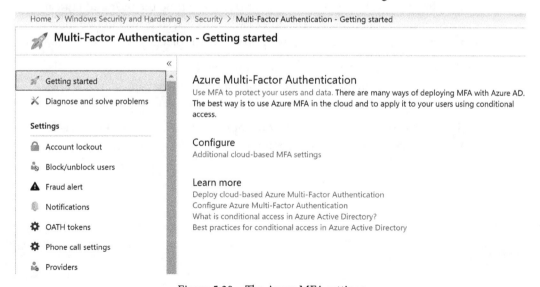

Figure 5.20 – The Azure MFA settings

> **Tip**
> From the **Getting started** screen, click on **Additional cloud-based MFA settings** to add or remove the specific verification options you want your users to use.

You can find additional details on the configurations shown in the preceding screenshot at `https://docs.microsoft.com/en-us/azure/active-directory/ authentication/howto-mfa-mfasettings`.

At the time of writing, there are several authentication methods supported in Azure AD. They include the following:

- Using the Microsoft Authenticator app on your Apple or Android device:

  - A verification code provided on the app (renews every 30 seconds)

  - A push notification where you click on **Approve** when prompted

  - Passwordless sign-in by entering the numbers provided on the screen in the app

- A verification code provided by a hardware token

- A call to your phone where you will need to press # (not recommended)

- A text message to your phone with a verification code (not recommended)

There are three supported methods in which MFA can be required by your users:

- Enabled at the user level, which will require MFA every time the user authenticates. This option takes precedence over the other two methods, so it may not be the most practical one to use. This option is more appropriate for privileged accounts.

- Enabled using Conditional Access policies, allowing the use of advanced criteria, such as the location, device state, device compliance, approved applications, and assigned roles. This requires an Azure AD Premium license.

- Enabled using Azure AD Identity Protection, applied when any detected sign-in attempt or user risk occurs. This also requires an Azure AD Premium license.

You can enable user-level MFA by following these steps:

1. Log in to `https://portal.azure.com`.

2. Click on the portal menu at the top left, and then click on **Azure AD**.

3. Click on **Users**, and then put a checkmark next to the user you want to enable MFA for.

4. Click on **Per-user MFA**, and a new window will open.

5. Put a checkmark next to the user you want to enable, and then click on **Enable**, as in the following screenshot:

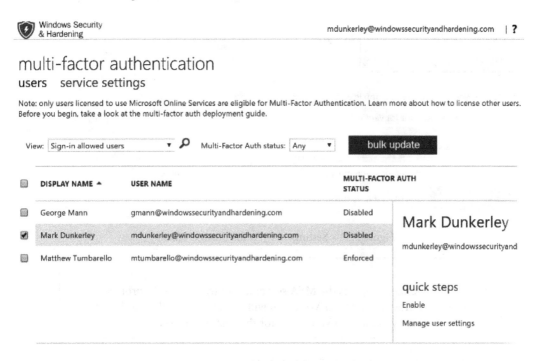

Figure 5.21 – Enabling MFA for a user

6. Click on **Enable multi-factor auth** on the pop-up window, and then click **Close**.

When the user you enabled accesses the portal for the first time, they will be required to set up MFA for their account. If the user is using the free version of an MFA license, they will only see the mobile app option.

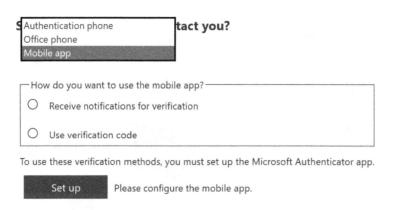

Figure 5.22 – Setting up MFA for the first time

> **Tip**
> Microsoft no longer supports the MFA server as of July 1, 2019. To protect your on-premises infrastructure with MFA, you will need to review the third-party options or ensure you have PAM in place for the best protection.

Next, let's look at what it means to go passwordless.

Transitioning to passwordless authentication

Going passwordless is the future, and Microsoft's making a big push in this direction as an authentication strategy. The technology is already available, and you can go passwordless today. Unfortunately, it may not be easy to get to a passwordless world straightaway, but you do need to understand and begin this journey sooner rather than later. The methods that are used to provide a passwordless world are currently much more secure for your users.

With the elimination of passwords, authentication is improved by using something you already have, such as a phone or a security key, in addition to something you are or know, such as biometrics or a PIN. Microsoft supports passwordless authentication with Windows Hello for Business, the Microsoft Authenticator app, or a **Fast ID Online (FIDO)** 2 security key.

> **Tip**
> FIDO is an alliance that works toward improving today's authentication challenges with passwords. They are looking to provide simpler and more secure authentication methods using open standards. You can view additional information about this at `https://fidoalliance.org/`.

To enable passwordless authentication in Azure, follow these steps:

1. Log in to `https://portal.azure.com`.
2. Click on the portal menu at the top left and choose **Azure AD**.
3. Click on **Security**, and then choose **Authentication methods**.
4. Click on **Policies**. Here, you can configure FIDO2 and the Microsoft Authenticator app for passwordless sign-in.

The following screenshot shows the configuration screen for passwordless authentication:

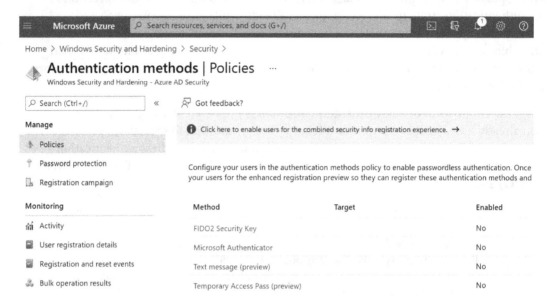

Figure 5.23 – Passwordless authentication in Azure AD

When configuring the Microsoft Authenticator app for passwordless authentication, a user does not need to enter their password to complete the sign-in. After entering the user ID, the sign-in notification asks the user to type the number seen on the screen into the Authenticator app. Once completed, this will log the user in and satisfy the need for MFA. To add additional security when logging in with passwordless authentication, or using the Authenticator app in general, it is recommended you enable the **Show additional context in notifications** option. Enabling this will provide you with a map of the location where the login is originating from, visually allowing you to validate whether the login is legitimate or not. The following screenshot is what the **Show additional context in notifications** option looks like:

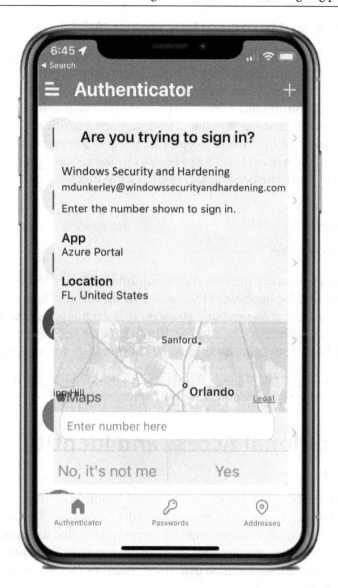

Figure 5.24 – Additional context in notifications

Next, let's look at using a form of biometric authentication to log in to our Windows devices, known as Windows Hello.

Passwordless authentication using Windows Hello

Windows Hello is Microsoft's biometric or PIN authentication feature for Windows devices. This technology can replace the traditional password authentication methods that have always been used on Windows devices. Windows Hello for Business supports cloud-only, hybrid, and on-premises deployments and FIDO v2.0-certified authenticators, such as a security key. When using Windows Hello-supported hardware, biometric data and PINs are only stored on the local device and are not available externally. This is a significant security enhancement that helps ensure an attacker can't get a copy of your biometric data. Today, the two available options for Windows Hello biometrics are as follows:

- Facial recognition
- Fingerprint recognition

Depending on your deployment method, the prerequisites will vary. We will cover enabling Windows Hello for Business in the cloud-only deployment later in *Chapter 8, Keeping Your Windows Client Secure*, but if you want to learn more, visit the following link: `https://docs.microsoft.com/en-us/windows/security/identity-protection/hello-for-business/hello-identity-verification`.

Next, let's review Azure AD Conditional Access and Azure AD Identity Protection and how to leverage the power of the cloud for advanced identity protection.

Using Conditional Access and Identity Protection

Two tools available in Azure AD that provide powerful identity-based security are known as Azure AD Conditional Access and Azure AD Identity Protection. Both solutions leverage identity-based signals that can help you create policies that define access controls as well as automate protection mechanisms when an anomaly occurs, based on machine learning and behavioral analytics used to determine risk. Let's review how to enable Azure AD Conditional Access.

Enabling Azure AD Conditional Access

Conditional Access is an Azure cloud policy tool that enforces compliance based on conditions for your users. The Conditional Access policies allow you to specify criteria against your users that will trigger specific requirements or exceptions, such as the sign-in location, device platform or type, application, and group membership. For example, if a user is not on a compliant managed device or from a trusted IP and is trying to access their email via Exchange Online, they can be required to use MFA. This is a simple example, but the possibilities of this scenario are extensive and allow increased fine-grained security controls. Conditional Access is a necessity within the cloud world, and you need to use it. If not, prioritize and enable it immediately.

You can learn more about Conditional Access at `https://docs.microsoft.com/en-us/azure/active-directory/conditional-access/overview`.

The following is a list of several use cases to get you started today with Conditional Access:

- Require MFA for your admins (privileged roles).

- Require compliant mobile devices before allowing access to company resources.

- Enable MFA for all guest users.

- Block all legacy authentication protocols.

- Allow users on trusted company devices to access resources without MFA.

- Block any access outside a specific region – for example, a United States-based company can block any connection from outside the country.

- Enforce MFA on any app that contains PII, whether on a trusted device or not.

- Only allow specific applications to be accessed from trusted devices.

To set up a Conditional Access policy, follow these steps:

1. Log in to `https://portal.azure.com`.

2. Click on the portal menu at the top left and choose **Azure AD**. Click on **Security**, then **Conditional Access**, then **New Policy**, and then **Create new policy**.

> **Tip**
> By default, Microsoft enables security defaults, which are a set of standard security features that it enforces. To set up your own Conditional Access policies, you will need to disable this by going to the portal menu at the top left. Click on **Azure AD | Properties | Manage Security Defaults**. Change this setting to **No**, and then click **Save**.

The following screenshot shows where you will need to set up your new Conditional Access policy:

Home > Security > Conditional Access >

New ...
Conditional Access policy

Control access based on Conditional Access policy to bring signals together, to make decisions, and enforce organizational policies. Learn more

Name *

New CA Policy ✓

Assignments

Users or workload identities ⓘ

  0 users or workload identities selected

Cloud apps or actions ⓘ

  No cloud apps, actions, or authentication contexts selected

Conditions ⓘ

  0 conditions selected

Access controls

Grant ⓘ

  0 controls selected

Session ⓘ

  0 controls selected

Control access based on signals from conditions like risk, device platform, location, client apps, or device state. Learn more

User risk ⓘ

  Not configured

Sign-in risk ⓘ

  Not configured

Device platforms ⓘ

  Not configured

Locations ⓘ

  Not configured

Client apps ⓘ

  Not configured

Device state (Preview) ⓘ

  Not configured

Filter for devices ⓘ

  Not configured

Enable policy

(Report-only   On   Off)

Create

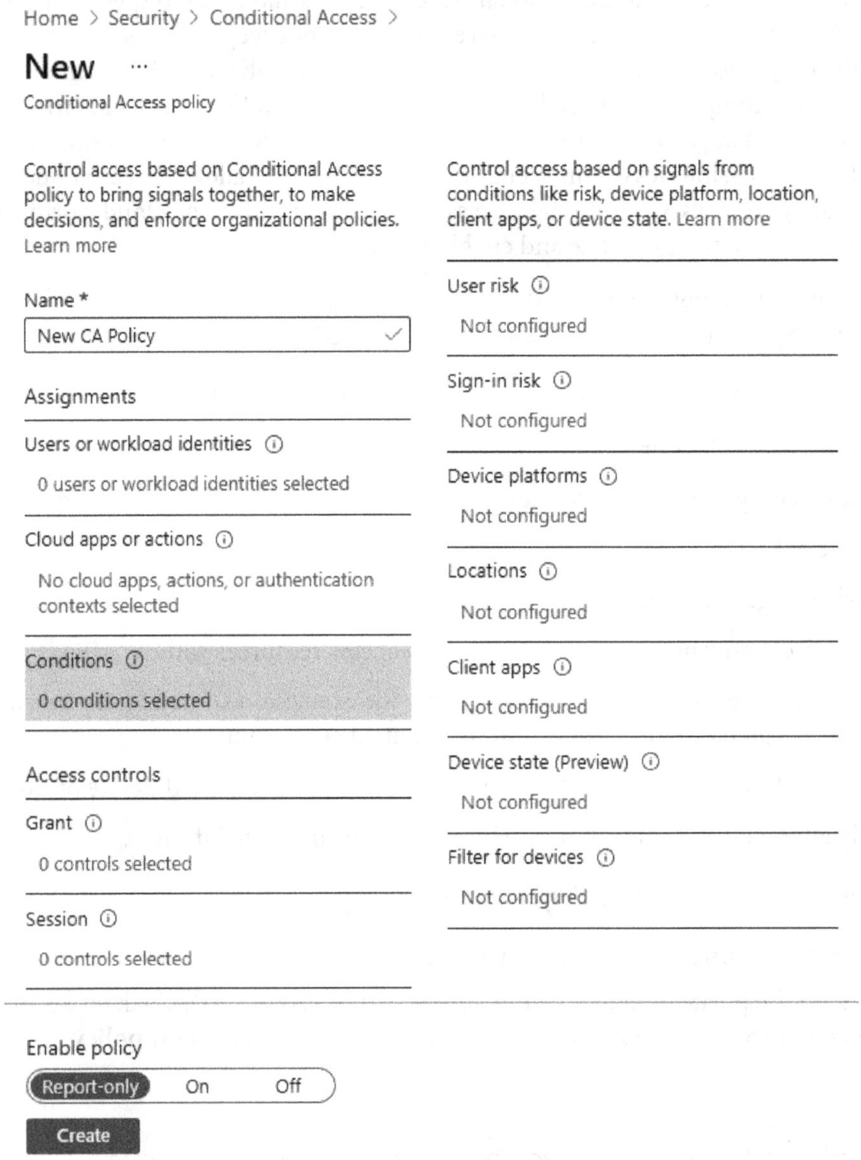

Figure 5.25 – A new Conditional Access policy

Within the new policy, configure your assignments and the following configurations:

- **Users or workload identities** is where you can include and/or exclude users and groups.

- **Cloud apps or actions** is where you can include or exclude apps as part of the policy, as well as any user-specific actions or authentication contexts.

- **Conditions** is where you configure specific conditions for the policy. These include **User Risk**, **Sign-in risk**, **Device platforms**, **Locations** (using IP ranges or countries/regions), **Client apps**, **Device state**, and **Filter for devices**.

3. Next, you will configure your access controls:

- **Grant** is where you can grant or block access. If you grant access, you can select multiple requirements to comply with – for example, requiring MFA, devices to be marked as compliant, or using an approved client app.

- **Session** applies session limits to cloud apps such as sign-in frequency and using Conditional Access app control to send sessions into Defender for Cloud Apps.

4. Finally, select **Report-only** to review the policy or select **On** to enable them immediately, and then click on **Done** to complete the policy setup.

> **Tip**
> Thoroughly test the Conditional Access policy before enabling it for the organization. There are scenarios where you can lock yourself out or cause major disruption to your users if you don't test and validate.

The following screenshot shows the **Conditional Access | Insights and reporting** page, which can be found by logging in to `https://portal.azure.com`, searching for and clicking on **Security**, clicking on **Conditional Access**, and then navigating to **Insights and reporting**.

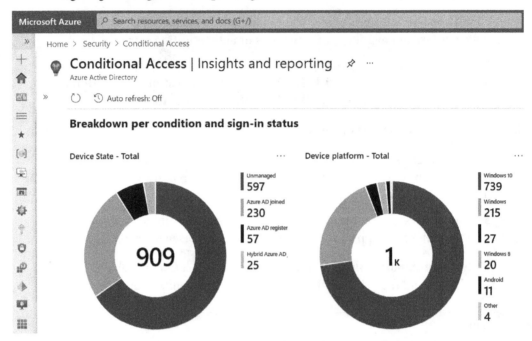

Figure 5.26 – Conditional Access | Insights and reporting

Location conditions can also be used to block anyone from accessing the organization outside of a specific country. To do this, you can build locations within the **Named locations** option of the **Conditional Access policy** menu and add a new location based on IP ranges or country/region.

Next, let's review Azure AD Identity Protection and how to build conditions for automated risk remediation.

Configuring Azure AD Identity Protection

Azure Identity Protection, like Conditional Access, is an Azure identity-based security tool. It works by analyzing identity-based signals from user activity to generate a user risk level, as well as identifying risky sign-ins as they occur in real time. With the advancement of the cloud and its collective customers, Microsoft currently analyzes 6.5 trillion signals per day to help better protect its customers and support services such as Identity Protection. By analyzing these signals against your user's activity to identify detection markers, Identity Protection can provide automated remediation for identity-based risks. Detection markers are used to determine the level of risk, and examples include atypical travel, anonymous IP addresses, malware-linked IP addresses, and a variety of others. Risks are split up into low, medium, and high categories for both user and sign-in risk. User risk detection helps to determine whether an account has been compromised, and sign-in risk detection helps determine whether the authentication request is from the real owner. By creating policies based around both user risk and risky sign-ins, automated protection mechanisms can apply controls in real time to block access to high-risk sign-ins or use the risk-level tag as a condition in a Conditional Access policy for fine-grained control.

> **Tip**
>
> To learn more about the risks, visit `https://docs.microsoft.com/en-us/azure/active-directory/identity-protection/concept-identity-protection-risks`.

To access Identity Protection and enable the policies, follow these steps:

1. Log in to `https://portal.azure.com`.

2. Go to the portal menu at the top left, click on **Azure AD**, choose **Security**, and click on **Identity Protection** to access the Identity Protection features.

3. Within this management console, you can enable **User risk policy, Sign-in risk policy**, and **MFA registration policy** (required to protect against sign-in risks unless you block access).

4. To enable the user risk policy, select **User risk policy** on the left and configure each of the following options:

   - **Users**: Decides who will be included or excluded from the policy

   - **User Risk**: Where you select the likelihood the user is compromised (**Low and above, Medium and above**, or **High**)

   - **Access**: To block or allow access and require a password change if allowed

5.   Once configured, change the policy to **On** and click **Save**. The following screenshot shows the **User risk policy** configuration screen:

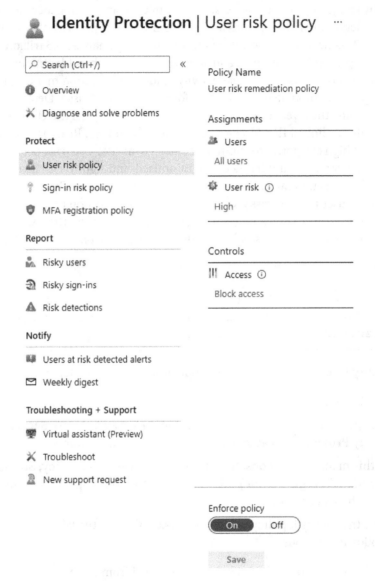

Figure 5.27 – Enabling the user risk policy

Complete the preceding steps again for the sign-in risk policy and ensure you enable the MFA registration policy as well.

> **Tip**
>
> To best protect your users, it is highly recommended you configure both the
> user risk policy and the sign-in risk policy to **High**, with the **Block access**
> control enabled.

Summary

In this chapter, we covered identity and access management and its importance to ensure
you provide the best protection for your users. We started with an overview, covering
the foundation of IAAA and each of its components. We then reviewed account and
access management, which covered the account life cycle, as well as external access and
privileged access.

Next, we covered authentication methods, including securing your passwords and
password management, enabling SSPR, and authenticating with Azure AD using SSO. The
remainder of this section included a review of passwordless authentication and enabling
Windows Hello for Business. We finished this chapter by covering advanced cloud identity
and access protection tools, known as Conditional Access and Identity Protection.

In the next chapter, we will look at the administration and policy management of our
Windows environment. We will review device management evolution and how to
remotely manage devices in an enterprise environment.

Part 2: Applying Security and Hardening

This section will move into the core of the book, applying controls and administering Windows endpoints. First, you will be introduced to the modern administration methods used for Windows environments and how to implement and manage security policies and configurations. After that, we will provide details about securely deploying Windows endpoints in an enterprise environment. You will then be provided with guidance on applying security baselines to harden endpoints and review mitigating common attack vectors should an intruder get inside your network. The section will finish off with a focus on securing a Windows Server infrastructure and how best to protect your Windows servers.

This part of the book comprises the following chapters:

- *Chapter 6, Administration and Policy Management*
- *Chapter 7, Deploying Windows Securely*
- *Chapter 8, Keeping Your Windows Client Secure*
- *Chapter 9, Advanced Hardening for Your Windows Clients*
- *Chapter 10, Mitigating Common Attack Vectors*
- *Chapter 11, Server Infrastructure Management*
- *Chapter 12, Keeping Your Windows Server Secure*

6
Administration and Policy Management

In this chapter, we will be covering the administration and policy management of Windows devices within an organization. This is a very important task and one that requires ongoing interaction, support, and reviews to stay up to date with changing trends in technology. The foundational direction of your administration strategy should largely be driven from your established framework and baselines, as discussed in *Chapter 2, Building a Baseline*. This should help direct admins on how to administer your Windows environment and provide clear direction during policy configuration to meet the defined requirements that have been created and agreed upon.

When it comes to recommendations for enterprise-grade solutions for managing Windows clients, **Microsoft Endpoint Configuration Manager (MECM)** and Intune are the industry-leading standards. In the following sections, we will dive into an overview of both solutions, as well as reviewing the tools available in each to help ensure the successful administration of your baselines, configurations, and policies.

To accomplish these tasks, we will review the following topics:

- Understanding device administration
- Managing devices with Configuration Manager
- Managing devices with Intune
- Administering a security baseline

Technical requirements

To follow along in this chapter, you will need the following technical requirements as a minimum:

- MECM
- Microsoft Intune
- Access to the Group Policy Management Console

Understanding device administration

Ensuring devices remain hardened and secured is unrealistic without a proper device management solution. Unless you administer a very small number of devices, you are going to need some form of management to effectively apply policies, distribute software, manage security updates, and continuously service endpoints after they are handed over to your employees. Without it, your organization will be putting itself in an extremely vulnerable situation.

First, let's look at the evolution of device management and the progression to unified endpoint management throughout the years.

Device management evolution

Using a device management model, many large organizations have adopted MECM, formerly known as **System Center Configuration Manager (SCCM)**, as the standard for on-premises device management. Configuration Manager is a fully mature solution whose capabilities range from image deployment to software distribution, patch management, and policy enforcement. To operate effectively, a Configuration Manager hierarchy requires resources and the deployment of infrastructure either on-premises or in IaaS. As new PC hardware is purchased and new Windows builds are released, a lengthy and complex life cycle process to support the new requirements will typically follow. Additionally, the underlying operating system and physical hardware, if on-premises, must also be maintained. Throughout the book, we may refer to Configuration Manager as MECM, SCCM, ConfigMgr, and Endpoint Manager. All are common forms used to reference Configuration Manager. MECM is a highly complex and robust solution, and therefore hundreds of reference docs are available through Microsoft docs and other third parties. In the sections that follow, we will try to reference only what we consider relevant to the material in this book, with the understanding that there may be other ways to complete these tasks.

Recently, we have seen disruption to the previous model and a shift that is changing the dynamics of device management. In recent years, there has been increased adoption of **Mobile Device Management (MDM)** tools that evolved in parallel with the growth of mobile platforms such as iOS and Android. This growth has sometimes led to multiple administration environments within enterprises, one for phones and tablets, and the other for desktops and laptops. This generates a lot of overhead and requires a broader skillset to support, manage, and operate multiple platforms. It also adds overhead to your security strategy as both need to meet the security requirements of your policies. Validating security within multiple environments can create challenges and adds its own complexity.

A major advantage of using an MDM solution is an out-of-the-box approach. The ability to take an **Original Equipment Manufacturer (OEM)** device out of the box, power it on, log in with your corporate account, and begin receiving policies, configurations, software, and security settings is a game-changer. This approach has been well received and is now widely adopted by corporate-owned and bring-your-own Windows, iOS, and Android devices. Using MDM helps companies shift away from traditional imaging, complex on-premises management solutions, and the overhead it brings. More recently, Microsoft has made significant enhancements with Intune and Configuration Manager integrating both solutions into one unified management approach for your device management program with Intune and co-management.

As the model continues to evolve, we are slowly seeing a transition to unified endpoint management, bringing together all endpoints into one management solution, as shown in the following diagram:

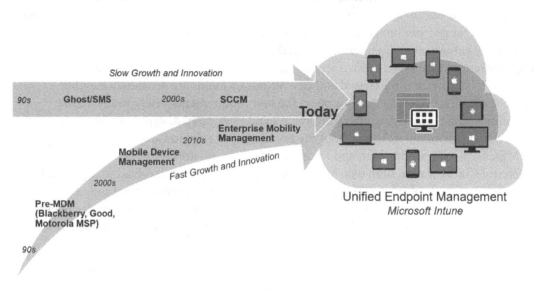

Figure 6.1 – The evolution of device management

For most organizations, this shift isn't going to happen overnight, but the good news is that Microsoft has built a solid foundation and avenue to make the journey a reality.

As a review and precursor to device management, we are going to discuss joining a Windows device to your organization. Traditionally in on-premises deployments, devices are joined to your domain in Active Directory and managed through group policy. Today, we have flexibility, and devices can be both domain and hybrid **Azure Active Directory (AD)** or fully Azure AD joined. MDM with Intune is supported for hybrid, domain, and fully Azure AD-joined devices.

Let's look at an overview of each model.

Differences between domain join, hybrid, and Azure AD-joined devices

Domain join is the traditional model and allows the central administration of your corporate end user devices and Windows servers on-premises. When joining your device to a domain, you are joining that machine to your **Active Directory Domain (ADD)**. Once joined, you can use **Group Policy Objects (GPOs)** to configure and enforce your policies based on your defined baselines. You will also most likely supplement Group Policy with an additional management tool such as Configuration Manager. This method has served as a very mature and reliable model.

If you want to use a cloud-only solution to administer your devices, then you will need to Azure AD join your devices. This method directly joins your device to the cloud environment and Azure AD where you can manage and administer your devices using an MDM solution such as Intune. This model also supports the ability to use co-management with MECM. Moving to the cloud model provides a far more powerful and robust deployment, allowing for the improved provisioning of your devices and increased flexibility for your users. Azure AD-joined devices will require you to log in with your Azure AD cloud credentials, which can be synced from an on-premises AD.

> **Tip**
> There is also the concept of Azure AD-registered devices that supports the use case of **Bring Your Own Device (BYOD)**. This allows Windows, iOS, Android, and macOS devices access to your corporate resources without being required to sign in to your device with your work account.

Many organizations may face dependencies that still require an on-premises Active Directory infrastructure. If this is still a requirement, hybrid Azure AD join provides the benefits of both Azure AD and Active Directory. With this method, your devices will still be joined to your on-premises Active Directory, but will also be registered in Azure AD. This allows for management using Group Policy, Configuration Manager, or the option to use co-management with Microsoft Intune. Windows 7 and later and Windows Server 2008/R2 and later all support this method.

The following diagram shows these three different options for your devices:

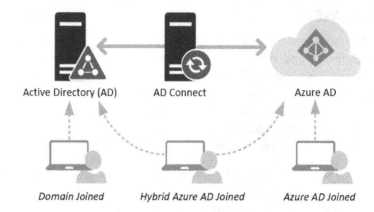

Figure 6.2 – Domain-joined, hybrid, and Azure AD-joined examples

Next, let's look at one of Microsoft's device management solutions – Microsoft Endpoint Configuration Manager. In the following section, we will provide details on how to effectively manage devices and policies using this enterprise-grade solution.

Managing devices with Configuration Manager

MECM is a world-class enterprise device management solution for workstations, laptops, and Windows servers. Its robust set of tools and centralized management are invaluable for ensuring that device security and configurations are compliant. It is highly scalable and can deploy operating systems, manage the application life cycle, push security updates, configure security baselines, manage antivirus, collect telemetry, and define compliance policies with a robust set of built-in reporting.

The Configuration Manager infrastructure usually contains a hierarchy of servers, including a **Central Administration Site** (CAS), primary sites, and additional secondary sites based on the organization's scale configured with system roles. Depending on the number and location of users and devices, having multiple servers to act as **Management Points (MPs)** and **Distribution Points (DPs)** for hosting content is not uncommon. Even with a complex deployment, your entire ecosystem of devices can be controlled from a single administrative console. To connect to the hierarchy, Windows clients have deployed an agent that creates the connection point to Configuration Manager.

Software Center is the familiar user-facing application installed on client machines that allows users to request and install applications and software updates. For companies with remote workers without an office or possibly VPN, clients can be managed directly over the internet using an Azure **Platform as a Service (PaaS)** resource known as a cloud management gateway or through internet-based client management through internet-facing site systems. Configuration Manager integrates with many other Microsoft cloud-based solutions, including Intune for co-management, Endpoint Analytics with tenant attach, and Windows Autopilot. Together, these tools create a robust, unified endpoint management solution branded as Microsoft Endpoint Manager. Configuration Manager is a key component of unified endpoint management and we want to review a few concepts regarding managing devices. In the next few sections, we will be covering the following topics:

- Client collections, settings, and communications
- Securely deploying clients for Configuration Manager
- Connecting to the Azure cloud and Intune co-management

A Configuration Manager deployment can become highly complex, and we won't go into detail about configuring a site hierarchy. There are many great blogs available over the internet, and the Microsoft docs are a great place to get started at the following link: `https://docs.microsoft.com/en-us/mem/configmgr/`.

Let's have a look at managing client settings and communicating between site components.

Client collections, settings, and communications

Monitoring the health and status of your clients is an important aspect of maintaining a well-managed device administrative ecosystem. If a production client is active but not sending a heartbeat, then the device could be at risk and unable to be managed or patched remotely. Monitoring client activity and status is also very helpful for maintaining an accurate device inventory.

In Configuration Manager, resources (or clients) are organized by being placed in collections by users or devices. Resources can be added to collections manually, by building custom rules, using WQL queries, and through other methods. For more information about collections in Configuration Manager, visit this link: `https://docs.microsoft.com/en-us/configmgr/core/clients/manage/collections/introduction-to-collections`.

Built-in security roles

Role-based access control (**RBAC**) can be applied to limit the scope of resources that administrators have access to manage. There are many built-in security roles, and they can be assigned directly to individual users or Active Directory security groups. A few examples of security roles include the following:

- The **Application Administrator** built-in security role grants permission to the author and deploys applications.

- The **Software Update Manager** role grants permission to define and deploy software updates.

Role-based administration is also helpful for hierarchies with multiple sites that have different administrators to create a separation of duties between the clients that they can manage. To assign security roles, go to **Administration | Security | Administrative Users**.

For more information on role-based administration in Configuration Manager, visit this link: `https://docs.microsoft.com/en-us/mem/configmgr/core/understand/fundamentals-of-role-based-administration`.

Now that we have reviewed collections, we need to define the settings the agents will use to communicate with the site systems. Let's look at the client settings options for Windows clients.

Client settings

Client settings are used to specify agent configurations such as caching settings, policy polling, compliance settings, and Software Center customizations, such as company branding. These settings are grouped into client setting policies and are assigned to device collections. Multiple policies can be assigned to devices, and they are evaluated by a priority weight set during policy creation. The default custom client setting is set to 1,000. In the following screenshot, you can see all the different settings available for configuring client settings:

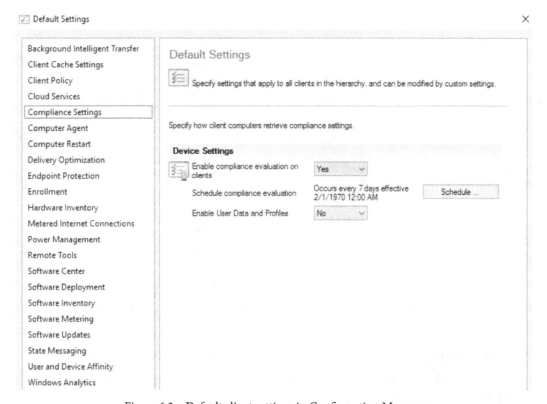

Figure 6.3 – Default client settings in Configuration Manager

Client settings can be configured under **Administration | Client Settings** and include both device and user-based options.

Remote device actions

Configuration Manager also has the ability to set remote device actions, such as collecting device inventory telemetry or forcing a policy sync. This helps force a policy sync if the request needs to be expedited sooner than the polling interval cycle configured in the client settings. The following screenshot shows some examples of the available options for device actions:

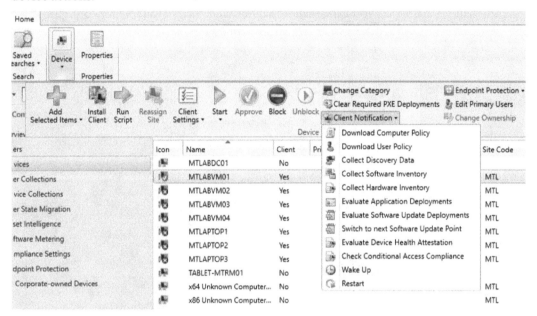

Figure 6.4 – Remote actions in Configuration Manager

In addition to client notification remote actions, there are other remote menu options available, including the following:

- **Client Settings** will show you a view of all client settings applied to a device.

- **Start** can initiate a remote desktop session and open **Resource Explorer** for viewing system information.

- **Run Script** will let you run a PowerShell script against a device.

- **Block** will prevent a client from receiving a policy and communicating with the hierarchy.

For more details on client management operations, visit this link: `https://docs.microsoft.com/en-us/configmgr/core/clients/manage/manage-clients`.

Now that we have covered configuring collections and client settings, let's discuss how clients communicate with the site systems.

Client communications

Clients communicate with different site systems such as MPs and DPs using the most secure method available. An MP site system provides location information to clients, such as the location of distribution points, and delivers policy data. The DP site system contains application source files and software updates and is where the client connects to download content. Client authentication uses HTTP or HTTPS methods, and authorization can occur for both the user and device.

> **Tip**
>
> Microsoft recommends using HTTPS or enhanced HTTP for all client communications. HTTP communications are deprecated in version 2103 for client communication with site systems.

For secure client communications, and typically seen in on-premises Active Directory, clients use an internal certificate issued through their certificate authority. Client authentication can also use a self-signed certificate and authenticate by verifying an Azure AD user or device token. In the following screenshot, the Configuration Manager control panel applet displays the client certificate as self-signed for an Azure AD-joined co-managed device over the internet:

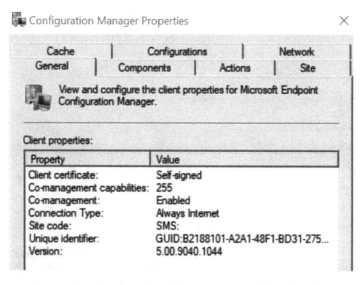

Figure 6.5 – Configuration Manager applet client properties

To read more about the communications between clients and Configuration Manager, visit this link: `https://docs.microsoft.com/en-us/mem/configmgr/core/plan-design/hierarchy/communications-between-endpoints`.

> **Tip**
>
> MECM supports CNG v3 certificates. To read more about *cryptography and next-generation certificates*, visit this link: `https://docs.microsoft.com/en-us/mem/configmgr/core/plan-design/network/cng-certificates-overview`.

To configure client communication settings for your MPs and DPs, go to **Administration | Site Configuration | Servers and Site System Roles**. Each site system you select will list all the installed roles that are configured. Double-click them to view their properties. In the following screenshot, the MP properties for this site server require client communications to use HTTPS:

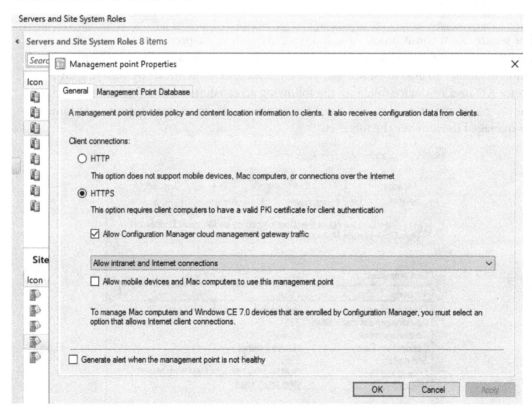

Figure 6.6 – MP client communication settings in Configuration Manager

Configuration Manager encrypts the data between the clients and hierarchy to prevent traffic from being analyzed using SHA-256 as the primary algorithm. To read more about the cryptographic functions in detail, visit this link: `https://docs.microsoft.com/en-us/mem/configmgr/core/plan-design/security/cryptographic-controls-technical-reference`.

Securely deploying clients for Configuration Manager

There are a few different methods that can be used to deploy the Configuration Manager client. For an in-depth analysis, Microsoft docs has a great write-up that can be found at this link regarding best practices for client deployments: `https://docs.microsoft.com/en-us/configmgr/core/clients/deploy/plan/best-practices-for-client-deployment`.

Let's look at several different methods used to deploy clients today:

- **Group Policy** installation is supported for Windows only. Using a GPO, clients can be deployed on computer objects. The installation bootstrapper and commands are configured in the software installation node in GPMC under **Computer Configuration | Policies | Software Settings**, as shown in the following screenshot:

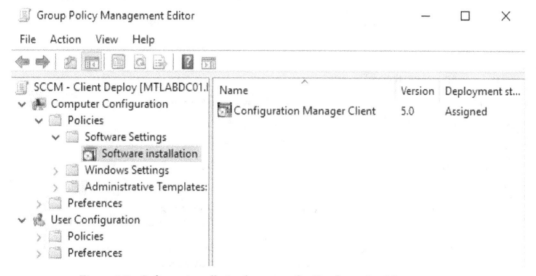

Figure 6.7 – Software installation bootstrap for Configuration Manager agent

- **Software update-based** installation also uses Group Policy and can be used for both new installations and client upgrades. Software update-based installations will require a server with the **Software Update (SUP)** site system role installed.

- **Logon script** installations can be done with GPO using the `CCMSetup.exe` file located in the `Program Files/Microsoft Configuration Manager/Client` directory from a UNC path or network share.

- **Client push** installation methods can be used to deploy clients automatically on computers that are discovered by site discovery within Configuration Manager.

- **Image** installation is an option to preinstall the client during device imaging.

- **Manual** or packaged installations are useful for boots-on-the-ground installations or for targeting a collection of devices for client upgrade scenarios.

- **Intune MDM installation** can be used for devices that are managed by Intune.

Using Intune, `CCMSetup.msi` is configured in an app deployment. The `ccmsetup` command line can be customized to support on-premises and Azure AD authentication scenarios. Additional resource requirements are needed if using the Azure AD method with the cloud management gateway. The following screenshot shows an example of the Intune app installation properties of the `ccmsetup.msi` bootstrapper:

Figure 6.8 – Microsoft Endpoint Manager view of the ConfigMgr client in Intune

> **Tip**
>
> To find the CCMHOSTNAME value for the Azure AD method, run the
> following query in SQL Server Management Studio for the main site database:
> `select * from proxy_settings`. The `ConnectorInfo` column
> contains the necessary information.

For more information about deploying clients with Azure AD Identity, visit `https://docs.microsoft.com/en-us/configmgr/core/clients/deploy/deploy-clients-cmg-azure`.

Regardless of the method used to deploy agents, Microsoft recommends the following security best practices:

- Use PKI certificates for clients that communicate with **Internet Information Services (IIS)** site systems by using HTTPS.

- If managing computers in Active Directory, only auto approve clients from trusted domains.

- If deploying the agent on an image, remove the certificates before Sysprep to ensure clients don't impersonate each other.

- Use a trusted root key to validate MP for your clients.

- Always update software update points to include the latest agent version.

- Disable WINS lookup to ensure that only Active Directory Domain Services is used to find valid management points.

- Use HTTPS or enhanced HTTP for secure communication between site systems.

To view a full detailed list of security best practices from Microsoft, visit this link: `https://docs.microsoft.com/en-us/configmgr/core/clients/deploy/plan/security-and-privacy-for-clients`.

Next, let's review the connection of Configuration Manager to the Azure cloud and enabling co-management with Intune.

Connecting to the Azure cloud and Intune co-management

As part of ongoing investment from Microsoft in Configuration Manager, you can now connect your hierarchy to Azure services and create a better-unified endpoint management experience. Some of the advantages of cloud attachment include the following:

- User and device-based Azure AD authentication.
- Support for the cloud management gateway and internet-based client management.
- Azure AD user and group discovery.
- Connections to Log Analytics.
- Deploy Microsoft Store for Business apps.
- Azure AD tenant attach for Configuration Manager actions and tools from the Microsoft Endpoint Manager admin center.

Depending on which services you configure, you can use the built-in Azure Services Wizard to deploy the resources in Azure AD needed to create the connections. For details on how to use the Azure Services Wizard and more information about Azure services in general, visit this link: `https://docs.microsoft.com/en-us/mem/configmgr/core/servers/deploy/configure/azure-services-wizard`.

In addition to connecting with Azure services, when enabling Configuration Manager for co-management, you can cloud-attach your hierarchy and manage devices both with Configuration Manager and Intune. Inside the administration console, you can control which workloads are managed by each solution. Some of the major benefits of enabling co-management include the following:

- Configuration Manager actions from Microsoft Endpoint Manager.
- Manage device configurations, apps, compliance policies, and endpoint protection from the solution of your choice.
- Better RBAC and central visibility for all your co-managed devices.
- Support for Windows Autopilot.
- Support for Conditional Access with device compliance conditions.
- Client health details in Microsoft Endpoint Manager.

The following screenshot shows the workload sliders for switching between Intune and Configuration Manager. It can be found under **Administration | Cloud Services | Co-management**:

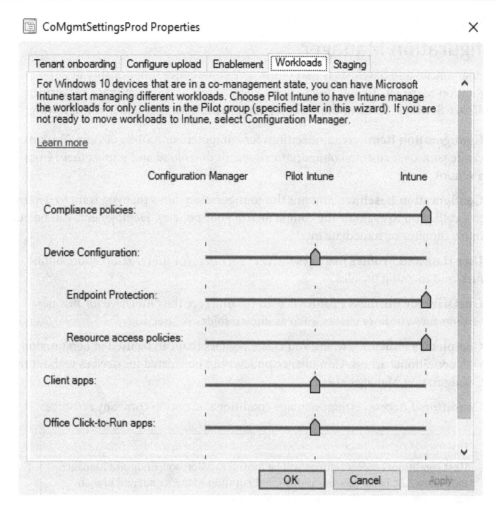

Figure 6.9 – CoMgmtSettingsProd in Configuration Manager

Co-management can be enabled on existing Configuration Manager devices, or set up with modern provisioning using the Intune MDM installation discussed earlier. You will need to have at least Azure AD, Microsoft Intune, and Windows 10 or later with hybrid Azure or fully Azure AD-joined devices.

For more information about enabling co-management, visit this link: `https://docs.microsoft.com/en-us/mem/configmgr/comanage/overview`.

Next, let's look at how to manage policies and baselines using Configuration Manager.

Managing policies and baselines in Configuration Manager

Managing policies and baselines is part of the core compliance functionality in Configuration Manager. To view them, navigate to **Assets and Compliance | Overview | Compliance Settings**. Some of the available compliance settings include the following:

- **Configuration Items** contains settings for computers or mobile devices. You can create your own custom configuration items or download and import them from a vendor.

- **Configuration Baselines** contains the configuration items that you want to deploy to a collection to evaluate the compliance of your policies. Deployments can be put into a monitor or remediate mode.

- **User Data and Profiles** manages the user's settings for folder redirection, offline files, and roaming profiles.

- **OneDrive for Business Profiles** is used to configure the OneDrive for Business settings for Windows clients, such as known folder redirection.

- **Compliance Policies** is where you create policies that can be used in conjunction with conditional access. Compliance policies can be created for devices without the Configuration Manager client.

- **Conditional Access** settings manage conditional access to company resources.

> Tip
> Most conditional access settings will be moved to Microsoft Endpoint Manager and Azure AD. Later versions in the Configuration Manager current branch servicing model now direct you to use this portal.

- **Company Resource Access** allows you to create VPN, Wi-Fi, and certificate profiles to deploy to your devices.

- **Microsoft Edge Browser Profiles** contains custom settings to create an Edge browser policy in Windows 10.

Let's look at what configuration items are and how they are used to evaluate settings on Windows 10 devices and servers.

Creating a configuration item

A **configuration item** contains a list of settings and compliance rules that are used for evaluation against a device's current settings. There is a lot of flexibility when choosing a setting type, and the options will vary depending on the platform selected during policy creation.

Some of the available setting types in a configuration item include the following:

- Active Directory query
- An assembly
- A filesystem
- IIS metabase
- A registry key and registry value
- Using a custom script
- SQL query
- XPath query

The following screenshot shows a configuration item using the integer registry value as the setting type for the detection method. The specific hive, key path, and value name are all specified:

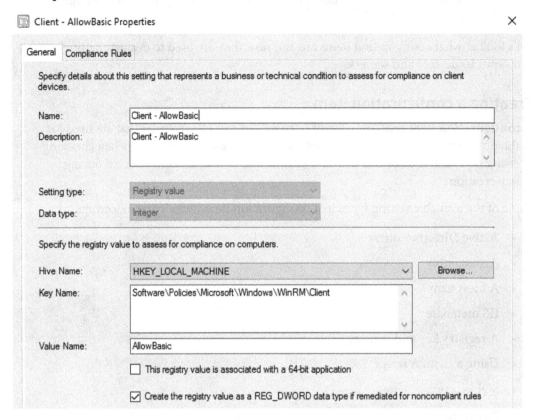

Figure 6.10 – Configuration item using the Registry value setting type

> **Tip**
> The checkbox for **Create the registry value as a REG_DWORD data type if remediated for noncompliant rules** is selected. This allows the REG_DWORD data type to be created or updated if the configuration baseline is set to remediate non-compliant rules.

The **Compliance Rules** tab in the **Configuration Item** setting is where you can specify the conditions for evaluation against the device. Each setting can have more than one compliance rule.

The following screenshot shows a compliance rule with a condition that says the value name must be set to zero to be compliant. Choose the **Remediate noncompliant rules when supported** checkbox to force this setting onto the PC if found to be non-compliant.

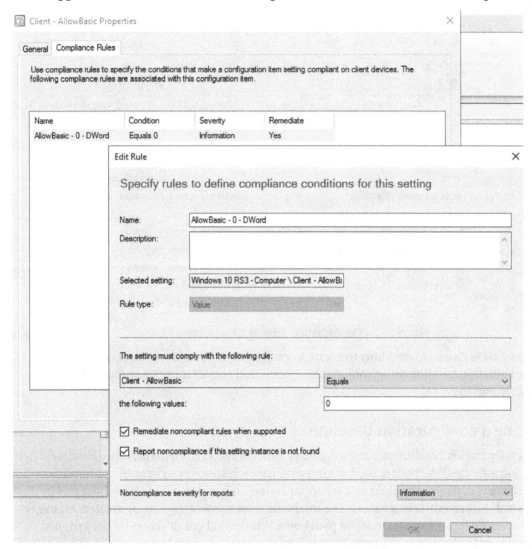

Figure 6.11 – Compliance rule options for a configuration item

You can specify many different settings in a configuration item and then assign it to a configuration baseline for deployment to devices.

The following screenshot shows an example of different configuration items that can be created and assigned to baselines:

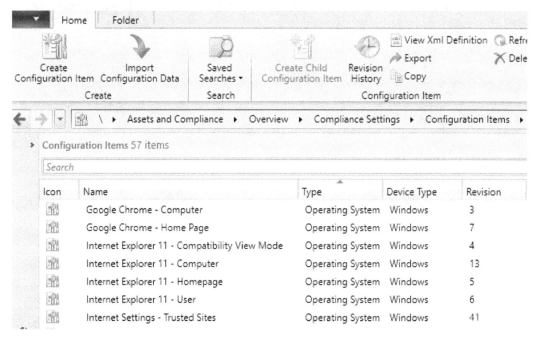

Figure 6.12 – Configuration items in Configuration Manager

Now that we have covered how to create a configuration item, let's look at building a configuration baseline using configuration items and remediate any non-compliant configurations.

Using a configuration baseline

A configuration baseline can contain one or many configuration items, other baselines, or software updates that are used during a policy evaluation on a client. A configuration baseline is deployed to collections where each device then downloads it and assesses it against the current device values. The following screenshot is the **Configuration Manager Properties** applet from the control panel on a Windows client. It shows all the assigned configuration baselines, the revision number, the last evaluation time, and the compliance status for each baseline:

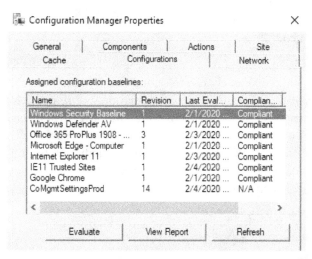

Figure 6.13 – Configurations in the Configuration Manager control panel applet

The client will receive any new baselines during the next *machine policy retrieval and evaluation cycle*. When organizing baselines, it's not uncommon to group them by their functions, as seen in the preceding screenshot. For example, you could create a baseline for Windows security settings and another for Google Chrome security and each one would contain all relevant configuration items.

To create and view configuration baselines in the administrative console, navigate to **Assets and Compliance | Compliance Settings** and choose **Configuration Baselines**. In the following screenshot, the properties of **Google Chrome Baseline CIS** contain two configuration items that will be used for evaluation against the device:

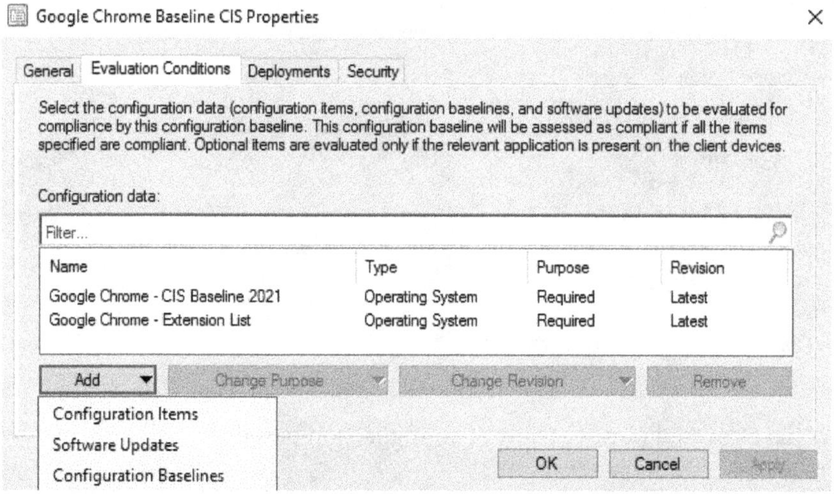

Figure 6.14 – Configuration baseline properties

Configuration baselines can also be imported from other sources if you don't want to create them from scratch.

After the baseline has been built, you can deploy it to collections of devices. When scheduling a deployment, you can choose to remediate any non-compliant rules, generate alerts, and schedule the frequency of evaluation. For remediation to work, the configuration item must have the **Remediate noncompliant rules when supported** checkbox selected in the compliance rules, as discussed previously when creating a configuration item. Baselines deployed without the remediation option selected will be in monitor mode only. The following screenshot shows the deployment properties for the Google Chrome baseline:

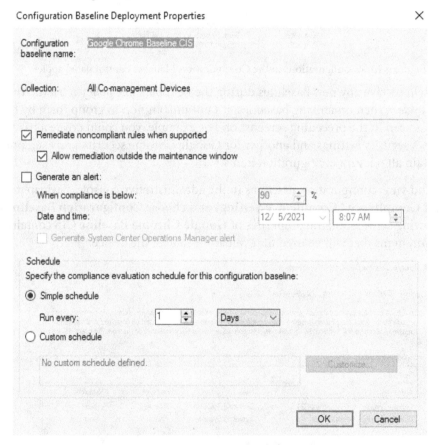

Figure 6.15 – Configuration baseline deployment properties

Now that we have covered our overview of configuration baselines, let's look at how to run reports to view device compliance.

Reporting on a configuration baseline

Configuration Manager is filled with many built-in reports for a wide range of collected data, from software inventory to resource usage to baseline deployment compliance. To find the reports, navigate to **Monitoring | Reporting | Reports** and browse through the list. Let's look at the overall compliance status of a configuration baseline deployment. Follow these steps to run the **Summary compliance of a configuration baseline for a collection** report:

1. In the **Configuration Manager** console, go to **Monitoring**, click on **Reporting**, and choose **Reports**. Expand the **Reports** folder in the left-hand navigation to view the expanded tree view.

2. Select the **Compliance and Settings Management** folder and choose the **Summary compliance of a configuration baseline for a collection** report to open the menu.

3. Click on **Values** next to **Configuration Baseline Name** and select a configuration baseline. We chose **Google Chrome Baseline CIS**, created previously in this example.

4. Click on **Values** next to **Collection** and choose a collection that you wish to report on. In this case, we will choose the collection we deployed earlier.

5. Click on **View Report** to open the report.

    The following screenshot shows the compliance status of the Google Chrome baseline. Click on the disk icon in the toolbar to view the export options, such as exporting to CSV:

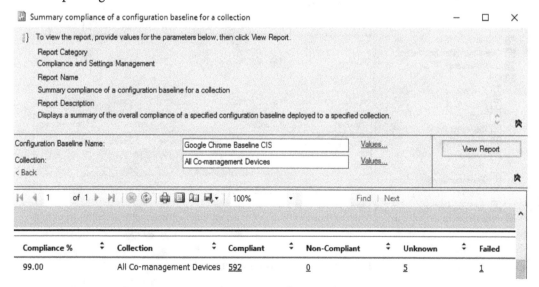

Figure 6.16 – Compliance report for a configuration baseline

Some values that are returned on the report contain clickable links. You can click the number to dig deeper into the baselines and view details such as which clients are non-compliant for each different configuration item, or even to the individual rule level that makes up the underlying configuration item.

Now that we have covered reporting on a configuration baseline deployment, let's look at using CMPivot to query devices in real time.

Querying devices with CMPivot

The CMPivot tool in Configuration Manager can be used for querying against an individual device or an entire device collection and return real-time data. This is extremely helpful if you need to troubleshoot an issue or view up-to-the-minute data on a client as there is a delay in the results returned from reporting. To use CMPivot, you build queries against the entities table using the **Kusto Query Language** (**KQL**). The KQL syntax is comparable to building SQL statements or, if you are familiar with WMI, using the WQL language. It is also the query language used for Azure Log Analytics. To start CMPivot, open a device collection, right-click a device, and choose **Start CMPivot**. Note that only one instance of CMPivot can be running at a time. To run a basic query to return the operating system, double-click on the OS entity table to populate it in the query editor and choose **Run Query**. If the device is online, the results will be returned as in the following screenshot:

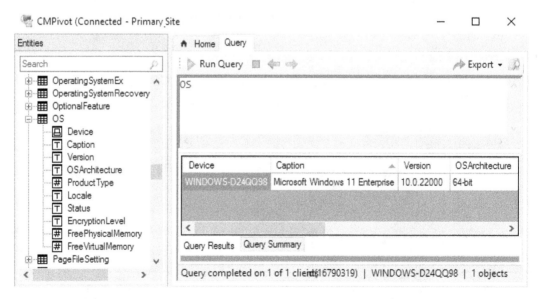

Figure 6.17 – CMPivot query results

Part of the convenience of using CMPivot when used for querying multiple devices is the option to run scripts directly against the query results right from the interface. You can build complex KQL queries to narrow down a selection of devices sourced from a larger device collection based on criteria and use **Run Scripts** to send a remediation action such as stopping a service or closing an application.

If the Configuration Manager hierarchy is tenant-attached to Microsoft Endpoint Manager, you can also run CMPivot queries directly from the Microsoft Endpoint Manager portal. To do this, the account running CMPivot from MEM must also have the appropriate permissions to do so in Configuration Manager. To learn about the requirements and limitations of CMPivot, visit this link: `https://docs.microsoft.com/en-us/mem/configmgr/core/servers/manage/cmpivot`.

We have just covered an overview of Configuration Manager, including how to create and manage policies and baselines for your clients. Next, let's dive into Microsoft's cloud-based device management solution, Intune, and learn how to use MDM to manage devices and enforce policies.

Managing devices with Intune

Microsoft Intune is an Azure-based device management solution and is a standard for MDM and **Mobile Application Management (MAM)**. Intune has a full feature set for deploying apps and configurations with support for Windows, macOS, iPadOS, and mobile platforms such as iOS and Android. It is a single solution to enforce security baselines and device configurations, manage endpoint security, deploy applications and layer protection policies, and much more.

As mentioned earlier, Intune can be integrated with Configuration Manager for co-management, as well as connected to solutions such as Microsoft Defender for Endpoint and other third-party mobile threat defense partners. Before we jump into enforcing policies, let's review some of the core concepts of Intune MDM. In the next section, we will cover the following topics as they relate to features in Intune:

- **Configuration Service Provider (CSP)**
- MDM versus MAM or **Intune App Protection Polices (Intune APP)**
- Windows enrollment methods

First, let's discuss how Intune manages the configuration of your MDM-enrolled devices using CSPs to drive policy configuration.

CSP

A CSP is the layer that MDM providers use to configure custom settings on mobile devices. More specifically, Windows version 10 and later use the **Open Mobile Alliance Uniform Resource Identifier (OMA-URI)** to identify and configure these settings. Configurations for CSPs are typically formed using a **Synchronization Markup Language (SyncML)** XML-formatted message and enforced on the client using the **Open Mobile Alliance Device Management (OMA-DM)** protocol. Just like traditional Group Policy, the CSP acts as a medium to configure settings such as a registry key, for example. It is important to understand CSPs for MDM as this is what configures settings on your devices using Intune compared to using GPOs and the traditional domain Group Policy processing engine.

CSPs are typically represented graphically in tree form. In the diagram that follows, the root node is at the top, followed by rounded elements and rectangular elements. The rounded elements are the different nodes, and the rectangular elements represent a setting that can be configured with a value. The first element after the root dictates the identifiable name of the CSP, and the first rectangular element is the setting. When put together, they formulate the full URI path. The following diagram demonstrates the tree view of a CSP:

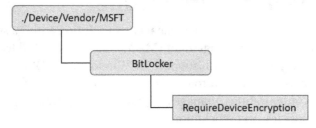

Figure 6.18 – CSP diagram for the BitLocker RequireDeviceEncryption setting

For this CSP, MSFT (Microsoft) is the vendor root and BitLocker is the CSP node. The full URI path for the CSP would be as follows in XML format:

```
./Vendor/MSFT/BitLocker/RequireDeviceEncryption
```

Setting the Data integer to 0 in the SyncML XML format will disable this policy:

```
<SyncML>
    <SyncBody>
        <Replace>
            <CmdID>$CmdID$</CmdID>
            <Item>
                <Target>
                    <LocURI>./Device/Vendor/MSFT/BitLocker/
```

```
RequireStorageCardEncryption</LocURI>
                </Target>
                <Meta>
                    <Format xmlns="syncml:metinf">int</Format>
                </Meta>
                <Data>0</Data>
                </Item>
        </Replace>
    </SyncBody>
</SyncML>
```

Documentation on the BitLocker CSP can be found here: `https://docs.microsoft.com/en-us/windows/client-management/mdm/bitlocker-csp`.

When building device configuration profiles in Intune, choosing the **Settings Catalog** or **Templates** profile type contains mostly CSP-backed policies. If the policy you're looking for doesn't exist in Intune, choosing the custom setting from **Templates** allows you to manually configure a CSP and specify the OMA-URI, data type, and value. The following screenshot shows the custom profile type for configuring a Microsoft Defender removable storage access control policy:

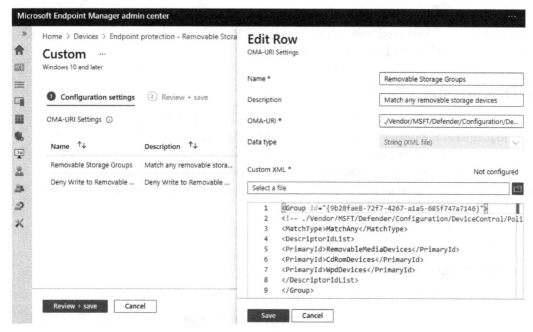

Figure 6.19 – Custom template in Intune

This policy maps to registry keys on the Windows client. After the profile has been assigned to a group of users or devices, you can confirm this setting has been configured by opening the registry on the Windows client and going to `HKEY_LOCAL_MACHINE\SOFTWARE\Policies\Microsoft\Windows Defender`. The XML data type in the screenshot has been populated under the data for the `PolicyGroups` string value.

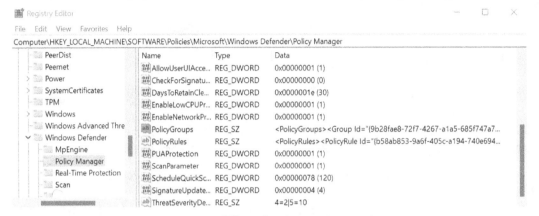

Figure 6.20 – CSP configuring a registry setting

For a full list of the available CSP, visit this link: `https://docs.microsoft.com/en-us/windows/client-management/mdm/configuration-service-provider-reference`.

Now that we have discussed how Intune uses a CSP for policy definitions on your devices, let's look at the difference between MDM and MAM.

MDM versus MAM

In the managed device, or MDM, scenario, devices are enrolled in Intune to become fully managed at the device level. Using MDM on a Windows device allows admins to administer it in a similar fashion as Group Policy by deploying applications and assigning user and device-based settings using device configuration profiles. Some examples of policies and remote actions that are available with Intune MDM include the following:

- PIN and password requirements for unlocking your device
- VPN configurations and settings
- Deploying certificates
- Configuring Wi-Fi settings and profiles
- Distributing line-of-business applications and other apps
- Assigning compliance policies such as a minimum required Windows version

- Configuring security baselines and administrative-backed templates (ADMX)

- Remote device wipe with factory reset

- Custom configurations for CSPs not available natively

- Windows update and delivery optimization settings

Having control at the device level allows device-specific remote actions such as forcing a policy synchronization, initiating a wipe and factory reset, locating a device, collecting diagnostic logs, and managing BitLocker key rotation. Intune also provides a comprehensive set of inventory and reporting data such as device attestation reports and hardware inventory.

In the MAM scenario, corporate data is controlled at the application level only by using Intune app protection policies. Intune app protection policies can be targeted at Microsoft 365 apps as well as custom line-of-business and other third-party applications that have integrated with the Microsoft Intune App SDK. MAM can be used in conjunction with MDM to provide an additional security layer over corporate data. Although many device-specific policies and remote commands can't be enforced using MAM, certain actions can be used to selectively wipe company data from apps or restrict how data can be interacted with between the app and host device. A few examples include applying an Intune app protection policy for Microsoft Outlook that doesn't allow company data to be copied and pasted outside the app, requiring a PIN on app launch, or blocking access to company data from a jailbroken or rooted device. Assigning Intune app protection policies for applications running on Windows is called **Windows Information Protection (WIP)**. To learn more about WIP, visit this link: `https://docs.microsoft.com/en-us/windows/security/information-protection/windows-information-protection/protect-enterprise-data-using-wip`.

Next, let's discuss the two types of enrollment methods for Windows devices to become Intune-managed.

Windows enrollment methods

To manage Windows devices with Intune, they need to be connected to the service. The two main methods of enrollment are user self-enrollment and administrator-based enrollment, and each has different methods and scenarios for completing the enrollment process. User self-enrollment scenarios include the following:

- Windows Autopilot in user-driven mode

- Intune enrollment (MDM only)

- Azure AD join with auto-enrollment
- BYOD

Administrator-based enrollment scenarios include the following:

- Windows Autopilot for pre-provisioned deployment
- Bulk enrollment methods for many devices
- Using a device enrollment manager to enroll kiosks, point-of-sale systems, or shared PCs
- Using Configuration Manager-based enrollment

Each enrollment scenario has its own unique capabilities, and Microsoft has outlined the best practices for when to use each one. It is recommended to review this list to determine which method may best suit your needs. It's not uncommon to use more than one, especially if your company has a mix of user-assigned devices and shared devices such as kiosks or point-of-sale systems. You can review the best practice recommendations by visiting this link: `https://docs.microsoft.com/en-us/mem/intune/enrollment/enrollment-method-capab`.

Now that we have covered the basics of MDM and Intune, in the next section, we will discuss how to use Intune and Microsoft Endpoint Manager.

Using Intune and Microsoft Endpoint Manager

MEM is the unified endpoint management solution that brings Intune and Configuration Manager together. The MEM admin center is used with Intune for configuration management of both MDM and MAM devices, as well as for deploying apps, creating compliance policies, configuring software updates, and managing endpoint security. MEM provides unified reporting views to gain insights into the compliance of your Windows devices from both Intune and Configuration Manager telemetry. Through Endpoint Analytics, you can analyze the productivity of your devices by identifying slow hardware, boot times, or app crashes. Using software updates, Windows updates for business policies can be scheduled and assigned directly from the Microsoft Endpoint Manager console, including feature update deployments or expedited security updates for out-of-band patch releases. There are many features worth exploring with Microsoft Endpoint Manager and you can log in to the admin center at this link: `https://endpoint.microsoft.com`.

Intune also integrates with other solutions across the Microsoft stack, such as Defender for Endpoint and additional third-party solutions. To view the connectors, log in to the admin center, go to **Tenant Administration**, and select **Connectors and tokens**. Here you can also connect the Microsoft Store for Business, view your connection to Configuration Manager, and configure certificate connectors to your on-premises certificate authority.

Next, let's look at an overview of device actions that are available for MDM-managed devices in Microsoft Endpoint Manager.

Devices and remote actions

To view the available remote device actions, go to the device overview by clicking on **Devices**, choose **Windows** under **By platform**, and then search for or select a device. The remote actions will be available in the top toolbar and include the following options:

- **Retire** will remove managed app data.
- **Wipe** restores the device to the factory Windows OS.
- **Delete** removes the device from Intune management.
- **Sync** forces a check-in to the Intune service for policy.
- **Restart** forces a device to restart.
- **Fresh Start** removes any apps (pre-installed OEM) installed on a Windows 10 PC.
- **Autopilot Reset** reapplies a device's original settings, bringing it back to a business-ready state.
- **Quick scan/Full scan** will run Defender antivirus scans.
- **Update Windows Defender security intelligence** will update the Defender malware definitions.
- **BitLocker key rotation** will rotate BitLocker keys.
- **Locate device** will send a location signal for tracking geolocation.
- **Sync machine policy/Sync user** policy actions will force a policy refresh for MECM-connected clients.

The following screenshot shows the available device actions in the toolbar from Microsoft Endpoint Manager:

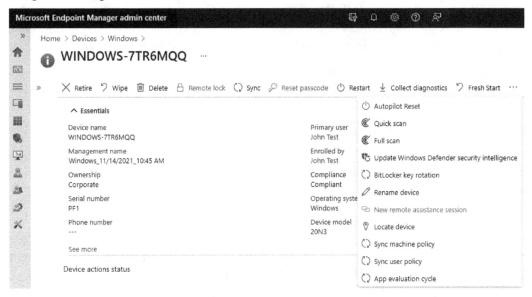

Figure 6.21 – MEM remote device actions

The **Remote device**, **Sync machine policy**, **Sync user policy**, and **App evaluation cycle** actions are only available for Configuration Manager tenant-attached devices.

Now that we have looked at an overview of Intune and Microsoft Endpoint Manager, let's cover how to manage policies and baselines using Intune.

Managing policies and baselines in Intune

There are a few areas in MEM for configuring policies, and sometimes that can lead to confusion on where to configure them and increase the potential for policy conflicts on the device. It's key to understand what the policies are doing, what method to use to apply them, as well as when to assign them to users or devices. It's particularly important when moving away from GPO-based policy management into modern management to be aware of conflicts and which setting takes priority. Luckily, there are a few options for how to handle this, which we will discuss later in *Chapter 8, Keeping Your Windows Client Secure*. Let's review the options to apply policies and configurations for Windows. They include the following:

- Intune security baselines
- Endpoint security
- Device configuration profiles

- Device compliance polices
- PowerShell scripts
- Windows Update for Business policies

To leverage the CSPs for MDM and modern management using Intune, the goal is to create the lightest-weight policy payload possible to deliver to the device. When defining how to apply your baseline configurations, it is recommended to use the following approach:

1. Deploy Intune security baselines first. They provide recommended best practices and security settings are preconfigured.
2. Deploy device configuration profiles or endpoint security policies for security-related settings to supplement baselines.
3. Deploy ADMX-based policy templates for Win32 apps or policies not yet available in Intune.

Let's review each configuration option based on the recommendation.

Using Intune security baselines

When building policies in Intune, it is recommended to start here first. Intune's security baselines are preconfigured groups of Microsoft's recommended security settings. They are the MDM equivalent of the GPO security baselines found in the Windows Security Compliance Toolkit. Security baselines are a collective grouping of individual device configuration profiles put together to form one policy that can be targeted at both users or devices for Windows 10 and later platforms. Microsoft frequently releases updated versions of these baselines and allows you to compare the new release to your current baseline for version control directly in Microsoft Endpoint Manager. When you are ready to append your baselines with the newly released policies, all you need to do is change the instance version and the deployment is automatic.

The available security baselines include preconfigured policies for the following:

- Security baseline for Windows 10 and later
- Microsoft Defender for Endpoint baseline
- Microsoft Edge baseline
- Windows 365 security baseline

To view the security baselines in Microsoft Endpoint Manager, log in to the admin center at `https://endpoint.microsoft.com`, click on **Endpoint Security**, and choose **Security Baselines**. The following screenshot is what the portal looks like for managing and monitoring Intune security baselines:

Figure 6.22 – Intune security baselines in MEM

To learn more about Intune's security baselines, visit this link: `https://docs.microsoft.com/en-us/mem/intune/protect/security-baselines`.

Next, let's look at device configuration profiles in Intune.

Device configuration profiles

Device configuration profiles are where most of the policy settings exist for Windows in Intune. It is recommended to use these settings to supplement the Intune security baselines. In fact, security baselines are just a collection of device configuration profiles already preconfigured by Microsoft. There are a few profile types available when creating a new device configuration profile that include the following:

- **Settings catalog** for searching all available policies

- **Templates**, which contain preconfigured groups of settings by function

- **Administrative templates** for ADMX-backed policies such as Group Policy preferences that exist today

- **Custom settings** for deploying CSPs yet to be mapped in Intune

The administrative templates and custom settings are options listed under the **Templates** profile type. After building a profile and selecting the assignment, you can choose to configure scope tags or applicability rules depending on the profile type. These tags help define a criterion that a device must meet for the settings to apply. A **scope** tag is a custom attribute assigned to a device to help identify it. An **applicability** rule specifies a property value in relation to the OS edition and version to determine whether a device meets the minimum requirements to apply the profile. The following screenshot demonstrates an applicability rule checking for Windows 10/11 Enterprise and the client must be within the build number range:

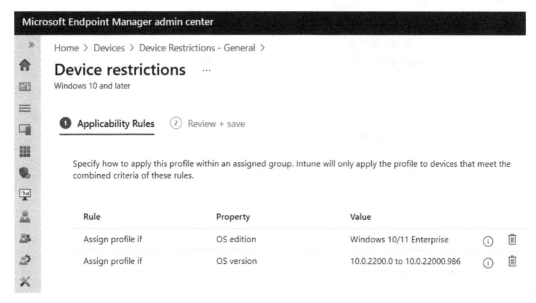

Figure 6.23 – Applicability Rules in Intune device configuration profile settings

First, let's look at using the settings catalog to build a custom policy.

The settings catalog

The settings catalog contains all the available MDM-backed settings configurable in Intune. It allows you to build your own policy from scratch and choose from any available mapped MDM setting to design the policy in a way that makes sense for your administrative needs. Using the settings picker, you can search for a variety of settings, including administrative templates, and add them to your profile. As with all device configuration profiles, it also includes reporting capabilities directly in the profile overview after you assign it to users or devices. The following screenshot shows the **Settings picker** option to add settings to your profile:

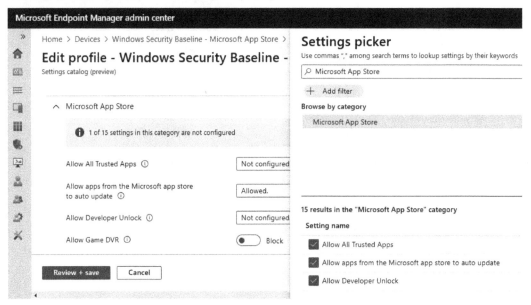

Figure 6.24 – Settings picker in Intune

Next, let's look at using templates when creating a device configuration profile.

Using templates

The **Templates** profile type contains settings that are grouped and organized by specific functions for ease of use. Unlike the settings catalog, where you must build a profile from scratch, you can use templates to build policies based on a category or functionality. A few examples include **Delivery Optimization** settings for software updates, or **Device Restrictions** settings for controlling a wide range of settings, from the Windows App Store to password requirements. Templates contain both CSP-backed policies, including the custom profiles type, and administrative templates for ADMX-backed policies. We covered the custom profile type in the *Managing devices with Intune* section, including a link to the CSP references. Let's review the **Administrative Templates** options.

Administrative templates in Intune contain thousands of settings structured in a similar way to Group Policy in Active Directory. In fact, they are built around the XML configurations of a GPO. Aside from Windows-specific settings, administrative templates include ADMX-backed policies for Microsoft Office and Microsoft Edge, and support configuring policies for other Win32 and third-party applications, such as Google Chrome, through a process called ADMX ingestion.

Policy assignments can be applied to user and device groups depending on whether the policy is scoped to user-specific or computer-specific settings. When browsing through administrative templates, follow a familiar path of a Group Policy setting. For example, in the following screenshot, policies found traditionally in `Computer Configuration \ Administrative Templates \ Windows Components \ Windows Remote Management` can be found by using the same path, starting with **Windows Components**:

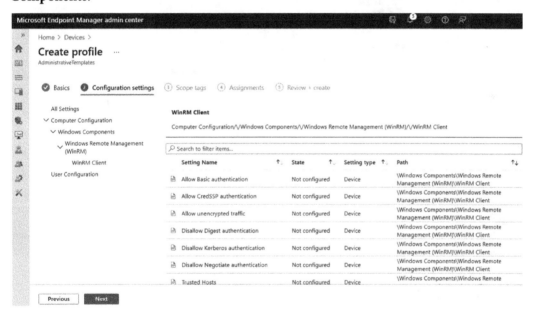

Figure 6.25 – Administrative templates in Intune

Administrative templates provide the familiarity and flexibility many administrators are used to when working with Group Policy. Next, let's look at configuring security settings using endpoint security.

Managing endpoint security

Just like a device configuration profile, endpoint security policies allow you to configure MDM-backed settings, but with a specific focus on Windows device security. This provides flexibility by allowing security administrators using Intune to focus more on security-based policies. We covered Intune security baselines earlier in this section, but you can find additional device security features under the **Manage** section. These include specific policies for Defender Antivirus, disk encryption, and advanced Defender for Endpoint features such as attack surface reduction rules. These policies can be used on their own or to supplement the Intune security baselines. The following screenshot shows the antivirus summary from **Endpoint security** and lists any policies that are created:

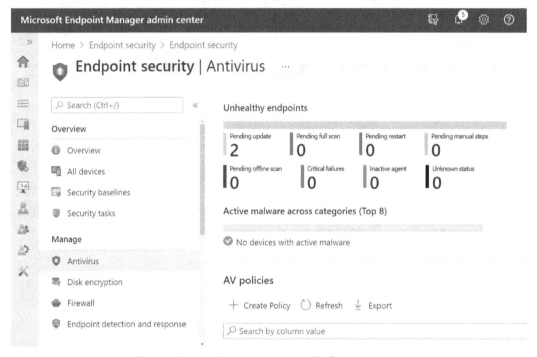

Figure 6.26 – Antivirus summary in Endpoint security

One advantage of endpoint security over device configuration profiles is the ability to assign policies to Configuration Manager tenant-attached devices. This includes non-Intune-managed Windows Server and Windows 10 and later clients.

Next, let's look at using a PowerShell scripts policy to deploy custom scripts.

Deploying PowerShell scripts

The PowerShell scripts node can be found in MEM under the **Devices | Policy** section. To deploy a policy, the device must be enrolled in Intune. When configuring a PowerShell script policy, you can choose to run the script under the currently logged-on user's credentials, enforce script signature checks, and run the script using 64-bit PowerShell. On the local device, the PowerShell script's function relies on the Intune Management Extension service to check for and run scripts. The process can be found in the service's **Microsoft Management Console (MMC)**. It is recommended to review the documentation in detail to understand all the prerequisites and behavior when using a PowerShell script policy. You can find more information at this link: `https://docs.microsoft.com/en-us/intune/apps/intune-management-extension`.

Next, let's look at deploying apps in Intune.

Deploying apps in Intune

Deploying apps in Intune follows a similar app deployment methodology to many other device management solutions. Some of the app types supported include the following:

- MSI
- APPX and MSIX
- Web apps
- Store links
- Office **click-to-run (C2R)** apps
- Microsoft Edge
- IntuneWin file extensions

In addition to custom line-of-business applications, you can connect Intune to Microsoft Store for Business and sync over any purchased modern apps and assign them to your users. In Windows 11, business store apps assigned to your private store have moved to the Company Portal app.

> **Tip**
> If creating a policy to block public apps, but while allowing the private store, users will see a blocked message when opening the Microsoft App store on Windows 11.

For handling complex app installations that are not packaged in a single MSI file, you can use the **Microsoft Win32 Content Prep** tool to package it into an `IntuneWin` file and deploy it through Intune. For more information on packaging Win32 apps for Intune, visit this link: `https://docs.microsoft.com/en-us/mem/intune/apps/apps-win32-prepare`.

For more information regarding application deployment in general using Intune, visit this link: `https://docs.microsoft.com/en-us/mem/intune/apps/apps-windows-10-app-deploy`.

Next, let's look at an overview of device compliance policies and how they can be used to control access to company resources.

Overview of device compliance policies

One great benefit of managed devices is the ability to set compliance requirements that must be met to ensure security settings are in place to gain access to company resources. When a Windows device is evaluated, Intune leverages Azure AD to keep a compliance status. This status, coupled with conditional access, allows administrators to block or apply restrictions to devices until they meet the rule requirements set by the policies.

The following screenshot is the device compliance overview of a managed Intune device. There are six separate compliance policies assigned to this device used in the compliance evaluation:

Figure 6.27 – Device compliance state of an Intune managed device

Actions can be configured for devices that are non-compliant, and these are as follows:

1. Mark a device as non-compliant.
2. Send an email to the end user.
3. Retire the non-compliant device.

The **Device compliance** status can then be used as an access condition for a conditional access policy; for example, a conditional access policy assigned to all users that states that access to SharePoint Online requires multi-factor authentication or a compliant device. If the user's device falls out of compliance, then they must use multi-factor authentication to access SharePoint Online even if on a company-issued device.

We recommend configuring Windows compliance policies for the following evaluations:

- Windows device health attestation rules requiring BitLocker, Secure Boot, and code integrity.
- Windows device properties setting a minimum OS version. This will help ensure that devices are patched or otherwise marked as non-compliant.
- Windows Security policies to block simple passwords, require a minimum length of 6 characters, and lock a device after 15 minutes of inactivity.
- Require data storage encryption.
- Windows device security policy that requires a firewall, antivirus, and antispyware solution to be enabled.
- Require a **Trusted Platform Module** (**TPM**) to be present.
- Require the Microsoft Defender for Endpoint solution to be enabled with real-time protection and up-to-date signature definitions.
- Require a machine risk score to be medium or lower from Microsoft Defender for Endpoint.

> **Tip**
> Co-management customers can also leverage Configuration Manager compliance as an evaluation condition for devices.

At the time of writing, Microsoft has added a custom compliance policy setting for use with a discovery script for added flexibility.

When building your compliance policies, it will help keep you organized by separating them by function in the event you need to troubleshoot issues with evaluation. In the following screenshot, the compliance settings are categorized by functions that can serve as a guide when building your own policies:

Figure 6.28 – Compliance settings in Intune

Tip

Be careful when assigning policies to both user and device groups. This could cause a policy to try and be evaluated in both contexts and could cause false compliance evaluations.

Now that we have provided an overview of Microsoft's unified endpoint management solutions, let's discuss different scenarios for administering security baselines for Windows clients and servers.

Administering a security baseline

Once we have chosen our framework and designed and built our policies and standards, we need to start applying our baselines to managed devices. There are many scenarios and device states that play a factor in your administration decision and more than one way to deploy baselines. Technical limitations aside, each method will have its own pros and cons and you might find taking more than one approach to complement the next is preferred depending on your environment.

Deploying managed configurations

The following chart can act as a reference when deciding what solution best suits your environment. This list is not all-inclusive, but these are the most common methods for administering baselines on enterprise-managed PCs:

	Hardened Image	Group Policy	Intune Security Baseline	Intune Device Configuration	ConfigMgr Security Baselines	Intune Endpoint Security
Workgroup Only	✓				✓	
Domain Joined	✓	✓			✓	
Hybrid / Azure AD Intune	✓	✓	✓	✓		✓
Hybrid / Azure AD ConfigMgr	✓	✓			✓	
Hybrid / Azure AD Co-Managed	✓	✓	✓	✓	✓	✓
Hybrid / Azure AD Tenant Attach	✓	✓			✓	✓
Azure AD Intune	✓		✓	✓		✓
Azure AD Co-Managed	✓		✓	✓	✓	✓

Figure 6.29 – Scenarios for administering a security baseline

Be mindful when choosing the application method and simplification is best to avoid creating conflicts.

Summary

In this chapter, we covered administration and policy management and the importance of managing devices. We started with an overview of device administration and the different ways in which Windows devices can connect and register to your domain. We then learned about Configuration Manager and how to securely deploy clients and manage policies and configurations.

Next, we walked through managing devices with Intune. We discussed the differences between MDM and MAM and covered how to use Intune with Microsoft Endpoint Manager. We talked about managing policies and baselines in Intune and recommendations on how to apply policies when using MDM. Finally, we covered different scenarios for how to administer security baselines to managed devices.

In *Chapter 7*, *Deploying Windows Securely*, we will cover different ways to build and deploy hardening images, covering different tools and techniques for Windows clients, servers, and virtualized environments.

7
Deploying Windows Securely

Over the years, methods for deploying Windows have remained consistent, with little change to the overall approach. Before the advancement of unified endpoint management tools and **Infrastructure as a Service (IaaS)**, companies were challenged with efficiently deploying devices with a consistent set of configurations to their users.

Historically, companies may have relied on **Original Equipment Manufacturer (OEM)** images, applied Group Policy, or built provisioning scripts to layer configurations onto devices. For many, the standard is to use imaging tools that allow companies to build and capture pre-configured images that could be pushed out to new devices or provisioned by a third-party partner or local IT company. While effective, these methods are time-consuming and resource-intensive, but still serve a valuable purpose for hardening Windows systems for deployment. Now, with the advancements in Azure AD and Intune, organizations can deploy Windows with very little or zero-touch provisioning using the Windows Autopilot service.

In this chapter, we're going to cover the tools that can be used to build and distribute a hardened Windows image. We will cover using the Windows Autopilot service for Windows clients, as well as deploying images for Azure Virtual Desktop and Windows 365 Cloud PC. It is important to understand the available tools for accomplishing these tasks since the approach may vary, depending on what types of resources your organization deploys.

In this chapter, we will cover the following topics:

- Device provisioning and upgrading Windows
- Building hardened Windows images
- Provisioning devices with Windows Autopilot
- Deploying images to Azure Virtual Desktop
- Deploying Windows 365 Cloud PC

Technical requirements

In this chapter, we will be referencing the following tools and services:

- Windows Assessment and Deployment Kit
- Windows Deployment Services (Windows Server roles and features)
- Microsoft Deployment Toolkit

Additionally, we will be covering the following tools that require additional licensing:

- Microsoft Intune licensing
- Windows Autopilot licensing
- Azure Virtual Desktop
- Windows 365 Cloud PC

> **Tip**
> Most Intune licenses also cover the use of System Center Configuration Manager and don't require a standalone purchase.

Device provisioning and upgrading Windows

It is strongly recommended that you update to the latest version of Windows as soon as possible. Windows 11 is rich with both hardware and security features, a great productivity tool for users, and is optimized for modern management, including support for **bring your own device (BYOD)** scenarios. BYOD has started to become a popular trend that's particularly aligned with the rapid adoption of remote work and can help significantly reduce costs and overheads for the IT department.

While this is an effective approach in some scenarios, it comes with unique challenges, including from a security lens and support and operational perspective. Depending on the organization's security and hardware requirements, it may not be suitable to support a BYOD program. Factors that may influence this decision could also include compliance requirements, the complexity of application configurations, and delivery, such as bandwidth requirements for large application payloads.

From a security perspective, you will not have as much fine-grained control over device security configurations and there may be other privacy and legal concerns if corporate data resides on users' devices. Until apps become more modernized and built with the appropriate app protection policy SDKs for modern management, it'll likely be necessary to enforce restrictions at the device level. This is where maintaining hardened images and purchasing certified hardware has a clear advantage in an enterprise scenario. Although maintaining images and deploying hardware comes with additional overheads and costs, MDM technologies such as Intune and Windows Autopilot have made these processes much lighter and more effective. Even if maintaining an image is still required due to heavy payloads from apps and policies, Intune and Autopilot can layer additional policies and enforce device configurations.

The decision of determining what provisioning model suits your needs depends on your deployment logistics, environment, and compliance requirements. Therefore, it's important to understand the technologies that are available to accomplish these tasks.

Upgrading Windows

When you're upgrading a major version of Windows, the two most common upgrade paths are known as in-place upgrades or migrations. For Windows clients, this can mean moving from Windows 7 or 8.1 to Windows 10/11 or converting an existing client into an Autopilot device. For Windows Server, upgrade options could also include rolling upgrades and license conversions to expand its feature set. In-place upgrades are supported for Windows Server and are typically limited to only one or two versions above the current version. However, it's worth noting that the migration path for Windows Server is typically recommended as the cleanest method. When you need to upgrade multiple assets, it's common to build a task sequence to assist with this process. Task sequences are a set of XML instructions that define a series of events to help automate the installation and upgrade process. They are created using tools such as MDT and orchestrate or *sequence* the installation of the device drivers, operating system, custom scripts, and software installations. Let's review the main upgrade paths.

In-place upgrades

With this option, Windows is *upgraded* from the source version to the desired version. Applications, settings, and user data are retained. Typically, these upgrades are managed by deploying task sequences using Windows Deployment Services or Configuration Manager, though they can also be controlled using Feature Updates in Intune, WSUS, or Windows Update. Manual options also exist when you're using an enterprise ISO, media creation tool, or Microsoft Update Assistant. With this upgrade option, a new operating system is installed on top of the existing one, which can cause some inconsistencies and challenges due to over-layering the operating system and setting translations. There may also be limitations that apply depending on the current source version and the target version you're looking to upgrade to.

Migration

With this option, Windows is completely reinstalled and can be migrated from basically any operating system, assuming there's support for the underlying hardware. The most common scenarios for migration include PC refresh or replacement or moving to new virtual machine hardware. Enterprise tools can be used to manage migrations such as in-place upgrades by using task sequence deployments. Applications, settings, and user data will need to be accounted for separately in this upgrade option as they will be deleted during the migration process. Use tools such as the **User State Migration Tool** (**USMT**) to migrate application and user data. If only user data is important, a cloud-based storage solution such as OneDrive can be used. For an overview of USMT, go to `https://docs.microsoft.com/en-us/windows/deployment/usmt/usmt-overview`.

Upgrading to Windows 11

Microsoft released Windows 11 for general availability on October 5, 2021, and recommends upgrading as soon as possible. Just like Windows 10, Windows 11 also supports both the in-place and migration upgrade path options and using similar methods to the ones we outlined previously. There may be some limitations if you're using the in-place path for versions before Windows 10 and it's worth understanding whether migration may the best option for you in this scenario. Windows 11 also has specific hardware compatibility requirements to support an upgrade. Specifics around system requirements can be found at `https://www.microsoft.com/en-us/windows/windows-11-specifications`.

For organizations that manage devices using Windows Updates for Business, notably with Intune, Windows 11 will not be pushed automatically like Feature Updates were in Windows 10. To accomplish this, use the **Feature Updates for Windows 10 and later** feature, and set Windows 11 as the feature update to deploy.

> **Tip**
> If you're using update rings to control Windows Update for Business settings, it is recommended that you set the feature update deferral window to 0 days if you're using Feature Updates to upgrade Windows.

After setting a feature update deployment, you can view its progress through the **Reports** feature in Intune:

Figure 7.1 – Windows Feature Update report in Intune

To view the reports in the **Microsoft Endpoint Manager** (MEM) admin center, go to `https://endpoint.microsoft.com`, click on **Reports**, and choose **Windows updates** under **Device management**.

Microsoft has also prepared methods to determine the upgrade-readiness of your devices through various upgrade scripts and solutions in Azure, such as Endpoint Analytics or the Update Compliance solution in Log Analytics. Next, let's look at modern options for backing up user data and settings.

Backing up user data and settings

Many modern cloud-based storage solutions are available to help you securely back up and restore user data both automated and manually. Fortunately, with modern operating systems and application architectures, it's becoming a less common need to back up app data-specific settings for client applications. This helps simplify migrations, especially during a PC replacement scenario. With cloud-based storage, organizations can empower users to take ownership of their data and use cloud storage such as OneDrive for Business.

Using OneDrive for Business

The OneDrive client is part of the Microsoft 365 suite of applications and has been designed to run seamlessly on Windows for backing up and syncing user files. It has enhanced security controls to protect company-owned data, with the ability to restrict external sharing and work with information protection solutions to apply policies for sensitive information types, data retention, and data loss prevention. For additional information about how OneDrive safeguards data, go to `https://support.microsoft.com/en-us/office/how-onedrive-safeguards-your-data-in-the-cloud-23c6ea94-3608-48d7-8bf0-80e142edd1e1?ui=en-us&rs=en-us&ad=us`.

OneDrive's settings can also be controlled through Group Policy or Intune. Some settings that are worth noting for PC migrations include configuring known folder redirection to back up important folders such as your desktop, documents, and pictures. This will automatically back up common local locations where users save files, which may help you avoid a future headache. On company-owned devices, you can set which organization accounts are allowed to sync and prevent a user from syncing data to other accounts. For PC replacement scenarios, use the **Files On-Demand** feature to avoid users' data from automatically being downloaded once they've signed in to OneDrive.

Organizations can keep track of how users are handling OneDrive data by reviewing the audit logs in the Microsoft 365 Compliance Center and monitoring file usage using a cloud app security broker such as Defender for Cloud Apps. The following screenshot is of Microsoft Defender for Cloud Apps, where you can view details about the usage, files, and activities in OneDrive:

Figure 7.2 – OneDrive for Business activity in Defender for Cloud Apps

OneDrive is great for backing up user files, but let's look at another feature in Azure AD for backing up user-specific settings in Windows known as Enterprise State Roaming.

Enabling Enterprise State Roaming

Enterprise State Roaming is a feature of **Azure Active Directory** (**AAD**) joined Windows 10/11 devices that can synchronize specific settings such as theme, language preferences, and even mouse settings to the cloud. By configuring this feature, these preferences can roam with the user from device to device without them having to be backed up manually. Enterprise State Roaming is available for organizations with an Azure AD Premium or comparable license. To configure Enterprise State Roaming, log in to `https://portal.azure.com`, go to **Azure Active Directory**, and choose the **Devices** blade, as shown in the following screenshot:

Figure 7.3 – Enterprise State Roaming in Azure AD

When Enterprise State Roaming is enabled, the user's settings are backed up to Azure, where the data is encrypted both in transit and at rest using the Azure Rights Management service. The data is retained indefinitely and determined stale by Microsoft after a user's Azure AD account has been deleted.

To view what settings are synced on an Azure AD joined device on Windows 11, go to **Settings**, then **Accounts**, and choose **Windows Backup**. In the following screenshot, you can see that sync is disabled and is being managed by the organization through an MDM policy:

Figure 7.4 – Windows backup in Windows Settings

Tip

Be mindful of setting Enterprise State Roaming in BYOD scenarios. Setting this policy with MDM or GPO will override any personal settings and can potentially have unwanted effects on users' devices.

Next, let's look at backing up Microsoft Edge browser settings using Microsoft Edge enterprise sync.

Configuring Microsoft Edge enterprise sync

The Microsoft Edge chromium-based browser allows you to sync browser-specific settings across all devices where a user has logged in with an Azure AD account. Some settings that are supported include favorites, passwords, addresses, settings, extensions, history, and more. Microsoft Edge enterprise sync is available for Azure AD Premium license subscribers or equivalent and the policies can be configured through Group Policy and Intune. If you're using Intune App Protection policies to manage Microsoft Edge for iOS and Android, users can access some of the available settings, such as history and favorites, right from their mobile phone. This is an extremely convenient feature and a great way to be productive using the browser both on work and personal or company-owned devices.

Because Microsoft Edge is a chromium-based browser, many of the same security settings that are available in the Google Chrome security baselines also apply to Edge. An example of a few policies that can be configured for Microsoft Edge sync are as follows:

- Enabling sync and restricting sign-ins to certain domains
- Setting which data or sync types to disable, such as passwords, extensions, and open tabs
- Forcing sync and not requiring a user to consent

To view the configured settings in the Edge browser, go to `edge://settings/profiles/sync`. The **Settings** option that shows the briefcase icon depicts the settings that are managed by the policy, as shown in the following screenshot:

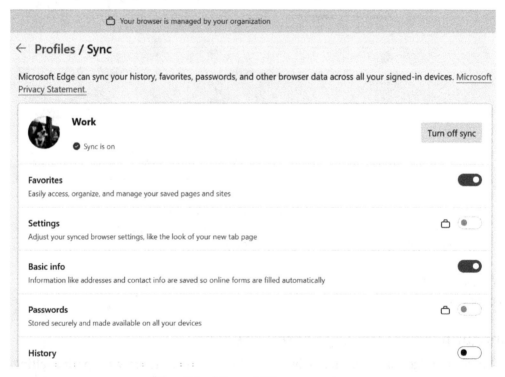

Figure 7.5 – Microsoft Edge sync settings

When combined with other cloud technologies such as Enterprise State Roaming for user preferences and OneDrive for Business for user files, these solutions cover a large percentage of the necessary user data and settings to account for during a migration. Once they have been configured, they also require little administrative effort to maintain and can significantly reduce the hands-on keyboard time a technician would need to spend with an end user during a PC replacement scenario.

These methods can significantly help in simplifying your upgrade deployments securely and cost-effectively. Next, let's discuss the tools we can use to build and deploy a hardened image.

Building hardened Windows images

Even with the increased adoption of the BYOD model and the advancement of self-provisioning technologies such as Windows Autopilot, building hardened images is still a relevant and effective method of device provisioning. Many organizations may have shifted to a lighter imaging process as these technologies can complement the traditional image deployment process and automate some provisioning tasks. Use cases for building and maintaining images still exist, whether it's for deploying laptops, workstations, or virtual desktops, and having a hardened image might be the best option for balancing security with ease of deployment. In the next few sections, we will provide an overview of the following device provisioning tools:

- **Windows Assessment and Deployment Kit (Windows ADK)**
- **Windows Configuration Designer** (WCD)
- **Microsoft Deployment Toolkit (MDT)**
- **Windows Deployment Services (WDS)**

Let's start by looking at Windows ADK.

Windows ADK

Windows ADK contains a variety of tools that can be used to build a custom Windows image. Some of the tools that are included with ADK are the Windows Configuration Designer, **Deployment Image Servicing and Management tool (DISM)**, and the User State Migration Tool. Windows ADK is typically seen as a prerequisite for using the Microsoft Deployment Toolkit. Here is a list of what Windows ADK can do and why it's used:

- You can create custom boot images or WIM files using the **Windows Preinstallation Environment (WinPE)**. WinPE is used for deployment and recovery and can later leverage a **Preboot Execution Environment (PXE)** or support boot from USB. Windows PE is now available as an add-on to the ADK.

- You can create an Unattended XML Windows Setup answer file. The answer file is used to automate the selection of options during the Windows setup process. **Unattend XML** can apply custom configurations and security features such as setting registry keys, disabling Windows features, and running custom scripts.

- You can back up user data with the USMT for migration or break-fix scenarios.

- You can create a reference *hardened* image that's used for deployment.

- You can create a **Provisioning Package** (**PPKG**) using Windows Configuration Designer. This can be applied after the image to enforce security settings, automate Azure AD Join, add certificates, install applications, and set network configurations such as Wi-Fi.

To download and install the latest ADK for Windows 11, go to `https://docs.microsoft.com/en-us/windows-hardware/get-started/adk-install`.

Next, let's look at Windows Configuration Designer, which helps configure shared devices or a smaller subset of devices that may not require you to maintain a fully customized image.

Windows Configuration Designer (WCD)

Windows Configuration Designer is an installation option of Windows ADK that's used to create provisioning packages for configuring clients through a friendly, easy-to-use interface. After selecting your project type and setting customization options using the tool, WCD exports the project as a provisioning package or PPKG file. Using a provisioning package may be useful for smaller organizations that don't have the resources, infrastructure, or skillset to build and deploy images at scale. They could also be useful for field IT to deploy shared devices, such as point of sale clients or kiosks. Provisioning packages can be applied directly to an out-of-the-box Windows installation or through the **Accounts | Access work or school** options in Windows settings. Some of the available configurations that can be set in WCD are as follows:

- Bulk MDM enrollment and Azure AD join

- Add applications

- Certificate enrollment

- Network profiles such as Wi-Fi and proxies

- Security restrictions such as password, device lock, encryption settings, update settings, and other privacy settings

- Customizations such as start tiles

If you're looking to automate Azure AD join through the provisioning package, a **Device Enrollment Manager** (**DEM**) account is required in Intune. To do this, log in to MEM at `https://endpoint.microsoft.com`, go to **Devices**, choose **Enroll devices**, and select **Device enrollment managers**. The following screenshot shows the available device enrollment managers in the Intune subscription. You can enroll up to 1,000 devices per account in a single AAD:

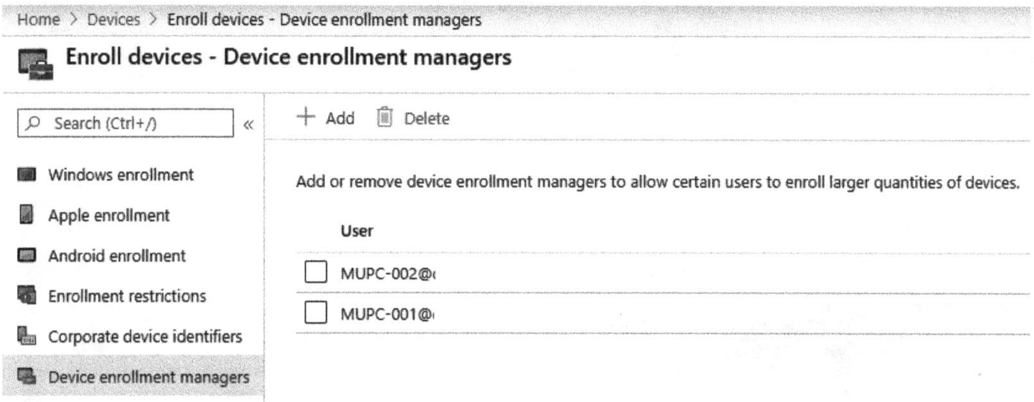

Home > Devices > Enroll devices - Device enrollment managers

Enroll devices - Device enrollment managers

Search (Ctrl+/)	«	+ Add 🗑 Delete
Windows enrollment		Add or remove device enrollment managers to allow certain users to enroll larger quantities of devices.
Apple enrollment		**User**
Android enrollment		☐ MUPC-002@
Enrollment restrictions		☐ MUPC-001@
Corporate device identifiers		
Device enrollment managers		

Figure 7.6 – Device enrollment managers in MEM

> **Tip**
> A DEM account must be assigned an Intune license to work properly.

To find out more about the DEM, go to `https://docs.microsoft.com/en-us/mem/intune/enrollment/device-enrollment-manager-enroll`.

To complete the Azure AD join process, WCD requires a token to be acquired using the DEM account. This will have a validity of 30 days. After the token expires, a new provisioning package must be created to successfully join the Azure AD client. In the following screenshot, you can see the **Bulk AAD Token** option. Clicking on **Get Bulk Token** brings up an AAD login page, where you can use one of the device enrollment manager accounts to acquire the AAD token:

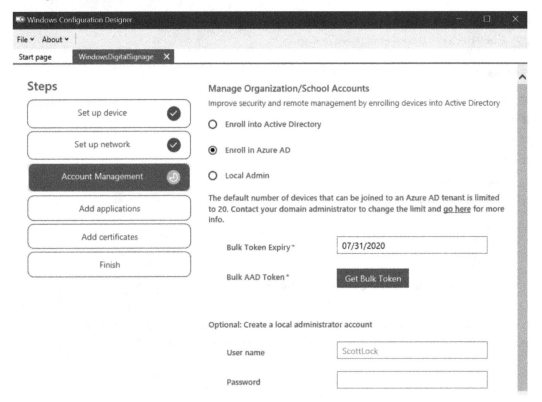

Figure 7.7 – WCD Azure AD token

For a full list of what can be configured in WCD, go to `https://docs.microsoft.com/en-us/windows/configuration/provisioning-packages/provisioning-packages`.

Now that we have covered the WCD tool, let's discuss **Microsoft Deployment Toolkit (MDT)**. MDT is the recommended tool for building custom images and offers the greatest flexibility for applying baselines and maintaining images to be used for larger deployments.

Using MDT to build custom images

MDT is an enterprise standard for building boot images, automating deployments, and maintaining reference images. Using MDT, a task sequence is created to automate the installation of drivers, operating systems, applications, security configurations, and device settings. It is commonly used to create the process known as the **Lite Touch Installation** (**LTI**) method for Windows device provisioning, requiring little interaction by a technician. By integrating with Configuration Manager, this method can be taken a step further and almost fully automated using the **Zero-Touch Installation** (**ZTI**) method. For the LTI deployment model, boot media is created by using the Deployment Workbench application, which is mapped to a deployment share containing all the source files, drivers, and applications needed to build the image. Once the operating system, applications, and configurations have been set up, boot media is generated that can be used to deploy systems with a standalone USB or through network-based deployments. MDT comes with some default task sequence templates configured out of the box. Some of these include a Sysprep and Capture task sequence, Standard Client task sequence, Custom task sequence, Lite Touch OEM task sequence, Post OS Installation task sequence, Deploy to VHD task sequence, and Standard Client Upgrade task sequence.

> **Tip**
> System preparation (**Sysprep**) is the process of *generalizing* an operating system that's used to capture a reference image where computer-specific information is removed.

MDT helps reduce the complexity of using the standalone Windows ADK (which is a prerequisite to using MDT). The following features of MDT can be used when you're building a hardened image:

- Apply local Group Policy objects
- Support for PowerShell scripts
- Install applications such as Microsoft 365 Office Click-to-Run apps
- Enable BitLocker in the WinPE environment (offline BitLocker)
- Customize the **Windows Recovery Environment** (**WindowsRE)**
- Add or remove local administrators to/from devices
- Support for Windows clients, servers, and VHDs
- Add or remove Windows Server roles and features
- Support for UEFI and the GPT partition table
- USMT support to automate the process of backing up user data and settings

To learn more about getting started with MDT and information about the features listed here, go to `https://docs.microsoft.com/en-us/windows/deployment/deploy-windows-mdt/get-started-with-the-microsoft-deployment-toolkit`.

The following screenshot is of a task sequence in MDT. It contains the list of step-by-step instructions that were used during the LTI process to deploy the operating system and install various applications and customizations:

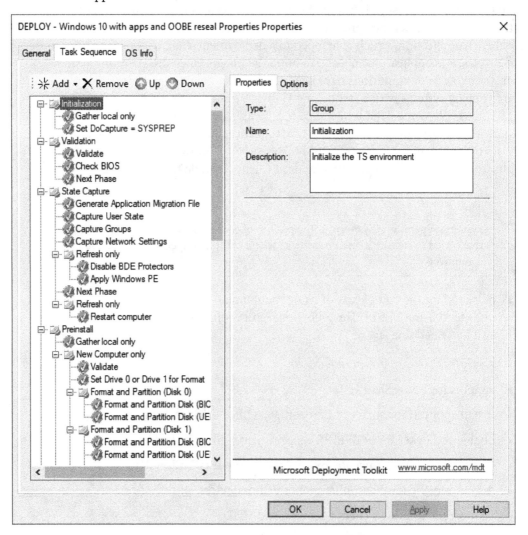

Figure 7.8 – Custom MDT task sequence

MDT can be used to deploy custom images over an Active Directory network with deployment share replication using **Distributed File System Replication (DFSR)**. For more information on configuring an MDT deployment environment and DFS-R replication, go to `https://docs.microsoft.com/en-us/windows/deployment/deploy-windows-mdt/deploy-a-windows-10-image-using-mdt`.

MDT is a free tool that you can install from Microsoft, and the latest downloads can be found at `https://docs.microsoft.com/en-us/configmgr/mdt/index?redirectedfrom=MSDN`.

Now that we have covered MDT, let's discuss using **Windows Deployment Services (WDS)** to deploy an image.

Deploying images with WDS

WDS is a feature of Windows Server that's used to deploy customized or MDT-based images using network-based installation via PXE boot. To use the PXE method, you need a DHCP server role and a TFTP server on your network that's available to retrieve the boot images. Computers on the network listening for PXE can boot using the network adapter first in the boot order and initiate the image. WDS technology still exists but is not as heavily invested in today. If you're still using WDS, here are a few recommendations for securing your image deployments:

- Set a UDP port range to limit the communication protocols that are used to connect to WDS.
- Require that *F12* is pressed to enter PXE boot.
- Use Active Directory authentication by specifying a domain controller to use for authentication and authorizing your WDS through DHCP.
- Use NTFS permissions on boot images and security groups on image groups to lock them down. `Read` and `Execute` are the only required permissions to deploy images.
- Require administrative approval to PXE from unmanaged AD computer objects using the pending device's snap-in.
- Specify the format of the computer name and **organizational unit (OU)** placement of the computer object in Active Directory. If necessary, use a staging OU that requires an administrator to move the PC object once the image is validated. This can also help prevent an overwrite of an already active and existing PC object.

WDS is a common method that's used by organizations to deploy images. Since it requires standalone infrastructure, additional upkeep and maintenance are required to keep it secure. It's important to note that Microsoft has deprecated some functionality in WDS with Windows Server 2022 and they recommend using alternatives to deploy Windows images, such as Configuration Manager. Workflows for Windows 11 and Windows Server that rely on boot.wim from installation media are now blocked, but PXE boot to custom boot images is still supported.

Next, let's look at deploying images with Configuration Manager.

MDT and Configuration Manager

MDT integration with Configuration Manager is commonly used in medium to large-sized organizations to support the deployment of operating systems and task sequences at scale. If you already have a Configuration Manager hierarchy deployed, depending on the architecture, your existing infrastructure may already be sufficient to support image deployments. Network-based installations can be supported by using distribution points that have been enabled for PXE support and using the Cloud Management Gateway for in-place upgrades over the internet. Support for task sequence deployments over the internet was introduced in version 2006. More information about these features can be found at `https://docs.microsoft.com/en-us/mem/configmgr/osd/deploy-use/deploy-task-sequence-over-internet`.

By integrating MDT with Configuration Manager, task sequences can be built directly in the console and then deployed to users and devices, just like any other software application, compliance policy, or configuration baseline. In addition to network-based installations, deployment methods include using Software Center and creating bootable media and standalone media. There is also no requirement that clients must have been previously managed by Configuration Manager or pre-imported if you're using bootable media or network-based options.

For backing up user state data, a task sequence leveraging the user state migration tool can grab the data and store it locally, remotely, or on a site server with the state migration point role enabled. Using this method and by creating a state migration point, the user data can then be restored on another computer by creating a computer association in the console. Computers, such as new hardware purchases, can be pre-staged or imported and associated with the user for PC replacement type scenarios. Once the task sequence starts on the new PC, the user state data can be restored automatically.

For more information on integrating MDT with Configuration Manager, go to `https://docs.microsoft.com/en-us/windows/deployment/deploy-windows-cm/prepare-for-zero-touch-installation-of-windows-10-with-configuration-manager`.

If you're using MDT and Configuration Manager to deploy images, the following recommendations are ways to protect your task sequences:

- Use security scopes and built-in security roles to define what objects administrators have access to. Security scopes can include boot images, driver packages, collection software updates, and applications, to name a few.

- Target only the necessary collections to receive your deployment.

- If you're leveraging PXE or media-based deployments, password-protect the task sequence.

- Modify the default permissions on deployment shares. This could be done to safeguard a user from accidentally kicking off a re-image if they unknowingly enter PXE.

MDT with Configuration Manager is a robust way to build and distribute images. Just like other on-premises tools, the underlying hardware and infrastructure require constant upkeep to ensure systems remain secure and run smoothly.

Next, let's discuss using the Windows Autopilot service and how it supports device provisioning by leveraging cloud technologies such as Azure AD and Intune.

Provisioning devices with Windows Autopilot

Windows Autopilot is used to automate and customize the device setup, deployment, and life cycle of a Windows client while minimizing the effort that's needed by IT staff in the device provisioning process. This is a fundamental change compared to the traditional model and enables a self-service scenario that can be fully completed by the end user without IT intervention. Imagine shipping a device directly from the supplier to the user, where they only need to turn it on and log in with their corporate credentials and be in a business-ready state. The remainder of the provisioning happens automatically over the internet, remains mostly transparent to the user, and can be used instead of a custom OEM image that's been pre-applied by a vendor.

For the Autopilot service to work, devices need to be registered in the service through your Azure AD tenant. This can be accomplished directly through your partnerships with device vendors or as a step by IT before shipping the device to the user. Once registered in the tenant, a profile is assigned to the device that determines the settings and configurations the user sees to control the onboarding process directly from the Windows **Out Of Box Experience (OOBE)**. Windows Autopilot can help reduce costs and with features such as Autopilot reset, it can increase efficiencies compared to the traditional model. Registering the device into company management is simplified too through automating Azure AD join and automatic enrollment into Intune.

To use Windows Autopilot, you will need a supported version of Windows 10 or later, AAD, and Intune licenses. Windows Autopilot also supports hybrid domain join. For more information about the Windows Autopilot requirements, go to `https://docs.microsoft.com/en-us/mem/autopilot/software-requirements`.

First, let's look at the different deployment scenarios for Windows Autopilot.

Deployment scenarios

Windows Autopilot supports different deployment scenarios, including user-driven mode, self-deployment mode, pre-provisioned mode, and a reset feature called Windows Autopilot reset:

- **User-driven mode** is designed for the user self-service model requiring minimal IT interaction. A device can be shipped directly to the user, powered on, connected to a network, and fully provisioned by making a few selections in the **OOBE** menu.

- **Self-deploying mode** is useful for shared devices, kiosks, or digital signage. In self-deploying mode, all that is needed is an active network connection. The provisioning process will join the device to Azure AD and Intune, and then apply all the assigned policies, configurations, and applications through an enrollment status page.

> **Tip**
> Self-deploying mode requires the device to have a minimum of TPM 2.0, which supports TPM attestation to provision successfully and authenticate to Azure AD with a device token.

- **Pre-provisioning** is available in Windows 10, 1903, and later, and allows the IT department or vendor to pre-provision a device by installing targeted policies, settings, and applications before logging in as the user. Formerly called **white glove**, pre-provisioned mode gets the device into a business-ready state and helps minimize the wait time for application and policy payloads to be downloaded and installed after the user logs in for the first time. The requirements for pre-provisioning mode are similar to the ones for self-deploying mode and require a TPM 2.0 with attestation support and an ethernet connection to complete the setup.

- **Windows Autopilot Reset** can be used to reset the device and bring it back to a business-ready state by removing user-specific data and reapplying original device settings. This mode can be used to transfer assets to other associates and doesn't require IT to clear the device records in Azure AD and Intune or re-image the device. The Azure AD device identity and connection with Intune remains active and the next user that logs in will become the primary user.

> **Tip**
> Autopilot reset can be started locally by configuring a local policy and remotely through the Microsoft Endpoint Manager admin center.

Next, let's learn how to register a device with the Windows Autopilot service.

Registering devices with the Autopilot service

The first step to using Windows Autopilot is to register the device in your tenant by exporting the unique device identity, known as the **hardware hash**, and uploading it to the service. Once it's registered in your Azure AD tenant, an Autopilot profile is assigned to the device and a profile is assigned that drives the OOBE by pre-selecting options for the user. Registering the hardware to the Windows Autopilot service helps secure the device by creating an ownership association with your organization. This helps protect against theft and resale if the PC is stolen and wiped and comes back online. The registration process can be done manually, by an OEM, through a reseller or partner, or automatically for existing devices.

The Windows Autopilot cycle happens in three steps:

1. Devices are registered to your organization by the IT department or through an OEM or vendor.

2. Device profiles or Autopilot profiles are created and assigned to the devices.

3. The device is powered on, connected to a network, and checks the Windows Autopilot deployment service for registration.

Collecting the hardware hash can be done automatically with Configuration Manager or by using the `Get-WindowsAutoPilotInfo.ps1` script, which is located in the PowerShell gallery at `https://www.powershellgallery.com/packages/Get-WindowsAutoPilotInfo/3.5`.

After running the script locally on a device, a CSV file is generated that contains details about the manufacturer, model, and device serial number. The file can be imported manually into MEM or uploaded automatically through the scripted process using the online parameter and logging in with an account that has the appropriate permissions to import a device. This script can be used as a step during a task sequence installation to automate how the device hardware hash is collected and uploaded. The following screenshot shows an example of the CSV file that will be generated by the script:

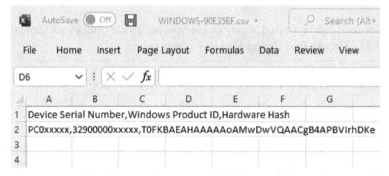

Figure 7.9 – Device hardware hash

To manually upload the hash into Intune, log in to Microsoft Endpoint Manager at `https://endpoint.microsoft.com` and choose **Devices**. Then, select **Windows**, select **Windows enrollment**, and click **Devices** under **Windows Autopilot Deployment Program**. Here, you can view your existing devices or import new ones. For more information on registering devices in the Windows Autopilot service, go to `https://docs.microsoft.com/en-us/mem/autopilot/registration-overview`.

Next, let's look at configuring an Autopilot profile to automatically provision a Windows device from OOBE.

Configuring an Autopilot profile

To configure an Autopilot profile and apply it to a device, a profile needs to be assigned to a device group, and the client needs to be added as a member after it's been registered. This membership assignment can be done manually or dynamically by using a dynamic group membership expression. After the device has been added, Intune will periodically evaluate the group memberships and apply any assigned profiles.

> **Tip**
> Autopilot registered devices have a **Zero Trust Device ID (ZTDID)** value assigned to them that can be used to dynamically add devices to a device group. You can also use the `GroupTag` operator in the script and assign a value for even greater flexibility.

For more information on creating a dynamic device group, go to `https://docs.microsoft.com/en-us/mem/autopilot/enrollment-autopilot`.

To create the deployment profile, go to the MEM admin center and choose **Devices**, select **Windows**, choose **Windows Enrollment**, and click **Deployment profiles**. Select the **Create profile** dropdown and choose **Windows PC**.

Give the profile a friendly name and description. Select **Yes** regarding **Convert all targeted devices to Autopilot** if you wish to automatically register any corporate-owned devices in the assigned device collection in the Autopilot service. Click **Next** to configure the OOBE settings. In the following screenshot, we have configured the OOBE experience to use user-driven mode, chosen the standard user account type, and selected **Yes** for **Allow White Glove OOBE**. Choosing **Allow White Glove OOBE** will add support for the pre-provisioning deployment mode:

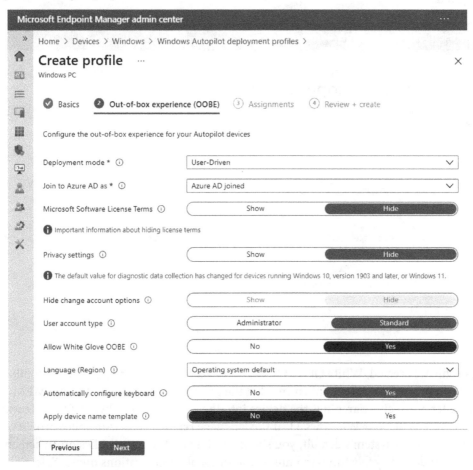

Figure 7.10 – Autopilot deployment profile in Intune

A welcomed security feature of Windows Autopilot is the ability to allow a standard user to enroll in Azure Active Directory and Intune. Without it, the user must be a local device administrator.

After the OOBE settings have been selected, click **Next** to assign the profile to a device group. Once the profile has been assigned, you can view the status of the device and what profile was assigned to it through Windows Autopilot devices in MEM. It's important to note that the Autopilot profile must be assigned before the device comes online during OOBE for it to apply correctly. Otherwise, Windows will download a blank profile and none of your configurations will apply. From the MEM admin center, go to **Devices**, choose **Windows**, select **Windows enrollment**, and click **Devices** under **Windows Autopilot Deployment Program**. Search for the device by serial number to view the profile assignment status. In the following screenshot, the **Profile status** column shows as **Assigned**:

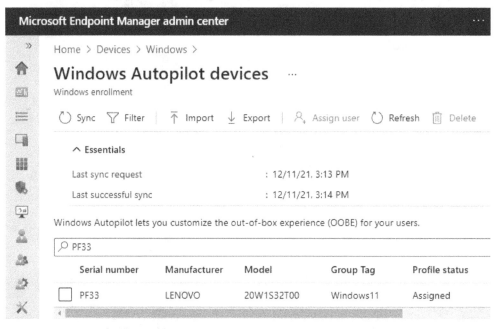

Figure 7.11 – Profile status

In the profile we created, White Glove OOBE was selected, enabling Windows Autopilot pre-provisioning mode. After confirming that the profile has been applied in the MEM admin center, the technician plugs the device into the power and hardwires Ethernet with internet connectivity. For Windows 11, if the language region on the Autopilot profile was left as the operating system's default, you should end up on the Azure AD sign-in screen. Press the Windows key five times to enter the additional configurations menu. Select **Windows Autopilot provision** and click **Continue** to begin the setup process:

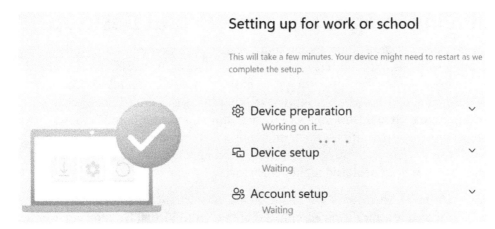

Figure 7.12 – Windows Autopilot pre-provisioning mode

With both pre-provisioning and the self-deploying mode, Autopilot relies on TPM attestation to confirm that the device is trusted by validating it against the device that's been registered with the Autopilot service in Azure AD. If this validation is successful, a device token will be issued, and the client can join Azure AD without the need to enter credentials.

For user-driven scenarios, you can configure an **Enrollment Status Page** (ESP) that will display the configuration process to the user, as shown in the preceding screenshot. This will allow you to set requirements to ensure the device is at a business-ready state before the user can access the desktop. A few options for the ESP include requiring the installation of specific applications, allowing users to collect troubleshooting logs if the enrollment fails, and actions to allow or block a user from continuing to the desktop if there are issues during the configuration process. To learn more about the ESP, go to `https://docs.microsoft.com/en-us/mem/autopilot/enrollment-status`.

For more information about configuring Windows Autopilot, go to the official documentation at `https://docs.microsoft.com/en-us/mem/autopilot/windows-autopilot`.

Next, let's look at deploying images for virtualized desktops running in Azure using the Azure Virtual Desktop service.

Deploying images to Azure Virtual Desktop

With the decreased costs of cloud computing, the adoption of remote work, and collaborating with business partners or vendors from all over the world, the virtualized desktop is a great solution for supporting your business needs. In *Chapter 3, Hardware and Virtualization*, we provided an overview of the features that are available when you're using **Azure Virtual Desktop** (**AVD**), including support for both Windows 11 multi-session persistent and non-persistent hosts. Let's review a few concepts for managing hosts in Azure and ways to build custom images that can be used for virtual machines that have been deployed to the AVD service.

There are many reference materials available on the internet, including dedicated groups and forums focused specifically on administering the Azure Virtual Desktop environment. If you're using the AVD service in your environment, you're strongly encouraged to join these groups as the technology is rapidly changing and can be highly complex. Staying up to date could offer opportunities to not only keep the environment secure but improve inefficiencies, provide a better experience for users, and ultimately save you money.

Managing hosts in AVD

The AVD administrative console can be found by logging into the Azure portal and searching for AVD. The structure of the AVD environment consists of the following resource types:

- Host pools
- Application groups
- Workspaces

Host pools contain the virtual machines that are deployed to make up the backend session hosts that a user connects to. When you're building your image, each virtual machine in the host pool should have the same images to maintain consistency. Host pools can contain multi-session hosts for pooled connections that have been load balanced across the hosts, or personal hosts that are dedicated and assigned to an individual user.

Application Groups contain the grouping of applications that are installed on the virtual machines in the host pools. The two main types are **RemoteApp** and **Desktop**. A RemoteApp group is used to publish applications individually for users to run instead of connecting to the full desktop experience. The Desktop app group grants access to the full desktop experience.

Workspaces contain collections of application groups (RemoteApp and Desktop app groups) and organize the resources in the Remote Desktop app that users can choose to connect to.

The following screenshot shows an example of both the RemoteApp and Desktop app groups being assigned to a workspace:

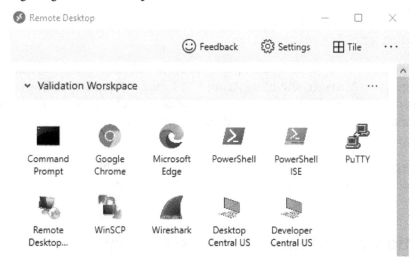

Figure 7.13 – Remote Desktop app for Azure Virtual Desktop

Next, let's look at the approach to building a master or *golden* image in Azure Virtual Desktop.

Building a master image

The master image, or golden image, acts as a base template that consists of the operating system and all the apps and configuration settings you want to use for your session host deployments. Azure has many pre-configured Windows multi-session images available in the marketplace, including images with Microsoft 365 apps already installed. Just like with Hyper-V and other virtual machines, the underlying image is created from a VHD file that is sysprepped, captured, and stored as an image resource in Azure and used for host deployments. Your master VHD image can be maintained directly on Azure VMs or downloaded and customized locally using tools such as Hyper-V or MDT to build customized task sequences. The high-level steps for creating a master image in Azure are as follows:

1. Create the base template from an Azure VM. It is recommended that you start with one of the available multi-session host images from the marketplace.

2. Create a snapshot of the VM in case you need to revert and make any changes.

3. Customize the VM with apps and settings. Apply local security configurations and run security updates.

4. Create another snapshot as a fallback if Sysprep fails.

5. Sysprep the VM.

6. Capture the VM as a managed image or use Azure Compute Gallery to provide replication for highly available images.

There is more than one way to maintain images for AVD and some of these include MDT, Configuration Manager, and using Azure VM Image Builder. Depending on which tool you use, the steps could vary. For more information about building images with Azure VM Image Builder, go to `https://docs.microsoft.com/en-us/azure/virtual-machines/windows/image-builder-virtual-desktop`.

Once the image has been created, it needs to be stored and available to deploy virtual machines. Let's learn how to use Azure Compute Gallery to store highly available images in Azure.

Replication with Azure Compute Gallery

Azure Compute Gallery is a solution in Azure that's used for storing and maintaining images. The gallery acts like a repository that organizes groups of image versions into image definitions. The gallery is highly available, uses premium storage, replicates globally, and allows images to be available for use across subscriptions and Azure AD tenants. The core concept of image management using the gallery consists of grouping images into image definitions. These definitions provide descriptive information about the collection, such as the operating system's type, VM generation, publisher, and SKU. For example, you may have various images for different lines of business where in each, you have different versions that include various flavors of apps or even different security patch levels for compatibility reasons. These versions are stored inside an image definition that can be used for cataloging purposes. Each image version contains the image source, which is typically a managed image, snapshot, or VHD, and virtual machines can be deployed directly from these versions across the replicated Azure regions. This is helpful for disaster recovery planning as only one image needs to be stored and it automatically replicates and is available in other Azure regions in the event of an outage. The following screenshot shows an image definition that contains different versions, the published date, the replicated regions, and the provisioning status:

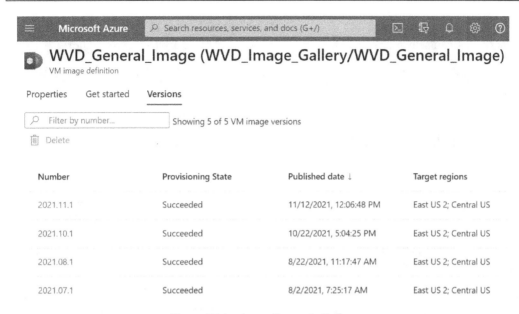

Figure 7.14 – Azure Compute Gallery

Once your image has been created and stored, there are a few methods that can be used for deployment. Let's look at some of the options that are available for deploying images to Azure Virtual Desktop.

Deploying images in Azure

As we mentioned earlier, host pools are the resources that contain a collection of virtual machines that users connect to when accessing VMs on the AVD service. To add virtual machines to an existing host pool, the golden image must contain the Azure Virtual Desktop Agent and a registration key must be generated to securely add the VM to the host pool. The full image deployment process can be automated end to end using the following methods:

- PowerShell using ARM templates
- Using a CI/CD pipeline with Azure DevOps
- The Azure portal through the AVD service
- Using templates in Azure
- Using Azure Compute Gallery and manually registering the VM

The following screenshot shows a session host deployment using the Azure portal. The image that's been selected from the dropdown is one from Azure Compute Gallery:

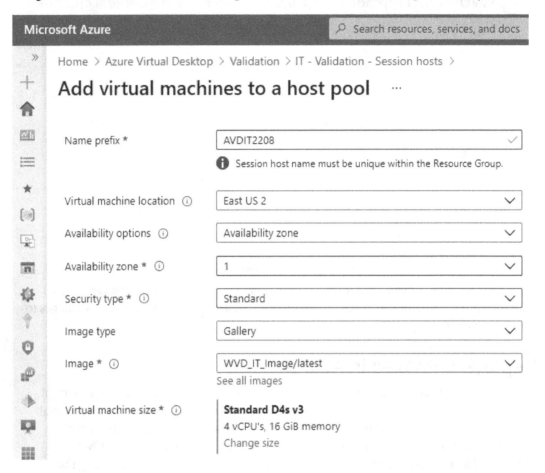

Figure 7.15 – Adding a session host to AVD

When you're deploying resources with ARM templates through PowerShell or Azure DevOps, the image reference is defined by the Azure resource ID of the image source. To get the image source resource ID, open Azure Compute Gallery, select your image definition, choose the target version, and click **Properties**. In the following screenshot, the imageReference parameter has the resource ID listed in the parameters.json file in Azure DevOps:

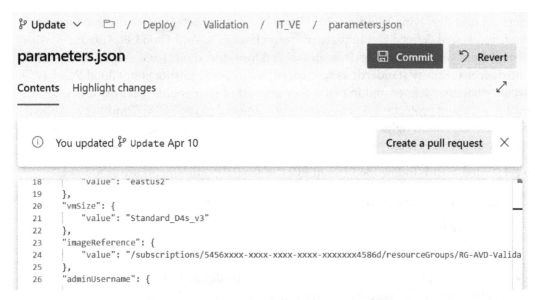

Figure 7.16 – Azure DevOps deployment for AVD

Deploying AVD can be highly complex and require a great understanding of how virtualization technologies work to be efficient. If you're considering deploying AVD in your environment, you are strongly encouraged to read the official documentation carefully to fully understand all the requirements. The official Microsoft documentation can be found at `https://docs.microsoft.com/en-us/azure/virtual-desktop/`.

Next, let's review another offering from Microsoft for streaming Windows in Azure using a simplified deployment model called Windows 365 Cloud PC.

Deploying Windows 365 Cloud PC

With the need to simplify technology for the consumer comes Windows 365 Cloud PC. In *Chapter 3, Hardware and Virtualization*, we introduced Windows Cloud PC with several benefits as to why you would want to adopt the technology. In short, Windows 365 Cloud PC is simple to set up, manage, and consume since it has security built into the service. It is considered a SaaS in which Microsoft manages the platform for you, taking away many of the complexities of managing traditional virtual platforms or physical devices.

Before we start deploying Windows 365 Cloud PC, it is important to ensure your security foundation is well defined and in place for your cloud PCs. Your Cloud PCs are no different from your end user devices, virtual desktops, and servers and should follow the same rigorous security standards as referenced in this book. Ensure your Cloud PCs have a security baseline defined and in place, they are part of your routine security and OS updates, endpoint protection is active, users are required to use MFA, Conditional Access policies are in place to tighten access controls, and compliance policies are enforced to ensure your Cloud PCs are healthy.

To get started with Windows 365 Cloud PC, the general requirements are as follows:

- Access to an Azure tenant with an active subscription.

- An active Intune and Azure AD tenant.

- On-premises Active Directory with Azure AD Connect to sync computer accounts.

- Intune device type enrollment restrictions are configured to allow the Windows MDM platform for corporate enrollment.

- Hybrid Azure AD join is configured.

- A **virtual network** (**VNet**) within Azure that acts as a gateway to the internet that can route to a DNS server for AD DS record resolution and allows direct access to your Domain Controller.

- A license for Windows, Intune, Azure AD, and Windows 365.

You can review the latest detailed requirements here: `https://docs.microsoft.com/en-us/windows-365/enterprise/requirements`.

To get started with Windows 365, you will need to assign the required licenses to your users. Once licensed, the next step is to create your on-premises network connection to allow your Cloud PCs to be provisioned, joined to your domain, and configured with connectivity to your internal network. To create the on-premises network connection, log in to `https://endpoint.microsoft.com/`. Then, browse to **Devices** > **Windows 365** > **On-premises network connection** and click **Create connection**. Here, you can follow the steps to create and complete the on-premises network connection for your Cloud PCs.

Next, we will review the options for deploying images with your Windows 365 service.

Deploying customized or gallery images

When you're deploying your Cloud PCs with Windows 365, you have the option to use the Azure marketplace gallery images. Alternatively, if you prefer, you can use a custom-created image. You can use the same images that were created and replicated with Azure Compute Gallery, as we mentioned earlier in the *Deploying images to Azure Virtual Desktop* section. You need to meet the following requirements, regardless of whether you're using the Azure marketplace or a custom image:

- Windows 10 Enterprise 1909 or later (Excluding 2004) or Windows 11 Enterprise 212H

- A generation 2 virtual machine that has been generalized

- It must be a single-session VM as multi-session VMs are not supported at this time

- A default OS disk size of 64 GB

- Custom images must be replicated to Azure as managed images

You can view and create a marketplace gallery Cloud PC through a provisioning policy, or you can navigate directly to the marketplace in Azure by logging into `https://portal.azure.com`, searching for and accessing **Marketplace**, searching for **Windows 365**, selecting **Windows 365 Enterprise – Cloud PC**, and then clicking **Windows 365 Enterprise – Cloud PC**. From here, you will be provided with all the available Cloud PC images that can be deployed to the service:

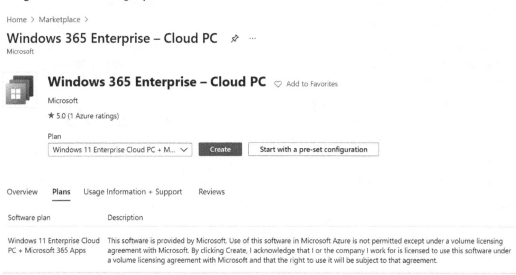

Figure 7.17 – Windows 365 Enterprise Cloud PC marketplace images

The available images include the option of an image with pre-installed Microsoft 365 apps fully optimized for Teams or an image with a vanilla or non-customized Windows installation.

If you would like to use a custom image, you will need to ensure the image meets the requirements provided earlier in this section. To add a custom image within Windows 365, log in to `https://endpoint.microsoft.com/`, browse to **Devices** > **Windows 365** > **Device images**, and click on **Add**. When the image pane appears, enter the image's name, version, and the source of the image from the available custom images. Click **Add** to make the image available.

The following article provides details on creating a custom image within Azure: `https://docs.microsoft.com/en-us/azure/virtual-machines/windows/capture-image-resource`.

Now that the image is ready, the last step is to create a provisioning policy.

Provisioning policies for Cloud PC

Provisioning policies allow you to create Cloud PCs and user assignments. To create a provisioning policy, log in to `https://endpoint.microsoft.com/`, then browse to **Devices** > **Windows 365** > **Provisioning policies**, and click **Create policy**. Then, complete the following steps:

1. On the **General** page, enter your provisional policy's **Name** and **Description** and select your **On-premises network connection**. Then, click **Next**.

2. Within the **Image** page, select whether you would like to use a gallery image or a custom image under **Image type**. Highlight the image you would like to use and click **Next**.

3. Click **Select groups** to add the group you would like to include and apply the provisioning policy on the **Assignments** page. Click **Next** once all the groups have been selected. Only users with a Windows 365 license applied will receive a provisioned PC.

4. On the **Review + create** page, ensure all the information is correct and click **Create**.

5. You will now see the provisioning policy listed within the **Provisioning policies** section. It can take up to an hour for your provisioning policy to finalize and your Cloud PCs to be ready for your users to use. You can also edit or delete your provisioning policy.

Once the provisioning policy is complete and ready, your assigned users will be able to access their Cloud PC. Next, let's review the available options for users to access Cloud PCs.

Accessing Windows 365 Cloud PCs

There are two options available for accessing Cloud PCs:

- Via the web at `http://windows365.microsoft.com`

- Using the **Microsoft Remote Desktop Client (MSRDC)**

When you're accessing your Cloud PCs via the web, simply navigate to `https://windows365.microsoft.com` and follow the instructions to log in. Once you've logged in, you will be able to select your Cloud PC from within your browser. Here, you can also restart, rename, or troubleshoot your Cloud PC within the management console. To access your Cloud PC from a web browser, you will need to meet the following requirements:

- You will need a modern browser such as Microsoft Edge and Google Chrome

- You will need Windows OS, macOS, Chrome OS, or Linux OS

To use the Microsoft Remote Desktop client to access a Cloud PC, you will need to complete the following steps:

> **Tip**
> The Microsoft Remote Desktop client is a new remote desktop client for Windows. **Remote Desktop Connection (MSTSC)** is the client that is built into Windows.

1. Download and install the Remote Desktop client from `https://docs.microsoft.com/en-us/windows-server/remote/remote-desktop-services/clients/remote-desktop-clients`.

2. Once installed, you will need to open the Remote Desktop client and select **Subscribe**.

3. Enter your Azure AD credentials.

4. Double-click to launch your Cloud PC from the Remote Desktop client.

The Remote Desktop client can be downloaded using Windows Desktop, Microsoft Store, iOS, macOS, and Android. To find out more about the Windows 365 Cloud PC service, visit `https://docs.microsoft.com/en-us/windows-365/enterprise/`.

Summary

In this chapter, we provided an overview of Windows device provisioning and how to upgrade Windows. We reviewed the different upgrade paths, including in-place upgrades, migrations, and how to upgrade to Windows 11 using Intune Feature Updates. We also reviewed options for backing up user data and settings. Then, we covered the tools that are available for building hardened images, such as Windows ADK, WCD, MDT, WDS, and MDT with Configuration Manager.

Next, we learned how to provision devices with Windows Autopilot. This included an overview of the different deployment scenarios, how to register devices with the Autopilot service, and how to configure an Autopilot profile. Then, we reviewed managing hosts in Azure Virtual Desktop. This included building a master image, replicating those images with the Azure Compute Gallery, and using different methods to deploy them in Azure. We finished off this chapter by providing an overview of Windows 365 Cloud PC, including how to deploy customized or gallery images, provisioning policies for Cloud PC, and how to access your Cloud PC.

In the next chapter, we will discuss securing your Windows clients. We will cover how to stay updated using Windows Update for Business, enforce policies and configurations with Intune, and enable BitLocker encryption to prevent data theft. We will review how to configure biometric authentication and passwordless sign-in with Windows Hello for Business, how to approach deploying Windows security baselines, and how to configure many of the available Windows security features.

8
Keeping Your Windows Client Secure

In this chapter, you will learn about the current best practices that are used to keep Windows clients secure. So far, we have covered many foundational topics that must be in place to have a robust and well-rounded security program, including administration and management tools, hardware-based security, network security, and the importance of protecting your identity. This chapter will cover specific features, policies, and configurations that will directly secure and harden the endpoints that make up some of your most vulnerable assets.

Securing the Windows client is a critical task, and it requires ongoing operational support to provide the latest available protection for devices as vulnerabilities continue to evolve. We will cover how to keep your Windows devices updated using Windows Update for Business, enforcing policies with Configuration Manager and Intune, enabling encryption with BitLocker, using Windows Hello for Business, enforcing security baselines, and configuring Windows Security features such as Defender Antivirus. Then, we will finish this chapter by setting up the Windows Security experience.

In this chapter, we will cover the following topics:

- Securing your Windows clients
- Staying updated with Windows Update for Business
- Enforcing policies and configurations
- Enabling BitLocker to prevent data theft
- Going passwordless with Windows Hello for Business
- Configuring a device compliance policy
- Deploying Windows Security baselines
- Configuring Windows Security features

Technical requirements

To follow the instructions in this chapter, you will need an Azure AD tenant and rights to manage resources in Azure and Microsoft 365. There will also be references to Group Policy in Active Directory. You will need the following deployments and licensing or equivalent SKUs from Microsoft:

- Microsoft 365 E3, E5, or equivalent
- Configuration Manager Hierarchy
- Microsoft Security and Compliance Toolkit with Policy Analyzer
- Windows 10/11 Pro or Enterprise

Let's start by learning how to secure and service Windows clients.

Securing your Windows clients

As a critical first step, it's important to keep Windows up to date and remain aligned with the modern life cycle policy from Microsoft to retain support. Once a product has been marked as "end of life," Microsoft no longer invests time in releasing features or security updates for these products and they can quickly become vulnerable and out of compliance. Depending on the servicing and licensing model you use, extended support is available in some instances and can be purchased at an additional cost. However, generally, it's a good practice to remain up to date by following the life cycle policy.

According to the modern life cycle policy guidelines, Windows remains in support if the following criteria are met:

- Windows remains current as per the servicing and system requirements that have been published for the product or service.

- Windows must be licensed.

- Microsoft must offer support for the product.

More information about the modern life cycle policy can be found at `https://docs.microsoft.com/en-us/lifecycle/policies/modern/`.

The modern Windows operating system functions as what's referred to as **Windows as a Service** (**WaaS**). Depending on the edition, updates are released using a servicing channel that allows organizations to determine when new features are available. To remain up to date, you can use many of the traditional tools to deploy security updates such as Windows Server Update Services, Configuration Manager, and Windows Update. To help automate this task, you can use Windows Update for Business and configure deployment rings using Intune or Group Policy.

> **Tip**
> Microsoft has recently released the Windows Autopatch service to fully automate Windows and Microsoft 365 app updates. This places the entire patching process in the hands of Microsoft and IT admins can take a *hands off* approach to update deployments. To learn more, visit the official docs at `https://docs.microsoft.com/en-us/windows/deployment/windows-autopatch/`.

Beginning with Windows 11, Microsoft has announced feature upgrades will be reduced to an annual cadence compared to the twice-a-year release schedule for Windows 10. Cumulative updates or quality updates will remain on the standard "patch Tuesday" releases and Microsoft has put in significant effort to reduce the size of these updates moving forward to increase the efficiency of delivery. In the next few sections, we will learn how to manage both feature and quality updates in more detail using **Windows Update for Business** (**WufB**). WufB can significantly help reduce the IT administration overhead of managing update deployments by automating delivery through deferral schedules and simplifying the experience for end users and IT admins alike.

> **Tip**
>
> You can learn more about Windows as a Service at `https://docs.`
> `microsoft.com/en-us/windows/deployment/update/`
> `waas-overview.`

Let's look at the Windows Update for Business model and how to plan deployments and configure update rings.

Staying updated with Windows Update for Business

If you are familiar with traditional deployment methods that use **Windows Server Update Services** (**WSUS**) or Configuration Manager, there are two main differences when it comes to adopting Windows Update for Business as the servicing method. In the **WufB** deployment, clients establish a direct connection to the Windows Update deployment service instead of a software distribution point or WSUS server. Also, updates are not reviewed and approved like they are with WSUS but managed based on "when" clients should receive them by configuring deferral schedules. This is accomplished by creating a Windows Update deployment ring policy. Let's look at a few key concepts around the servicing model for Windows Update for Business as it relates to Windows 11:

- **Feature updates** are now released annually, usually in the second half of the year. These are the major build releases in Windows, and they offer the latest updates in security and UI enhancements.

- **Quality updates** are the traditional monthly cumulative updates that include critical updates, security patches, drivers, updates to the servicing stack (Windows Update Service), and other Windows products.

- **Driver updates** for Windows and third parties can be delivered through Windows Update and toggled to be on or off.

- **Microsoft product updates** are updates that target Windows Store apps and Office products, such as the Camera or Photos apps.

There are two types of release schedules, known as servicing channels, to choose from when you're configuring update policies:

- The **General Availability Channel** or retail channel is the default servicing channel for all Windows devices.
- The **Insider Program** is for pre-release builds and includes options to join the beta or dev channel release programs.

Now that we've covered the main components of the servicing model, let's look at deployment planning using the Windows Update for Business service.

Planning for deployment

Windows Update for Business settings are configured on devices by assigning an update ring profile. The update ring profile includes configurations for setting the servicing channel, allowing Microsoft product and driver updates, setting deferral periods for both feature and quality updates, and configuring deadlines when clients will be required to comply with update installations. The deferral periods specify the time in days before the latest available update is offered to the device from the Windows Update service. The maximum deferral periods that can be configured are as follows:

- The maximum deferral period for feature updates is 365 days.
- The maximum deferral period for quality updates is 30 days.

There is no *best practice* approach when you're planning out your deployment rings, but as a general recommendation, update rings can be configured by department or based on the total number of devices and should include pilot rings for early adopters. This will give the pilot group time to validate and report to the IT team that there are no issues after updates are released. Organizing rings by department may help you troubleshoot issues with an update release as business departments may have unique software needs that could interact with security patches differently. When you're choosing early adopters, there must be equal representation across the different lines of business and app personas that exist in your organization. If an issue is discovered, update rings can be paused for a period of up to 35 days to stop additional devices from receiving the updates until the issue can be resolved.

The following diagram demonstrates how to use update rings to roll out Windows updates. Each ring has incrementally more devices with increased deferral periods and installation deadline requirements:

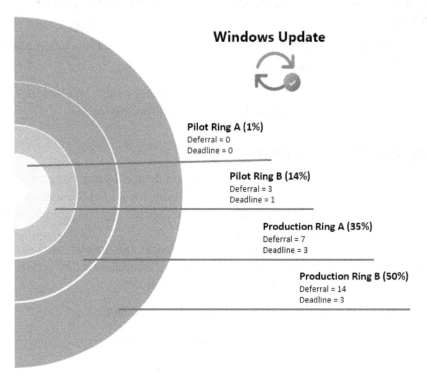

Figure 8.1 – Update ring deployment example

In addition to configuring the deferral periods, setting an installation deadline will help ensure updates get installed in a timely fashion. Deadlines can help drive greater patch compliance by forcing users to action any pending installation, schedule restarts, or automatically reboot to finish an installation during non-active hours. Deadlines can be configured for both feature and quality updates, and a grace period can be set that warns the user of an impending reboot. This behavior and recommended configuration are outlined as follows:

- Set the feature update deadline, in days, to 7 with a grace period of 2. The countdown for Feature Updates begins after the update has been installed and the device is pending a restart.

- Set quality update deadlines to 3 with a grace period of 2. The countdown for quality updates begins as soon as the update is offered to the device, but the grace period countdown doesn't begin until the device is pending a restart.

We can conceptualize what this looks like from a deferral and deadline perspective by looking at the following calendar. In this example, the following settings have been configured on the update ring profile:

- **Quality update deferral period (days)**: 7
- **Deadline for quality updates**: 3
- **Grace period**: 2:

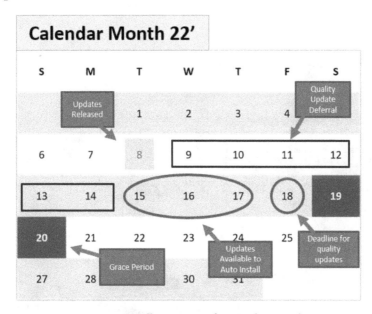

Figure 8.2 – Deadline settings for Windows Update

Now that we have reviewed deferral periods and installation deadlines, let's learn how to configure update rings in Intune.

Configuring update rings for Windows clients

In this section, we will configure a Windows Update for Business ring. We will be configuring a policy for an early adopters pilot group with a zero-day deferral period for feature and quality updates and specify user experience and installation deadline options. Follow these steps:

1. Log into **Microsoft Endpoint Manager** at
   `https://endpoint.microsoft.com`.
2. Click on **Devices** and choose **Update rings for Windows 10 and later** under **Policy**.
3. Click **Create Profile**.

4. Give it a friendly name, such as IT Users - Pilot Ring A, and a description. Click **Next**.

5. Set up **Update settings**, as shown in the following screenshot:

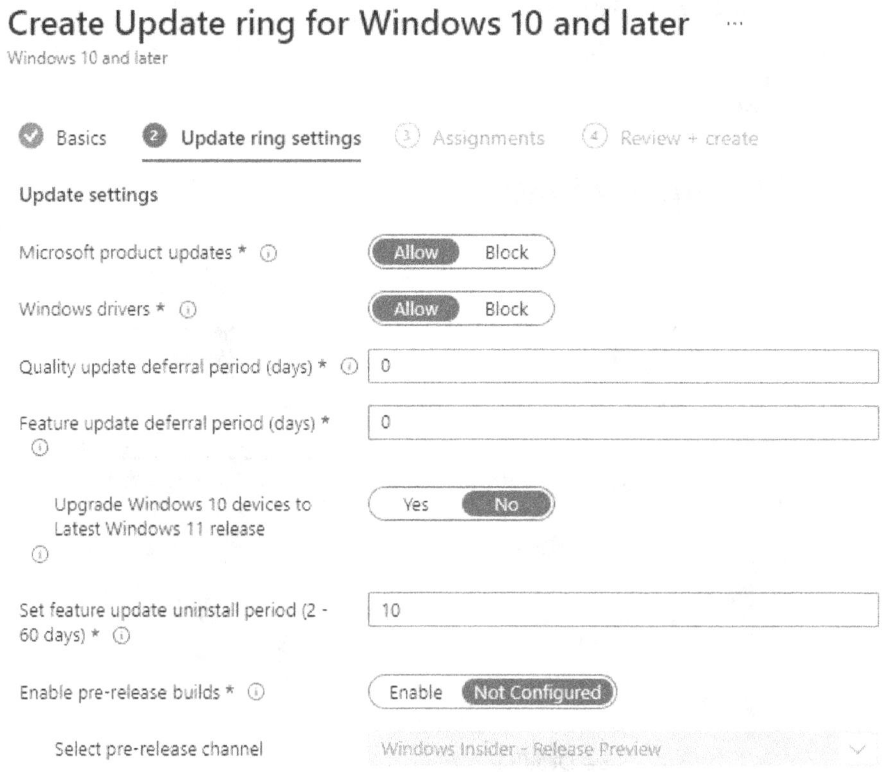

Create Update ring for Windows 10 and later

Windows 10 and later

✓ Basics   ② Update ring settings   ③ Assignments   ④ Review + create

Update settings

Microsoft product updates * ⓘ	(Allow Block)
Windows drivers * ⓘ	(Allow Block)
Quality update deferral period (days) * ⓘ	0
Feature update deferral period (days) * ⓘ	0
Upgrade Windows 10 devices to Latest Windows 11 release ⓘ	(Yes No)
Set feature update uninstall period (2 - 60 days) * ⓘ	10
Enable pre-release builds * ⓘ	(Enable Not Configured)
Select pre-release channel	Windows Insider - Release Preview ⌄

Figure 8.3 – Update settings

> **Tip**
> Managed devices will *not* automatically upgrade to Windows 11 unless you use a feature update deployment policy or change your update ring to allow it.

6. Set up **User experience settings**, as shown in the following screenshot:

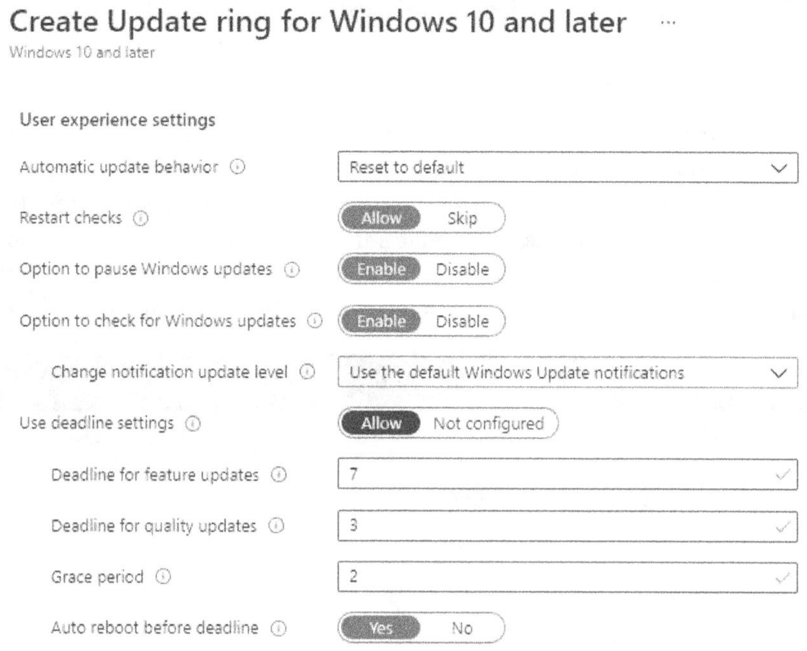

Create Update ring for Windows 10 and later ⋯
Windows 10 and later

User experience settings

Automatic update behavior ⓘ	Reset to default ⌄
Restart checks ⓘ	(**Allow** Skip)
Option to pause Windows updates ⓘ	(**Enable** Disable)
Option to check for Windows updates ⓘ	(**Enable** Disable)
Change notification update level ⓘ	Use the default Windows Update notifications ⌄
Use deadline settings ⓘ	(**Allow** Not configured)
Deadline for feature updates ⓘ	7 ✓
Deadline for quality updates ⓘ	3 ✓
Grace period ⓘ	2 ✓
Auto reboot before deadline ⓘ	(**Yes** No)

Figure 8.4 – User experience settings

Microsoft recommends setting the **Automatic update behavior** option to **Reset to default**. This provides the best option to ensure the success of update installations balanced with end user experience. Using **Reset to default** uses the Windows default update behavior, which uses intelligent active hours based on device usage to auto-install updates and notifies users of reboots. In addition to this update behavior, it is recommended to leave the other user experience settings (restart checks, pause updates, check for updates, and notification update level) as is.

7. Click **Next**. Choose a group to assign the policy to and click **Next**.

8. Click **Review + Create** and choose **Create** to build the policy.

Windows Update policies can be assigned to both user and device groups. It is recommended to assign them to device groups to avoid policy conflicts that can occur from both user and device policy evaluations. This is particularly true if you have shared devices or kiosks where multiple users will sign in. To view the configured Windows Update settings on a Windows client, open **Window Settings**, choose **Windows Updates**, click **Advanced options**, and then choose **Configured update policies**.

Pausing update deployments

Once your update rings have been assigned, updates can be paused if any issues were found during the rollout. Both feature and quality updates can be paused up to 35 days from a specified start date independently from each other.

To view the pause settings, in Microsoft Endpoint Manager, select **Devices**, choose **Update rings for Windows 10 and later**, and select a profile. In the profile overview, click the **Pause** dropdown and choose to pause either feature or quality updates. There are additional dropdowns for uninstalling updates if you need to roll back or if you need to extend the pause period for an additional 35 days. The following screenshot shows an overview of an updating ring profile and the state of both the feature and quality updates:

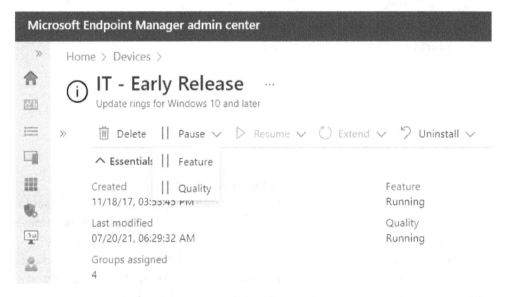

Figure 8.5 – Pause updates

Now that we have reviewed creating an update deployment ring, let's look at managing feature updates and expedited quality updates in Intune.

Managing feature updates and expedited quality updates

Intune has two additional features for managing update deployments for both feature and expedited quality or out-of-band updates. These features allow for more fine-grained control over when major updates are rolled out and in scenarios where an out-of-band security patch must be deployed for clients to immediately be brought into compliance. There are a few prerequisites to using these policies, including the following:

- Devices must be Intune enrolled and Hybrid or Azure AD joined.

- A telemetry policy must be enabled to send Windows Update data.

- Devices must run a supported version of Windows 10/11.

- The **Microsoft Account Sign-In Assistant** (**wlidsvc**) service must be allowed to run and not set to disabled.

> Tip
>
> The required telemetry policy can be deployed by using an Intune **Device Configuration profile**, selecting **Windows 10 and later** as the platform, and selecting the **Templates** profile type. Choose the **Windows Health Monitoring** template name and make sure that you enable Windows Updates under the **Scope** setting.

Now, let's look at configuring a feature updates policy.

Feature updates for Windows 10 and later

To configure a feature update deployment, log in to Microsoft Endpoint Manager, click on **Devices**, choose **Feature updates for Windows 10 and later**, and click **Create Profile**. Give it a name and a description and choose the feature update to deploy from the dropdown. In the following screenshot, we've selected **Windows 11**:

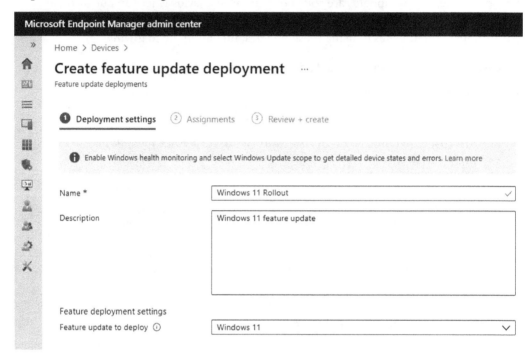

Figure 8.6 – Feature update deployment in Intune

Under the rollout options, you can choose to make them available as soon as possible, or gradually roll the feature update out based on the dates you set in the policy. By making the update available, it will gradually distribute the update to different subsets of devices in the assignment group at different times. This is helpful to reduce the load on the network and for troubleshooting issues during the deployment. For more information about Windows rollout options, go to https://docs.microsoft.com/en-us/mem/intune/protect/windows-update-rollout-options.

> **Tip**
> For Windows 11 deployments, when a device reaches out to the Windows Update Deployment service, it will be evaluated to see if it meets the minimum compatibility requirements to run Windows 11. An additional policy can be enabled from the **Settings Catalog** area to allow diagnostic data to be processed by the deployment service to increase the efficiency of this capability check. To deploy this policy, search for **Allow Wufb Cloud Processing** in the **Settings picker** section of the **Settings Catalog** area and deploy it to your devices.

Click **Next** and choose a group to assign the policy to. Finally, review and create the policy.

It is possible to use both a feature update policy and an update ring policy. The feature update deployment policy will take precedence over the update ring feature deferral setting, and it's recommended to set the feature update deferral to **0** days in the update ring to avoid complications. After a device receives the targeted version from the feature update policy, it will remain at that version until a later version is targeted through future feature update deployments. There is a slight delay before a device checks into Intune to become aware of a new policy, so if you're transitioning feature updates from update rings, be careful not to set the zero-day deferral too soon; otherwise, it could trigger an unexpected installation. You can determine if a device is ready to be transitioned by using the **Windows 10 and later feature updates report** feature and verifying that the **OfferReady** status is shown in the **Update Substate** area, as shown in the following screenshot:

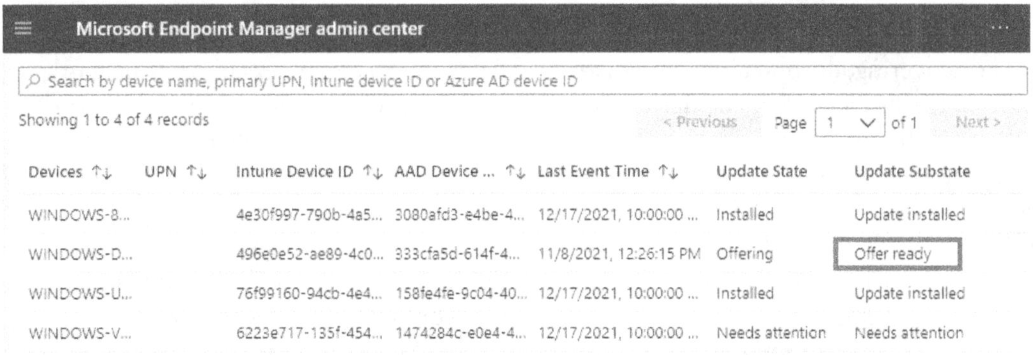

Devices ↑↓	UPN ↑↓	Intune Device ID ↑↓	AAD Device ... ↑↓	Last Event Time ↑↓	Update State	Update Substate
WINDOWS-8...		4e30f997-790b-4a5...	3080afd3-e4be-4...	12/17/2021, 10:00:00 ...	Installed	Update installed
WINDOWS-D...		496e0e52-ae89-4c0...	333cfa5d-614f-4...	11/8/2021, 12:26:15 PM	Offering	Offer ready
WINDOWS-U...		76f99160-94cb-4e4...	158fe4fe-9c04-40...	12/17/2021, 10:00:00 ...	Installed	Update installed
WINDOWS-V...		6223e717-135f-454...	1474284c-e0e4-4...	12/17/2021, 10:00:00 ...	Needs attention	Needs attention

Figure 8.7 – Feature update readiness

To view this report in Microsoft Endpoint Manager, choose **Reports**, click on **Windows updates**, and select **Windows Feature Update Report**.

For more information about transitioning to a feature update policy, go to https://docs.microsoft.com/en-us/mem/intune/protect/windows-update-for-business-configure.

This report is also helpful when you're tracking the rollout status of a Windows feature update.

Quality updates for Windows 10 and later

Configuring an expedited quality update policy can be done similarly to what we described in the preceding section regarding the feature update policy. It is useful for deploying out-of-band security updates as quickly as possible if there is a need to mitigate an immediate security risk for a zero-day threat. Unlike with an update ring policy, expedited quality updates are only used to deploy out-of-band security updates and are not recommended as an alternative to managing regular quality updates. Once a policy has been assigned, devices at a lower build of the targeted patch will immediately start the download and installation process as soon as possible. Any quality update deferral periods specified in an update ring policy will be ignored and restarts will be enforced. This means that there is a chance that a client could restart in the middle of the workday. Read the documentation carefully at `https://docs.microsoft.com/en-us/mem/intune/protect/windows-10-expedite-updates` if you're planning to use expedited quality updates in Intune. There are a few prerequisites to meet to ensure your devices can receive the expedited update successfully.

Next, let's look at the delivery optimization settings in Windows for configuring how clients download and distribute Windows updates.

Using delivery optimization

Delivery optimization (DO) is a cloud-managed service that helps manage settings to handle delivering, downloading, and distributing updates to computers on your network and over the internet. It can be used to help reduce bandwidth consumption and network bottlenecks by configuring the distribution mode and limiting the number of available bandwidth clients that can be used to download updates. The delivery optimization service can be used with WSUS, Windows Update for Business, and expedited quality updates through Intune. DO is also not limited to just security updates but also delivers Windows Store files, Win32 apps from Intune, Windows Defender definition updates, and Microsoft 365 apps and updates. DO settings can be configured natively in Group Policy or Intune using the **Templates** device configuration profile type. Some of the available distribution modes that can be configured in DO are as follows:

- **HTTP only, no peering (0)** uses the internet only.

- **HTTP blended with peering behind the same NAT (1)** uses the internet and other clients on the same NAT IPs.

- **HTTP blended with peering across a private group (2)** uses the internet and peers connected to the same Active Directory domain and NAT IPs.

- **HTTP blended with Internet peering (3)** uses the internet and other computers on your network.

- **Simple download mode with no peering (99)** uses the internet directly from Microsoft and bypasses the DO cloud service.

- **Bypass mode (100)** uses the **Background Intelligent Transfer Service (BITS)** to get updates, not DO.

Other DO settings that can be configured on the client include setting active business hours, limiting the available bandwidth percentage for uploads and downloads, specifying the minimum battery level to upload, and setting the maximum download cache size.

To find a full list of configurable DO settings, go to `https://docs.microsoft.com/en-us/mem/intune/configuration/delivery-optimization-settings`.

In Intune, the Delivery Optimization template settings have plenty of configurations to get you started, but you can always use **Settings Catalog** to configure more fine-grained policies if they don't meet your needs. To view the DO settings that have been configured on a client, open **Windows Settings**, choose **Windows Update**, click **Advanced Options**, and choose **Delivery Optimization** under **Additional Options**.

> Tip
>
> The **Update Compliance** solution in Azure Log Analytics can be used to evaluate DO efficiencies across all your Intune-managed devices.

The following screenshot shows a report on the DO performance for devices that have been onboarded in the **Log Analytics Update Compliance** solution in Azure:

DEVICE CONFIGURATION

CONFIGURATION	DEVICE COUNT
HTTP + Group (2)	0
HTTP + LAN (1)	583
HTTP + Internet (3)	0
HTTP Only (0)	1
Simple (99)	0
Bypass (100)	0
Unknown	8

CONTENT DISTRIBUTION (%)

CONTENT	BW SAV %	DEVICE COUNT (USED P2P)
Feature Updates/Flights	8.5%	28
Quality Updates	0.8%	399
Apps	1.7%	572
Driver Updates	0%	345
Office	2.2%	574
Other	1.3%	578

Figure 8.8 – Delivery optimization performance

As you can see, the configured distribution mode is shown across all your clients, as well as for insights into content distribution across different types of updates. In company networks, even more flexibility is provided for configuring DO, including specifying update groups that clients can be peers with or restricting peers based on subnets to help reduce bandwidth consumption. Combining DO with the **quality of service** (**QoS**) filtering capabilities that are available in most modern-day networking solutions will ensure Windows Updates can still be delivered effectively without slowing down your company's network.

Next, let's learn how Configuration Manager and Intune can be used to enforce policies and configurations. These enterprise-grade solutions are critical to ensuring compliance with your security baseline requirements.

Enforcing policies and configurations

In *Chapter 6*, *Administration and Policy Management*, we covered the differences between domain, hybrid, and Azure AD join for devices. Traditionally, when devices join an Active Directory domain, they connect to a domain controller and receive specific user and computer configurations controlled by Group Policy. This is still applicable in the hybrid Azure AD join model, but the approach to managing policies may begin to change, depending on your administration preferences. When a device becomes fully Azure AD joined, Group Policy is no longer an option, but it opens new opportunities to administer configurations and enforce policies. For companies not starting with fresh configurations or are greenfield, this can present a challenge as many organizations have years' worth of GPOs they rely on to harden their Windows systems and enforce baseline controls. The question becomes how to move and enforce these policies if GPO isn't an option. The answer is to use an MDM solution with unified endpoint management such as Microsoft Endpoint Manager, Intune, or Intune co-management. Unfortunately, for companies moving to full Azure AD join, there is no clear lift-and-shift path, and part of the challenge will involve auditing and evaluating what currently exists and determining the best path forward to apply your baselines.

In this section, we are going to learn about creating and enforcing policies with Configuration Manager and Intune. First, we will walk through creating policies with Configuration Items and assigning them to a Configuration Baseline in Configuration Manager. Then, we will discuss assigning MDM policies using Intune and how to handle policy conflicts between CSPs and GPOs. Toward the end of this section, as a practical example, we will walk through creating a custom policy using an OMA-URI to manage memberships to the local administrator's group using Restricted Groups. Let's start by creating policies for a security baseline in Configuration Manager.

Creating security baselines in Configuration Manager

In *Chapter 6*, *Administration and Policy Management*, we provided an overview of Configuration Items and Configuration Baselines. These two compliance types are the nuts and bolts to building and enforcing security baselines with **Microsoft Endpoint Configuration Manager** (**MECM**). A Configuration Item contains a collection of policies, settings, and scripts that are used for evaluation and policy enforcement, and it is assigned to users and devices through its relationship with a Configuration Baseline. Let's look at this in practice by creating a Configuration Item using the Configuration Manager console. In this example, we will create a single policy for controlling the password manager feature of Google Chrome, which is controlled by setting a registry value. This process can be repeated for any additional settings that will build the collection of policies inside the Configuration Item.

To create a Configuration Item, follow these steps:

1. In the Configuration Manager console, go to **Assets and Compliance**, choose **Compliance Settings**, and select **Configuration Items**. Click the **Create Configuration Item** button in the toolbar.

2. Give it a friendly name, such as `Google Chrome - Security Baseline`. Enter a description and choose **Windows Desktops and Servers (custom)** under the settings for devices that are managed with the Configuration Manager client.

3. Keep all the versions selected under **Supported Platforms**.

4. Choose **New** under **Specify settings for this operating system** and set the following options:

    *   **Name**: `Chrome - PasswordManagerEnabled`
    *   **Description**: `Disable the Google Chrome password manager`
    *   **Setting type**: `Registry Value`
    *   **Data type**: `Integer`
    *   **Hive Name**: `HKEY_LOCAL_MACHINE`
    *   **Key Name**: `Software\Policies\Google\Chrome`
    *   **Value Name**: `PasswordManagerEnabled`

5. Select the **Create the registry value as a REG_DWORD data type if remediated for noncompliant rules** option.

6. Click on the **Compliance Rules** tab and choose **New**. Set the following options in the **Specify rules to define compliance conditions for this setting** menu:

    *   **Name**: `PasswordManagerEnabled - 0 - Dword`

7. In the **Setting must comply with the following rule** section, choose **Equals** from the drop-down list and enter 0.

8. Select the **Remediate noncompliant rules when supported** and **Report noncompliance if this setting instance is not found** options.

9. Choose **Information** for **Noncompliance severity for reports**.

10. Choose **OK** and then **OK** again. Click **Next** to see your compliance rule.

11. Click **Next** to view the summary and create the Configuration Item.

Repeat this process for any additional policies you want to organize within the Google Chrome security baseline. In the following screenshot, multiple policies have been defined in the Configuration Item, as recommended by the **Center for Information Security (CIS)** benchmark recommendations for Google Chrome:

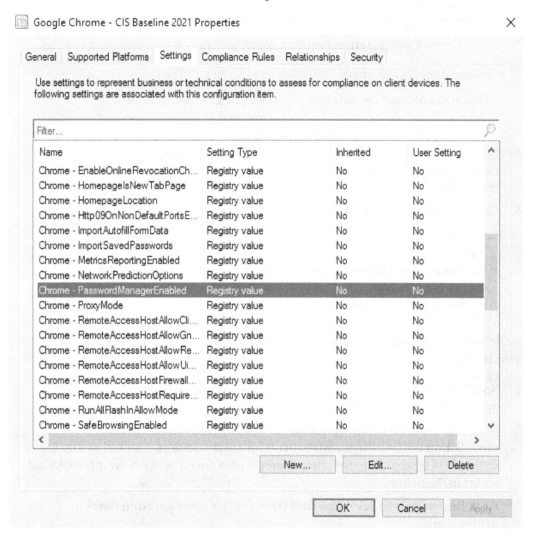

Figure 8.9 – Configuration Item settings

Once the Configuration Item has been created, the next step is to add it to a Configuration Baseline that can be deployed to a collection of devices. To do this, follow these steps:

1. In the Configuration Manager console, go to **Assets and Compliance**, then **Compliance Settings**, and choose **Configuration Baselines**. Click **Create Configuration Baseline** from the top toolbar.

2. In the **Create Configuration Baseline** dialog box, provide the following settings:

    -   **Name**: `Google Chrome CIS Security Baseline`

    -   **Description**: `Google Chrome CIS Security Baseline`

3. Click **Add** and choose **Configuration Item**.

4. In the **Add Configuration Items** dialog box, find the **Google Chrome – Security Baseline** Configuration Item that we created earlier. Select it and click **Add**. Then, select **OK**.

5. Click **OK** to commit these changes.

> Tip
>
> The **Always apply this baseline even for co-managed clients** option will force the baseline to be applied to devices whose workloads are managed in Intune in co-management scenarios.

Next, let's look at deploying this policy to a collection of devices. Follow these steps to deploy the baseline:

1. Select **Google Chrome CIS Security Baseline** and click **Deploy** in the toolbar to open the deployment menu.

2. Select the **Remediate noncompliant rules when supported** and **Allow remediation outside the maintenance window** checkboxes. This will ensure the settings will be applied to the client if they are not found or are different from what we set in the policy.

3. Click **Browse** next to **Select the collection for this Configuration Baseline deployment**.

4. Click on the dropdown and choose **Device collections**. Then, select the collection you wish to target.

5. Under **Schedule**, keep **Simple schedule** selected and change the **Run every** option to **1** day. Click **OK**.

Tip

For mission-critical baselines, choosing between 1 and 3 days is a good middle-ground for policy evaluations. Keep in mind that the shorter the time interval, the more network traffic and strain will be placed on the site server and its services.

The following screenshot shows the menu options that are available when you're scheduling a deployment:

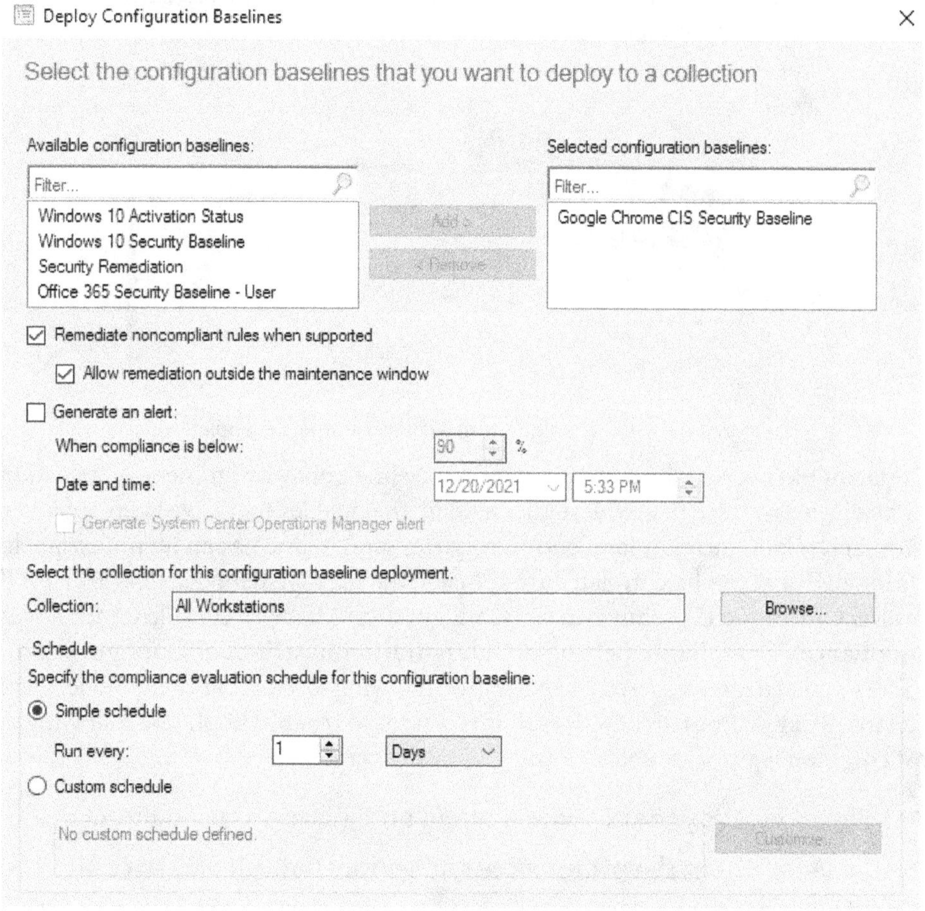

Figure 8.10 – The Deploy Configuration Baselines menu

Once the baselines have been deployed, the client will be notified of a new policy assignment during the next **Machine Policy Retrieval & Evaluation Cycle**. This policy cycle can be triggered manually on the client by opening the **Configuration Manager** applet in **Control Panel**, clicking the **Actions** tab, selecting **Machine Policy Retrieval & Evaluation Cycle**, and clicking **Run Now**. You can also click **Sync Policy** from **Software Center** under **Options | Computer maintenance**. In the **Configuration** tab, you can view the assigned baselines and their compliance statuses. If there are conflicts, choose **View Report** to see an HTML output of the conflicting settings. The following screenshot shows the compliance state of the Google Chrome baseline we deployed earlier from the **Configurations** tab in the applet:

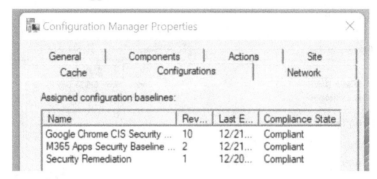

Figure 8.11 – The Configuration Manager Properties applet

Once the client has been notified about the new baseline policy assignment, a collection of logs can be viewed to trace the request and validate the policies that have been applied to the device. Each Configuration Item and Configuration Baseline has an identifier labeled as **CI Unique ID** that can be correlated from inside the management console. To view this ID, show the **CI Unique ID** column in the Configuration Manager console under **Assets and Compliance | Compliance Settings | Configuration Baselines** or **Configuration Items**. Then, you can match these IDs inside the following logs on the client device, which is located in `%SystemRoot%\CCM\Logs` if you need to troubleshoot. The following are the most common logs for troubleshooting policy delivery:

- `PolicyAgentProvider.log` will record policy changes on the client.

- `PolicyAgent.log` shows the requests for policies through the Data Transfer Service.

- `PolicyEvaluator.log` shows the details about the evaluation of the policies compared to the existing device settings.

Configuration Manager is a robust and extremely powerful tool with a lot of flexibility for deploying security baselines to your Windows devices instead of Group Policy. The settings can be as simple as a registry key or creating complex evaluation and remediation scripts using VBScript or PowerShell. You can also automate additional actions based on the client's compliant status, such as adding the non-compliant device to a collection and targeting apps and policies based on your evaluation rules.

Finally, most policy settings for Google Chrome will exist in the `HKLM:\SOFTWARE\Policies\Software\Google\Chrome` path in the registry of the client. Just like with Group Policy, anything in the `HKLM:\SOFTWARE\Policies` node will take effect after the next Group Policy update cycle on the client. Let's confirm that the password manager is disabled by opening Google Chrome and typing `chrome://settings/passwords` in the address bar. In the following screenshot, the **Offer to save passwords** setting has been disabled. There's an icon next to it that indicates that this setting is managed by the company policy:

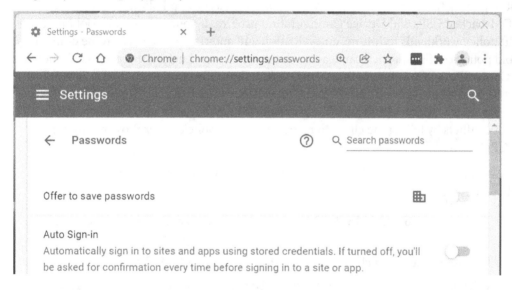

Figure 8.12 – Chrome Password Manager

Now that we have discussed deploying a Configuration Baseline with MECM, let's look at deploying policies using MDM in Intune.

Deploying MDM policies in Intune

In *Chapter 6*, *Administration and Policy Management*, we provided an overview of the available methods to apply policies to Intune-managed devices using **configuration service providers** (**CSPs**). Over the past few years, the Intune team has done a great job of mapping CSP-backed policies, which were traditionally handled by the Group Policy to Windows settings recommended by CIS and Microsoft Security Baselines. Most settings are now available through the Intune UI via the Settings Catalog, Endpoint Security, or Templates profile types. In *Chapter 6*, *Administration and Policy Management*, we also discussed the recommended approach to layer policies using Intune to create the lightest policy payload possible to be processed by the device to increase efficiencies. To recap, when configuring a policy, it's best to check if the setting exists in Security Baselines; otherwise, it can be supplemented using the Settings Catalog or Templates profile type. If the setting is not available in the UI, a custom policy can be created using the CSP's OMA-URI. If your company is starting fresh with Intune management and not migrating from an existing policy workload such as Group Policy, then you must use MDM-backed CSPs and Device Configuration profiles. If your company is looking to shift policy workloads to Intune, an evaluation and auditing phase needs to be done to avoid policy conflicts, which can lead to client-side issues. Luckily, there is a CSP known as **ControlPolicyConflicts/MDMWinsOverGP** that allows you to control which policy will be used whenever both MDM and Group Policy have been configured for conflicting settings. This CSP can significantly help expedite the policy shift into Intune and help avoid conflicts by telling the client to prioritize Intune policies over those configured by GPO.

We can configure the control policy conflicts of CSP by using the Intune Settings Catalog. Follow these steps:

1. Log into **Microsoft Endpoint Manager** at `https://endpoint.microsoft.com`.

2. Choose **Devices** and click **Configuration Profiles** under **Policy**.

3. Click **Create Profile**, choose **Windows 10 and later** as your platform, select **Settings Catalog** as the profile type, and click **Create**.

4. Give the profile a name, such as `Windows Control Policy Conflicts`, and click **Next**.

5. Under **Configuration Settings**, click **Add Settings** to open the **Settings picker** area.

6. Type `control policy conflict` in the search bar and click **Search**.

7. Select the **Control Policy Conflict** result to show the **MDM Wins Over GP** setting and click **Select all these settings** to add them to the profile.

8. Close the **Settings picker** area.

9. Click the dropdown and choose **The MDM policy is used and the GP policy is blocked**. Then, click **Next**.

10. Add a user or device group under **Included groups** to create the assignment and click **Next**.

11. Click **Next** under **Scope tags** and click **Create** under **Review + Create** to finish creating the policy.

Intune-managed devices typically check into the Intune service every 8 hours, but this can be expedited either through the MEM admin center remote device actions or manually on the device. To run this on a client, open **Settings**, choose **Accounts**, and then **Access work or school**. Click the dropdown on the desired work or school account and click the **Info** button. Under **Device sync status**, choose **Sync**.

> **Tip**
> A sync can also be initiated through the Intune Company Portal if the app is deployed to devices, restarting the Intune Management Extension service, or rebooting the PC.

To view the status of the profile in Intune, open the Device Configuration profile overview pane. The following screenshot shows the device and user check-in status for a Settings Catalog profile:

Figure 8.13 – Device Configuration profile status

Click **Per setting status** to view the status of each setting as it's applied across users and devices. To validate the settings that have been applied on the client's PC, look at the policies that have been applied through the **Access Work or School** information pane we navigated to earlier to run a device sync. In the following screenshot, you can see that the **ControlPolicyConflict** policy has been applied, along with other CSP-based policies that have been assigned by Intune:

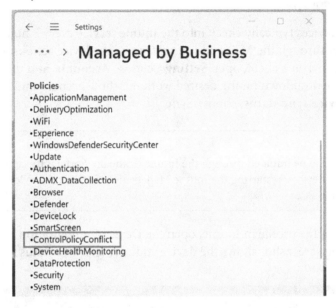

Figure 8.14 – Access Work or School policies

Let's dig a little deeper and do a registry analysis to understand where the CSP settings have been applied. Open the registry editor, navigate to the HKLM:\SOFTWARE\ Microsoft\PolicyManager\current\device path, and find the following information to confirm that the setting has been configured correctly:

- **Path**: HKLM:\SOFTWARE\Microsoft\PolicyManager\current\device
- **Key**: ControlPolicyConflict
- **Setting**: MDMWinsOverGP

- **Value**: 1:

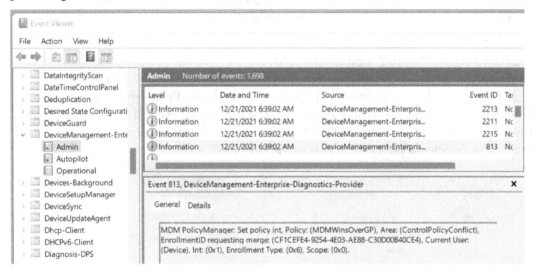

Figure 8.15 – ControlPolicyConflict in the registry

Next, let's review the event logs on the client by opening **Event Viewer**. Navigate to **Applications and Services Logs | Microsoft | Windows | DeviceManagement-Enterprise-Diagnostics-Provider** and open the **Admin** logs. Look for event ID 813. In the following screenshot, event ID 813 shows that the MDM policy manager configured the **MDMWinsOverGP** policy in the device context to an integer value of 1, as shown in the registry's DWORD type. This was enforced through the policy provider's GUID identifier, which correlates with **MDMWinsOverGP_WinningProvider**, as shown in the preceding screenshot:

Figure 8.16 – The DeviceManagement-Enterprise-Diagnostics-Provider log

Now that we have walked through how to deploy the ControlPolicyConflicts CSP using Settings Catalog, let's learn how this CSP works when dealing with group policy conflicts.

Controlling policy conflicts with MDM

By default, in Windows, Group Policy will win over any MDM CSP settings unless we configure the ControlPolicyConflicts CSP mentioned in the previous section. Setting this policy will ensure that any CSP that's mapped to the PolicyCSP node will win over the corresponding Group Policy setting. We just covered how to deploy this policy using the UI through Settings Catalog, but for reference, let's look at the OMA-URI. For other policies, if the option to configure the CSP doesn't exist in the Intune UI, a custom profile type can be used to configure it through the OMA-URI path and set the appropriate data type and value:

- **OMA-URI**: `./Device/Vendor/MSFT/Policy/Config/ControlPolicyConflict/MDMWinsOverGP`

- **Data type**: Integer

- **Value**: 1

To determine which MDM policies will win over Group Policy, the OMA-DM **device description framework (DDF)** contains tags that are used to identify the corresponding Group Policy. A reference to the list of policies in PolicyCSP that are supported by Group Policy can be found at `https://docs.microsoft.com/en-us/windows/client-management/mdm/policies-in-policy-csp-supported-by-group-policy`.

After enabling the ControlPolicyConflicts CSP, if the equivalent Group Policy has been configured on the device, a blocking record will be created to ensure that the CSP will take precedence. This can be seen in the **DeviceManagement-Enterprise-Diagnostics-Provider | Admin** logs we mentioned earlier while looking for event IDs 2211 and 2213. The following screenshot shows event ID 2211 creating a blocking record for a Windows update feature deferral policy that exists in the PolicyCSP node:

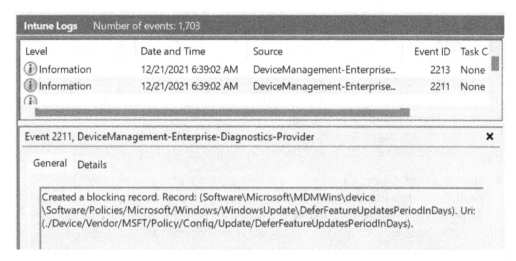

Figure 8.17 – MDM blocking record

Intune has a reporting and diagnostic tool built into Windows that can generate logs to provide a report of blocked Group Policies based on conflicting CSPs. To run the report on a client, open **Settings**, choose **Accounts**, and then **Access work or school**. Click the dropdown for the desired work or school account and click the **Info** button. Scroll to the bottom of the page and choose **Create report** under **Advanced Diagnostics Report**. The output will store an HTML file located in **%SystemDrive%\Users\Public\Documents\ MDMDiagnostics**. Open the **MDMDiagReport.html** file and scroll down to the **Blocked Group Policies** section, as shown in the following screenshot:

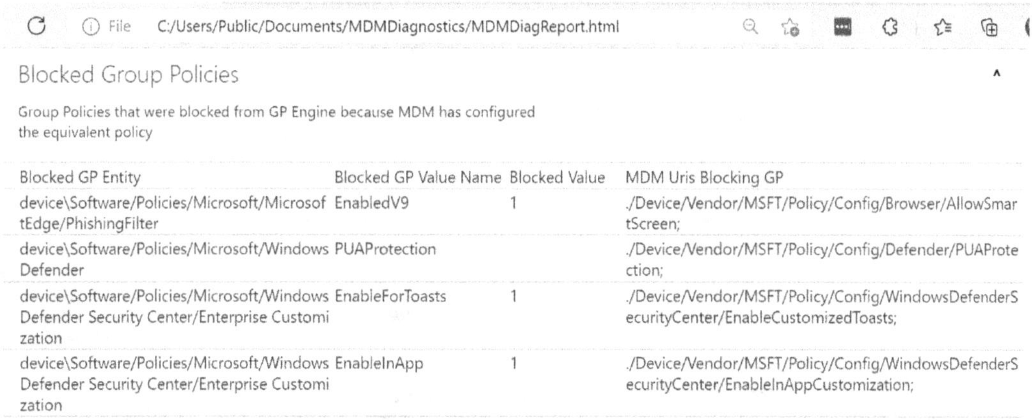

Figure 8.18 – MDM diagnostics report

Next, let's look at an example of using the custom profile type to control a local administrator's group memberships on Windows through Restricted Groups.

Managing Azure AD local device administrators

In *Chapter 5*, *Identity and Access Management*, we discussed the inherent security risk with users retaining local administrative rights on to their computers. We also discussed the risk of having inactive accounts remaining on the system as they can become an attack target that's used for privilege escalation and lateral movement through your environment. When a device becomes fully Azure Active Directory joined, group membership to the local administrator's group is no longer managed by Active Directory Group Policy and any on-premises password rotation solutions may no longer work.

By default, a standard Azure AD user account can register a device in Azure AD, but local administrative privileges are required to enroll an already configured device into Intune. To avoid assigning administrative privileges, users will need to enroll in Intune by using a new device with an assigned Autopilot profile or one that was enrolled through the bulk enrollment method. If an Autopilot profile isn't applied correctly or the **User account type** profile setting isn't set to **Standard**, then any user that completes the enrollment process will become a member of the Administrators group. Assuming our policies and security controls have been applied correctly, let's look at the default members of the **Administrators** group on an Azure AD joined device, as shown in the following screenshot:

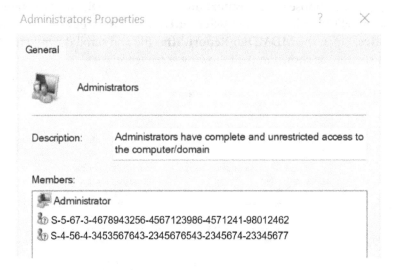

Figure 8.19 – The Administrators group in Windows 11

In this preceding screenshot, we can see two **Security Identifiers** (**SIDs**) that are added after the device joins Azure AD. One is for the **Global Administrators** Azure AD role, while the other is for the **Azure AD Joined Device Local Administrator** role. Any users that have been assigned those roles in Azure will have local admin permissions on all Azure AD joined devices. These SIDs are unique to your tenant and consistent across all the devices in your organization. It is possible to manually elevate a user to the Administrators group on a specific device by running the following command in an administrative command prompt:

```
net localgroup administrators AzureAD\username@company.com /add
```

The `AzureAD` prefix to the username in the preceding command is used across all the tenants for Azure AD users.

The challenge with Azure AD joined devices is that there is no graceful way to manage local administrator access to just a single client. Adding a user to these groups extends the range of their elevated access beyond just their device. The two main methods for managing memberships and governing access to local groups using Azure AD and Intune are as follows:

- Use the **Azure AD Privileged Identity Management** policy and allow users to request access to the **Azure AD Joined Device Local Administrator** role through eligible assignments. Configure an approval workflow with a short expiration time. Users will become administrators to all Azure AD joined devices using this method. Once a user has been removed from an assignment, it may take up to 4 hours before those rights are revoked.

- Use the **Restricted Groups** or **Local Users and Groups** policy to manage memberships to the Administrators group through an OMA-URI to ensure no unapproved members are added.

When a Restricted Groups policy is applied, any current member that is not on the member's list will be removed. Additionally, any user that is included in the member's list will be added. Enforcing this policy is a good method of governing rogue memberships if users are added manually or may have remained an admin during Intune enrollment. We can create a Restricted Groups policy using a custom profile type in Intune by following these steps:

1. Log into **Microsoft Endpoint Manager** at `https://endpoint.microsoft.com`.

2. Choose **Devices** and click **Configuration Profiles** under **Policy**.

3. Click **Create Profile**, choose **Windows 10 and later** as the platform, select **Templates** as the profile type, choose **Custom** under **Template name**, and click **Create**.

4. Give the profile a name, such as **Windows Restricted Groups Policy**, and click **Next**.

5. Under **OMA-URI Settings**, click **Add** and enter the following information:

   - **Name**: `Administrators`

   - **Description**: `PolicyCSP - RestrictedGroups`

   - **OMA-URI**: `./Vendor/MSFT/Policy/Config/RestrictedGroups/ConfigureGroupMembership`

   - **Data type**: `String (XML file)`

6. Under **Custom XML,** you will need to upload a file that contains the parameters to apply, as shown in the following code. Remember, each SID will be different, depending on your tenant:

```
<groupmembership>
    <accessgroup desc = "Administrators">
        <member name = "Administrator"/>
        <member name = "S-1-12-1-5498761259-1234456789-
1212684579-2455789161"/>
        <member name = "S-1-12-1-4193365479-1304569875-
3141135495-4254897561"/>
    </accessgroup>
</groupmembership>
```

> **Tip**
> Ensure that you include the default Administrator, Device administrator, and Global administrator SIDs as members in the access group to avoid errors.

7. Click **Save** and then **Next**. Add any group assignments you wish to target.

8. Click **Next** under **Applicability Rules**. Finally, click **Create**.

> **Tip**
>
> You can review more about this CSP and copy the formatting for the XML file by going to `https://docs.microsoft.com/en-us/windows/client-management/mdm/policy-csp-restrictedgroups`.

Let's verify this on the local client device by running an Intune Sync. Then, open the event viewer and navigate to **Applications and Services Logs | Microsoft | Windows | DeviceManagement-Enterprise-Diagnostics-Provider**. Open the **Admin** log and look for event ID 814, as shown in the following screenshot:

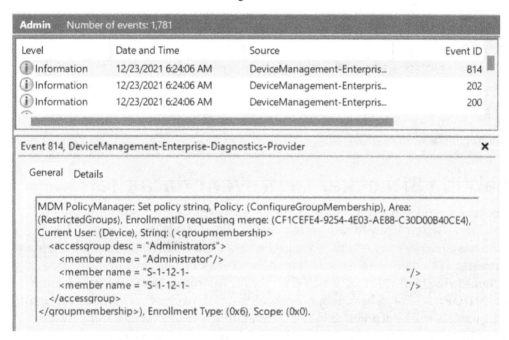

Figure 8.20 – The Restricted Groups policy in Event Viewer

After confirming that the policy has been applied, open the **Local Users and Groups** management console by clicking **Run** and typing `lusrmgr.msc`. Then, click **Groups** and open **Administrators**. The membership should match what was specified in the XML that was sent to the device. Setting this policy is the equivalent of configuring Restricted Groups using Group Policy, which can be found in the GPMC under **Computer Configuration | Windows Settings | Security Settings | Restricted Groups**.

To avoid complications with enforcing Restricted Groups, Microsoft recommends using the Local Users and Groups policy instead for Windows 10 versions 20H2 and above. This will allow you to add users incrementally and avoid removing any existing members by accident. You can deploy a Local Users and Groups policy using the custom profile type or by going to **Endpoint Security | Account Protection** and choosing the **Local user group memberships** profile type under the **Windows 10 and later** platform. You should not deploy both a Restricted Groups and Local Users and Groups policy to the same device. More information about this CSP can be found at `https://docs.microsoft.com/en-us/windows/client-management/mdm/policy-csp-localusersandgroups`.

> **Tip**
> Local user group memberships can also be configured using a built-in template in Endpoint Security.

Now, let's learn how to configure a BitLocker encryption policy for an Intune-managed device.

Enabling BitLocker to prevent data theft

BitLocker encryption is a common technology that's used for encrypting data on disk drives. It is an effective way to help protect data if a device is stolen or the hard drive is removed as a recovery key is required to gain access. Historically, BitLocker was deployed and managed through Group Policy using the **Microsoft BitLocker Administration and Monitoring (MBAM)** tool, which is part of the **Microsoft Desktop Optimization Pack (MDOP)**. Microsoft has announced that MBAM development ended in 2019 and that its services will be deprecated in 2024. They strongly recommended using Azure AD and Intune to deploy and manage BitLocker drive encryption as soon as possible. If you're planning a deployment of BitLocker, visit the official Microsoft docs to read the deployment guide: `https://docs.microsoft.com/en-us/windows/security/information-protection/bitlocker/bitlocker-basic-deployment`.

So far, we've covered deploying Intune policies using Settings Catalog and the Custom profile type. Now, let's learn how to configure a BitLocker policy by using the Endpoint Protection template in an Intune Device Configuration profile.

Configuring BitLocker with Intune

The best way to configure BitLocker in Intune is by using Endpoint Security or an Endpoint Protection template in the Templates profile type. Both methods are acceptable, but the Endpoint Protection template does provide the option to silently enable BitLocker to avoid end user notifications. Let's look at creating a Windows encryption profile using the Endpoint Protection template. In this example, we will configure the standard BitLocker settings for OS drive encryption, allow silent enablement of BitLocker encryption, and configure the recovery key information to be stored in Azure AD:

1. Sign into **Microsoft Endpoint Manager** at `https://endpoint.microsoft.com`.

2. Go to **Devices** and choose **Configuration Profiles** under **Policy**.

3. Create a profile and choose **Windows 10 and later** from the **Platform** dropdown, select the **Endpoint Protection** template name, and click **Create**.

4. Give it a name, such as **Windows BitLocker Encryption**, and a description. Click **Next**.

5. Expand **Windows Encryption** to view the BitLocker configurable settings.

> **Tip**
> There are many available configurations for Windows encryption. By enabling BitLocker, a recommended set of standard BitLocker encryption methods are configured unless specified. Each organization may have unique encryption standards and policies.

6. Change **Encrypt Devices** to Require.

7. Under the BitLocker base settings, change **Warning for other disk encryption** to **Block**.

8. Change **Allow standard users to enable encryption during Azure AD Join** to **Allow**.

> **Tip**
> Leave **Configure encryption methods** set to **Not configured** to use the Windows default BitLocker encryption algorithms.

9. Under the BitLocker OS drive settings, change **Additional authentication at startup** to Require.

10. Change **OS drive recovery** to Enable.

11. Change **Save BitLocker recovery information to Azure Active Directory** to **Enable**.

> **Tip**
> Configuring a certificate-based data recovery agent and allowing users to create recovery keys is a business decision that should be documented in your policies. We will skip these settings for these instructions.

12. Change **Save BitLocker recovery information to Azure Active Directory** to **Enable**.

13. Ensure **BitLocker recovery information stored to Azure Active Directory** is set to Backup recovery passwords and key packages.

There are several additional settings worth considering in your BitLocker policy. Let's review what each is so that you can decide whether they should be configured:

- **Client-driven recovery password rotation** allows you to use Intune device actions to remotely rotate the BitLocker recovery keys if needed. **BitLocker recovery information stored to Azure Active Directory** and **Store recovery information in Azure Active Directory before enabling BitLocker** must be configured.

- **Store recovery information in Azure Active Directory before enabling BitLocker** ensures that recovery information is saved to Azure AD before BitLocker is enforced.

- **Recovery options in the BitLocker setup wizard** will allow you to block users specifying recovery key options if you're enabling BitLocker manually.

- **BitLocker fixed data-drive settings** lets you apply similar configurations to fixed data drives, as we covered regarding the OS drive settings.

- **BitLocker removable data-drive settings** allows you to block write access to removable data drives that are not protected by BitLocker and prevent writing to devices that have been configured in other organizations.

Once you have configured the profile so that it meets your needs, assign it to a group of users or devices. If some devices have already been encrypted with BitLocker and they match the policy configurations, the profile will report a success message. If there is a configuration mismatch with the algorithm, the device needs to be decrypted first before it can be re-encrypted using the configured settings in the profile.

Once these settings have been applied to the client, you can view the BitLocker policies that have been applied via Intune and the winning provider GUID in the registry at HKLM:\SOFTWARE\Microsoft\PolicyManager\current\device\ BitLocker.

To view the status of all the volumes on a client using PowerShell, run the Get-BitLockerVolume command. The output should look as follows:

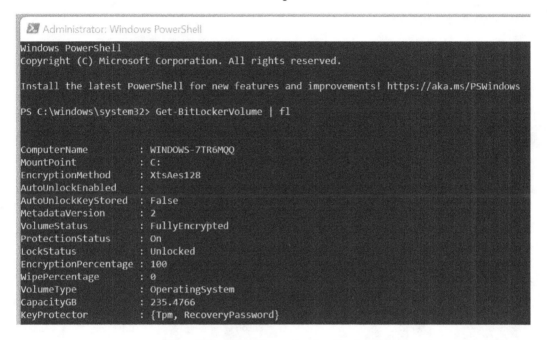

```
Administrator: Windows PowerShell

Windows PowerShell
Copyright (C) Microsoft Corporation. All rights reserved.

Install the latest PowerShell for new features and improvements! https://aka.ms/PSWindows

PS C:\windows\system32> Get-BitLockerVolume | fl


ComputerName         : WINDOWS-7TR6MQQ
MountPoint           : C:
EncryptionMethod     : XtsAes128
AutoUnlockEnabled    :
AutoUnlockKeyStored  : False
MetadataVersion      : 2
VolumeStatus         : FullyEncrypted
ProtectionStatus     : On
LockStatus           : Unlocked
EncryptionPercentage : 100
WipePercentage       : 0
VolumeType           : OperatingSystem
CapacityGB           : 235.4766
KeyProtector         : {Tpm, RecoveryPassword}
```

Figure 8.21 – BitLocker status for all volumes

Now that we have enabled a BitLocker profile, if the device has encrypted successfully, the recovery keys should be stored in Azure AD. Let's look at a few ways to access these keys using Azure AD, including by using a self-service option that's available for users.

Viewing BitLocker recovery keys

Once encryption has been applied to the drives, the recovery key will be stored in Azure AD and can be recovered by the end user or an IT admin with the Intune Service Administrator, Security Reader, or equivalent role. IT administrators can view BitLocker recovery keys in the following ways:

- Go to the device's overview in Microsoft Endpoint Manager and select **Recovery keys** under **Monitor**.

- From Azure Active Directory, choose **Devices**, choose **BitLocker keys**, and type in the 32-digit BitLocker key ID.

The following screenshot is of Azure Active Directory after searching for a BitLocker key ID:

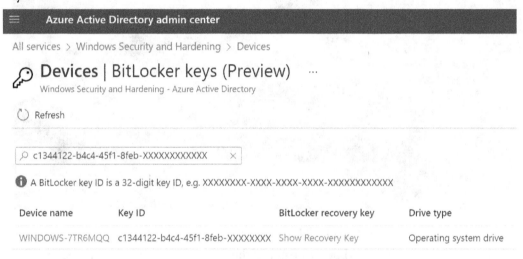

Figure 8.22 – BitLocker keys in Azure AD

For the self-service option, users can recover their BitLocker key by logging into `https://myaccount.microsoft.com/device-list`, selecting their device, and choosing **View BitLocker Keys**, as shown in the following screenshot:

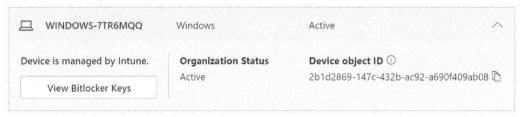

Figure 8.23 – Self-service recovery key for MyAccount

With that, we've learned how to enable BitLocker encryption using the Endpoint Protection template type. Now, let's learn how to configure PIN and biometric sign-in for Windows using Windows Hello for Business. We will accomplish this by configuring tenant-wide Windows enrollment settings from within Microsoft Endpoint Manager.

Going passwordless with Windows Hello for Business

Enabling biometric and device PIN sign-in to Windows is a great security enhancement and a crucial step in any passwordless strategy. Windows Hello for Business allows you to use fingerprint and facial recognition biometric sensors in addition to configuring a local device PIN to support passwordless sign-in. These methods are documented as significantly more secure than using a standard account password for the following reasons:

- Biometrics are unique to the user and hard to spoof. This requires a user to be physically present to unlock the device.

- Device PINs, although they can be less complex than passwords, are only tied to the device they have been configured on. They cannot be used elsewhere to gain access to your account.

The Windows Hello for Business technology can be backed by using either an asymmetric public/private key pair or through certificate-based authentication. It supports authentication with Azure AD, on-premises AD, and other identity providers that support **Fast ID Online v2 (FIDO)**. When configured with Azure AD, a user goes through a registration process. The credentials (PIN, fingerprint, and face ID) are bound to the device, and the public key is mapped to the Azure AD tenant. The validation of this pairing occurs through a two-step **multi-factor authentication (MFA)** verification process during the user registration process and key generation is supported using a hardware **Trusted Platform Module (TPM)** or through software. Windows Hello for Business supports the following types of authentication mechanisms:

- Device PIN

- Fingerprint or face biometric authentication

- Phone sign-in when using a Windows Hello for Business PIN on an Azure AD joined companion device

For more information about using Windows Hello biometrics in an enterprise, go to https://docs.microsoft.com/en-us/windows/security/identity-protection/hello-for-business/.

Now, let's learn how to configure Windows Hello for Business during Intune enrollment using Microsoft Endpoint Manager.

Enabling Windows Hello for Business

Windows Hello for Business settings can be managed by both Group Policy and MDM. To avoid policy conflicts, it is recommended that you use one or the other as Windows Hello for Business does not follow the ControlPolicyConflict CSP we configured earlier for dealing with conflicting MDM and Group Policy settings. Using Intune, settings can also be managed using security baselines, Endpoint Security, and Settings Catalog. In the following example, we will use Microsoft Endpoint Manager to configure organizational-wide settings to enable registration during Windows enrollment:

1. Log in to `https://endpoint.microsoft.com`.

2. Choose **Devices**, then select **Windows** and choose **Windows Enrollment**.

3. Select **Windows Hello for Business** under **General**. Change the **Configure Windows Hello for Business** dropdown to **Enabled**.

> **Tip**
> The default settings are a good starting point and offer a balance of security and usability. Modify any settings as necessary to meet your baseline requirements.

4. Click **Save** to apply and assign the policy to all users.

During device registration, the user will be prompted to configure Windows Hello for Business. The available biometric methods (fingerprint and facial recognition) will depend on what methods the underlying hardware can support. After the registration is complete, standard users can modify or configure additional sign-in methods by going to the **Windows settings** app, selecting **Accounts**, and choosing **Sign-in Options**. The following screenshot shows the Windows sign-in screen configured with Windows Hello for Business sign-in methods:

Figure 8.24 – Windows Hello for Business – Sign-in options

In addition to a password (key icon), you will see that the icons for fingerprint and PIN are available.

The following settings are suggested recommendations for configuring a Windows Hello for Business policy:

- Set **Use a Trusted Platform Module (TPM)** to Required to ensure hardware security is enforced.

- Set **Minimum PIN length** to no less than 6 characters.

- Configure lowercase, uppercase, and special character requirements to add complexity.

- Set a PIN expiration time to require PIN change after a set number of days.

- Set **Remember PIN history** to 5 or more to avoid PIN reuse.

- Allow biometric authentication.

- Enable **Use enhanced anti-spoofing, when available**.

For additional information about creating a Windows Hello for Business policy, go to `https://docs.microsoft.com/en-us/mem/intune/protect/windows-hello`.

If you are planning for an Azure AD cloud-only deployment or on-premises, take a look at the deployment guide at `https://docs.microsoft.com/en-us/windows/security/identity-protection/hello-for-business/`.

Now that we have covered different methods for setting and enforcing configurations with Intune, let's look at configuring a device compliance policy. Device compliance policies in Intune help perform attestation against certain conditions before marking a device as one that is compliant with company policy.

Configuring a device compliance policy

In *Chapter 6, Administration and Policy Management*, we reviewed device compliance policies in MEM and how they help attest to conditions as part of the zero-trust strategy and ensure a device meets the company requirements before being marked as compliant. We covered the actions that can be taken for devices marked as non-compliant and provided recommendations for which conditions to evaluate against. In this example, we will create a policy that evaluates if hardware security features are enabled on a device by checking if BitLocker encryption is enabled, a TPM is present, and Secure Boot is enabled. Let's get started:

1. Log into **Microsoft Endpoint Manager** at `https://endpoint.microsoft.com`.

2. Choose **Device** and click **Compliance policies**.

3. Click **Create Policy**, select **Windows 10 and later** as the platform type, and click **Create**.

4. Give it a name, such as **Windows Device Health Compliance**, and a description and click **Next**.

5. Choose the following compliance settings under **Device Health** and **System Security**:

    -   **Require BitLocker** must be set to Require.

    -   **Require Secure Boot to be enabled on the device** must be set to Require.

    -   **Trusted Platform Module** must be set to Require.

6. Click **Next**.

7. Under **Actions for noncompliance**, leave the default settings as-is and click **Next**.

8. Assign the policy to a group and click **Next**.

9. Click **Create** after reviewing the summary to finish creating the policy.

> **Tip**
> Be mindful if you're assigning to user groups as this can cause a conflict of policies if multiple users log into the same device with different assigned policies.

Once the policy has been evaluated, we can review its status by clicking **Device status** under the **Monitor** section from within the compliance policy overview. From here, you can click further into the device to validate each setting. The following screenshot shows the compliance state of each setting on a device in the scope of the policy assignment:

Figure 8.25 – Device compliance status

Now that we have created a policy that can be used for evaluating a device's compliance, we can leverage the device's state as a condition in a resource access policy. In this example, we will use the **Require device to be marked as compliant** condition in an **Azure AD Conditional Access** policy to allow access to the Azure portal. Using these controls can ensure that only devices that meet strict compliance evaluation rules are allowed access to the portal; otherwise, they will be blocked.

Log into the Azure Active Directory portal at `https://portal.azure.com`, search for **Azure AD Conditional Access**, and create a new policy with the minimum following settings:

- **Users or workload identities** must be scoped to some users or groups.

- **Cloud apps or actions** includes an app that contains the **Microsoft Azure Management** resource.

- **Grant controls** includes the **Require device to be marked as compliant** and **Require all the selected controls** resources.

> **Tip**
> We learned how to create a Conditional Access policy in *Chapter 5, Identity and Access Management*.

After saving the policy and attempting to log into `https://portal.azure.com` from a non-compliant device, the sign-in attempt will be blocked and will look as follows:

Figure 8.26 – Access blocked to the Azure portal

The Azure Activity Sign-in log will show the following details for the sign-in failure:

- **Sign-in error code**: 53000

- **Failure Reason**: Device is not in required device state: {state}. Conditional Access policy requires a compliant device, and the device is not compliant. The user must enroll their device with an approved MDM provider like Intune.

This example demonstrated the power of using device compliance evaluation with conditional access for building strong access controls. For less strict access requirements, multiple controls can be selected to provide a better experience for your end users. An example would be requiring a compliant device or requiring MFA to access a resource such as SharePoint Online. If the user was on a compliant company-issued device, they would not be required to use MFA, while if the client was Azure AD joined, the sign-in experience could be seamless for the user.

Next, let's learn how to build and deploy baseline controls for Windows clients. We will be referencing both the Microsoft Security Baselines and the CIS benchmark controls, and we will enforce them using Group Policy, Configuration Manager, and Intune.

Deploying Windows Security Baselines

Security baselining is the practice of implementing a minimum set of standards and configurations to apply to your Windows systems across the environment. More specifically, as a critical component of the zero-trust strategy, it ensures systems and devices are deployed consistently in the desired state according to the baseline. Two main frameworks to reference when defining these configurations are Microsoft Security Baselines and the **Center for Internet Security (CIS)**. Earlier in *Chapter 2, Building a Baseline*, we provided an overview of these frameworks and how to use them. We also referenced a tool that's available in the Microsoft Security and Compliance toolkit called Policy Analyzer, which can be used to compare recommended policies against the pre-configured settings locally on the system.

In the next section, we are going to learn how to apply these recommendations by leveraging tools such as Group Policy, Configuration Manager, and Intune. Then, we will talk about how to approach migrating policies from on-premises Group Policy into MDM to take a cloud-first approach to device management. Even if you're only enforcing policies through MDM, understanding how to implement these baselines using other methods provides a great roadmap for cloud-only deployments. Let's start by building a GPO using Microsoft Security Baselines.

Building a GPO using Microsoft Security Baselines

In *Chapter 2, Building a Baseline*, we referenced downloading the Microsoft Security and Compliance toolkit and comparing baselines with Policy Analyzer. Even if your clients are no longer domain joined, you can still build these GPOs to act as a reference source and use them for analysis later against your MDM-based policies. For this exercise, we will need to download the Microsoft Security and Compliance toolkit with the Windows 11 Security Baseline and Policy Analyzer. You can find the download for this at `https://www.microsoft.com/en-us/download/details.aspx?id=55319`.

Once the downloads are complete, extract the files to a location on your computer that has access to open and edit Group Policy using the **Group Policy Management** console.

Before applying policies, we must compare the MSFT Windows 11 – Computer baseline using Policy Analyzer to see the effective state of a non-hardened Windows 11 Enterprise operating system. In total, there are 176 total policies with 128 total combined differences and conflicts from the Microsoft recommendations.

Let's create a GPO from the MSFT Windows 11 – Computer baseline template that's supplied in the toolkit by following these steps:

1. Open the **Group Policy Management** console.

2. Select an **organizational unit (OU)** that contains the systems you wish to target. Right-click on it and choose **Create a GPO in this domain, and Link it here…**.

3. Give it a friendly name, such as `MSFT Windows 11 - Computer`.

4. Click the **Group Policy Objects** folder and find the policy that was just created. Right-click on it and choose **Import Settings**.

5. Click **Next** a few times until you get to the **Backup location** section, where you can choose a backup folder. Browse to the GPOs folder inside the downloaded and extracted baseline and click **Next** and then select **MSFT Windows 11 – Computer** from the list of backed-up GPOs.

6. Click **Next** until you get to the **Migrating References** menu and select **Copying them identically from the source**. Click **Next** and then **Finish**.

If Group Policy was imported successfully, you will see **GPO: MSFT Windows 11 - Computer…Succeeded**. Now that the policy has been imported, let's modify the default settings. We are going to change the behavior of the user account control elevation prompt for standard users to prompt for credentials instead of blocking them:

1. Right-click on the GPO and choose **Edit**.

2. Navigate to **Computer Configuration | Window Settings | Security Settings | Local Policies | Security Options**.

3. Find the **User Account Control: Behavior of the elevation prompt for standard users** policy and double-click on it.

4. Set the dropdown to **Prompt for credentials on the secure desktop** and click **OK**.

This setting will allow the user account control window to prompt for credentials on the secure desktop when a process requires administrative privileges. The secure desktop helps protect against input and output spoofing by locking down the credential box so that it can only be accessed by trusted system processes. If we left this policy as recommended, a standard user would receive a blocked notice.

> **Tip**
> Right-click on **GPO** and choose **Save Report** to export an HTML report containing all the settings that will be applied.

Now that we have applied the GPO to an OU, let's connect to a client that is in the scope of the new baseline to validate it. Follow these steps to run a **Resultant Set of Policies (RSOPs)**:

1. From a client connected to the domain network, open the Command Prompt and run gpupdate /force to check for policy updates.

2. Next, open the Command Prompt as an administrator and type in gpresult /r /scope computer. If **GPO** was applied, you will see the **MSFT Windows 11 – Computer** policy listed, as shown in the following screenshot of **Applied Group Policy Objects**:

```
Applied Group Policy Objects
-----------------------------
    MSFT Windows 11 - Computer
    MECM - Client Authentication Certificate
    Default Domain Policy
    Client Push Policy Settings
    SQL Ports for SCCM Replication
    Local Group Policy
```

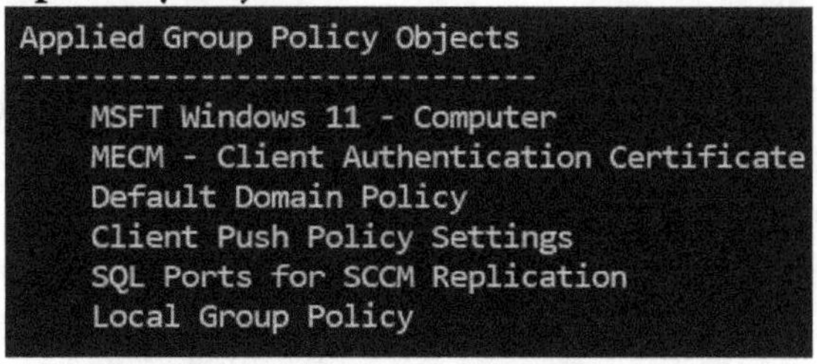

Figure 8.27 – Applied Group Policy Objects

After making the GPO change in the preceding steps, running the comparison between the Windows 11 – Computer baseline template from the toolkit and the effective settings now only shows one difference, as shown in the following screenshot:

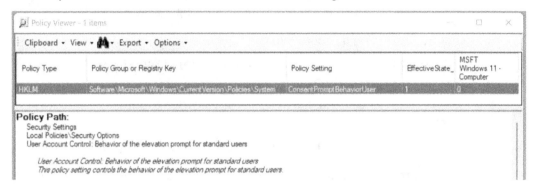

Figure 8.28 – Policy Analyzer comparison

We strongly recommend that you test these settings before deploying them in a production environment. It's important to understand the effects they will have on your systems and the end user's experience.

Next, let's compare the policies we just assigned from the Windows Security Baseline to the CIS recommendations.

Reviewing CIS recommendations

In addition to the Microsoft Security Baselines, it's recommended that you compare them to the CIS benchmark recommendations. CIS benchmarks are developed by a consensus of people that are actively engaged in the security of a given technology area (such as Windows) and contribute testing and feedback to help build these recommendations. The goal of the benchmark is to provide practical guidance that creates a generally applicable baseline for most organizations and maintains operational effectiveness for people to perform their jobs. CIS benchmarks typically contain one defined profile type that provides basic recommendations for each technology. For Windows clients, there are two levels of profile types, as follows:

- **Level 1** profiles recommend essential basic security requirements that can be configured on any system and should cause little or no interruption to services or inhibit and reduce functionality. They are practical and provide clear security enhancements to harden Windows clients.

- **Level 2** profiles recommend security settings for environments that require greater security controls. They typically extend upon and contain all Level 1 recommendations and are for organizations that must maintain a higher level of security, such as government agencies. Level 2 profiles could result in some reduced functionality.

We can run a comparison by using the **CIS-CAT Pro Assessor** tool to validate the Windows Security Baselines against the L1 and L2 CIS profiles. To download and use this tool, you will need a license that is available with a CIS SecureSuite membership. More information can be found at `https://www.cisecurity.org/cis-securesuite/`.

The CIS Windows benchmark contains 10 different profiles that you can use for analysis. We will choose **L1 – Corporate/Enterprise Environment (general use)** and **the L2 – High Security/Sensitive Data Environment (limited functionality)** for comparison. Depending on your licensing, you can export the results as HTML, CSV, Text, ARF XML, or JSON. The following table shows the results of the analysis:

Benchmark	Total	Pass	Fail	Score
Level 1 (L1) – Corporate/Enterprise Environment	331	211	120	64%
Level 2 (L2) – High Security/Sensitive Data Environment	424	215	209	51%

Figure 8.29 – CIS benchmark comparison

As you can see, there are a considerable number of recommendations in both the L1 and L2 profiles compared to the Microsoft Security Baselines. There are a few options regarding how to approach combining the recommendations from CIS with the already created baseline for Windows 11, as follows:

- Use the Build Kit that is included with a CIS SecureSuite membership. All Group Policy objects from the 10 assessed profile types can be imported directly into a new or existing group policy object.

- Manually review, assess, and compare the recommended policies with the Windows 11 Microsoft Baselines and add them to your GPOs.

For more information about CIS-CAT Pro, go to `https://www.cisecurity.org/blog/introducing-cis-cat-pro/`.

Now that we have learned how to create a GPO for Windows Security Baselines and compared it against the CIS recommendations, let's look at a tool that can help convert our reference GPO into a Configuration Baseline for use with Configuration Manager if Group Policy is not an option. This can significantly help automate the creation of policies when Configuration Manager is your main policy engine.

Converting a GPO into a Configuration Baseline

In the *Enforcing policies and configurations* section of this chapter, we learned how to manually create a Configuration Item and add it to a Configuration Baseline using Configuration Manager. This is an efficient way of enforcing policies, especially for devices that are fully Azure AD joined and lack the presence of Group Policy. Many GPO settings can be mapped to entries in the Windows registry that can be shifted into Configuration Manager and deployed as a baseline. Since we want to use the recommendations from CIS and Microsoft and we can't just import the GPO template into ConfigMgr, the process of identifying all the associated registry keys and creating a CI is very manual and time-consuming. Fortunately, a free tool is available that can help map GPOs and their associated settings to significantly expedite the creation of a Configuration Item using PowerShell.

For this task, we are going to be using the `Convert-GPOtoCI` PowerShell script from the **Sam Roberts** GitHub repository.

> **Tip**
> While this project is no longer under development or support, it still works well.

To download the files, go to `https://github.com/SamMRoberts/Convert-GPOtoCI`, click on the **Code** dropdown, and choose **Download ZIP**. It is recommended that you view the `README.md` file for full usage instructions, including syntax.

In this example, we are going to use the tool to convert the MSFT Windows 11 – Computer GPO we created earlier into a Configuration Item. Follow these steps:

1. Find and extract the ZIP file to a system that can manage **Group Policy Objects** and connect to Configuration Manager.

2. Open PowerShell as an administrator and go to the extracted directory by typing `CD "PATH/TO/Extracted/Folder"`, replacing `PATH` in quotes with the actual file path of the extracted script, and pressing *Enter*. In our example, we will be using `"C:\Convert-GPOtoCI"`.

We are going to target the **MSFT Windows 11 – Computer** policy that we created earlier. You will need your **Fully Qualified Domain Name** (**FQDN**) and the three-letter site code of your Configuration Manager site to create the Configuration Item.

3. In PowerShell, type in the following code and replace the parameters so that they fit your site and domain settings:

```
.\Convert-GPOtoCI.ps1 -GpoTarget "MSFT Windows 11 -
Computer" -DomainTarget mydomain.prod.com -siteCode ABC
```

> **Tip**
>
> Use the `-remediate` parameter to set the **Remediate noncompliant rules when supported** flag in the compliance rules.

The script will process and return some results. As shown in the following screenshot, it found 116 keys and 200 values for registry items to be created in the Configuration Item:

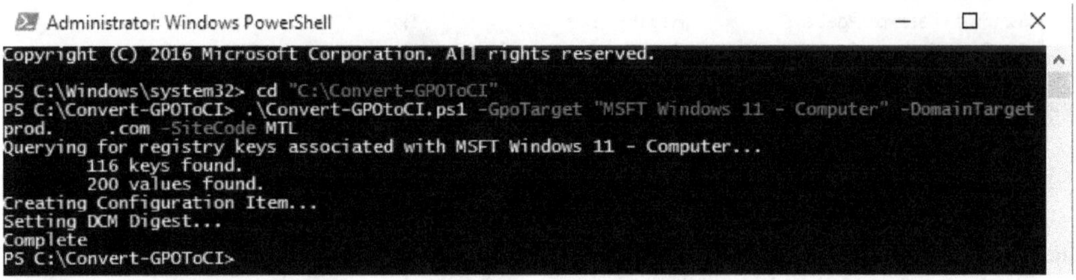

Figure 8.30 – PowerShell output for creating a Configuration Item

4. Open the **Configuration Manager** console and navigate to **Assets and Compliance | Compliance Settings | Configuration Items**. You should see the **MSFT Windows 11 – Computer** Configuration Item that was created by running the script.

If you open the properties of the Configuration Item and click on the **Settings** tab, you can see all the different settings that have been mapped to the registry values that were imported, as shown in the following screenshot:

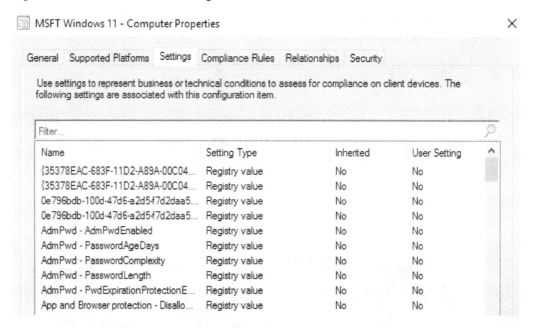

Figure 8.31 – The Settings tab of a Configuration Item

Click the **Compliance Rules** tab and open a rule. **Report noncompliance if this setting instance is not found** should be selected.

> **Tip**
> The severity of the non-compliant items can be changed by specifying the severity setting in the PowerShell command. The default value is `Information`.

Next, let's assign this CI to a Configuration Baseline and validate the settings against our client. In the *Creating security baselines in Configuration Manager* section, we learned how to create a Configuration Baseline and add a CI. Complete these steps to add this CI to an existing baseline or create a new one. We would consider this baseline as critical and recommend setting the compliance evaluation schedule to **Simple schedule** and the **Run every** option to 1 day to ensure that the policy is being evaluated frequently. Finally, assign the baseline to a collection of devices. We deployed it to our **All Windows Clients** collection.

> **Tip**
>
> Select **Remediate noncompliance rules when supported** if you're using the
> `-Remediate` parameter when running the script in the Configuration
> Baseline settings. In this example, we are deploying in monitor mode only,
> which is useful for reporting on GPO compliance.

Now that the Configuration Baseline has been deployed, let's head over to our client
device and check its status:

1. Open **Control Panel** and find the **Configuration Manager** control panel applet.

2. Click the **Configurations** tab and check for **Windows 11 Security Baseline**. Select
   it and click **Evaluate**.

> **Tip**
>
> From the Configuration Manager control panel applet, choosing the **Actions**
> tab and running **Machine Policy Retrieval & Evaluation Cycle** will force the
> PC to check into Configuration Manager and download any pending policies.

As shown in the following screenshot, the policy is reported as **Non-Compliant**. This is
due to a mismatch in the settings between our baseline and what's been configured locally
on the device. The Configuration Manager applet has a **View Report** button that you can
click on to export an HTML file and see what settings are mismatched:

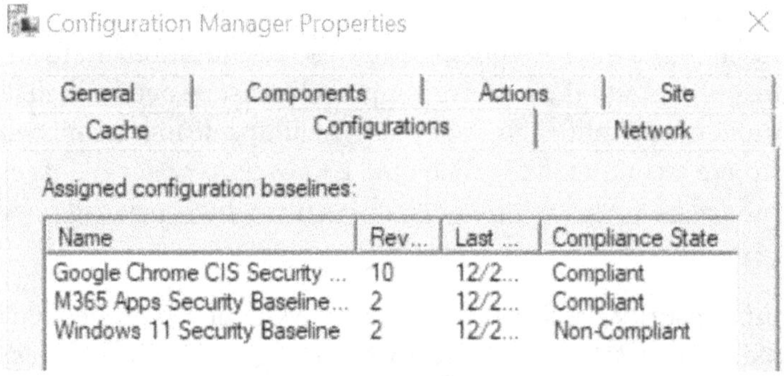

Figure 8.32 – Non-compliant evaluation of a security baseline

During the GPO import process, some settings may have a `null` value and need to be
manually updated in the CI to accurately reflect the settings in your policy or what the
client expects to read in the policy path. You can validate this in the HTML report as the
expression column under non-compliant rules will only display `Equals` in absence of
a value.

If you have configured the CI to remediate non-compliant rules and there is a conflict or issue is enforcing the policy setting, **Compliance State** may show as **Error** in the Configuration Manager applet.

If you're using Configuration Baselines to deliver your Windows Security Baseline payload, we recommend always updating the reference GPO whenever a policy needs to be added or changed. This will act as the source and helps create reports that can easily be read due to friendly names and descriptions of the policy settings.

Using Configuration Baselines is a great way to enforce policies, even in Intune Co-managed scenarios, but a lot of investments are being made by Microsoft to easily map these settings to CSPs through the Microsoft Endpoint Manager portal and Intune. Next, let's look at using Intune and MEM to deploy security baselines.

Deploying security baselines with Intune

Intune includes pre-configured groups of Microsoft's recommended best practices and security settings, known as Intune **Security Baselines**. These recommendations are created from the same security team that creates the Microsoft Security Baselines that we used in the toolkit earlier. In *Chapter 6*, *Administration and Policy Management*, we provided an overview of Intune Security Baselines and covered the different types of profiles that are available. These pre-configured settings are great, especially if your Intune deployment is new or if you're starting greenfield with security policies as it provides a framework to build from and can be quickly deployed. Part of the challenge that IT security administrators face is that not all the policy settings from the CIS or Microsoft Security Baselines are exposed in the Intune UI or are CSP-backed. Additionally, there are multiple places to configure policies and the preferred approach must be determined if you wish to avoid policy conflicts and confusion, especially if multiple teams are managing Intune settings. As a recap, we recommend the following approach to policy management as it will help maintain a level of organization and deliver the lightest payload possible to your devices:

1. Deploy Intune Security Baselines first. They provide the recommended best practices and security settings are already preconfigured in the policies.

2. Deploy Device Configuration profiles or Endpoint Security policies for security-related settings to supplement policies. Not all CIS or Microsoft Security Baseline recommendations will be available in the Intune Security Baselines.

3. Use Device Configuration Templates. They contain logical groupings of settings with friendly names specifically designed for managing Windows devices in Intune.

4. Deploy ADMX-based policy templates for win32 apps or policies not yet available in Intune. Use custom profile types with ADMX ingestion for third-party apps that are not natively supported in Intune.

If you're building your baseline controls using CIS or Microsoft Security Baseline recommendations, there is no quick method to convert the Group Policy recommendations into the Intune UI or expedite the configuration of the Device Configuration profile. Following the preferred approach for policy management, you could use a spreadsheet to track each policy and reference where each setting will be applied to help stay organized. There is a tool in Intune called Group Policy analytics that can help you analyze on-premises GPOs and determine how they map to CSPs to get you started. Let's look at an example of how to use this.

Using Group Policy analytics

In this example, we are going to analyze the MSFT Windows 11 – Computer baseline we created earlier and import it into Group Policy analytics for analysis. We are going to export the GPO from a computer with access to the Group Policy Management console:

1. Find the **MSFT Windows 11 – Computer** Group Policy, right-click it, and choose **Save Report**. Click the dropdown and choose **XML** file.

2. Sign into **Microsoft Endpoint Manager** at `https://endpoint.microsoft.com`.

3. Select **Devices | Group Policy analytics** under **Policy**.

4. Select **Import** and choose the saved XML file to be analyzed by Intune.

In the following screenshot, you can see that there is 80% support for policies that are backed by MDM and ready to be migrated:

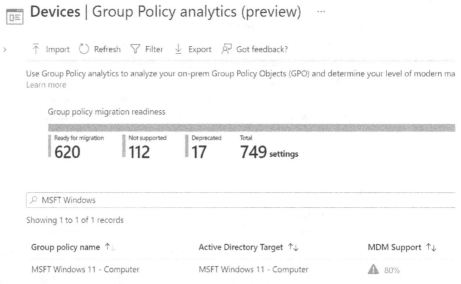

Figure 8.33 – Group Policy analytics in Intune

Under the **MDM Support** column in the records, click on the percentage to view more details about each policy. It will show information about the setting's name, GPO setting, value, CSP name, and the CSP mapping. Choose **Export** to save the report as a CSV file. We will use this report as our reference policy template as we start migrating policies to Intune.

Migrating policies to Intune

Now that we have used Group Policy analytics to identify which settings are mapped to CSPs, they will need to be individually evaluated against your security requirements. We must also decide on how to effectively enforce them. We can use the CSV file that we exported from the analysis against our GPO baseline as a reference point and follow the preferred approach. Let's get started:

1. Use Intune Security Baselines for the bulk of your settings.
2. Layer additional settings using Settings Catalog or with Endpoint Security where applicable.
3. Use Device Configuration templates.
4. Use ADMX-based policy templates for Windows settings, win32 apps, and custom profiles for third-party apps that don't have CSP support.

Append the CSV report and add additional headers for the MDM policy workload, as shown in the following screenshot, and build a spreadsheet to track where the settings will be applied:

H	I	J	K	L
Intune Security Baselines	Settings Catalog	Endpoint Security	Templates	ADMX Backed
	X			
	X			
X				
X				
		X		
X				
X				
X				

Figure 8.34 – MDM CSP mapping report

Since Intune Security Baselines are already pre-configured with recommended values, it may make sense to create the profile, go through the settings list from the top down, and mark each corresponding setting that was found in the CSV file. To easily match the settings, compare the category title in Intune with the CSP mapping column in the CSV file to find the relationship in the path. For example, the **Above Lock** settings in the Intune Security Baseline would match the **./Device/Vendor/MSFT/Policy/Config/AboveLock** OMA-URI path in the CSP mapping column of the CSV file. After the analysis is complete, you may find additional settings listed in the Intune Security Baseline that are not shown in the CIS or Microsoft security baselines. Once you've done this, you can supplement the baselines with Settings Catalog or other methods to round out the baseline.

This method might be time-consuming but it's also a great way to learn about many of the policies that exist in Windows and understand the relationship between Group Policy and CSPs. Just like with any profile that's deployed using Intune, these settings can be validated on the device using the same processes that were discussed in the *Enforcing policies and configurations* section. In *Chapter 9, Advanced Hardening for Windows Clients,* we will look at deploying security baselines for Microsoft Edge, Google Chrome, and Microsoft 365 productivity apps. We will also take a deeper dive into creating baselines using ADMX-backed policies through a process called ADMX ingestion.

> **Tip**
>
> Microsoft has recently added a migrate feature that takes the imported GPO and converts it to settings found in the **Settings** catalog. More information about migrating Group Policy can be found at this link `https://docs.microsoft.com/en-us/mem/intune/configuration/group-policy-analytics-migrate#review-and-migrate-your-gpos-to-a-settings-catalog-policy`

Next, let's look at the built-in security features that are available in the **Windows Security** app and how to configure them on Windows clients using Intune.

Configuring Windows Security features

Windows Security settings can be found in the **Windows Security** app, which acts as the control interface for many Windows Security-related features. Windows Security contains multiple categories for controlling different security products built into Windows, such as Defender Antivirus, Defender Firewall, and hardware-based security options. Inside each category, additional advanced features are available, many of which can be enabled if certain prerequisites are met, such as meeting certain hardware requirements or appropriate licensing for Microsoft Defender for Endpoint features. Many features that are found in the Windows Security app can be managed and configured using MDM and Group Policy. The main categories that are available in the Windows Security app are as follows. We've covered a few of these already throughout this book:

- Virus and threat protection with Defender Antivirus
- Account Protection (using Windows Hello for Business)
- Firewall and network protection with Defender Firewall
- App and browser control using Microsoft Defender SmartScreen
- Device security using advanced hardware-based security features
- Device performance and health to check the status of your device

To access the Windows Security app, search for it in the *Start* menu or use Windows search. In the following sections, we will cover features that we haven't discussed yet. First, let's look at configuring a Windows Defender Antivirus baseline for virus and threat protection features.

Configuring a Defender Antivirus baseline

Windows Defender Antivirus is a next-generation feature of Microsoft Defender for Endpoint and a subset in the Microsoft Defender suite of products. In Windows, protection is enabled by default, and virus definitions are updated automatically through Windows Updates. We can enforce a baseline for Defender Antivirus and manage additional settings such as scanning behavior using Microsoft Endpoint Manager. Intune Security Baselines has a pre-configured baseline for Defender Antivirus, but we want to keep some security configurations separated in MEM, so we will configure it using Endpoint Security. In this example, we will configure real-time protection, scanning options, and remediation settings using Endpoint Security in Intune to enforce a Windows Defender Antivirus baseline. Let's get started:

1. Sign into **Microsoft Endpoint Manager** at `https://endpoint.microsoft.com`.

2. Click **Endpoint Security**, choose **Antivirus**, and select **Create Policy**. Choose **Windows 10 and later** as your platform and select **Microsoft Defender Antivirus**.

3. Give it a friendly name, such as `Microsoft Defender Antivirus Baseline`, and a description. Then, click **Next**.

In the configuration settings, you are presented with seven categories. We recommended configuring these settings based on the corresponding screenshots provided:

- **Cloud protection** settings will enhance real-time protection by allowing Windows Defender to send information to Microsoft for rapid analysis and verification, which is faster than relying on definition updates. By setting the extended timeout, Defender Antivirus can control how long a file won't open until a result is returned from the cloud protection service:

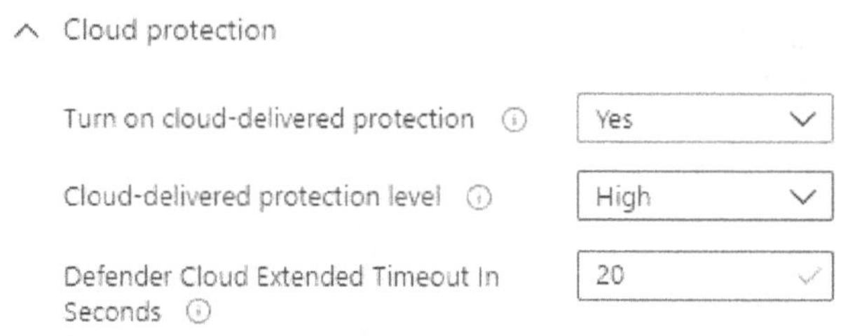

Figure 8.35 – Cloud protection settings

- **Microsoft Defender Antivirus Exclusions** will exclude processes, file extensions, and folder paths from being scanned. Setting **Disable local admin merge** to **Yes** will prevent any rules that have been added manually from being honored:

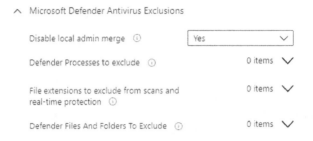

Figure 8.36 – Antivirus Exclusions settings

- **Real-time protection** settings enable real-time monitoring features. We left **Scan network files** disabled as this is handled better by the file servers instead of by a client:

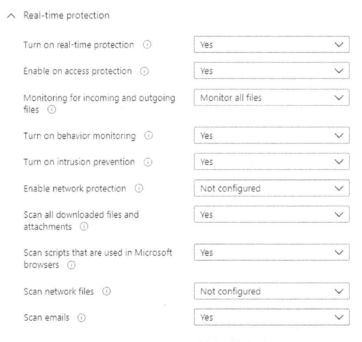

Figure 8.37 – Real-time protection settings

- **Remediation** settings control the action to take on detected threats from real-time protection. Setting **Action to take on potentially unwanted apps** to **Enable** will block unwanted applications such as advertising software, and software bundles containing low-reputation software:

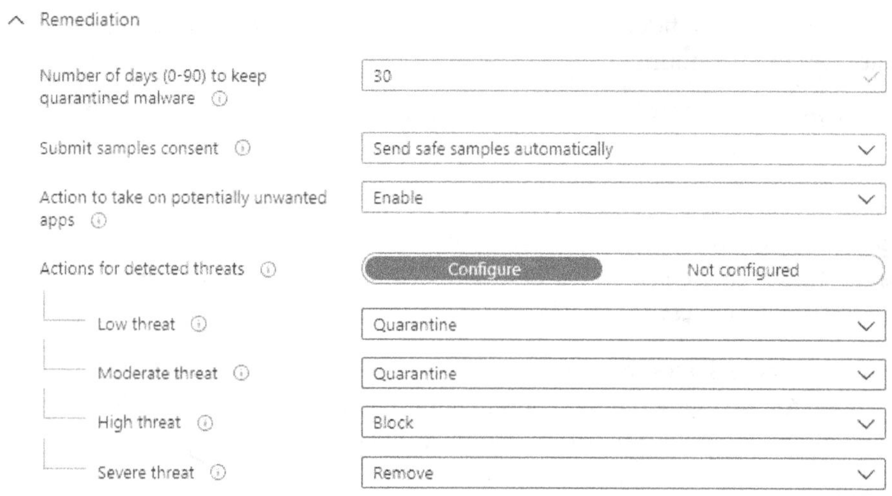

Remediation

Number of days (0-90) to keep quarantined malware ⓘ	30 ✓
Submit samples consent ⓘ	Send safe samples automatically ⌄
Action to take on potentially unwanted apps ⓘ	Enable ⌄
Actions for detected threats ⓘ	**Configure** Not configured
└── Low threat ⓘ	Quarantine ⌄
└── Moderate threat ⓘ	Quarantine ⌄
└── High threat ⓘ	Block ⌄
└── Severe threat ⓘ	Remove ⌄

Figure 8.38 – Remediation settings

- **Scan** settings configure the conditions that scans will run under. Keep in mind that setting a full scan during business hours could affect PC performance:

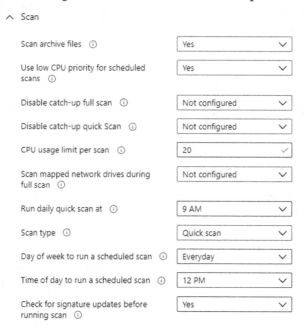

Figure 8.39 – Scan settings

- **Updates** settings configure how often to check for security definition updates. The default setting is 4 hours.

- **User Experience** controls if a user can access the Windows Defender app and suppress notifications. The default setting is to allow user access.

When you're configuring the profile settings, they can be left as **Not configured**, which will accept the default Windows Defender configurations. There are a few additional settings that can't be configured in the Endpoint Security profile but can be supplemented through other profiles. These are as follows:

- **Scan removable drives during full scan** should be set to **Yes**. This is the default Defender Antivirus setting, but if it's not enforced, a user can disable it. This policy can be set using Intune Security Baselines, or a Device Configuration Profile in the Settings Catalog or Templates profile type.

- **Expedite telemetry reporting frequency** can be set to **Yes** to increase the reporting frequency from the client to Defender for Endpoint. This can be configured in **Endpoint Security** under the **Endpoint detection and response** profile.

- **Tamper Protection** can be enabled to prevent malicious processes from attempting to disable security features and avoid detection. Tamper protection can be enabled organization-wide through the Microsoft 365 Defender portal, the Endpoint Protection template type in MEM, or through the Windows Security Experience policy type in Endpoint Security.

Now, let's learn where we can configure other Windows Security features.

Account protection features

Account protection features manage security settings for account and sign-in settings, such as configuring Windows Hello for Business. We covered how to enable Windows Hello for Business earlier in the *Going passwordless with Windows Hello for Business* section. Using MEM, account protection features can also be configured in Endpoint Security with an **Account protection** profile, using Settings Catalog, or an **Identity protection template type** under **Templates**. Using the **Account protection** profile, you can also enable Credential Guard and configure a local user group membership policy to add, remove, and replace members of local groups on Windows.

Firewall and network protection

Firewall and network protection settings allow you to manage firewall and network settings on Windows. In *Chapter 4, Networking Fundamentals for Hardening Windows*, we provide an overview of Defender Firewall and network settings on the Windows client. We also discuss how to deploy a security baseline for Defender Firewall using MEM.

App and browser control

App and browser control settings include managing various features, including Microsoft Defender SmartScreen, Defender Application Guard, and Exploit protection. These features provide reputation-based protection against malicious apps, files, and websites, isolated browsing to protect against infected websites, and exploit protection to apply mitigations against exploits so that they can't use operating system processes and apps. We will cover enabling some of these features in *Chapter 9, Advanced Hardening for Windows Clients*.

Device security

Device security shows the hardware-based security features that are available on your device. They include core isolation features that use virtualization-based security, information related to the TPM security processor, and a status indicator for Secure Boot. We covered the TPM, enabling Secure Boot, and VBS security features in *Chapter 3, Hardware and Virtualization*. Instructions for enabling BitLocker encryption can be found earlier in this chapter in the *Enabling BitLocker to prevent data theft* section.

Now that we have covered the main categories that are available in the Windows Security app, let's look at configuring the experience settings. Some of these configurations include setting notification levels, hiding specific features, and adding support contact information for use in notifications and inside the app.

Setting the Windows Security experience

Now that we have reviewed the features in the Windows Security app, we can configure the experience to control how users can interact with the app and add support information using MEM. To create the profile using Endpoint Security, go to **Devices Endpoint Security | Antivirus** and choose **Create Policy**. Choose **Windows 10 and later** as the platform and select **Windows Security experience** as the profile type.

Choose the settings that best fit your organization's needs, but at a minimum, we recommend the following settings:

- Set **Enable tamper protection to prevent Microsoft Defender being disabled** to **Enable**. If this setting has been configured in Microsoft 365 Defender, it does not need to be configured here.

- Set **Hide the Family options area in the Windows Security app** to Yes. For most organizations, these settings can be hidden.

- Set **Disable the Clear TPM option in the Windows Security app** to Yes. Users should not have the option to clear the TPM without contacting IT support.

- Set **Organization's support contact information** to Display in app and in notifications. This will provide a clear message about how to contact support if an app or website is blocked using Defender's next-generation security features.

Once the profile has been created, assign it to groups of users or devices. The following screenshot shows an example of the support information that can be seen from the Windows Security app:

Figure 8.40 – Organizational contact information

This support contact information will also be included in the Windows notification area on some triggered events.

Summary

In this chapter, we discussed a wide range of topics around securing Windows clients. We started by discussing the guidance in the Microsoft life cycle policy and the importance of staying updated. We covered how to plan for and deploy Windows Updates for Business using MEM and configuring Delivery Optimization for downloading updates.

Next, we reviewed enforcing policies and configurations using Configuration Manager and Intune. We reviewed how to handle conflicts between Group Policy and MDM when shifting policies into Intune cloud management. After that, we covered how to enable encryption on Windows clients using BitLocker and configuring passwordless sign-in with Windows Hello for Business. Then, we reviewed how to create and deploy Windows Security baselines using both CIS and Microsoft Security Baseline recommendations and how to migrate these policies into Intune for cloud-based management.

Finally, we discussed how to configure the Windows Security features that are listed in the Windows Security app. In the next chapter, we will look at applying advanced hardening to secure enterprise browsers, Microsoft 365 apps, and configuring the advanced protection features of Microsoft Defender for Endpoint.

9

Advanced Hardening for Windows Clients

Implementing a Windows baseline is a great step in your overall hardening, but it's also important to consider other areas of risk to tighten controls and reduce the attack surface. In this chapter, we will cover some of those areas, including securing enterprise web browsers, protecting Microsoft 365 apps, and enabling advanced features of Microsoft Defender for Endpoint as part of your baseline controls. In *Chapter 8*, *Keeping Your Windows Client Secure*, we covered enforcing a foundational security baseline that provides a robust layer of protection, but other apps, such as web browsers and the Microsoft Office suite, remain consistently vulnerable to exploits. The first part of this chapter will cover building baselines to protect these apps and how to apply a policy using MDM. We will provide examples using Intune security baselines, the settings catalog, the Office cloud policy service, and ADMX ingestion for third-party apps such as Google Chrome.

Next, we will cover enabling advanced features of Microsoft Defender that directly apply zero-trust security principles. This includes protection from unwanted software, attack surface reduction rules, enabling cloud-based protection, and leveraging hardware-based security to secure endpoints through virtualization-based security. It's important to think about protecting Windows from a holistic lens, as there are many areas to consider in your hardening efforts and not just one solution that can fit all these needs. Attack methodologies and techniques are constantly shifting and it's important to stay informed and learn about implementing new defense measures as threats evolve.

This chapter will include the following topics:

- Securing enterprise web browsers
- Securing Microsoft 365 apps
- Advanced protection features with Microsoft Defender

Technical requirements

To follow along with the instructions, you will need a Microsoft 365 subscription and rights to manage resources and policies in the Microsoft 365 admin center. In addition, to enable certain features with Microsoft 365 Defender and manage devices with **Microsoft Endpoint Manager** (**MEM**), you will need the following licenses or equivalent SKUs:

- Microsoft 365 E3 and/or E5
- Windows 10/11 Pro or Enterprise
- Hardware with support for VBS security

Let's start by discussing how to secure enterprise web browsers.

Securing enterprise web browsers

The modern-day web browser powers more applications than ever and plays a pivotal role in supporting many business functions. Part of this adoption can be attributed to the increase of SaaS-based apps coupled with the flexibility of cross-platform compatibility. Developers no longer need to build complex thick client applications to create a powerful and robust user interface full of features. Therefore, enterprise web browsers must receive the utmost attention when it comes to security controls and hardening. The following figure is the worldwide browser market share as reported by StatCounter, and can be found at the following reference link: `https://gs.statcounter.com/browser-market-share`.

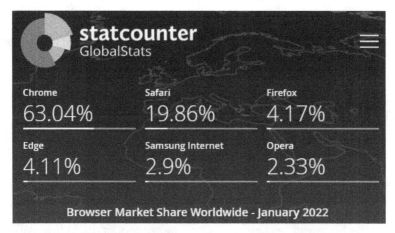

Figure 9.1 – StatCounter browser market share

As represented in *Figure 9.1*, Google Chrome dominates the market share followed by Safari, Firefox, and Microsoft Edge. In the following sections, we will discuss how to build and enforce security baselines for both Microsoft Edge and Google Chrome using Intune MDM based on recommendations from Microsoft and CIS. It's worth mentioning that support for Internet Explorer on Windows ended on June 15, 2022, and companies still relying on it will need to migrate as soon as possible. For more information about migrating from Internet Explorer, visit this link: `https://techcommunity.microsoft.com/t5/windows-it-pro-blog/proven-tools-to-accelerate-your-move-to-microsoft-edge/ba-p/2504818`.

First, we will review how to configure a Microsoft Edge security baseline.

Configuring a Microsoft Edge security baseline

Microsoft Edge is now a Chromium-based browser and supports many of the same functions and settings that are configurable in Google Chrome. Just like Windows, security baseline recommendations for Edge are available from the Microsoft Security and Compliance Toolkit and CIS. We can use the following methods to enforce the baseline controls:

- MDM
- Group Policy
- Microsoft Endpoint Configuration Manager

Let's review the recommendations from CIS and Microsoft to create the baseline. To do this, we will download and import the latest ADMX templates. Then, we will create a GPO from the Microsoft recommendations, compare it to CIS, and combine them into one GPO. Once the policy is final, we will create and deploy it using Intune.

Building the baseline

To complete the analysis, we followed the high-level steps listed as follows. We will go into more detail on each step throughout the section.

First, let's download the latest Microsoft Edge ADXM templates by visiting `https://www.microsoft.com/en-us/edge/business/download`:

1. Download the Microsoft Security and Compliance toolkit, including Policy Analyzer and the Microsoft Edge security baseline: `https://www.microsoft.com/en-us/download/details.aspx?id=55319`.

2. Create a new GPO named **MSFT Edge Security Baseline – Computer** and import the settings from the Microsoft Edge baseline GPO template. Apply the GPO to a Windows client to be used in the analysis.

3. Assess the Microsoft recommendations against **CIS Level 1 (L1) – Corporate/Enterprise Environment** by downloading the CIS Microsoft Edge benchmark PDF file. The Microsoft toolkit includes a spreadsheet of the Edge baseline policies, or you can create one using Policy Analyzer locally on the client. To speed this process up, with a CIS SecureSuite membership, you gain access to the CIS-CAT Pro Assessor tool that can automatically analyze the benchmarks and create an assessment report. CIS benchmarks can be downloaded here: `https://www.cisecurity.org/cis-benchmarks/`.

4. After the analysis is complete, manually combine the recommendations into the GPO to build the enhanced baseline.

> **Tip**
> CIS SecureSuite membership also includes access to build kits, which contain all the recommended policies pre-configured into a GPO.

5. Export the GPO into XML and upload it for analysis using Group Policy analytics in MEM.

6. Create the profiles in Intune and assign the baselines.

During our analysis, the current Microsoft Edge security baseline version 96 from the toolkit contains 21 policies. As mentioned in the preceding steps, we created a GPO named **MSFT Edge Security Baseline – Computer** from the recommendations and applied it to a client for analysis. If you don't have an Active Directory domain, you can use the LGPO tool that is part of the toolkit. The `.zip` file includes full documentation for usage on how to apply computer-based settings to a **Local Group Policy (LGP)**. After the GPO was applied, we opened the administrative command prompt and ran `gpupdate /force` followed by `gpresult /r /scope computer` to view the **Resultant Set of Policy (RSoP)** output and confirmed that our GPO had been successfully applied.

Each setting maps to a registry entry, and to view them, we have to open **Registry Editor** as an administrator and navigate to `HKLM:\SOFTWARE\Policies\Microsoft\Edge`. It should be populated with settings, as shown in the following screenshot:

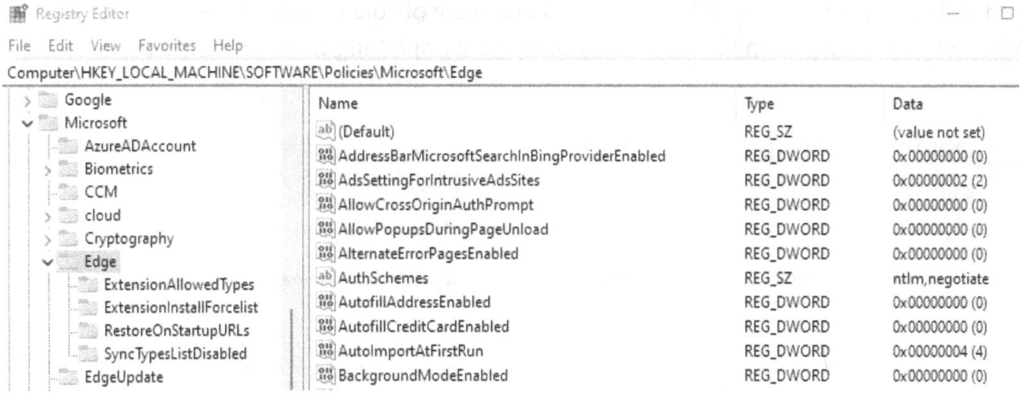

Figure 9.2 – Microsoft Edge policies in the registry

You can also view the managed policies directly in the Edge browser by typing `edge://policy` in the address bar.

Next, we are going to run the CIS-CAT Pro Assessor to compare the Microsoft Edge security baseline against the Microsoft Edge benchmarks from CIS. After running the assessment, we came back with the results shown in *Figure 9.3*. We ran it against both the Level 1 and Level 2 profiles. On comparing Microsoft recommendations to CIS, there are only 7 policies out of 60 that pass the assessment. This isn't assuming the Microsoft baselines are inadequate, it's just to show the differences between the two.

Benchmark	Total	Pass	Fail	Score
Level 1 (L1) – Corporate/Enterprise Environment	60	7	53	12%
Level 2 (L2) – High Security/Sensitive Data Environment	85	8	77	9%

Figure 9.3 – CIS comparison to the Microsoft Edge security baseline

After carefully reviewing and testing the CIS recommendations, we added the policies to the GPO created earlier to finalize the baseline that will be used as our reference template. Next, we are going to export the GPO to an XML file and analyze it using Group Policy analytics in MEM.

To assess our baseline to determine MDM support, follow these steps:

1. Find the **MSFT Edge Security Baseline – Computer** group policy, right-click, and choose **Save Report**. Click the dropdown and choose the **XML** file.
2. Sign in to MEM at `https://endpoint.microsoft.com`.
3. Select **Devices | Group Policy Analytics** under **Policy**.
4. Select **Import** and choose the saved XML file.

In the following screenshot, you can see there is a 97% support for policies that are backed by MDM and ready for migration. Some policies were obsolete to newer versions of Edge, so we removed them from the GPO template before importing them into MEM:

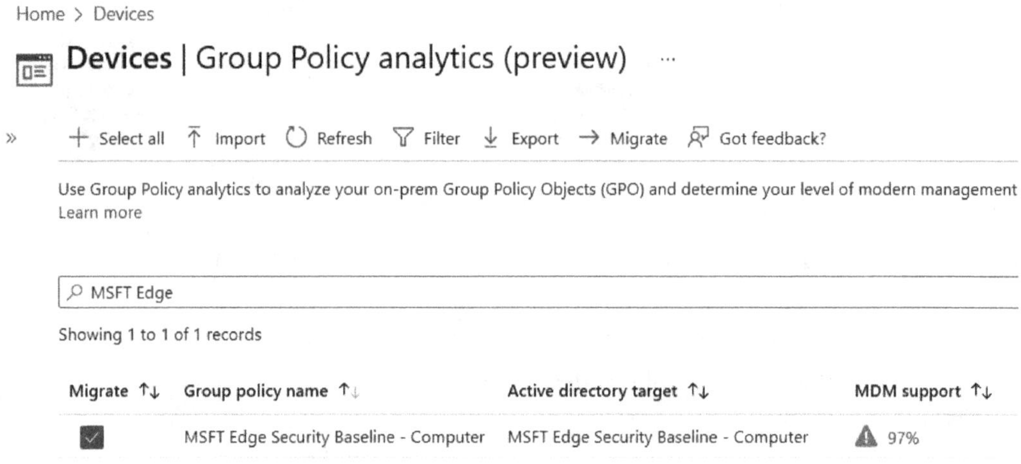

Home > Devices

Devices | Group Policy analytics (preview)   ...

»   + Select all   ↑ Import   ○ Refresh   ▽ Filter   ↓ Export   → Migrate   ⅋ Got feedback?

Use Group Policy analytics to analyze your on-prem Group Policy Objects (GPO) and determine your level of modern management Learn more

🔎 MSFT Edge

Showing 1 to 1 of 1 records

Migrate ↑↓	Group policy name ↑↓	Active directory target ↑↓	MDM support ↑↓
✓	MSFT Edge Security Baseline - Computer	MSFT Edge Security Baseline - Computer	⚠ 97%

Figure 9.4 – Group Policy analytics analysis

On the **MDM Support** column, click on the percentage to view more details about each individual policy. Here, we can choose **Export** to save the report as a CSV file. We will use this report for our documentation as we work through building policies in Intune. This will help us reference policies in the future and know where they were applied in Intune without having to hunt them down later.

> **Tip**
> Microsoft recently added a migrate option to automatically create a Device Configuration profile from the Group Policy Settings.

Now that we have our report, we are going to follow a similar approach already demonstrated for building policies in Intune:

1. Use the Microsoft Edge baseline in Intune security baselines for the initial policy payload. These baselines contain the bulk of the Microsoft recommended security baselines from the toolkit.

2. Supplement additional settings using the settings catalog.

In *Chapter 8, Keeping Your Windows Client Secure*, we walked through how to append the CSV report and add additional headers to the MDM policy workload to track where settings are being applied. After running the review, we ended up creating two profiles in Intune that apply the Edge policy payloads, as follows:

- Using Intune security baselines, a security baseline named **Microsoft Edge Baseline**

- Using the settings catalog, a policy named **Microsoft Edge Security Baseline**

To easily find the policies using the settings catalog, search for the setting name in the spreadsheet using the Settings Picker. Most settings can be found listed in alphabetical order by choosing the **Microsoft Edge** category too. Repeat the process for the remainder of the policies until the profile is complete.

Once the client syncs into Intune, the Edge policies are imported into the registry and a mapping is created to know how to associate policies settings with their corresponding registry keys, just like with Group Policy. This process is called **ADMX ingestion** and happens conveniently all through the Intune UI. If the **ControlPolicyConflicts** CSP has been enabled on the client, a blocking record event is created to gracefully handle any potential conflicts with Group Policy settings. We will cover how to manually ingest ADMX policies in Intune for Google Chrome, but we can see this in action by opening **Registry Editor** and navigating to HKLM:\SOFTWARE\Microsoft\ PolicyManager\AdmxDefault\{GUID}. In the following figure, you can see the Microsoft Edge ADMX ingested policies that include the following taxonomy as a child item underneath the GUID registry key: **Ingested name | Namespace | Category | Policy name:**

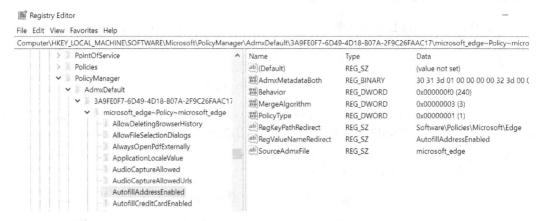

Figure 9.5 – ADMX ingestion of Microsoft Edge policies

This taxonomy can be seen in the full OMA-URI path for enforcing Microsoft Edge policies as follows:

```
./Device/Vendor/MSFT/Policy/Config/microsoft_
edge~Policy~microsoft_edge.
```

In this example, most of the Microsoft Edge security settings are configurable through the settings catalog, and the mapping all happens conveniently through the Intune UI. Next, let's look at how to manage Edge browser extensions.

Managing Edge browser extensions

Brower extensions are a great way to add additional features and functionality to the web browser and are typically developed for all major browsers. A few examples of what an extension is used for could include password managers, copy-to-clipboard functions, auto-filling information, ad blockers, reading email, and even games. Although useful, browser extensions can present a significant security risk and must be managed effectively in an enterprise environment. Malicious extensions can be used to steal data, install malware, create persistent backdoors, log keystrokes, and send information to command-and-control servers. In an enterprise scenario, we can manage extensions in Microsoft Edge through the following policies:

- **Configure allowed extension types** to control the extension types that can be installed in the browser. Examples include browser themes, edge extensions, and hosted apps.

- **Block external extensions from being installed** will block any extensions that are not available through the Microsoft Edge Add-ons website.

- **Control which extensions cannot be installed** creates a blocklist. All extensions can be blocked by setting the policy using an asterisk.

- **Allow specific extensions to be installed** can be used as a safelist to allow extension installation if the **Control which extensions cannot be installed** policy is set to block all.

- **Control which extensions are installed silently** allows admins to push extensions directly to users without requiring interaction. The user cannot manually remove these through the browser.

- **Configure extension management settings** allows for configuring extension-specific permissions or for auto-pinning to the toolbar. An example of extension permissions could be blocking access to USB, preventing script injection, or modifying web requests.

By default, both the CIS and Microsoft baseline recommendation is to set the **ExtensionInstallBlocklist** policy to block all extensions. This is the most secure configuration but will require additional IT administration if there is a business requirement to install or allow an extension. Let's look at using the **Control which extensions are installed silently** policy to silently install the Office browser extension in Microsoft Edge.

To create the policy, we will need the extension ID and update URL of the Microsoft store extension repository, to ensure it updates automatically. The extension ID can be found by visiting the Microsoft Edge Add-ons website and copying the end of the URL. For example, the current URL for the Office browser extension is as follows: `https://microsoftedge.microsoft.com/addons/detail/office/gggmmkjegpiggikcnhidnjjhmicpibll`.

The app ID or extension ID is `gggmmkjegpiggikcnhidnjjhmicpibll`. Since this is hosted in the Microsoft Add-ons store, we can use the default Microsoft update URL, so the browser knows where to check for updates:

```
https://edge.microsoft.com/extensionwebstorebase/v1/crx.
```

Let's combine the two together to create the formatting string to be used by the policy, as in this example:

```
extension_id;update_url.
```

> **Tip**
> You can also use the Google Chrome Web Store update URL if the Microsoft Add-on store does not have the extension.

Next, let's update the **Microsoft Edge Security Baseline** policy we created earlier in MEM:

1. Sign in to MEM at `https://endpoint.microsoft.com`, choose **Devices | Configuration profiles**, and click the policy created earlier.

2. Click **Edit** next to **Configuration settings**.

3. Once in **Configuration settings**, click on **Add settings** to open the *Settings Picker*.

4. Search for `Microsoft Edge`. Double-click the **Microsoft Edge\Extensions** category to view the available settings.

5. Select the **Control which extensions are installed silently** setting.

6. Close the Settings Picker and find the **Extensions** subcategory and enable the **Control which extensions are installed silently** setting.

7. Enter the following string in the input box: `gggmmkjegpiggikcnhidnjjhmicpibll;https://edge.microsoft.com/extensionwebstorebase/v1/crx.`

8. Click **Review + Save** and finally **Save** to update the profile.

The following screenshot shows adding the extension ID and the update URL to the policy in MEM:

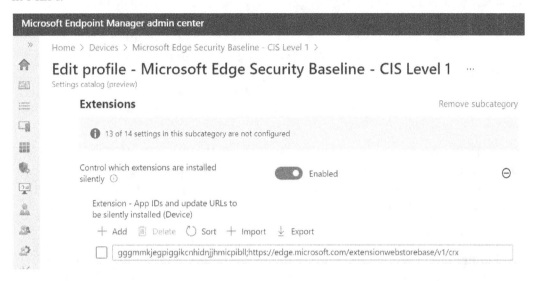

Figure 9.6 – Microsoft Edge extension policy

After the next Intune policy sync, the client should create a blocking record and add the new **ExtensionInstallForcelist** policy with the configured extension ID. To confirm the policy applied, open a command prompt and run the following command: `gpupdate /force`. The Office browser extension should install automatically. You can also view settings in the following locations in the registry:

```
HKLM:\SOFTWARE\Microsoft\PolicyManager\current\device\
microsoft_edge~Policy~microsoft_edge~Extensions
```

```
HKLM:\SOFTWARE\Policies\Microsoft\Edge\
ExtensionInstallForcelist
```

Multiple extensions can be added using this method, and creating a blocklist is done in a similar manner. Next, let's look at configuring enterprise sign-in mode to add sync capabilities for your users.

Configuring enterprise sign-in mode

A great feature available in Microsoft Edge is enterprise sync which can be used to sync favorites, passwords, and other settings to the Microsoft cloud. Earlier, in *Chapter 7, Deploying Windows Securely*, we covered Microsoft Edge sync and its uses for backing up user data and settings during device provisioning. Although the Level 1 CIS recommendation is to disable the synchronization of data using Microsoft sync services, this is required for Edge Sync to work. Using the **Configure the list of types that are excluded from synchronization** policy can limit the types of data that are allowed to sync if privacy is a concern.

Let's now look at configuring Edge sync using Intune by logging in to MEM at `https://endpoint.microsoft.com` and editing **Microsoft Edge Security Baseline** we created earlier. Choose **Add settings** to open the Settings Picker and filter for the **Microsoft Edge** category to view the settings. We will set the following policies:

- **Browser sign-in** is set to **Enabled | Enable browser sign-in.**

- **Configure whether a user always has a default profile automatically signed in with their work or school account** is set to **Enabled**. This will ensure that a user cannot sign out of their work or school account in Edge.

- **Restrict which accounts can be used as Microsoft Edge primary accounts** is set to **Enabled** and includes only your corporate domains. Use the following format to set specific domains: `.*@mycompany.com`.

- **Disable synchronization of data using Microsoft sync services** is set to **Disabled**. This will allow for cloud sync.

- **Configure the list of types that are excluded from synchronization** is set to **Enabled** and we recommend entering `passwords`, `extensions`, `openTabs`, and `settings`.

When a user first opens Microsoft Edge, they are asked whether they want to sync settings. If you wish to enforce synchronization, set **Force synchronization of browser data and do not show the sync consent prompt** to **Enabled**.

Review and save your policy and assign it to a group of users or devices. After the policies apply, the account pop-up menu in Edge should look something like the following screenshot:

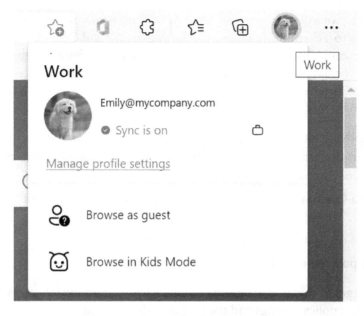

Figure 9.7 – Microsoft Edge enterprise sync

If you have applied all the recommendations in the previous sections, Microsoft Edge is hardened and ready for company use. Depending on the version of Windows your company has rolled out, it should be automatically installed for users as the legacy Edge is out of support starting March 9, 2021. Next, let's take a similar approach and build a Google Chrome security baseline using Intune.

Configuring a Google Chrome security baseline

The Google Chrome browser has quickly become the most widely used browser in the world, accounting for almost 64% of the browser market share according to *StatCounter*. The latest worldwide browser usage statistics can be found at the following link: `https://gs.statcounter.com/`.

It is fast, user-friendly, offers cross-platform compatibility, and contains a large set of configurable security features. For Windows-based systems, Google offers an enterprise bundle that allows admins to configure fine-grained security controls over policies, add-ons, and customizations that can be configured using Group Policy, Configuration Manager, or MDM. Just as with Microsoft Edge, CIS also releases recommended Level 1 and Level 2 benchmarks for Google Chrome. We will use these recommendations to configure a security baseline for Google Chrome using Intune. Until recently, Google Chrome policies were not available natively in Intune through the settings catalog or other device configuration profiles.

To accomplish this, we will use a process called ADMX ingestion by pushing the ADMX-backed definitions to a client using Intune via the Policy CSP and OMA-URI. Under the hood, this is very similar to how Microsoft Edge policies are configured with the settings catalog. Since Intune has minimal support for controlling third-party apps, we need to ingest the Google Chrome ADMX metadata, so the client is aware of how to map the registry values. To do this, we will complete the following steps:

1. Download the latest Google Chrome enterprise bundle that includes the ADMX templates from the following link: `https://cloud.google.com/chrome-enterprise/browser/download`.

2. Configure a **Custom template** device configuration profile and ingest the ADMX metadata.

3. Validate the data was ingested locally on a client.

4. Configure policies from the CIS recommendations.

First, let's look to get a basic understanding of how ADMX ingestion works. We will identify how to find policy settings and the way they should be structured for successful delivery to the client.

ADMX ingestion using Intune

For Windows to ingest an ADMX file, it needs to be processed through the Policy CSP on the client using the following OMA-URI:

`./Vendor/MSFT/Policy/ConfigOperations/ADMXInstall.`

Windows maps the ingested policy by parsing the application ADMX metadata file and storing the policy definition in the MDM Policy CSP client store. From there, a SyncML command is used to tell the client which registry keys to set based on the value configured in the payload. The formatting in the following screenshot can be used as our reference of the OMA-URI for the Google Chrome ADMX metadata:

Figure 9.8 – OMA-URI of ADMX ingestion

In Intune, we are going to create a custom template using the following steps:

1. From MEM, go to **Devices**, choose **Configuration Profile**, and click **Create Profile**.

2. Choose **Windows 10 and later** for **Platform**, select **Templates** for **Profile type**, choose **Custom**, and then click **Create**.

3.   Give the profile a name and description, such as `Google Chrome Security Baseline`. Click **Next**.

- On **Configuration Settings**, click **Add** to add a row, and fill out the following information:

- **Name**: `Google Chrome ADMX Ingestion`

- **OMA-URI**: `./Device/Vendor/MSFT/Policy/ConfigOperations/ADMXInstall/Chrome/Policy/ChromeAdmx`

- **Data type**: `String`

Follow these steps to copy the ADMX metadata and paste it into the value field:

1.   Download and extract the Google Chrome enterprise bundle.

2.   Open **Configuration | ADMX folder** and find the `chrome.admx` file.

3.   Right-click and open with Notepad.

4.   Click **Edit** and choose **Select All**. Click **Edit** and choose **Copy**.

5.   Go back to the **Add Row** settings and paste the contents into the value field.

The **Add Row** OMA-URI settings menu should look like the following figure:

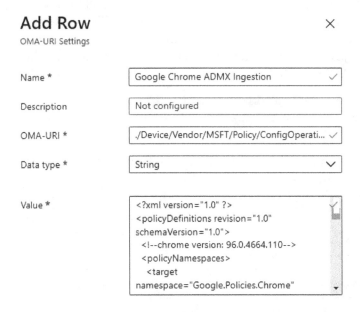

Figure 9.9 – ADMX ingestion for Google Chrome

6.   Click **Save** to commit the settings, and **Review + Save** to commit the setting changes.

The policy is now ready to assign to users or device groups. Let's look at a client that received the policy to confirm it applied correctly.

Open **Registry Editor**, navigate to `HKLM:\SOFTWARE\Microsoft\PolicyManager\AdmxInstalled`, and expand the GUID. If the ADMX is ingested correctly, you should see the folder tree layout **Chrome | Policy | ChromeAdmx,** as in the following screenshot underneath the GUID:

Figure 9.10 – ADMX ingested into the registry

We can confirm this in the event viewer under the `DeviceManagement-Enterprise-Diagnostics-Provider-Admin` logs looking for event ID `873` and `866`. This should show the ADMX ingestion started and completed matching the GUID as the provider we expanded upon in the registry. Back in the registry, expand the GUID under `PolicyManager\AdmxDefault` to view all the available policy names and categories. These paths will be referenced in the OMA-URI when configuring our required policy. Let's look at how to disable the password manager by using the `PasswordManagerEnabled` policy as an example. Once we identify the correct settings from the ADMX metadata, the final OMA-URI path will be as follows:

```
./Device/Vendor/MSFT/Policy/Config/
Chrome~Policy~googlechrome~PasswordManager/
PassworeManagerEnabled
```

The following figure can also be used as a formula reference when building out the OMA-URI path for additional policies:

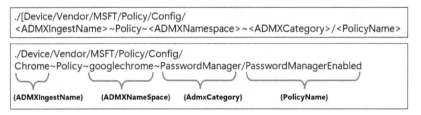

Figure 9.11 – OMA-URI reference for required settings

To understand exactly how the OMA-URI translates from the ADMX, let's find the password manager policy in the `chrome.admx` file:

1. Open the `chrome.admx` file in Notepad or your favorite text editor.
2. Search for `PasswordManagerEnabled`.

   In the XML snippet from the following figure, we can match child elements and attributes inside the code tree structure and find {`AdmxCategory`} and {`PolicyName`} to construct our OMA-URI path. The text highlighted in bold in the XML code shows how to find these mappings:

```
<policy class="Both" displayName="$(string.PasswordManagerEnabled)"

explainText="$(string.PasswordManagerEnabled_Explain)" key="Software\Policies\Google\Chrome"

name="PasswordManagerEnabled" presentation="$(presentation.PasswordManagerEnabled)"

valueName="PasswordManagerEnabled">

    <parentCategory ref="PasswordManager"/>
```

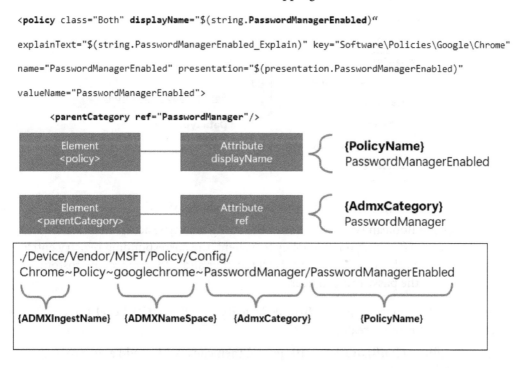

Figure 9.12 – ADMX XML tree structure

Once you get comfortable looking at the ADMX formatting, this will become apparent for any additional policies you want to configure.

Now that we have the OMA-URI path, the next step is to identify the possible values that can be accepted. In this case, it's a simple Boolean value, and the setting is either enabled (decimal value of 1) or disabled (decimal value of 0). Looking at the chrome.admx file, find the <enabledValue> element. It is highlighted with a border in the following screenshot:

```
<policy class="Both" displayName="$(string.PasswordManagerEnabled)"

explainText="$(string.PasswordManagerEnabled_Explain)" key="Software\Policies\Google\Chrome"

name="PasswordManagerEnabled" presentation="$(presentation.PasswordManagerEnabled)"

valueName="PasswordManagerEnabled">

        <parentCategory ref="PasswordManager"/>

        <supportedOn ref="SUPPORTED_WIN7"/>

        <enabledValue>

                <decimal value="1"/>

        </enabledValue>

        <disabledValue>

                <decimal value="0"/>

        </disabledValue>

</policy>
```

Set to disabled in Intune

Data type: String
Value:

Figure 9.13 – Identifying the policy value in ADMX

Since the policy is a Boolean type, the value of the string data type accepted in Intune will be . After applying the policy to a client, the corresponding registry entry will have a DWORD value of 0. Let's put this all together and add a row to our custom profile to configure the password manager:

1. From MEM, open the **Google Chrome Security Baseline – CIS Level 1** device configuration profile created earlier.

2. Choose **Edit** in **Configuration settings** and then click **Add** to add a new row. Enter the following values:

   - **Name**: Chrome - PasswordManagerEnabled

   - **OMA-URI**: ./Device/Vendor/MSFT/Policy/Config/ Chrome~Policy~googlechrome~PasswordManager/ PasswordManagerEnabled

   - **Data type**: String

   - **Value**:

3. Click **Save** and **Review + Save** to commit the setting changes.

After the client runs an Intune policy sync, let's confirm the setting applied. This time we will open Google Chrome and type `Chrome://Policy` in the address bar. You can see from the following screenshot that the **PasswordManagerEnabled** policy is set to `false`:

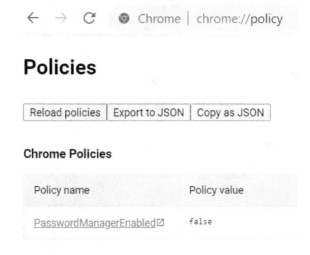

Figure 9.14 – Google Chrome applied policies

Next, let's look at expanding on ADMX ingestion and configuring Google Chrome extensions. Unlike the Boolean values for the password manager, there are a few additional steps to ensure the policy is configured correctly in Intune.

Managing Chrome browser extensions

Like Microsoft Edge, Google Chrome has an extensive catalog of add-ons and extensions available through the Chrome Web Store that can enhance user experience and add functionality. Because Microsoft Edge is a Chromium-based browser, similar policies are available for Chrome to manage these extensions. Both browsers are also aligned with the same CIS recommendations regarding their management of them. You can visit the *Configuring a Microsoft Edge security baseline* section in this chapter for more information about the different policy types.

In the following example, we are going to use ADMX ingestion to deploy an extension blocklist and silently install the Microsoft Defender Browser Protection and Windows 10 account extensions:

1. First, let's find the `ExtensionInstallBlocklist` policy inside the `chrome.admx` file and look at the available values. These values are located inside the `<elements>` XML, which is highlighted in the following screenshot:

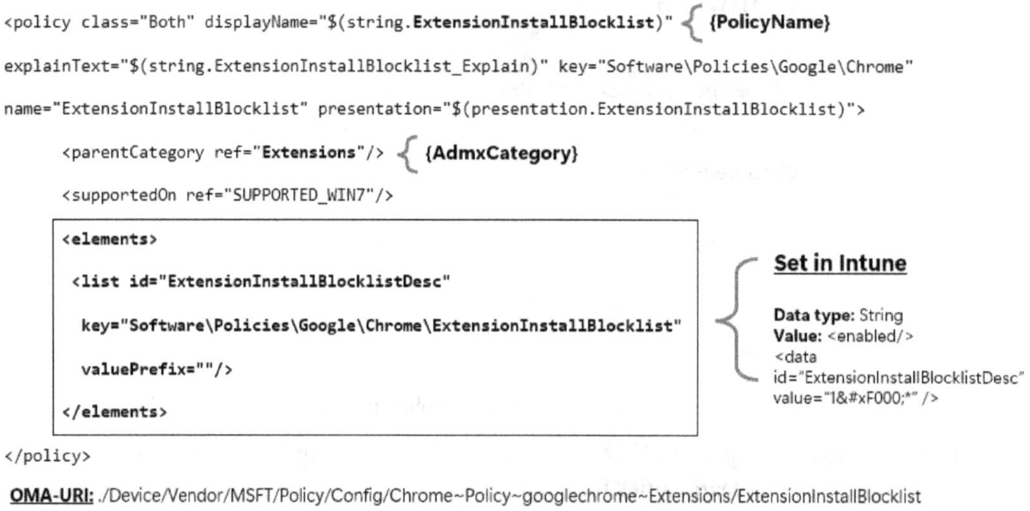

```
<policy class="Both" displayName="$(string.ExtensionInstallBlocklist)" { {PolicyName}

explainText="$(string.ExtensionInstallBlocklist_Explain)" key="Software\Policies\Google\Chrome"

name="ExtensionInstallBlocklist" presentation="$(presentation.ExtensionInstallBlocklist)">

    <parentCategory ref="Extensions"/> { {AdmxCategory}

    <supportedOn ref="SUPPORTED_WIN7"/>

    <elements>

    <list id="ExtensionInstallBlocklistDesc"

    key="Software\Policies\Google\Chrome\ExtensionInstallBlocklist"

    valuePrefix=""/>

    </elements>

</policy>
```

Set in Intune

Data type: String
Value:
<data
id="ExtensionInstallBlocklistDesc"
value="1*" />

OMA-URI: ./Device/Vendor/MSFT/Policy/Config/Chrome~Policy~googlechrome~Extensions/ExtensionInstallBlocklist

Figure 9.15 – Extension blocklist policy

Inside `<elements>`, there is a child element labeled `<list id>`. These elements map to `REG_SZ` registry strings as name/value pairs. When we input the value for the OMA-URI settings, `<data id>` must match the attribute shown inside `<list id>`. Each name/value pair is separated by using the Unicode character `0xF000` (encoded to `&#xF000`), and if the policy contains multiple entries, each entry needs to be separated by a semicolon. In the preceding example, the name/value pair is encoded as `1&#xF000;*`. In the registry, the name would be `1`, the type would be `REG_SZ`, and the value would be `*`.

2. Let's push this setting down to the client by modifying the device configuration profile created earlier and adding a row with the following settings:

   - **Name:** `Chrome - ExtensionInstallBlocklist`

   - **OMA-URI:** `./Device/Vendor/MSFT/Policy/Config/ Chrome~Policy~googlechrome~Extensions/ ExtensionInstallBlocklist`

- **Data type**: `String`

- **Value**: `<enabled/> <data id="ExtensionInstallBlocklistDesc" value="1&#xF000;*" />`

> **Tip**
>
> The OMA-URI and values are case-sensitive or else the policy will fail to apply.

3. Once the client receives the policy from Intune, we can confirm this setting is applied by opening **Registry Editor** and navigating to `HKLM:\SOFTWARE\Policies\Google\Chrome\ExtensionInstallBlocklist`.

   If the policy is applied successfully, it should look like the following screenshot at the registry path:

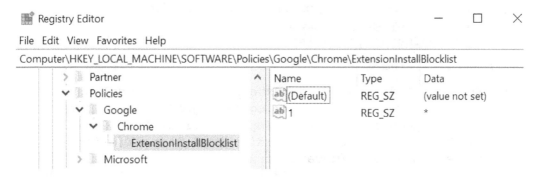

Figure 9.16 – ExtensionInstallBlocklist policy applied

4. Next, let's add the `ExtensionInstallForcelist` policy to silently install the Microsoft Defender Browser Protection and Windows 10 account extensions. Just like with the blocklist, this policy uses the `<list id>` element type and will have formatting like the blocklist policy.

   To format the string, we need to gather the extension ID from the Google Chrome store. Earlier, in the *Managing Edge browser extensions* section of this chapter, we covered how to find the extension ID and format the string with the Add-ons store update URL if you need a review. Extension IDs for Google Chrome can be found at the Chrome Web Store from the following link: `https://chrome.google.com/webstore/category/extensions?hl=en-US`.

5. The update URL is `https://clients2.google.com/service/update2/crx`. The following figure contains the formatted strings for both extensions:

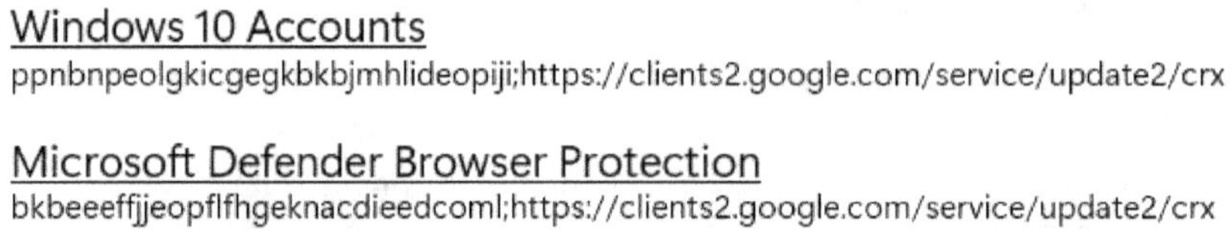

    <u>Windows 10 Accounts</u>
    ppnbnpeolgkicgegkbkbjmhlideopiji;https://clients2.google.com/service/update2/crx

    <u>Microsoft Defender Browser Protection</u>
    bkbeeeffjjeopflfhgeknacdieedcoml;https://clients2.google.com/service/update2/crx

    Figure 9.17 – Google Chrome extension IDs

6. Back in MEM, add a new row to the configuration settings and set the following values:

    - **Name**: `Chrome - ExtensionInstallForcelist`

    - **OMA-URI**: `./Device/Vendor/MSFT/Policy/Config/Chrome~Policy~googlechrome~Extensions/ExtensionInstallForcelist`

    - **Data type**: `String`

    - **Value**: `<enabled/> <data id="ExtensionInstallForcelistDesc" value="1&#xF000;ppnbnpeolgkicgegkbkbjmhlideopiji;https://clients2.google.com/service/update2/crx&#xF000;2&#x-F000;bkbeeeffjjeopflfhgeknacdieedcoml;https://clients2.google.com/service/update2/crx "/>`

7. After the client receives the next Intune policy update, let's confirm on the client by opening **Registry Editor** and navigating to `HKLM:\SOFTWARE\Policies\Google\Chrome\ExtensionInstallForcelist`.

8. If the policy is applied successfully, it should look like the following screenshot. Notice that multiple registry entries were added by separating the name/value pairs with a semicolon.

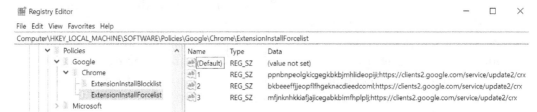

Figure 9.18 – ExtensionInstallForcelist policy applied

Using the blocklist and force list policies is an effective way to manage extensions for Google Chrome and a great way to keep users secure from accidentally installing malicious extensions. If there is a business requirement to allow additional extensions but you don't want to force-install them, use the `ExtensionInstallAllowlist` policy as a safelist.

Next, let's put this all together and complete the implementation of the CIS Level 1 profile recommendations to finish building the baseline.

Building the baseline

Unlike Edge, Google Chrome doesn't have a recommended security baseline from Microsoft. For our baseline, we will only use the recommendations from CIS. We can run an analysis using Policy Analyzer and compare the Level 1 and Level 2 CIS profiles. In the following figure, you can see there are 22 total additional policies in Level 2, with 21 differences in settings for overlapping policies:

Google Chrome Security Benchmark	Total	Differences
Level 1 (L1) – Corporate/Enterprise Environment	80	21
Level 2 (L2) – High Security/Sensitive Data Environment	102	--

Figure 9.19 – Analysis of Level 1 and Level 2 CIS profiles using Policy Analyzer

If only applying the Level 1 recommendations, we recommended disabling the following setting:

- **Browser Sign-in settings** are set to `Disabled`. Unless you are a G Suite Enterprise customer, leaving this setting enabled is a significant security risk as corporate data can leak and synchronize with non-managed accounts.

The following are the high-level steps we took to build the baseline using ADMX ingestion in Intune:

1. Download the Google Chrome enterprise bundle and import the latest ADMX files to the Group Policy Central Store.
2. Use the CIS Google Chrome Benchmark Build Kit or manually build a GPO based on the Level 1 recommendations.
3. Save the GPO report as an XML file and upload it to Group Policy analytics.
4. Export the analysis as a CSV file.

5. Audit each policy and update the CSP mapping column to reflect the OMA-URI path of the ADMX ingested policy specific to Google Chrome. By default, it will try to map them to Microsoft Edge.

6. Use the `chrome.admx` file to determine the required policy values and format the correct value for Intune.

7. Implement each policy by adding a row for each unique setting in our device configuration profile.

Unfortunately, the export will show **Migrate to Edge** in the MDM support column. We decided to use this CSV file as it's in a similar format to the Microsoft Edge baseline we audited and implemented earlier, but you can use whatever works best for you. In most instances, we can replace `microsoft_edge~Policy~microsoft_edge` of the OMA-URI path with `Chrome~Policy~googlechrome`. Use the registry entries under **PolicyManager\AdmxDefault** on a client that ingested the ADMX metadata as a reference for identifying the correct category and policy names. After updating the CSP mapping column, we had a reference file that looked like the following figure. We left a few Edge references to compare on rows 7 and 8.

	A	D	G H I J K L M N O P
1	Setting Name	Value	CSP Mapping
2	Allow download restrictions/Download restrict	1	./Device/Vendor/MSFT/Policy/Config/Chrome~Policy~googlechrome/DownloadRestrictions
3	Allow Google Cast to connect to Cast devices	Disabled	./Device/Vendor/MSFT/Policy/Config/Chrome~Policy~googlechrome~GoogleCast/EnableMediaRouter
4	Allow invocation of file selection dialogs	Disabled	./Device/Vendor/MSFT/Policy/Config/Chrome~Policy~googlechrome/AllowFileSelectionDialogs
5	Allow or deny audio capture	Disabled	./Device/Vendor/MSFT/Policy/Config/Chrome~Policy~googlechrome/AudioCaptureAllowed
6	Allow or deny screen capture	Disabled	./Device/Vendor/MSFT/Policy/Config/Chrome~Policy~googlechrome~ScreenCapture/ScreenCaptureAllowed
7	Allow or deny video capture	Disabled	./Device/Vendor/MSFT/Policy/Config/microsoft_edge~Policy~microsoft_edge/VideoCaptureAllowed
8	Allow proceeding from the SSL warning page	Disabled	./Device/Vendor/MSFT/Policy/Config/microsoft_edge~Policy~microsoft_edge/SSLErrorOverrideAllowed

Figure 9.20 – Policy analysis from Group Policy analytics

Next, let's look at setting the **Allow download restrictions** policy. This policy is again slightly different to build the formatting of the value and is worth covering:

1. Open the `chrome.admx` file and search for `DownloadRestrictions` in the file. In this example, it contains `<enum id>`, which makes it an Enum type. This means that the policy settings can have one of the defined attributes as outlined in the XML tree. The CIS recommendations say to configure for `BlockDangerousDownloads`, which is a decimal value of 1. Once again, in the following figure, we can get an idea of how to configure the value for Intune:

```
<policy class="Both" displayName="$(string.DownloadRestrictions)"  { {PolicyName}
  explainText="$(string.DownloadRestrictions_Explain)"
  key="Software\Policies\Google\Chrome" name="DownloadRestrictions"
  presentation="$(presentation.DownloadRestrictions)">
      <parentCategory ref="googlechrome"/> { {AdmxCategory}
      <supportedOn ref="SUPPORTED_WIN7"/>
      <elements>
        <enum id="DownloadRestrictions" valueName="DownloadRestrictions">
          <item displayName="$(string.DownloadRestrictions_DefaultDownloadSecurity)">
            <value>
              <decimal value="0"/>
            </value>
          </item>
          <item displayName="$(string.DownloadRestrictions_BlockDangerousDownloads)">
            <value>
              <decimal value="1"/>
            </value>
          </item>
            . . .
        </enum>
      </elements>
</policy>
```

Set in Intune

Data type: String
Value:
<data id="DownloadRestrictions"
value="1" />

OMA-URI: ./Device/Vendor/MSFT/Policy/Config/Chrome~Policy~googlechrome/DownloadRestrictions

Data Type: String

Value: <data id="DownloadRestrictions" value="1" />

Figure 9.21 – DownloadRestrictions policy

Just as with `<list id>`, `<enum id>` will map to `<data id>` in the Intune value field.

2. Let's add a row to our device configuration profile for the `DownloadRestrictions` policy and set the following values:

- **Name**: `Chrome - DownloadRestrictions`

- **Description**: `Allow download restrictions`

- **OMA-URI**: `./Device/Vendor/MSFT/Policy/Config/ Chrome~Policy~googlechrome/DownloadRestrictions`

- **Data type**: `String`

- **Value**: `<enabled/> <data id="DownloadRestrictions" value="1" />`

3. After the client receives the next Intune policy update, confirm in **Registry Editor** by navigating to `HKLM:\SOFTWARE\Policies\Google\Chrome`.

4. If the policy is applied successfully, you should see `DownloadRestrictions` `REG_DWORD` with a value of 1.

ADMX ingestion isn't just limited to establishing apps that have ADMX configurations but can also be used for configuring your own custom policies. The following are a few additional resources that are helpful for understanding how ADMX-based policies work using MDM and Intune:

- `https://docs.microsoft.com/en-us/windows/client-management/mdm/win32-and-centennial-app-policy-configuration`

- `https://docs.microsoft.com/en-us/windows/client-management/mdm/understanding-admx-backed-policies`

Next, let's look at securing Microsoft 365 apps by creating a security baseline and enabling advanced features using Intune and the Microsoft 365 admin centers.

Securing Microsoft 365 apps

The Microsoft Office suite includes some of the world's most popular productivity apps used today, with programs such as Outlook, OneDrive, Word, Excel, PowerPoint, OneNote, SharePoint, and Teams. According to `https://www.statista.com/`, Microsoft Office 365 controls around 47.5% of the market share for major office suite technologies worldwide, and is used by over 731,000 companies in the United States alone. This wide usage rate makes it a great attack target to build exploits and should not be overlooked when building out security controls. According to `https://www.cvedetails.com`, there were 63 named CVE vulnerabilities for Microsoft Office in 2021 alone, with three having a score of 9.3 or greater. The following URL shows the total of all Office vulnerabilities since 1999: `https://www.cvedetails.com/product/320/Microsoft-Office.html?vendor_id=26`.

In addition to enforcing security controls through baselines, it's important to install security updates regularly and build a program to report on non-compliance. Modern versions of Microsoft 365 apps no longer use Windows Updates to download updates and instead rely on Microsoft Office's **Content Delivery Network (CDN)**. The update mechanism or maintenance tasks that check for updates on a client are separate from Windows Updates, so it's important not to assume an updated Windows client is also successfully receiving Office updates. Let's look at how to build a security baseline for Microsoft 365 apps.

Building a security baseline for M365 apps

Security baselines for Microsoft Office apps are available from CIS and through the Microsoft Security and Compliance Toolkit. Just as with Windows and Edge, we recommend you review, audit, and combine recommendations from both controls to meet your company's security requirements. One fundamental difference worth mentioning with enforcing Office policies compared to Windows or web browsers is that most policies are user-targeted. This is important to note because your policy delivery mechanism must be able to import settings for the currently logged-on user, which could present a challenge as many device management solutions use a system context for payload delivery. Once the recommendations have been reviewed, we can use the following methods to apply them:

- ADMX policies for user and device scopes through the MEM settings catalog

- The Office cloud policy service for user-based policies

- Configuration baselines and LGPO application packages in Microsoft Endpoint Configuration Manager for both user and computer policies

- Group Policy

The Microsoft security baselines set includes recommendations for both computer and user-based policies bundled together across all apps in the Office suite. There are six pre-built GPOs in the toolkit, but the main MSFT Microsoft 365 apps for both the user and computer scope provide the best combination for security and compatibility. The additional four policies cover security settings specifically for macros and legacy files that should also be tested and implemented. The reason they are not combined is most likely that macro-specific policies tend to have a larger impact on user experience and were called out for awareness. The CIS benchmark recommendations include a Level 1 profile separated among each app individually for Word, Excel, Outlook, Access, PowerPoint, and common elements shared across all apps.

Choosing an approach for policy management of Office apps will depend on the deployment method . For example, if the Office deployment uses the MSI or Click-to-Run installer, the Office version, and the licenses purchased for users are all factors that determine how policy can be managed. It's worth reviewing the deployment prerequisites to understand what best suits your needs. A link to the Microsoft documentation, with detailed information on deployment planning, can be found here: `https://docs. microsoft.com/en-us/deployoffice/plan-microsoft-365-apps`.

First, let's look at auditing the baseline from Microsoft by downloading the reference GPOs and importing them into Group Policy analytics in Intune for analysis.

Building the baseline

To compile the baseline, we are going to complete the following steps:

1. Download the latest Microsoft 365 Apps for enterprise security baseline. The baseline can be downloaded from the Security and Compliance toolkit at the following link: `https://www.microsoft.com/en-us/download/details.aspx?id=55319`.

2. Download and import the latest Microsoft 365 Apps administrative templates. The latest administrative template files can be found by visiting this link: `https://www.microsoft.com/en-us/download/details.aspx?id=49030`.

> **Tip**
> The administrative templates for the legacy JScript block policies can be found in the `SecGuide.admx` files located in the templates folder of the download.

3. Create a new GPO in the Group Policy Management Console for each of the baselines downloaded from the toolkit.

    We created the following six group policies:

    -   **MSFT Microsoft 365 Apps v2112 – Computer**
    -   **MSFT Microsoft 365 Apps v2112 – User**
    -   **MSFT Microsoft 365 Apps v2112 – DDE Block – User**
    -   **MSFT Microsoft 365 Apps v2112 – Legacy File Block – User**
    -   **MSFT Microsoft 365 Apps v2112 – Legacy JScript Block – Computer**
    -   **MSFT Microsoft 365 Apps v2112 – Require Macro Signing – User**

4. Save each report as an XML file and import them into Group Policy analytics for analysis.

    The following screenshot shows the analysis of MDM supported policies from the Microsoft security baseline recommendations:

Figure 9.22 – Group Policy analytics for M365 app baselines

5. Next, export each analyzed policy and consolidate them into a single CSV file. Then, we added a new header column to determine where each policy will be enforced for reference.

    The latest approach for policy enforcement is to use the Office cloud policy service if you meet the prerequisites. We will go into more detail in the next section but decided to use this as our first choice. If there are any gaps in policies not available in the cloud policy service, we will supplement them using Intune. In our exported CSV, we will add the following column headers to columns H through K.

    - **Office Cloud Policy Service**
    - **Settings Catalog**
    - **Templates**
    - **ADMX**

    > **Tip**
    >
    > We won't be covering supplementing policies in Intune in this section, but this can be accomplished through the settings catalog, as shown in previous examples in this chapter.
    >
    > Both user and computer based policies support the migrate option in Group Policy analytics.

    After combining the output of each analysis, we ended up with 358 unique policies across all M365 apps.

6. Audit each policy and build the baseline in the respective areas, making a note on the master spreadsheet where each policy is applied.

This will be time-consuming, but once it's complete, your Office apps will have robust security controls applied. It is strongly recommended to review and test each policy to understand the effect they could have on user experience before deploying it into a production environment. Many business-driven processes are automated using functions inside of Office apps, such as macros. We need to ensure to minimize interruption.

Next, let's review how to use the Office cloud policy service for policy management of your Office apps.

Office cloud policy service

The Office cloud policy service allows you to create policies, configure servicing options, customize office deployments, and monitor the health, inventory, and security updates of Office apps through the Microsoft 365 apps admin center portal. The service operates independently of device management, such as Intune, Configuration Manager, and Group Policy. It offers great flexibility because, unlike with device management methods, the client does not need to be under company management for policies to be enforced. Policy assignment is handled by Azure AD group memberships and applied directly onto the device or can be configured for anonymous access to Office web apps. On a Windows client, the policy engine that drives enforcement is the *Click-to-Run* service that syncs with the Office cloud policy service to check for any assigned policies for the user currently signed into Office apps. The intervals occur during first sign-in and on a regular frequency every 90 minutes, or on a 24-hour basis if no policy assignments were previously found for the signed-in user.

To use the Office cloud policy service, you must meet the following requirements:

- Microsoft 365 Apps for enterprise (Click-to-Run) and is not compatible with the MSI or perpetual versions.

- Version 1808 or later of Microsoft 365 apps for Windows only.

- The client must be able to communicate with the cloud policy service endpoints.

Depending on your needs, policy configurations are flexible, and some examples could include structuring them by application type or grouping them together by department, business function, or for the entire organization. Let's create an organization-wide policy from the security baseline recommendations by following these steps:

1. Sign in to the Microsoft 365 Apps admin center at `https://config.office.com`.

2. Click the **Customization** dropdown and choose **Policy Management**.

3. Click **Create** to open the **Create Policy Configuration** menu.

4. Give the policy a name (for example, `Microsoft 365 Apps Security Baseline Org Wide`).

5. Click **Select type** and choose **This policy configuration applies to users**.

6. Click **Select group** and choose an Azure AD security group to assign the policies.

7. Click **Configure Policies** to open the **Select policies** menu.

8. Click the recommendation column and uncheck (select all) and choose **Security Baseline**. Click **Filter**.

The results should now be filtered by Microsoft's recommended security baselines. Not all user-based policy recommendations from the Security and Compliance Toolkit will be available, but new policies are being added frequently.

To create our baseline, we further sorted by platform and policy name and then used the search function to locate policies based on the setting name in the CSV exports from Group Policy analytics using the search bar. To apply a configuration, select the policy and leave the Microsoft recommended security baseline option checked, and click **Save**. This will apply the recommended setting from the security baseline. We continued this process for the remaining policies until we had our profile built and saved it.

In **Policy Management**, evaluations occur based on the assigned priority order listed in the **Policy configurations** pane in the admin center. If a user is in the scope of multiple policy assignments, this order will determine which settings take precedence, with 0 being the highest priority. Our security baseline can be seen in the following figure with a priority of 0. The order can easily be changed as more configurations are added:

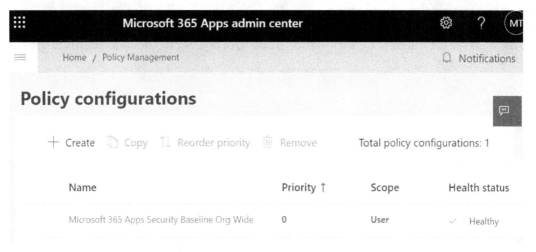

Figure 9.23 – Office cloud policy service

384 Advanced Hardening for Windows Clients

Now that the policy is created, let's confirm the client has checked into the cloud policy service and settings are applied. Open **Registry Editor** and navigate to `HKCU:\ SOFTWARE\ Microsoft\Office\16.0\Common\CloudPolicy`.

Click on the **LastUser** key and check the `PolicyHash` value. If it is blank or missing, then the client did not receive a policy from the Office cloud policy service. A successful policy check-in should look like the following screenshot, with a `PolicyHash` value populated:

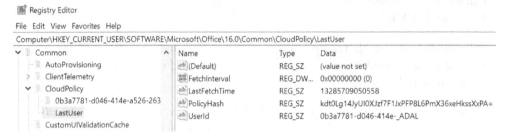

Figure 9.24 – Office cloud policy check-in

> **Tip**
>
> For testing purposes, if you need to force a check-in, delete the entire `CloudPolicy` key and close and re-open Office apps to download a fresh policy set.

Next, let's check whether the actual settings are applied from the policy configuration. The location in the registry is different than where we typically see from Group Policy or MDM. To view the applied settings from the Office cloud policy service, navigate to `HKCU:\SOFTWARE\Policies\Microsoft\Cloud\Office\16.0`. In the subkeys, you should see the different Office products listed with policy categories, as shown in the following screenshot:

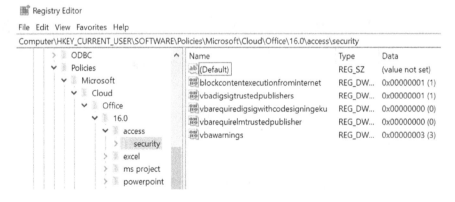

Figure 9.25 – Office cloud policy service policies

Remember, this service offers flexibility to manage Office policies for non-managed devices. If setting policies with the Office cloud policy service in addition to Group Policy or MDM, they will take precedence during conflict resolution even if more restrictive settings are configured. These configurations also *roam* with the user on other devices if the version of Office apps meets the requirements. This is an important consideration when determining whether the cloud policy service is right for your deployment.

For more information about the Office cloud policy service and using the Microsoft 365 Apps admin center features, visit this link: `https://docs.microsoft.com/en-us/deployoffice/admincenter/overview-office-cloud-policy-service`.

Next, let's look at enabling advanced features from the Microsoft Defender suite of protection capabilities.

Advanced protection features with Microsoft Defender

Microsoft Defender is the branding used by Microsoft for labeling their suite of security products that provide **extended detection and response (XDR)** capabilities. The individual products under the Defender moniker when combined are designed to protect, prevent, and respond to threats across multiple workloads as part of a holistic approach to security. Microsoft Defender products leverage Microsoft AI and **machine learning (ML)** to protect organizations through automation and real-time protection capabilities. Let's do a quick review of Microsoft Defender technologies to bring awareness around the different workloads it can protect and the admin portals where each solution can be accessed:

- **Microsoft Defender for Endpoint** powers client and server endpoint security and provides real-time protection and XDR capabilities to stop threats and persistence with a focus on zero-trust principles. This is the core feature set we will review in the following sections. The Microsoft 365 Defender portal can be found at the following link: `https://security.microsoft.com`.

- **Microsoft Defender for Office 365** provides protection for email and collaboration tools against malware, phishing, and business email compromise. These tools are also located in the Microsoft 365 Defender portal.

- **Microsoft Defender for Identity** helps to protect on-premises and hybrid identities from targeted attacks by monitoring user behavior for suspicious activity and reducing the attack surface. The Microsoft Defender for Identity portal can be found at the following link: `https://portal.atp.azure.com`.

- **Microsoft Defender for Cloud Apps** is a **cloud access security broker (CASB)** that provides visibility into workloads across SaaS apps, discovers shadow IT processes, protects sensitive information, and provides governance and compliance over your cloud apps. The Microsoft Defender for Cloud Apps portal can be found at the following link: `https://portal.cloudappsecurity.com`.

- **Microsoft Defender for Cloud** is a vulnerability and threat protection solution that provides insights into the security posture of cloud resources by assessing, securing, detecting, and resolving threats to those resources. Microsoft Defender for Cloud can be accessed in Azure at the following link: `https://portal.azure.com/#blade/Microsoft_Azure_Security/SecurityMenuBlade/0`.

Let's look at enabling advanced features in Microsoft Defender for Endpoint to protect Windows clients at the endpoint.

Defense evasion with tamper protection

A defense evasion tactic is something an attacker would do to try and avoid detection on a compromised system. A few techniques commonly seen in defense evasion include leveraging trusted processes to hide malicious software, obfuscating code, or disabling security software. This is where tamper protection comes in. With tamper protection enabled, Microsoft Defender is locked down and threats such as human-operated ransomware and other malware can no longer disable Defender's real-time protection capabilities and other features used to pick up threats through the registry, PowerShell, or by modifying Group Policy.

Earlier, in *Chapter 8*, *Keeping Your Windows Client Secure*, we recommended tamper protection as part of the Defender Antivirus baseline using endpoint security in Intune. It can also be configured in Intune security baselines, Group Policy, PowerShell, and the organizational level through the Microsoft 365 Defender portal. To enable tamper protection organization-wide, go to `https://security.microsoft.com` and select **Settings | Endpoints | Advanced Features | Tamper protection**.

Cloud-delivered protection is a prerequisite to enable tamper protection through the Microsoft 365 Defender portal. By default, in Windows, this setting is enabled, but it's also recommended as part of the Defender Antivirus baseline. To confirm tamper protection is enabled on a client, open **Windows Security | Virus & threat protection | Manage settings**. It will look like the following screenshot:

Figure 9.26 – Tamper Protection in Windows Security

For more information about tamper protection, visit this link: `https://docs.microsoft.com/en-us/microsoft-365/security/defender-endpoint/prevent-changes-to-security-settings-with-tamper-protection?view=o365-worldwide`.

Next, let's look at protecting against unwanted applications and malicious websites using Microsoft Defender's PUA protection and SmartScreen.

Protecting against untrusted applications and websites

Having protection in place from untrusted software and malicious websites is another zero-trust principle of always verify. Today, many freeware applications available on the internet often come bundled with add-ons as part of the installation. They may contain ads, which is why many software vendors are able to offer them free of charge. These software bundles may not necessarily be malicious in nature but are often unwanted, and having them installed can affect system performance and raise privacy concerns if they collect data about your computer and usage. In addition to protection from low-reputation software, users may inadvertently visit a malicious website facilitated by a phishing scam or by clicking on an unverified link. This can lead users to a credential harvesting website site or start a malware download. To provide protection against these common security risks, we can enable **Defender SmartScreen** and **potentially unwanted applications** (**PUA**) protection through Microsoft Defender features. PUA protection helps to detect and block unwanted software by verifying reputation against Microsoft's intelligence systems and safeguarding users through notifications and file quarantine. SmartScreen protects your users by cross-checking websites and files against a list of reported malware sites and validates reputation from intelligence sources collected from other Windows users. If the site or file in question is found, a warning notification will be presented to the user and the website or the file will be blocked depending on the level of enforcement. First, let's look at enabling PUA protection and review the user experience on a Windows client.

PUA blocking

In *Chapter 8, Keeping Your Windows Client Secure*, we recommended enabling PUA protection in block mode (or enabled) through the remediation section of the Defender Antivirus baseline in endpoint security. PUA protection can also be enabled in Intune security baselines, Group Policy, or with PowerShell. There are additional policies for PUA protection in Microsoft Edge, which are listed in the Microsoft and CIS recommendations.

If the policy is enabled for Microsoft Edge and a file is blocked during download, the user will receive a notification in the browser, as shown in the following screenshot:

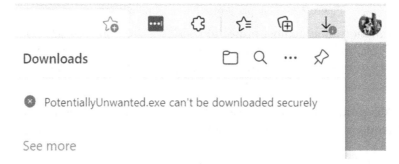

Figure 9.27 – PUA protection

If the download can succeed (as in a software bundle from a trusted reputation) and the installation program launches, PUA protection will quarantine any low-reputation files or additional software included in the bundle and show a notification like the following figure:

Figure 9.28 – PUA protection blocking an app

> **Tip**
>
> We configured the Windows security experience settings in *Chapter 8, Keeping Your Windows Client Secure,* to customize the text seen in the preceding notification.

For premium licensed customers, PUA events are logged in the Microsoft Defender for Endpoint portal and can be queried using an advanced hunting query. Automated workflows can be created from these alerts and trigger actions such as initiating an **IT Service Management (ITSM)** workflow into a ticketing system for follow-up. If a legitimate file is blocked from PUA protection, it can be added to the Defender exclusion list, which is configurable in the Microsoft Defender security baseline in endpoint security. To validate the fact that PUA protection is enabled on a client, open **Windows Security | App & browser control | Reputation-based protection settings** and look for **Potentially unwanted app blocking**. Enabling PUA protection does not require a premium license.

Next, let's enable Microsoft Defender SmartScreen features.

Checking apps and files with SmartScreen

Microsoft Defender SmartScreen provides protection against malicious apps and websites in Microsoft Edge, Windows, and the Microsoft App Store. When configuring a SmartScreen policy, there are a few recommended settings that should be enabled to provide the best level of protection for your users. First, let's look at enabling protection in Microsoft Edge. If you've implemented the Microsoft Edge recommended security baseline from Microsoft and CIS, then SmartScreen should already be enabled using Intune security baselines or through the settings catalog. If using the settings catalog, find these settings in the Settings Picker by searching for SmartScreen and selecting the **Microsoft Edge\SmartScreen** category to add the settings to your policy. We recommend enabling the settings seen in the following figure:

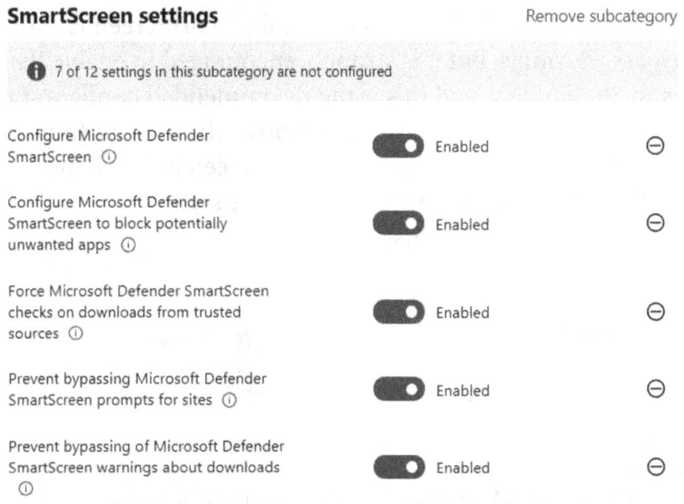

Figure 9.29 – SmartScreen browser settings

Once the policies have been applied, let's see what this looks like from an end user perspective on a client. In this example, we used Microsoft Edge to navigate to a potential phishing site. The user should see a notification as shown in the following screenshot and be blocked from proceeding through the warning:

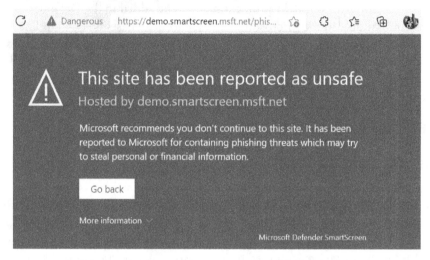

Figure 9.30 – SmartScreen protection in Microsoft Edge

> **Tip**
>
> SmartScreen protection can be enabled for Google Chrome by installing the Microsoft Defender Browser Protection extension.

Next, let's enable SmartScreen for Windows. By default, SmartScreen is configured to allow users to bypass warnings, but it is strongly encouraged to enable the **Prevent Override For Files in Shell** policy, and this is the recommended configuration from CIS and Microsoft. SmartScreen can be enabled for Windows through an Intune security baseline or using the settings catalog. To enable SmartScreen in block mode, search for Smart Screen in the Settings Picker and configure the settings shown in the following screenshot to follow the recommendations:

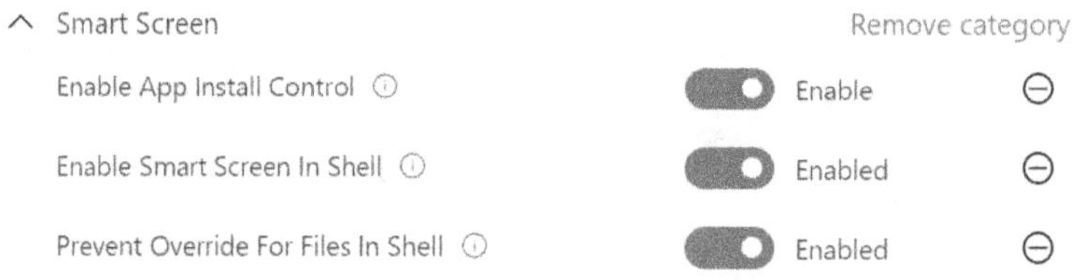

Figure 9.31 – Windows Defender SmartScreen policies

On the client, when a user tries to open an unverified or risky app that does not pass the SmartScreen check, they will be presented with the following notification:

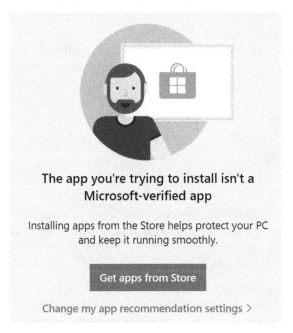

The app you're trying to install isn't a Microsoft-verified app

Installing apps from the Store helps protect your PC and keep it running smoothly.

Get apps from Store

Change my app recommendation settings >

Figure 9.32 – Windows Defender SmartScreen blocking a malicious app

> **Tip**
> If a false positive blocks an app from running, users can right-click the executable, open properties, and choose to unblock and allow the app to run.

To view the configured SmartScreen settings on the client, open **Windows Security | App & browser control | Reputation-based protection settings.**

Next, let's look at blocking access to the Microsoft Store to prevent users from installing unwanted consumer applications.

Block the Microsoft Store

The Microsoft Store is installed by default in Windows and contains many of the pre-installed applications available in Windows. By default, users are allowed to search for and install any additional applications from the app store without requiring admin rights. In an enterprise-managed environment, it may not be appropriate to allow users to install apps created for consumer purposes and that are not applicable for business functions. Applications installed through the app store increase the risk of installing vulnerable software that collects data and has privacy concerns. CIS recommends blocking access to the Microsoft Store completely or only enabling the private store for companies that distribute software through purchases from the Microsoft Store for Business and Education.

> **Tip**
> If you distribute online-licensed apps using the Microsoft Store, switch to the offline-license model otherwise required app assignments using Intune will fail.

Let's look at blocking access to the Microsoft Store with Intune and enabling access to the private store only. Once the store is disabled, the behavior in Windows 10 and 11 acts slightly differently. In Windows 10, the Microsoft Store will automatically open to the private store if enabled, but in Windows 11, users will need to use the Company Portal app to access the private store. This will require a separate deployment to ensure the Company Portal app is pushed to Windows devices if you plan on using the private store to distribute apps.

To configure this policy, we used the **Settings catalog** profile type for **Windows 10 and later**. We gave it the name **Windows Security Baseline – Microsoft App Store** and chose **Microsoft App Store** from the Settings Picker. The settings shown in the following screenshot are a good baseline recommendation for controlling the Microsoft App Store and will enable the use of the private store in the Company Portal app:

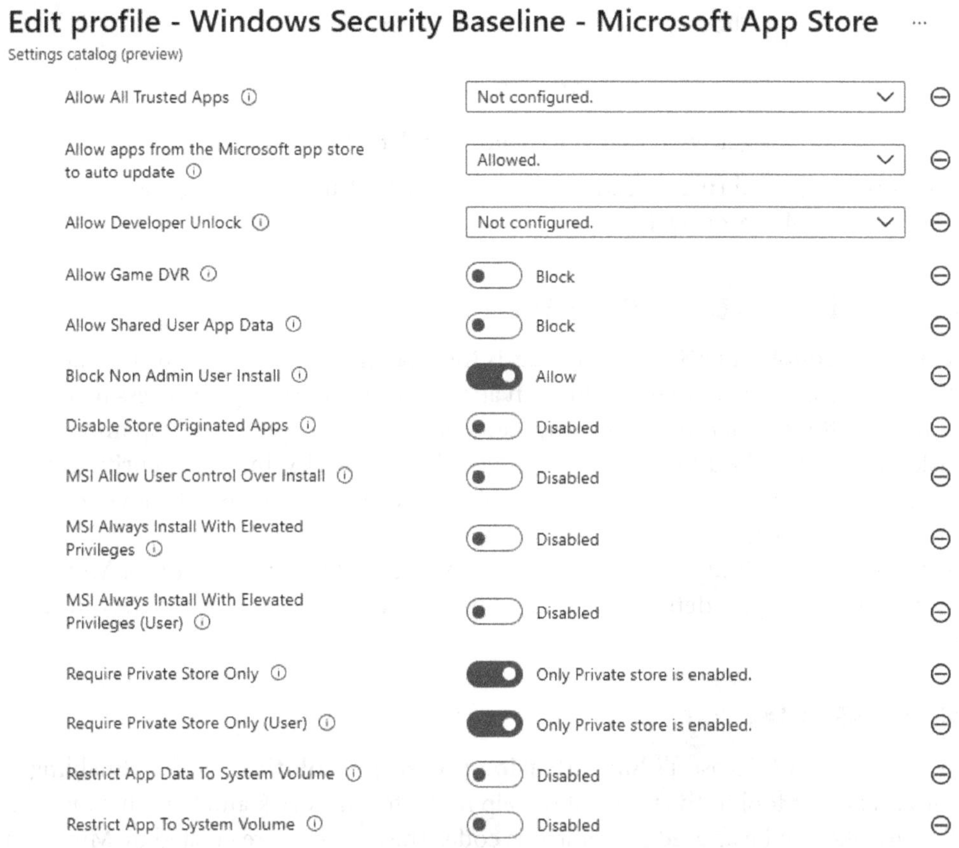

Edit profile - Windows Security Baseline - Microsoft App Store   ...

Settings catalog (preview)

Allow All Trusted Apps ⓘ	Not configured. ⌄
Allow apps from the Microsoft app store to auto update ⓘ	Allowed. ⌄
Allow Developer Unlock ⓘ	Not configured. ⌄
Allow Game DVR ⓘ	Block
Allow Shared User App Data ⓘ	Block
Block Non Admin User Install ⓘ	Allow
Disable Store Originated Apps ⓘ	Disabled
MSI Allow User Control Over Install ⓘ	Disabled
MSI Always Install With Elevated Privileges ⓘ	Disabled
MSI Always Install With Elevated Privileges (User) ⓘ	Disabled
Require Private Store Only ⓘ	Only Private store is enabled.
Require Private Store Only (User) ⓘ	Only Private store is enabled.
Restrict App Data To System Volume ⓘ	Disabled
Restrict App To System Volume ⓘ	Disabled

Figure 9.33 – Control Microsoft App Store

For instructions on how to push the Company Portal app to Intune devices, visit this link: `https://docs.microsoft.com/en-us/mem/intune/apps/store-apps-company-portal-app`.

After creating the policy and assigning it to a group of users and devices, when a user goes to open the Microsoft App Store, they will see the message displayed in the following screenshot:

Microsoft Store is blocked

Check with your IT or system administrator.

Report this problem

Figure 9.34 – Microsoft Store blocked in Windows 11

Next, let's look at enabled rules to reduce the attack surface and protect against ransomware using Microsoft Defender.

Reducing the attack surface

Attack surface reduction (ASR), in practice, is the enabling of security features that reduce the *attack surface* or areas in which an attacker can exploit to gain access to a system. Microsoft Defender includes many features that work together to help mitigate these risks and harden the attack surface. We already discussed Windows security features that enable these defenses, including hardware-based security with VBS, deploying a Windows Firewall security baseline, and enabling network and web protection for Microsoft Defender for Endpoint. Next, let's look at a few additional features of Microsoft Defender than can help harden our clients and further reduce the attack surface through ASR rules.

Enabling ASR rules

ASR rules are part of Microsoft's **host intrusion prevention solution (HIPS)**. Enabling ASR rules is another tool in the tool belt to help mitigate attack risk and harden common areas that are likely to be abused by malicious code. These rules were created by Microsoft based on an analysis of the top identified areas based on their telemetry that malicious actors would try to exploit. Many of the processes that ASR rules protect are otherwise seemingly standard functionality in Windows-based systems. A few examples of the protection types in ASR include blocking scripts from executing in email or preventing apps such as Office macros from creating child processes. Many times, attackers leverage these techniques as part of their evasion tactics to hide malicious actions through otherwise seemingly harmless processes.

Microsoft's ASR rules can be enabled through Group Policy, Configuration Manager, PowerShell, and Intune. In the following screenshot, we can see some of the available ASR rules in an endpoint protection profile. There is no recommended best practice as to which rules to enable and each rule will require careful planning to understand how they will affect processes used in your organization:

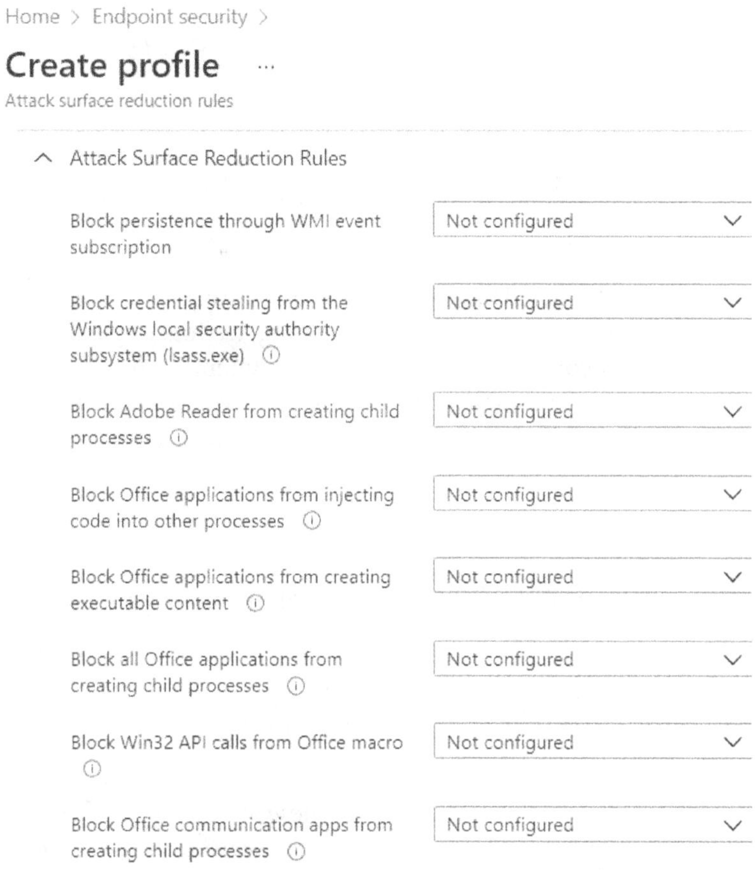

Home > Endpoint security >

Create profile   ···
Attack surface reduction rules

∧ Attack Surface Reduction Rules

Block persistence through WMI event subscription	Not configured ∨
Block credential stealing from the Windows local security authority subsystem (lsass.exe) ⓘ	Not configured ∨
Block Adobe Reader from creating child processes ⓘ	Not configured ∨
Block Office applications from injecting code into other processes ⓘ	Not configured ∨
Block Office applications from creating executable content ⓘ	Not configured ∨
Block all Office applications from creating child processes ⓘ	Not configured ∨
Block Win32 API calls from Office macro ⓘ	Not configured ∨
Block Office communication apps from creating child processes ⓘ	Not configured ∨

Figure 9.35 – ASR rules in endpoint security

For a full reference of ASR rules, including their GUID mapping, visit this link: `https://docs.microsoft.com/en-us/microsoft-365/security/defender-endpoint/attack-surface-reduction-rules-reference?view=o365-worldwide`.

In a policy where a rule is configured in *warn mode*, users will see a notification as displayed in the following screenshot, when an ASR rule is triggered on the client:

Figure 9.36 – ASR rule in warn mode

Microsoft documentation has detailed resources about how to plan for an ASR deployment, including incorporating phases for planning, testing, implementing, and operationalizing. We encourage reviewing these documents in detail by visiting this link: `https://docs.microsoft.com/en-us/microsoft-365/security/defender-endpoint/attack-surface-reduction-rules-deployment?view=o365-worldwide`.

While ASR rules can be enabled with minimal licensing requirements, to take advantage of the reporting and telemetry features in Microsoft 365 Defender as recommended in the deployment planning, Microsoft 365 Defender E5 or equivalent licenses are required.

Next, let's look at protecting files from ransomware with controlled folder access.

Prevent ransomware with controlled folder access

Ransomware attacks have become quite common as a way for attackers to steal data and extort businesses for monetary gain. Not only have these attacks targeted individual businesses, but more sophisticated human-operated ransomware groups have recently focused their sights on critical infrastructure sectors such as energy, water, and healthcare. Any interruption of operations in these sectors can have real-world detrimental effects on the commodities we depend on every day to survive. We covered ransomware preparedness in detail in *Chapter 1, Fundamentals of Windows Security*. But, as a reminder, building a ransomware playbook and having an incident response plan is a critical component of a well-rounded security program. An area of focus consists of enabling many tools to protect assets from ransomware attacks.

Typically, a ransomware attack starts with attackers gaining access to systems to gather information about what they can compromise. Once ready, they will detonate their encryption programs and encrypt files, documents, and processes that render them inaccessible, which can disrupt critical system processes. Threat actors will typically require a ransom be paid to decrypt files and restore operations or threaten to leak the data publicly or permanently delete it. As part of ASR planning, Microsoft Defender has a feature that can protect critical files and folders from unapproved modification using controlled folder access. Controlled folder access works by specifying directories on your clients that should be protected (such as desktop, documents, and system folders) from modification, except by identified processes or apps that are explicitly safe-listed. As a result, when a malicious program attempts to trigger encryption on these protected containers, the process will be stopped. Let's look at how to enable this feature.

For controlled folder access to work, Microsoft Defender Antivirus real-time protection must be enabled. If the baseline recommendations for Defender Antivirus are deployed as discussed in *Chapter 8*, *Keeping Your Windows Client Secure*, real-time protection is turned on and we can configure controlled folder access through an **Attack Surface Reduction Rules** policy in **Endpoint Protection** in Intune. The policy configuration will look like the following screenshot:

Figure 9.37 – Controlled folder access settings

The **Use advanced protection against ransomware** rule uses cloud-delivered protection to enhance real-time protection and help discover whether a file could be a product of ransomware.

The **Enable folder protection** setting enables the controlled folder access setting in the Windows Security app.

Once enabled, Windows system folders and common locations such as documents, pictures, and videos are protected by default. For a reference about which folders are protected by enabling the policy and for additional information about controlled folder access, visit this link: `https://docs.microsoft.com/en-us/microsoft-365/security/defender-endpoint/controlled-folders?view=o365-worldwide`.

Just as with the features we've covered, events are logged in the Microsoft 365 Defender portal. To view the status of controlled folder access on a client, open the **Windows Security** app and go to **Virus & threat protection** | **Manage settings** | **Manage Controlled folder access**. Click on **Block history** to view any protected folders that have blocked access from unapproved apps. If notifications are enabled on the client, users will see a Windows security notification when access to a protected folder was blocked, as seen in the following screenshot:

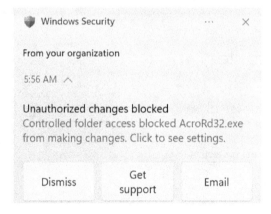

Figure 9.38 – Controlled folder access block

Next, let's look at enabling another intrusion prevention capability of Defender called Windows Defender exploit protection.

Enabling Defender exploit protection

Windows Defender exploit protection is another feature that is part of Windows Defender Exploit Guard. We've covered three of these features already: ASR rules, controlled folder access, and network protection. Exploit protection works to stop attackers from leveraging operating system processes and apps from malicious activities by applying specific mitigation tactics. A few examples of mitigation techniques that can be enabled in an exploit protection policy include the following:

- **Arbitrary code guard** helps by preventing memory protection flags from being marked as executable and allowing code not originating from the application to be loaded. Attackers try to leverage the permissions of trusted applications to execute malicious code.

- **Do not allow child processes** prevents applications from creating a new child application. This is an evasion technique that uses a trusted process to spin up a potentially malicious child process to remain under the radar.

- **Control flow guard** prevents memory corruption attacks that overwrite memory and change function pointers to be directed to execute malicious code instead of the expected application.

Exploit protection can apply many additional mitigation tactics and allow the creation of a policy audit mode to measure any potential compatibility impacts. To read more details about each mitigation control, visit the following link: `https://docs.microsoft.com/en-us/microsoft-365/security/defender-endpoint/exploit-protection-reference?view=o365-worldwide`.

Let's view the exploit protection features on a client by opening the **Windows Security** app and going to the **App & browser control** | **Exploit protection** settings. Exploit protection is automatically enabled in Windows at the recommended settings for both system processes and individual apps. Any additional mitigations you wish to add for protecting apps can be added in **Program settings** and will take precedence over preset system settings.

To deploy exploit protection settings to your clients, we can build a baseline template using the local Windows Security app. Once the mitigations have been added and tested, we can export the XML file and distribute it using Intune or Group Policy.

To export the policy for distribution, click on **Export settings** from the **Exploit protection** settings in the **Windows Security** app and save the XML file. In Intune, create an **Exploit protection** profile for Windows 10 and later using the **Attack surface reduction** profile type in **Endpoint security**.

Upload the XML file and choose Yes on the **Block users from editing the Exploit Guard protection interface** setting to prevent users from making changes. The profile should look like the following screenshot when creating it in Intune:

Figure 9.39 – Exploit protection profile in Endpoint security

Review and save your policy and assign it to the user or device groups to complete the deployment. Exploit protection settings should now be managed by Intune and match the exported baseline configurations.

> **Tip**
> If exploit protection settings were configured in the past for earlier Windows baselines, the Windows Security baselines from the SCT include a script to reset them back to the closest Windows default settings.

Just as with ASR rules and controlled folder access, when mitigation occurs on the client because of the exploit protection policy, the user will see a notification from Windows Security. To read more about testing policies in audit mode, visit the following link: https://docs.microsoft.com/en-us/microsoft-365/security/defender-endpoint/evaluate-exploit-protection?view=o365-worldwide.

Next, let's look at enabling Application Guard and use hardware-based process isolation to protect Windows from malicious code executed inside Office and Edge.

Zero trust with Application Guard

Application Guard is a technology available for Microsoft Office apps and the Edge browser to prevent untrusted files and websites from accessing resources on the host operating system through hardware-based isolation. This isolation, which is powered by virtualization-based security, provides zero-trust protection against zero-day threats. We covered the hardware virtualization technologies that drive Application Guard in *Chapter 3, Hardware and Virtualization*, and we recommend carefully reviewing the system requirements at the following link before enabling the feature. Currently, Application Guard is not officially supported in virtualized environments: `https://docs.microsoft.com/en-us/windows/security/threat-protection/microsoft-defender-application-guard/reqs-md-app-guard`.

First, let's look at enabling Application Guard for the Microsoft Edge browser and configuring a list of trusted cloud domains using enterprise-managed mode.

Application Guard for Microsoft Edge

Application Guard for Microsoft Edge opens non-trusted websites in isolation to help protect against malicious code and browser-based threats. Historically, admins may have enabled trusted sites or used firewalls to explicitly restrict the permissions websites had by placing them in trusted zones. With Application Guard, you can configure trusted domains and know that when a user visits a website outside of the safelist, they will be protected. Policies configurable through Intune can also be set to allow certain interactions with the host OS, such as allowing copying and pasting or printing. This protection can extend to other browsers, such as Google Chrome and Firefox, by installing an extension that will redirect untrusted sites to open in an Application Guard window of Microsoft Edge. Let's look at enabling an Application Guard policy with enterprise-managed mode using Endpoint Security in MEM.

To create a policy in MEM, use an **App & browser isolation** profile for Windows 10 and later using the **Attack surface reduction** profile type in **Endpoint Security**.

There are several settings in the profile that determine how websites inside Application Guard can interact with the host and we recommend configuring them to fit your company's needs. Use **Windows network isolation policy** to configure trusted network boundaries that safelist company IP ranges, cloud resources, and domains that can access the host computer. Any website not in the isolation policy will open in Application Guard and be subjected to the policies you configure in the profile. The following screenshot shows some of the configurable settings available for Application Guard:

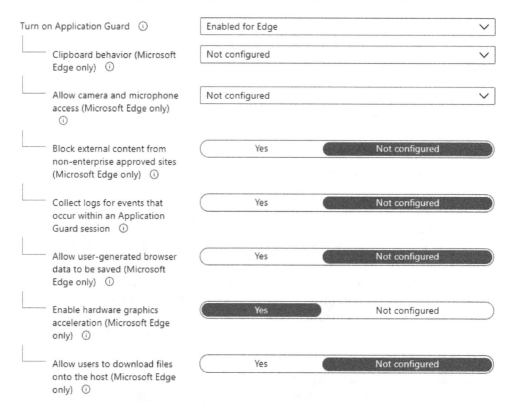

Figure 9.40 – Application Guard settings in Endpoint Security

In our profile, we are going to leave the defaults and configure the following settings:

- **Turn on Application Guard** is set to **Enabled for Edge**.

- **Enable hardware graphics acceleration (Microsoft Edge only)** is configured to **Yes**.

- **Windows network isolation policy** is set to **Configure**.

Reviewing the CSP reference documentation, typically, leaving the default setting of **Not configured** means that the functionality is turned off. More information about the Application Guard CSPs can be found at the following reference link: https://docs.microsoft.com/en-us/windows/client-management/mdm/windowsdefenderapplicationguard-csp.

> **Tip**
> Enabling Application Guard turns on new Windows features and a reboot may be required before the settings take effect.

Let's see what this looks like on a client after the policy applies. By visiting a website not listed in the network boundaries, a new window will open and load the web page in Application Guard, as can be seen in the following screenshot:

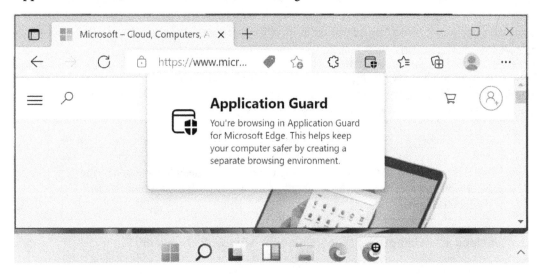

Figure 9.41 – Application Guard in Microsoft Edge

Notice how there are two instances of Edge open in the Windows taskbar, and the Application Guard window is depicted by an icon over it. Keep in mind that this isolation may have unwanted effects and could reduce the functionality of certain websites. For example, downloading files will be prevented unless specified in the policy, and be sure to test websites with redirects to make sure they behave as expected, especially if they are in your safe-listed domains. General troubleshooting can be done directly inside the Edge browser. For example, to determine why a site is opening in a container and not on the host, you can visit edge://application-guard-internals in the address bar and test a URL.

Application Guard policies are also supported for some third-party browsers. Next, let's review the requirements to enable Application Guard for Google Chrome and Firefox.

Third-party browser support

To extend Application Guard for use in Google Chrome, you will need to install the extension and companion app. You can find the Defender Application Guard extension at the following link: `https://chrome.google.com/webstore/detail/application-guard-extensi/mfjnknhkkiafjajicegabkbimfhplplj?hl=en-US`.

The Microsoft Defender Application Guard companion app can be installed from the Microsoft Store and pushed to managed devices using Intune app deployments. There are a few recommended policy settings for Google Chrome that should be reviewed and detailed usage instructions for both Google Chrome and Firefox can be found at this link: `https://docs.microsoft.com/en-us/windows/security/threat-protection/microsoft-defender-application-guard/md-app-guard-browser-extension`.

Once the prerequisites are configured and the extension and companion app is installed, whenever a user visits a website not in the configured trusted network boundary list, they will be redirected to a new Microsoft Edge Application Guard window automatically.

> **Tip**
> If blocking the Microsoft Store, the companion app will need to be added to the Microsoft Store for Business with an offline license to push as required to users' devices in Intune.

Next, let's look at enabling Application Guard for Microsoft Office apps.

Application Guard for Microsoft Office

Microsoft Application Guard technology is also extended to Office apps to help reduce the attack surface from Office-based exploits. Just like with Edge, Microsoft Office will open untrusted files in an isolated container that prevents interaction with the host OS. As mentioned previously, with Edge, policies can be configured to allow users the ability to print and copy/paste to minimize productivity loss. Application Guard for Office does require a Microsoft 365 E5 or equivalent license SKU. You can find the link to the reference documentation for additional prerequisites in the *Zero trust with Application Guard* section.

If you are familiar with the protected view feature in Office, documents will follow a similar behavior and will open in Application Guard if the following conditions are met:

- Files are downloaded from the internet.

- Files are not located in defined trusted locations.

- Attachments are opened from emails.

Since the interaction between the file and host is restricted, if a document is deemed trustworthy, the user can remove protection from the file by saving it locally and then re-opening it like any other trusted Office document. Since this is a fundamental change in how Office documents open and users interact with them, we recommend communicating these changes throughout your organization to help avoid confusion and to reduce service desk tickets.

Let's enable an Application Guard policy for Office using Endpoint Security in Intune. We can use the same profile created earlier or build a new profile using the **App & browser isolation** profile for Windows 10 and later and selecting the **Attack surface reduction** profile type in **Endpoint Security**.

To keep Application Guard enabled for both Edge and Office, set the **Turn on Application Guard** setting to **Enabled for Edge AND isolated Windows environments**.

After the policy applies, Office will open untrusted documents in the Application Guard container in a protected mode, as can be seen in the following screenshot:

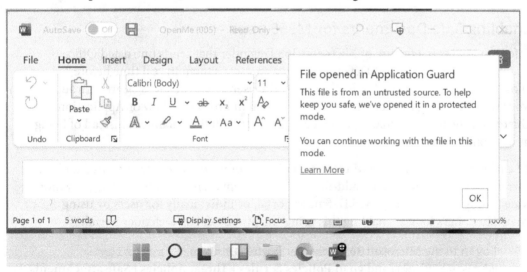

Figure 9.42 – Microsoft Office Application Guard

If the file does not contain malicious content, users can remove protection through the **File** menu and save the file locally if allowed by policy. Users will see this option, as shown in the following screenshot:

Figure 9.43 – Removing Application Guard protection

To enable Application Guard, Microsoft Office must be on version 2011 build *16.0.13530.10000* or later and on Current Channel or Monthly Enterprise Channel. Additional policies for how users can interact with files inside of Application Guard can be configured using the Office cloud policy service, and a reference can be found at this link: `https://docs.microsoft.com/en-us/microsoft-365/security/office-365-security/install-app-guard?view=o365-worldwide`.

Next, let's add an extra layer of protection before opening office files by enabling real-time document scanning with Microsoft Defender using Safe Documents.

Enabling Safe Documents for M365 apps

Safe Documents is a feature of Microsoft 365 Defender that sends untrusted Office documents to the Microsoft Defender service and scans them in real time before allowing them to be opened in edit mode by the user. Once the file is cleared through Microsoft Defender, the document will then open in Protected View or Application Guard and the user can enable content and interact with the document instead of being in read-only mode.

To enable Safe Documents, a Microsoft 365 E5 or equivalent license is required and is not currently available as a license add-on. Safe Documents can be configured organization-wide through the Microsoft 365 Defender portal, or individually for users by using PowerShell. Let's look at enabling this using the Microsoft 365 Defender portal:

1. Log in to the Microsoft 365 Defender portal at `https://security.microsoft.com` and go to **Policies & rules | Threat policies | Safe attachments**.
2. Click on **Global Settings** to open the pop-up menu. Toggle the option for **Turn on Safe Documents for Office clients** to enable the feature.

3. If a file is identified as malicious and you wish to allow users to proceed, enable the option to **Allow people to click through Protected View even if Safe Documents identifies the file as malicious**.

The global settings pop-up will look like the following screenshot:

Help people stay safe when trusting a file to open outside Protected View in Office applications.

Before a user is allowed to trust a file opened in a supported version of Office, the file will be verified by Microsoft Defender for Endpoint. Learn more about Safe Documents.

Turn on Safe Documents for Office clients. Only available with *Microsoft 365 E5* or *Microsoft 365 E5 Security* license. Learn more about how Microsoft handles your data.

Allow people to click through Protected View even if Safe Documents identified the file as malicious

Save   Cancel

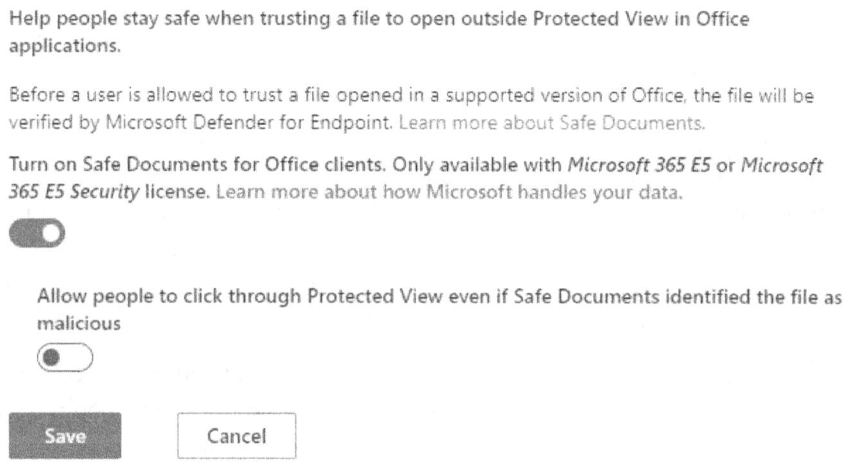

Figure 9.44 – Safe Documents in the Microsoft 365 Defender portal

After the policy has been applied, when a user opens an untrusted file using the Office client, the document will be scanned against the Microsoft Defender for Endpoint service. Depending on the size of the document, this may take several seconds. After the file is verified and found safe, the user will see an **Enable Editing** button, as in the following screenshot, and be able to work with the file as expected:

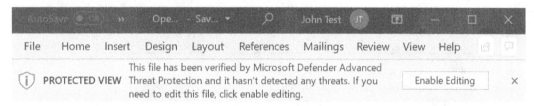

Figure 9.45 – Safe Documents banner notification

For more information on licensing requirements and how to enable using PowerShell for individual users, visit this link: https://docs.microsoft.com/en-us/microsoft-365/security/office-365-security/safe-docs?view=o365-worldwide.

For information about Microsoft Defender for Endpoint data storage and privacy, visit this link: https://docs.microsoft.com/en-us/microsoft-365/security/defender-endpoint/data-storage-privacy?view=o365-worldwide.

Next, let's review **Windows Defender Application Control (WDAC)** for creating application safelists. WDAC can be used to harden the attack surface by determining which applications are allowed to run in Windows.

WDAC

WDAC helps to harden the attack surface and mitigate risk from malicious applications by restricting what applications are allowed to run on a Windows client. To configure WDAC, a policy is created that acts as a safelist that lists trusted applications and file paths that apps and processes can run from. A WDAC policy can also be used to specify what **Component Object Model (COM)** objects can run and enforce script signing requirements to protect your system from unsigned or unauthorized script modifications. WDAC is fundamentally aligned with core zero-trust principles, as applications and processes must prove their trust before they are able to execute. If the application, process, or path is not on the safe list, it will be blocked or audited.

WDAC is the recommended approach by Microsoft for applying application control and is a successor to other technologies such as AppLocker. Unlike traditional application control policies such as AppLocker, WDAC policies are enforced at the device level and cannot be assigned to specific users and groups. WDAC policies can be protected from modification by enabling **hypervisor-protected code integrity (HVCI)** and VBS security for an added layer of protection. In the past, the combination of these two technologies deployed together was known as **Device Guard** and it provides full protection for both user and kernel-mode operations.

Later, in *Chapter 12*, *Keeping Your Windows Server Secure*, we will cover how to build a WDAC policy for Windows Server, but the same methodology applies to creating policies for clients. Once the baseline policy is created, it can be exported and distributed through Intune, Configuration Manager, or Group Policy.

> Tip
> WDAC policies are built in XML format. They are configured using PowerShell and are converted into binary before they are deployed. Windows has preconfigured baseline controls that can be used to build your own XML definitions around. To view the rules of these policies, use the `Get-CIPolicy` cmdlet in PowerShell. Example policies are stored in the following path: `%SystemDrive%\Windows\schemas\CodeIntegrity\ExamplePolicies`.

It is recommended that you thoroughly understand and test the effects that WDAC policies will have on your systems. As mentioned earlier, they are designed with zero-trust principles and improper configuration can block critical system components from running. We recommend deploying a policy in audit mode first and reviewing the logging to understand the effects on your Windows clients.

WDAC logs can be found in the event viewer under `Applications and Services\ Microsoft\Windows\CodeIntegrity\Operational` or by using advanced hunting in the Microsoft 365 Defender portal if your clients are onboarded to Defender for Endpoint.

Intune can natively deploy Application Control using **AppLocker CSP** policies through the **Endpoint protection** template in the **Templates** profile type. Support for customized policies and further fine-grained control can be distributed using a **Custom** template and by referencing the **ApplicationControl CSP** policies. The following screenshot shows the built-in Intune policy for Application Control set to `deploy` in **Audit Only** mode using **Endpoint protection**:

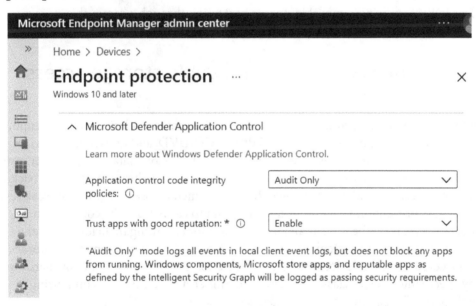

Figure 9.46 – Application Control in Intune

Considerations for WDAC or AppLocker

As previously mentioned, Microsoft recommends implementing WDAC over AppLocker if possible as they have committed to providing enhancements and support for WDAC over AppLocker in the future. We recommend reviewing the capabilities of each to determine what policy or combination of the two is appropriate for your organization. For more information about comparing WDAC to AppLocker, visit this link: `https://docs.microsoft.com/en-us/windows/security/threat-protection/windows-defender-application-control/wdac-and-applocker-overview`.

Configuring and testing custom policies can be extremely time-consuming and should be carefully considered if you're planning a deployment. Microsoft has created a lot of documentation around deployment planning and a link to the reference materials can be found at `https://docs.microsoft.com/en-us/windows/security/threat-protection/windows-defender-application-control/windows-defender-application-control-deployment-guide`.

Next, let's look at applying removable storage protection through a removable storage access control policy in Microsoft Defender.

Protecting devices with a removable storage access control policy

Removable storage can be defined as any storage device that is not built into the computer itself. This includes peripherals such as USB drives, CD/DVD, and non-volume devices such as cameras and smartphones. It is a convenient way to store data and quickly transfer files from one system to another. It is also portable and is a fast alternative for data backup comparable to network and cloud storage. Removable media poses a significant weakness in any data loss prevention program by allowing data to easily leave the company in an unauthorized way. It can be used as an attack surface for executing malicious code and for exfiltrating data from a system. Traditionally, to control the use of removable storage, IT admins could deploy a device installation restriction policy to block or allow hardware types by specifying a device class, instance, or identifier. While effective, configuring these policies can be complicated and impacts the use of other peripherals such as audio headsets, keyboards, and mice. This level of control is extremely fine-grained and, depending on your company's compliance and security policies, is an effective method for controlling peripherals.

For companies that don't need that fine level of control over hardware installation, Microsoft Defender allows you to control removable media through a removable storage access control policy. We can configure actions to audit, allow, or prevent the read, write, and execution of data to removable storage. These policies are designed to help complement any existing DLP program with significantly less overhead compared to an installation restriction policy by following these high-level steps:

1. Defining the removable media group or groups in which the policy should apply, including exclusion group.

2. Creating the policy and setting the enforcement action.

3. Deploying the policy to your endpoints using Intune or Group Policy

When defining your policies, a minimum of two XML files will be created. The first will define the removable storage groups for organizing devices, and the other will define the access controls that are applied to those groups. To simplify the creation of removable storage groups, devices can be scoped into groups by defining a `PrimaryId`. This can simplify the organization process instead of adding devices by listing their unique `HardwareId`, `SerialNumberId`, or `InstancePathId` in the policy. `PrimaryId` is defined as follows:

- `RemovableMediaDevices` includes USB data storage devices.

- `CdRomDevices` includes optical drives such as CD/DVD/Blu-Ray.

- `WpdDevices` are **Windows Portal Devices (WDPs)**, including mobile phones.

Building policies can get complex quickly and we strongly encourage reviewing the reference documentation in detail. For a full list of the available policy properties, visit the link at `https://docs.microsoft.com/en-us/microsoft-365/security/defender-endpoint/device-control-removable-storage-access-control?view=o365-worldwide`.

Let's build a policy that denies write and execute access to all removable media types by using `PrimaryId`. We will set the options to show a notification to the user when an action is blocked. The following screenshot shows the formatted XML file to define our removable media group. It includes `PrimaryID` of all supported removable media types as specified from the policy properties reference link shared previously:

```
<Group Id="{1c5d166a-9f15-4f7f-85c3-d429b8c919c3}">  { Unique GUID
    <MatchType>MatchAny</MatchType> { Match to any of the PrimaryId below
    <DescriptorIdList>
        <PrimaryId>RemovableMediaDevices</PrimaryId>  ⎧ Includes removable
        <PrimaryId>CdRomDevices</PrimaryId>           ⎨ data drives, optical
        <PrimaryId>WpdDevices</PrimaryId>             ⎪ drives, and Windows
    </DescriptorIdList>                               ⎩ Portable Devices
</Group>
```

Figure 9.47 – Removable storage group XML file

> **Tip**
>
> To create a unique GUID to be used as a group ID, use the following PowerShell command: `[guid]::NewGuid()`.

Next, let's look at the format of the access control policy. In the following screenshot, you can see it includes a unique GUID for `<PolicyRule Id>` and includes the GUID of our media group created previously in `<IncludedIdList>`. This tells the access control policy that every device in that removable media group will have the policy enforced. We will create a new GUID to define `<Entry ID>` and set the deny action for write and execute using `<AccessMask>` with the option to show notifications to the user.

```
<PolicyRule Id="{f582defe-506d-4492-9d70-dcd352cd2b56}"> { Unique GUID
    <Name>Block Write and Execute to removable media</Name>
    <IncludedIdList>
        <GroupId>{1c5d166a-9f15-4f7f-85c3-d429b8c919c3}</GroupId> { GUID of the
    </IncludedIdList>                                              removable media
    <ExcludedIdList></ExcludedIdList>                              Group
    <Entry Id="{4791c183-df01-46ef-87d0-db6adc2df3d7}">
        <Type>Deny</Type>              ⎧ Type: Deny action
        <Options>1</Options>           ⎨ Options: 1 = Show Notifications
        <AccessMask>6</AccessMask>     ⎩ AccessMask: 6 = Write and Execute
    </Entry>
</PolicyRule>
```

Figure 9.48 – Access control policy XML file

After creating the policies, we saved each XML file to be uploaded into Intune.

An audit mode is supported and allows you to analyze the telemetry data in Microsoft Defender for Endpoint to determine the effects that enforcing a policy will have on your devices. Use audit mode to start piloting the policy to help minimize the impact on end users. To leverage the streaming of telemetry to Microsoft Defender for Endpoint, a Microsoft 365 E5 license or equivalent is required. Let's look at delivering this policy to devices using Intune.

Enabling a removable storage access control policy

A removable storage access control policy can be deployed using Group Policy and Intune. It's important to note that the formatting of the XML file will differ slightly depending on the method. The reference links for building the XML file can be reviewed in the *Protecting devices with a removable storage access control policy* section.

In this example, we will use a **Custom** template from the **Templates** profile type for Windows 10 and later in Intune to deploy the policies.

First, we will add our removable media group by clicking **Add row** and entering the following details:

- **Name:** `All Removable Media`

- **OMA-URI:** `./Vendor/MSFT/Defender/Configuration/DeviceControl/PolicyGroups/%7b1c5d166a-9f15-4f7f-85c3-d429b8c919c3%7d/GroupData`

The OMA-URI path of our removable media groups should be in the following format and include the GUID for the Group ID in the path: `%7b**GroupGUID**%7d/GroupData`. Our Group ID in the example was `{1c5d166a-9f15-4f7f-85c3-d429b8c919c3}`.

- **Data type:** `String (XML file)`

The XML file will need to be uploaded. Next, we added a new row for the access control policy with the following details:

- **Name:** `Deny Write and Execute All Media`
- **OMA-URI:** `./Vendor/MSFT/Defender/Configuration/DeviceControl/PolicyRules/%7bf582defe-506d-4492-9d70-dcd352cd2b56%7d/RuleData`

The OMA-URI path of our access control policy should be in the following format and include the GUID for the PolicyRule ID in the path: `%7b**PolicyRuleGUID**%7d/RuleData`. Our PolicyRule ID in the example was `{f582defe-506d-4492-9d70-dcd352cd2b56}`.

- **Data type**: `String (XML file)`

The XML file for the access control policy rules will also need to be uploaded.

If you have multiple removable media groups and policies, each will need to be added as its own OMA-URI in Intune or the policy enforcement will fail. After you have added the rows, save, and assign the policy to a user or device group. Let's see what the behavior looks like for end users.

After the policy applies to a client and the user goes to write a file to removable media, they will see a notification as shown in the following screenshot:

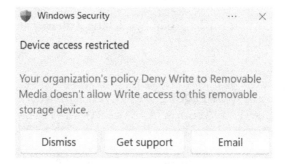

Figure 9.49 – User notification

Also, if the user opens File Explorer and attempts to add a new file by right-clicking on the removable media drive, the option for **New** will be missing. With auditing enabled, you can review the audit logs using advanced hunting in the Microsoft 365 Defender portal. KQL query examples are listed on the reference documentation page listed previously. The following screenshot shows an example of the parsed log file from the Microsoft 365 Defender portal for the triggered deny event:

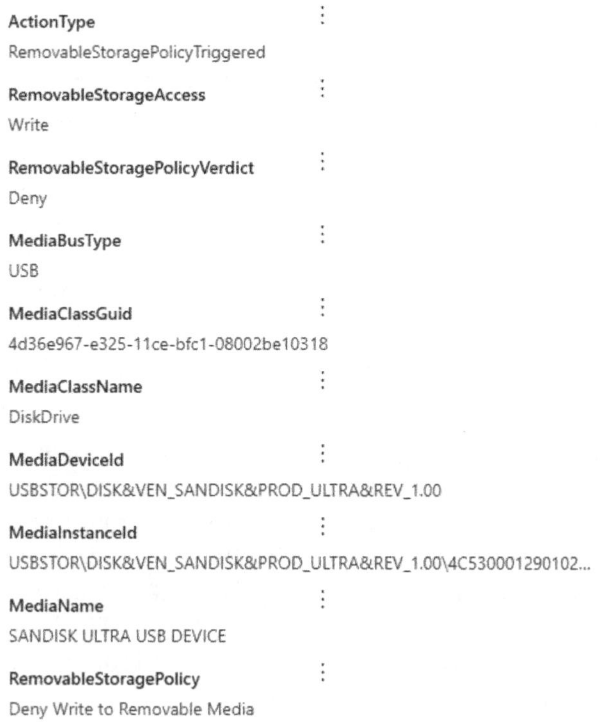

ActionType   ⋮
RemovableStoragePolicyTriggered

RemovableStorageAccess   ⋮
Write

RemovableStoragePolicyVerdict   ⋮
Deny

MediaBusType   ⋮
USB

MediaClassGuid   ⋮
4d36e967-e325-11ce-bfc1-08002be10318

MediaClassName   ⋮
DiskDrive

MediaDeviceId   ⋮
USBSTOR\DISK&VEN_SANDISK&PROD_ULTRA&REV_1.00

MediaInstanceId   ⋮
USBSTOR\DISK&VEN_SANDISK&PROD_ULTRA&REV_1.00\4C530001290102...

MediaName   ⋮
SANDISK ULTRA USB DEVICE

RemovableStoragePolicy   ⋮
Deny Write to Removable Media

Figure 9.50 – Microsoft 365 Defender advanced hunting

Deploying a removable storage access control policy is very powerful and quite flexible. It is a great way to lock down the use of removable storage and provides much of the same flexibility as an installation restriction policy. Using the reporting from the Microsoft 365 Defender portal, dashboards can be built to help analyze removable media usage and determine the impact before enforcing any deny actions on your users. Phased deployments can be accomplished by assigning the policy to different device groups in Intune.

Summary

In this chapter, we reviewed additional areas and next-generation features to incorporate into baseline controls that complement a Windows security baseline. Taking a holistic approach to security is key for improving your overall hardening and establishing a robust layer of protection for your endpoints. We covered baseline recommendations from CIS and Microsoft for Edge and Chrome, and how to manage sign-in settings and control extensions. We provided examples of the different methods for delivering policies using Intune with Intune security baselines, the settings catalog, and ADMX ingestion for managing third-party apps. We discussed the importance of protecting Microsoft 365 apps and provided an overview of using the Office cloud policy service in the Microsoft 365 Apps admin center.

Next, we reviewed many of the advanced protection features available in the Microsoft Defender suite of products and the importance of enabling them to reduce the attack surface beyond standard antivirus protection. This included protecting users from unwanted apps, enabling ASR rules, and protecting them from ransomware using controlled folder access. We then covered enabling Application Guard for Edge and Office to isolate processes and protect against zero-day threats. Finally, we finished the chapter by discussing application control policies and how to administer a removable storage access control policy to protect against data leakage.

In the next chapter, we are going to expand on ASR and discuss common attack vectors. We will review techniques such as Adversary-in-the-Middle, and discuss how to protect against lateral movement and privilege escalation.

10
Mitigating Common Attack Vectors

In this chapter, you will learn how to mitigate attack vectors that are commonly seen when standard computer communications protocols have been exploited. Once an attacker has gained access to your network, they will likely try to intercept communications and insert themselves in an attempt to gain a foothold. First, we will discuss different types of Adversary-in-the-Middle techniques and how they can be used to intercept communications, poison responses, capture user passwords, and relay authentication processes to access other systems. We will also discuss how network protocols such as mDNS, NetBIOS, LLMNR, WPAD, SMB, ARP, and IPv6 can be used to trick an unknowing victim into redirecting communications to the attacker's host and fool them into providing credentials.

Then, we will discuss protecting against lateral movement and privilege escalation. We will look at how a compromised standard user account can be used to identify systems and different ways to protect against reconnaissance activities. After that, we will review exploits that target Kerberos authentication and where attackers look to access credentials stored on the host system and in Active Directory. Finally, we will end this chapter by discussing Windows privacy settings. Many privacy permissions are allowed by default in Windows, and they could pose a privacy risk to your organization if they're not disabled.

In this chapter, we will cover the following topics:

- Preventing an Adversary-in-the-Middle attack
- Protecting against lateral movement and privilege escalation
- Windows privacy settings

Technical requirements

To follow this chapter's instructions, you will need a Microsoft 365 subscription and devices that are managed by Intune. We have included links to the official repositories for the pentesting tools that we will be using for demonstration purposes as appropriate. You will need the following licensing or an equivalent SKU to deploy policies with Intune:

- Microsoft 365 E3 and/or E5
- Windows 10/11 Pro or Enterprise

Let's start by learning about Adversary-in-the-Middle attacks.

Preventing an Adversary-in-the-Middle attack

For clients connected to corporate networks, intra-network communications are critical to the basic functionality of systems. This could be connectivity to the local intranet, printers, file servers, or accessing the internet. For attackers that have gained access to the internal network, several tools and techniques can be used to listen for and intercept these communications. If the attacker can place themselves in the middle of the communications path, they can gather information, manipulate, and modify traffic, and force users to unknowingly authenticate to them. If they're successful in their efforts, passwords can be captured, cracked, and forwarded to other systems in relay attacks to authenticate them against other systems. This technique is known as an **Adversary-in-the-Middle (AiTM)** or **Man-in-the-Middle (MiTM)** attack.

In the next few sections, we are going to review different network protocols that adversaries can use to intercept, spoof, and poison responses back to the victim to commit an **AiTM** attack. This includes the following:

- **Link-Local Multicast Name Resolution (LLMNR)**
- **NetBIOS Name Service (NBT-NS)**
- **Multicast DNS (mDNS)**
- **Web Proxy Auto-Discovery Protocols (WPAD)**
- **Server Message Block (SMB)** relay
- IPv6 DNS spoofing
- **Address Resolution Protocol (ARP)** cache

When a Windows system attempts to communicate with a destination host using protocols such as TCP/IP, UDP/IP, or ICMP/IP, it must discover the name or IP address of the destination system to establish communication. During the lookup process, Windows will first attempt to use a local host file or send a DNS request in the hopes that it finds an answer. If there is no valid answer, the Windows system will send additional broadcasts over the local network to garner a response and find the destination. These are known as **LLMNR** and **NBT-NS** requests. Additionally, clients will attempt to resolve hostnames to any available DNS server over the `.local` suffix using an **mDNS** request. These methods can become prone to name poisoning exploits and unknowingly receive responses from untrusted hosts.

If an attacker gains access to your network, they can set up tools and deceive clients by responding to these broadcasts and sending poisoned responses back. Depending on the malicious services that are running, attackers may present a fake login prompt to the user and coerce them into entering credentials. These activities can ultimately result in plaintext or hashed passwords being captured that can then be cracked or used in relay attacks. First, let's look at LLMNR.

LLMNR

When a DNS request fails, clients will fall back to using LLMNR for name resolution as a secondary protocol through an unauthenticated UDP broadcast over the local network. This allows for any system on the network to respond to the request when DNS services are not available or can't provide a valid response. Using a tool such as Responder (`https://github.com/SpiderLabs/Responder`), an attacker can answer these requests if LLMNR is not disabled on all the available network adapters. Let's look at an example of a poisoned response. In the following diagram, the client PC is attempting to connect to a network share and accidentally mistypes the name. Unbeknownst to them, an attacker sees this request and sends a poisoned response back, tricking the victim into thinking they found the desired destination:

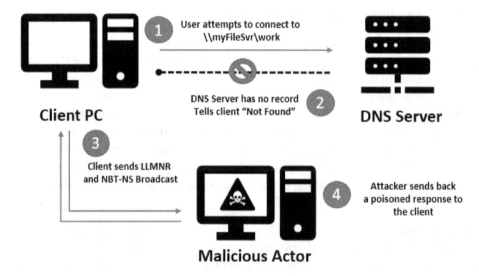

Figure 10.1 – Multicast response communication flow

Depending on the tooling, the user might be presented with an authentication prompt; otherwise, their credentials will be stolen automatically if the PC attempts to use Windows credentials to connect to the share. The following screenshot shows Responder sending a poisoned answer back to the user who was attempting to connect to a file server that doesn't exist on the network. As you can see, it captured the destination host, as well as the client's IP address, username, and password hash. In this example, Responder poisoned answers to the LLMNR, NBT-NS, and mDNS protocols:

```
[*] [NBT-NS] Poisoned answer sent to 192.168.1.111 for name WORKGROUP (service: Domain Master Browser)
[*] [NBT-NS] Poisoned answer sent to 192.168.1.111 for name WORKGROUP (service: Browser Election)
[*] [MDNS] Poisoned answer sent to 192.168.1.111  for name myfileserver.local
[*] [LLMNR]  Poisoned answer sent to 192.168.1.111 for name myfileserver
[HTTP] Sending NTLM authentication request to 192.168.1.111
[HTTP] Host         : myfileserver
[WebDAV] NTLMv2 Client   : 192.168.1.111
[WebDAV] NTLMv2 Username : \eyoung@mtlab.com
[WebDAV] NTLMv2 Hash     : eyoung@mtlab.com:::32ef0f5e0a9357ab:85966AD3F138540BBCB2AD5D9A73995D:010100000
ADB0125D605DD97EC426600000000200080056004B004E00470001001E00570049004E002D00440059004F003800590043003100
00140056004B004F0047002E004C004F00430041004C00030034005700490049004E002D00440059004F003800590043003100480005400
B004E0047002E004C004F00430041004C000500140056004B004E0047002E004C004F00430041004C00080030003000300000000000000
BC291C8F116A5200FA6D84E34662279A996CAA631D4410DD902E9BA16B48664F0A00100000000000000000000000000000000000000
050002F006D007900660069006C006500730065007200760065007200200000000000000000000
```

Figure 10.2 – Captured password hash

In a well-administered network, the LLMNR protocol can likely be disabled and can be configured using Intune, Group Policy, and Configuration Manager. To disable LLMNR using Registry, use the following setting:

- `HKEY_LOCAL_MACHINE\SOFTWARE\Policies\Microsoft\Windows NT\DNSClient`

- `EnableMulticast` and use a `REG_DWORD` value of `0`

To disable LLMNR with **Intune Settings Catalog**, from **Settings picker**, search for **Multicast** and select the **Administrative Templates\Network\DNS Client** category. Choose **Turn off multicast name resolution** and change the setting to **Enabled**.

Once the policy has been applied, the client should no longer broadcast messages using LLMNRR.

> **Tip**
> We strongly encourage you to understand how to use the LLMNR protocol and how it can adversely affect workstation communications before disabling it. Some organizations may need to consider additional network access controls and rely on strong password requirements as mitigation if this cannot be turned off.

Next, let's look at NBT-NS and how to disable NetBIOS over TCP/IP for network interfaces.

NBT-NS

NBT-NS is another broadcast protocol over TCP/IP that is used by the Windows system to find resources as a secondary to DNS. It is enabled by default for all the interfaces in Windows and is vulnerable to the attack techniques that are used by the LLMNR protocol. If name resolution cannot be satisfied using the Windows system's local hosts file or through a DNS request, then it will resort to using NBT-NS to locate the resource. Any system that's available on the network can respond to these broadcast messages and send a response.

> **Tip**
>
> As we mentioned previously regarding LLMNR, disabling NetBIOS can break communication with your systems if broadcast messages are being used to resolve NetBIOS queries. Please consider this before disabling it.

The NetBIOS settings of a network adapter can be viewed by opening the **Settings** app and clicking on **Network & Internet | Advanced network settings | More network adapter options**. Right-click on a network adapter and choose **Properties**. Then, select **Internet Protocol Version 4 (TCP/IPv4)** and choose **Properties**. Click on **Advanced** and choose the **WINS** tab. As shown in the following screenshot, the NetBIOS settings have been set to their defaults, which enables NetBIOS over TCP/IP:

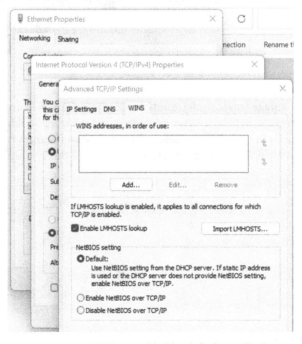

Figure 10.3 – NetBIOS is enabled by default on all adapters

To eliminate the risk of NetBIOS resolution poisoning, it must be disabled on all network adapters. This cannot be accomplished using a policy configuration alone. Let's look at how we can use PowerShell to detect and disable NetBIOS over TCP/IP for all interfaces. There are a few ways to deploy PowerShell scripts to managed devices, including Win32 app deployments or remediation scripts in Intune, Configuration Baseline with a discovery and remediation script in Configuration Manager, or even Group Policy. Let's look at using the discovery and remediation method in Configuration Manager:

1. Open the **Configuration Manager** console. Go to **Assets and Compliance | Compliance Settings | Configuration Items** and click on **Create Configuration Item**.

2. Give it a friendly name, such as `Disable NetBIOS Interfaces`, and click **Next**.

3. For **Support Platforms**, click **Next** to keep the default settings. Click **New** under **Specify settings for this operating system** and enter the following settings:

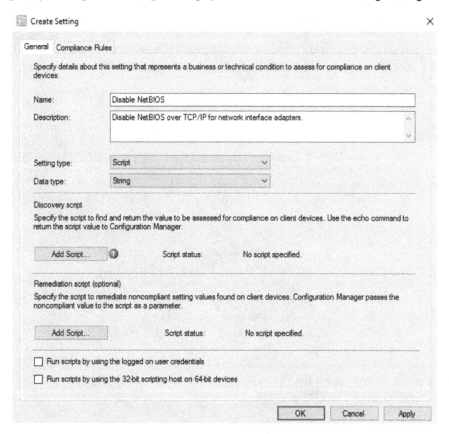

Figure 10.4 – Create Setting

4. Under **Discovery script**, select **Add Script…**.

5. Keep Windows PowerShell selected. Now, we can build the discovery script.

The first section of the code will define the $NBTNS variable. We will use the Get-ItemProperty cmdlet to look in the respective registry hive and enumerate all the TCPIP_GUID keys that start with tcpip using a * wildcard. We want to look specifically at the REG_DWORD value of NetbiosOptions:

```
$NBTNS = Get-ItemProperty HKLM:\SYSTEM\CurrentControlSet\
Services\NetBT\Parameters\Interfaces\tcpip* -Name
NetbiosOptions
```

The following values for REG_DWORD determine the effective state:

- A NetbiosOptions value of 0 is the default

- A NetbiosOptions value of 1 equates to enabled

- A NetbiosOptions value of 2 equates to disabled

In the following screenshot, you can see the amount of unique TCPIP_{GUID} values that must be potentially looked through as each is a unique network interface adapter in Windows:

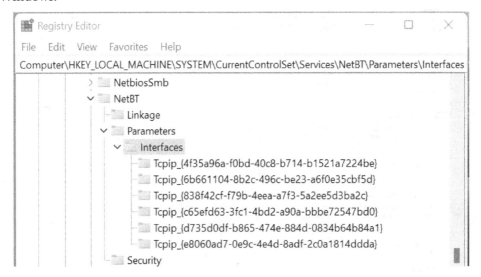

Figure 10.5 – The TCPIP_GUID values of network adapters

Next, we will create an `If` statement that states if the value of `NetbiosOptions` does not equal 2, set the `$NTBNSCompliance` variable to `No`. If it does equal 2, set it to `Yes`:

```
If (!($NBTNS.NetbiosOptions -eq "2")){ $NBTNSCompliance = "No"
} Else { $NBTNSCompliance ="Yes" }
```

Now, let's print the output of `$NBTNSCompliance` so that Configuration Manager can use it to evaluate for compliance. The discovery script should look as follows:

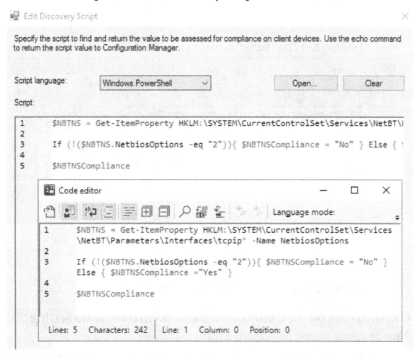

Figure 10.6 – Edit Discovery Script

Click **OK**. Then, click **Add Script** under **Remediation Script**. Keep **Windows PowerShell** set as the script language.

To remediate the settings for each unique network adapter, we will use the `Set-ItemProperty` cmdlet to change all the `NetbiosOptions` `REG_DWORD` values to 2 as it iterates through each entry:

```
Set-ItemProperty HKLM:\SYSTEM\CurrentControlSet\Services\NetBT\
Parameters\Interfaces\tcpip* -Name NetbiosOptions -Value 2
```

The following screenshot shows the **Edit Remediation Script** dialog box with the script code that was used for this task:

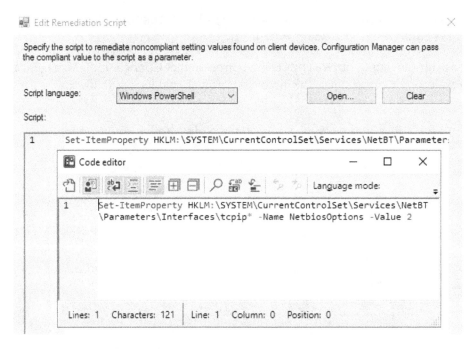

Figure 10.7 – Edit Remediation Script

Now, let's create a compliance rule to validate these settings:

1. Click **OK**, then click on the **Compliance Rules** tab in the **Create Setting** dialog box. Click **New** to create a new compliance rule.

2. Give it a friendly name, such as `Disable NetBIOS`. Regarding **The setting must comply with the following rule**, leave **Equals** selected and enter `Yes` in the box.

3. Select both **Run the specified remediation script when this setting is noncompliant** and **Report noncompliance if this setting instance is not found**.

4. Choose `Information` for **Noncompliance severity for reports**. Click **OK**, click **OK** again, and choose **Next** back on the **Create Configuration Item Wizard** screen.

5. Click **Next** a few times to complete the wizard and build the configuration item.

For this configuration item to be used to evaluate for compliance and remediation, you must create or assign it to a Configuration Baseline. We covered Configuration Baselines in *Chapter 6, Administration and Policy Management*.

Once the Configuration Baseline has been evaluated on the client, we can confirm that it's been enforced by checking the network interface settings, as we saw earlier. The NetBIOS setting should be set to **Disable NetBIOS over TCP/IP**. Upon rebooting, no more requests will be allowed to be made using NetBIOS over TCP/IP.

Next, let's learn how to use mDNS discovery to resolve hosts on a local link.

mDNS

Multicast DNS (mDNS) is used in zero-configuration networking as a way to locate devices, hosts, and services on local networks instead of a DNS server. The engineering specification for mDNS uses many of the same standards as regular DNS. Zero-configuration networking allows for devices connected on the same local network to locate each other without having to configure services such as DNS. mDNS was originally developed by Apple and called Bonjour for use with printer discovery. Instead of sending a unicast request to a DNS server, mDNS sends out a multicast message over the local network, looking for any available hosts to respond to the request and determine the best match for the location of the resource. This multicast broadcast is not private and is prone to the same type of response poisoning as LLMNR and NetBIOS to someone inside your network. The flow for mDNS can be seen in the following diagram, where the client PC sends out a multicast request, looking for a peripheral such as a printer or wireless display, and the malicious actor intercepts the request and poisons the response back to the victim:

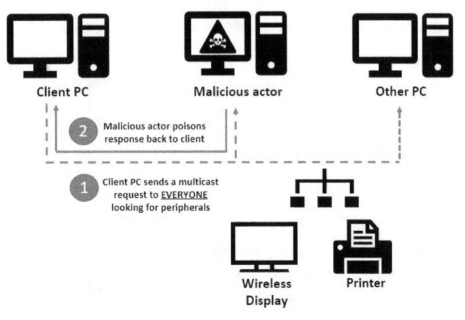

Figure 10.8 – Multicast DNS

An example of this type of request can be seen if a user types a word in their web browser instead of a formulated URL. In the following screenshot, the user typed the word `cookies` into the Microsoft Edge Browser. Using a tool such as **Responder**, an attacker can intercept the mDNS broadcast and send a poisoned response back to request NTLM authentication. In this case, it resulted in a password hash being captured, which remained transparent to the user:

```
[*] [MDNS] Poisoned answer sent to 192.168.1.114    for name qtnmawycyljdzfh.local
[*] [MDNS] Poisoned answer sent to 192.168.1.114    for name jmldyzwhyck.local
[*] [MDNS] Poisoned answer sent to 192.168.1.114    for name qrvsrrtiecqavu.local
[HTTP] Sending NTLM authentication request to 192.168.1.114
[HTTP] Sending NTLM authentication request to 192.168.1.114
[HTTP] Sending NTLM authentication request to 192.168.1.114
[*] [MDNS] Poisoned answer sent to 192.168.1.114    for name cookies.local
[HTTP] Sending NTLM authentication request to 192.168.1.114
[HTTP] Host          : cookies
[HTTP] NTLMv2 Client   : 192.168.1.114
[HTTP] NTLMv2 Username : mtlab\eyoung
[HTTP] NTLMv2 Hash     : eyoung::mtlab:ffdc4e965d16cb26:FDA5A6CE68713DB0B91DC8125F4
D801097537323DBB13EF00000000020008003300340034005200010001E00570049004E002D0054004F0
300040014003300340034005200520004C004C00430041004C0003003400570049004E002D0054004F00
002E0033003400340052002E004C004C004F00430041004C000500014003300340034005200520004F004
001000000002000009E24C955AF4B7714E7D7D440F529D38AA171AF5BAB606A07AB668C96CFC0AE610A
0000009001800480054005400540050002F0063006F006F006B00690065007300000000000000000000
```

Figure 10.9 – Poisoned mDNS response

Not only is mDNS enabled by default in Windows, but web browsers such as Microsoft Edge and Google Chrome use Google Cast to send out multicast requests to stream content to wireless devices. The Google Cast functionality is shown in the following screenshot. It can be accessed from the Chrome browser by clicking on **Cast**:

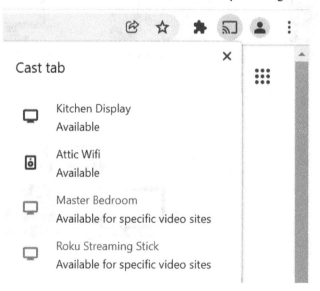

Figure 10.10 – Google Cast

Google Cast is generally unnecessary in a corporate environment, so we recommend disabling this functionality in your security baselines. In addition, you should block the Google Chrome mDNS inbound firewall rule that is created when the software is being installed, as shown in the following screenshot:

Figure 10.11 – Google Chrome mDNS firewall rule

If the rule isn't blocked, `chrome.exe` listens on port UDP `5353` inbound for all network profiles (domain, private, and public). We can demonstrate this by using `netstat` and process monitor (`ProcMon`). The Google Cast functionality will show that `chrome.exe` is listening over UDP port `5353`, as shown here:

```
[chrome.exe]
  UDP     0.0.0.0:5353              *:*
[chrome.exe]
  UDP     0.0.0.0:5353              *:*
[chrome.exe]
  UDP     0.0.0.0:5353              *:*
[chrome.exe]
  UDP     0.0.0.0:42000            *:*
[LTSVC.exe]
```

Figure 10.12 –Chrome.exe listening on UDP 5353

Note that just disabling this firewall rule once is not a permanent solution as each time Google Chrome updates, the rule will be re-enabled. To mitigate this, we recommend using a discovery and remediation compliance script through Configuration Manager or by using a Win32 app or remediation script in Intune.

Once this setting has been applied, the firewall rule will be blocked, as shown in the following screenshot:

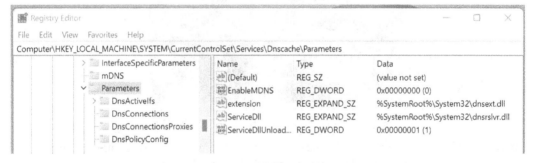

Inbound Rules				
Name	Group	Profile	Enabled	Action
⃠ Google Chrome (mDNS-In)	Google Chrome	All	Yes	Block
✓ Groove Music	Groove Music	Domai...	Yes	Allow
HomeGroup In	HomeGroup	Private	No	Allow

Figure 10.13 – Inbound firewall rules in Windows Defender Firewall

To disable mDNS on Windows-based systems, we can create a specific registry key, as follows:

- `HKLM:\SYSTEM\CurrentControlSet\Services\Dnscache\Parameters`

- **REG_DWORD**: `EnableMDNS`

- **Value**: 0

Once the value has been created, mDNS requests can no longer be poisoned. The following screenshot shows the aforementioned registry key:

Figure 10.14 – Disabling mDNS for Windows

At the time of writing, there is no group policy setting or CSP that controls this setting, so we recommend using a Configuration Manager baseline or deploying a PowerShell script to control this policy. To enable the registry setting, use the following PowerShell command:

```
Get-Item -path "HKLM:\SYSTEM\CurrentControlSet\Services\
Dnscache\Parameters" | New-ItemProperty -Name EnableMDNS -Value
0 -Force
```

Next, let's look at WPAD and how to prevent attacks by disabling the service.

The WPAD protocol

The WPAD protocol or **autoproxy** is a discovery protocol that's used by Windows to find web proxy settings located in a configuration file called wpad.dat. In an enterprise network, a network administrator typically configures DHCP and DNS to allow proxy settings to be fetched from a managed proxy server for the company's DNS suffix; the wpad.dat file is served to client PCs in a managed fashion. The security risk is that when a computer is configured to automatically look for proxy configurations on the network, no internal DNS or DHCP is configured. This allows anybody available to respond to the request. If an attacker is positioned inside the network, they can impersonate themselves as a proxy server, receive the connection requests, and offer the computer a malicious wpad.dat file. This can lead to an attacker redirecting all web traffic through their host for inspection and manipulation. Attackers can then modify websites to steal information, download malware-infected files, and force an HTTP-NTLM authentication request to entice users to supply credentials. The following diagram shows how this works:

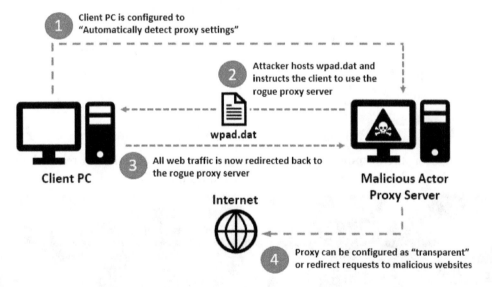

Figure 10.15 – Malicious proxy server

To view the Windows proxy settings, open **Settings** and go to **Network & internet |
Proxy**. The following screenshot shows that the default Windows settings for **Automatic
proxy setup | Automatically detect settings** have been set to **On**:

Figure 10.16 – Windows Proxy settings

Once again, using the **Responder** tool, you can see that the target computer reached out
for wpad.dat and received a poisoned response requesting NTLM authentication:

```
[*] [LLMNR]  Poisoned answer sent to 192.168.1.112 for name wpad
[*] [LLMNR]  Poisoned answer sent to 192.168.1.112 for name wpad
[*] [MDNS] Poisoned answer sent to 192.168.1.112    for name wpad.local
[*] [MDNS] Poisoned answer sent to 192.168.1.112    for name wpad.local
[*] [LLMNR]  Poisoned answer sent to 192.168.1.112 for name wpad
[*] [LLMNR]  Poisoned answer sent to 192.168.1.112 for name wpad
[HTTP] User-Agent        : Mozilla/5.0 (Windows NT 10.0; Win64; x64) AppleWebKit/537.36
.00.22976 Chrome/85.0.4183.121 Electron/10.4.3 Safari/537.36
[HTTP] User-Agent        : Mozilla/5.0 (Windows NT 10.0; Win64; x64) AppleWebKit/537.36
.00.22976 Chrome/85.0.4183.121 Electron/10.4.3 Safari/537.36
[HTTP] User-Agent        : Mozilla/5.0 (Windows NT 10.0; Win64; x64) AppleWebKit/537.36
.00.22976 Chrome/85.0.4183.121 Electron/10.4.3 Safari/537.36
[HTTP] User-Agent        : Mozilla/5.0 (Windows NT 10.0; Win64; x64) AppleWebKit/537.36
.00.22976 Chrome/85.0.4183.121 Electron/10.4.3 Safari/537.36
[HTTP] Sending NTLM authentication request to 192.168.1.112
```

Figure 10.17 – Rogue proxy server

As a result, the attacker was able to capture the password after serving the malicious WPAD file. This hash can be taken offline in an attempt to crack it in plaintext or forward it to other systems in a relay-style attack:

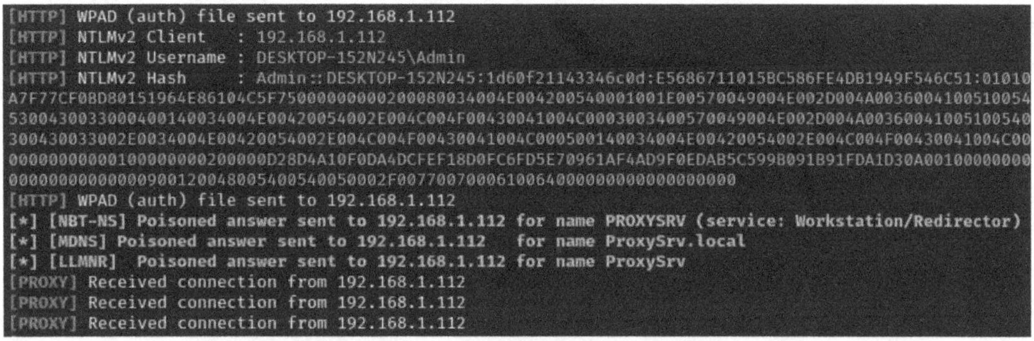

```
[HTTP] WPAD (auth) file sent to 192.168.1.112
[HTTP] NTLMv2 Client    : 192.168.1.112
[HTTP] NTLMv2 Username : DESKTOP-152N245\Admin
[HTTP] NTLMv2 Hash      : Admin::DESKTOP-152N245:1d60f21143346c0d:E5686711015BC586FE4DB1949F546C51:01010
A7F77CF0BD80151964E86104C5F75000000000200080034004E004200540040001001E00570049004E002D004A0036004100510054
5300430033000400140034004E00420050054002E004C004F0043004100054C000300340057004900049004E002D004A00360041005100540
30043003002E0034004E00420050054002E004C004F0043004100054C000500140034004E00420050054002E004C004F0043004100054C00
00000000000010000000200000D28D4A10F0DA4DCFEF18D0FC6FD5E70961AF4AD9F0EDAB5C599B091B91FDA1D30A0010000000
00000000000000900120048005400540050002F007700700006100640000000000000000000000
[HTTP] WPAD (auth) file sent to 192.168.1.112
[*] [NBT-NS] Poisoned answer sent to 192.168.1.112 for name PROXYSRV (service: Workstation/Redirector)
[*] [MDNS] Poisoned answer sent to 192.168.1.112  for name ProxySrv.local
[*] [LLMNR]  Poisoned answer sent to 192.168.1.112 for name ProxySrv
[PROXY] Received connection from 192.168.1.112
[PROXY] Received connection from 192.168.1.112
[PROXY] Received connection from 192.168.1.112
```

Figure 10.18 – Captured password hash

There are a few ways to protect against WPAD attacks and rogue proxy servers on a network, as follows:

- Disable WPAD completely or configure the proxy settings on client computers (recommended)

- On internal company networks, configure a WPAD server in the DNS lookup chain to avoid computers connecting to a rogue proxy server.

- Configure DHCP to point WPAD to a known host to control the traffic.

To disable WPAD on the client, you can use an Intune **Device Restrictions** profile | **Network proxy** template, Group Policy, or the registry. We also recommend that you disable the following service:

- **WinHTTP Web Proxy Auto-Discovery Service**: `WinHttpAutoProxySvc`

To disable the service through the registry, set the following key:

- `HKLM\SYSTEM\CurrentControlSet\Services\WinHTTPAutoProxySvc`

- **Start** REG_DWORD must be set to 4 (this sets startup mode to disabled)

After the service has been disabled and the PC restarts, the service will no longer be running. Additionally, if a user tries to re-enable WPAD, the setting will not be enforced unless the user physically starts the service using administrative rights. The following screenshot shows the LAN settings in **Internet Properties** once the service has been disabled:

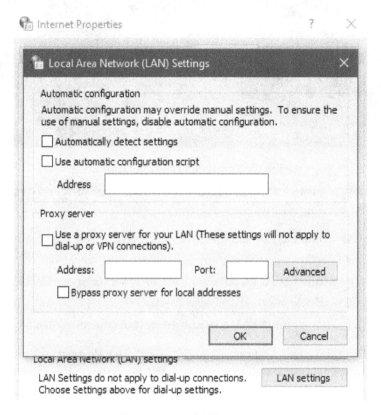

Figure 10.19 – LAN settings

With that, we have looked at several examples of how an attacker can become an AiTM by exploiting LLMNR, NBT-NS, mDNS, and WPAD. Ultimately, once a password has been captured, it may only be a matter of time before it's cracked or leveraged in a relay attack to connect to other systems. Next, let's look at an NTLM relay attack and how to mitigate its risks.

NTLM relay attacks

If an attacker has gained a foothold into your network and was able to successfully compromise a user and capture a password hash through MiTM techniques, they may not be able to crack the password into plaintext. This could be due to several things, such as corporate password complexity requirements or good password practices. One option is to relay them to systems and services that are vulnerable to relay attacks. Two services that are commonly targeted in relay-style attacks are SMB and LDAP. In this type of attack, the malicious actor sits in the middle of an NTLM authentication request and relays it to a target server of their choice. If the compromised user has administrative privileges on other systems, the attacker can move laterally and run arbitrary commands, launch a command shell, steal credentials, and manipulate the registry to gain a foothold. The following diagram shows how this works:

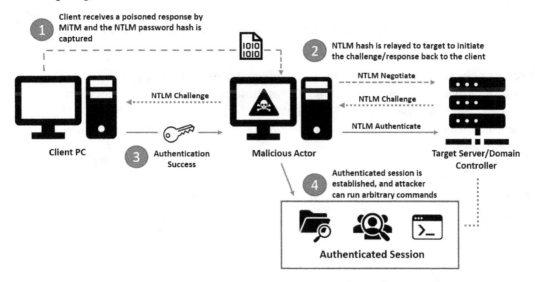

Figure 10.20 – NTLM relay attack

Let's look at a few recommendations that can help mitigate this type of attack.

Enable Kerberos authentication

In a network environment, you can use Kerberos authentication and disable NTLM completely. Kerberos is more secure than NTLM and offers better performance. To ensure that there will be no disruption if you're considering disabling NTLM, auditing policies can be enabled in Group Policy and events can be reviewed in Event Viewer or from a SIEM if you're streaming logs to a third-party service. For deployments that leverage the Azure cloud, system events can be sent to Log Analytics and queried using the KQL query language. The policies that you can use to enable NTLM auditing can be found by going to **Computer Configuration | Windows Settings | Security Settings | Local Policies | Security Options**, as follows:

- **Network Security: Restrict NTLM: Audit NTLM authentication in this domain** is set to `Enable all`

- **Network Security: Restrict NTLM: Audit Incoming NTLM Traffic** is set to `Enable auditing for domain accounts`

Events can be searched for in **Microsoft-Windows-Security-Auditing** with **Event ID 4624**. Under **Detailed Authentication Information,** you can view the protocol that was used under **Package Name** (LM, NTLMv1, or NTLMv2). In the following screenshot, we ran the following query in Azure Log Analytics to return all NTLM authentication requests:

```
SecurityEvent
| where EventID == "4624" and AuthenticationPackageName ==
"NTLM"
```

The output of this query will look as follows:

Activity	4624 - An account was successfully logged on.
AuthenticationPackageName	NTLM
ImpersonationLevel	%%1833
IpAddress	
IpPort	59623
KeyLength	128
LmPackageName	NTLM V1
LogonGuid	00000000-0000-0000-0000-000000000000
LogonProcessName	NtLmSsp
LogonType	3
LogonTypeName	3 - Network

Figure 10.21 – NTLM audit event

After carefully auditing the authentication requests and you are ready to disable NTLM, you can either disable it completely or use a server exception list by configuring the following policies:

- **Network Security: Restrict NTLM: Add server exceptions for NTLM authentication in this domain** and add any servers for which an exception must be listed.

- **Network Security: Restrict NTLM: NTLM authentication in this domain** must be set to Deny for domain servers or Deny all. Setting the policy to Deny for domain servers will allow those listed in the exemption list.

> **Tip**
> Adding users to the **Protected User** domain group will require them to authenticate using Kerberos only. This is a good way to confirm an account will work using Kerberos authentication before it's enforced.

Next, let's review enabling SMB signing.

Require SMB signing

SMB is a common protocol that's used for transferring files and facilitating **Microsoft Remote Procedure Calls** (**MSRPC**) in Windows. It is most notably used for its file-sharing capabilities on workstations and servers. By requiring SMB signing on both clients and servers, endpoints can verify the origin of the sender and the authenticity of the request. Devices that do not require SMB signing are suitable candidates for NTLM relay attacks. If an attacker intercepts the SMB packet flow with SMB signing enabled, the intended recipient will be able to detect that tampering occurred. Using a tool called **CrackMapExec**, an attacker can scan a subnet to find computers that do not require SMB signing, as shown in the following screenshot:

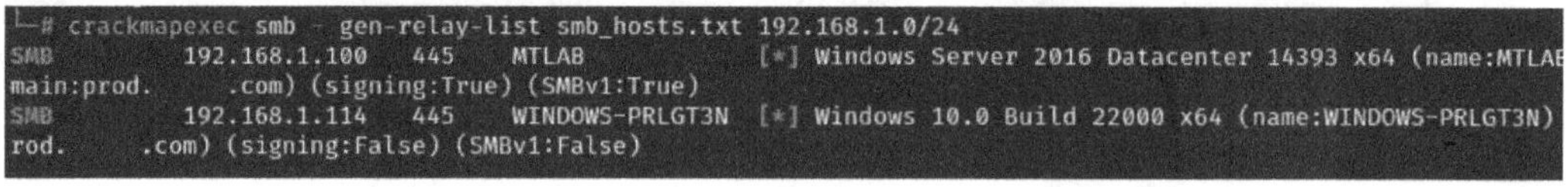

```
└─# crackmapexec smb - gen-relay-list smb_hosts.txt 192.168.1.0/24
SMB         192.168.1.100   445    MTLAB           [*] Windows Server 2016 Datacenter 14393 x64 (name:MTLAB
main:prod.    .com) (signing:True) (SMBv1:True)
SMB         192.168.1.114   445    WINDOWS-PRLGT3N  [*] Windows 10.0 Build 22000 x64 (name:WINDOWS-PRLGT3N)
rod.     .com) (signing:False) (SMBv1:False)
```

Figure 10.22 – SMB signing scan

Here's the link to CrackMapExec: `https://github.com/byt3bl33d3r/CrackMapExec`.

Once the systems have been identified, a combined list of targets can be exported and used in a coordinated multi-relay attack. To complete this, an attacker can use any combination of methods to capture user password hashes (LLMNR, NBT-NS, WPAD, and mDNS), and pass them along to a tool such as **NTLMRelayx** to try and authenticate to the target system with SMB. More information about NTLMRelayX can be found at `https://github.com/SecureAuthCorp/impacket/blob/master/examples/ntlmrelayx.py`.

To learn more about the basics of SMB signing, we strongly recommend reading this article on Microsoft docs: `https://docs.microsoft.com/en-us/archive/blogs/josebda/the-basics-of-smb-signing-covering-both-smb1-and-smb2`.

Enabling SMB signing can reduce network performance and should be reviewed carefully before it's enabled. Over the years, the policy guidance for configuring SMB signing has evolved alongside the evolution of the protocols (SMB1, SMB2, and SMB3). In Group Policy, these settings can be found by going to **Computer Configuration | Policies | Windows Settings | Security Settings | Security Options** and are as follows:

- **Microsoft network client: Digitally sign communications (always)**

- **Microsoft network client: Digitally sign communications (if server agrees)**

- **Microsoft network server: Digitally sign communications (always)**

- **Microsoft network server: Digitally sign communications (if client agrees)**

Typically, **always** means to require signing, while **if client/server agrees** means that signing is enabled, but it's up to the destination to decide the requirement. For SMB2 and above, the **if client/server agrees** setting is no longer applicable and SMB packet signing will never be negotiated. If you want to always require SMB signing, use the **Digitally sign communications (always)** policy setting for both clients and servers. In the latest **MSFT Windows 11 – Computer** security baseline, all four policies are set to `enabled` and SMB signing is a pre-configured policy recommendation in **Intune Security Baselines for Windows 10 and later**. It can also be configured through the Intune **Settings Catalog** under the **Local Policies Security Options** category or by using the Group Policy requiring path mentioned previously.

Next, let's look at protecting LDAP from relay attacks by enabling LDAP signing and LDAP channel binding.

Enable LDAP signing and LDAP channel binding

LDAP is an application protocol that can be used to query directory services such as Active Directory. This includes looking up data such as usernames, passwords, and computer objects. LDAP can also be used in authentication and to perform administrative actions such as resetting passwords and modifying security groups, which makes it a powerful and useful tool. LDAP is susceptible to MiTM and relay attacks if security measures aren't in place. To protect it, cryptographic protocols can be layered onto LDAP to establish secure connections and encrypt data in transit using SSL and TLS. This is known as **LDAPS** or **Secure LDAP**. LDAPS is essentially the same protocol as LDAP, but applications will require a certificate to create a trust chain, and communications in Active Directory use port `636` over `389`. It is strongly recommended to use LDAPS over LDAP whenever possible. Additional information about configuring it for Windows Server environments can be found at `https://techcommunity.microsoft.com/t5/sql-server-blog/step-by-step-guide-to-setup-ldaps-on-windows-server/ba-p/385362`.

In addition to using LDAPS, it is recommended that you enable both LDAP signing and LDAP channel binding to increase security against MiTM attacks. Enabling LDAP signing will require integrity verification through a digital signing signature, much like SMB signing, as described previously. LDAP channel binding will combine the transport and application layers through a *binding* to create a unique fingerprint for the LDAP communication to protect it against tampering. During a MiTM attack, if communication was intercepted, the fingerprint would change, resulting in the connection being dropped. For official guidance from Microsoft on enabling LDAP channel binding and LDAP signing, go to `https://msrc.microsoft.com/update-guide/en-us/vulnerability/ADV190023`.

The preceding link will help you enable these policies and refer to event logging that can be configured to minimize issues.

To enable signing, you can configure and set the following group policies, which are located at **Computer Configuration | Policies | Windows Settings | Security Settings | Local Policies | Security Options**:

1. Configure clients and member servers to negotiate signing by setting **Network Security: LDAP client signing requirements** to `Negotiate signing`. Ensure your clients received the updated policy before continuing.

2. Once the clients/servers have received the first policy, on your domain controllers, set the **Domain controller: LDAP server signing requirements** policy to `Require signing`.

3. Next, to enforce signing, change **Network Security: LDAP client signing requirements** to `Require signing` for clients and servers.

To configure channel binding on your domain controllers with Group Policy, go to **Computer Configuration | Policies | Windows Settings | Security Settings | Local Policies | Security Options** and set **Domain controller: LDAP server channel binding token requirements** to `Always`.

Next, let's look at how attackers can abuse IPv6 for MitM attacks and how we can help mitigate this risk.

Preventing IPv6 DNS spoofing

As the internet runs short of available IPv4 addresses, the solution is to use IPv6. Over the past few years, IPv6 usage has steadily increased and some countries have surpassed a 50% adoption rate. The following screenshot of 6lab shows the global IPv6 adoption by country:

Figure 10.23 – IPv6 worldwide adoption – 6lab

You can find the preceding chart and other great statistics about IPv6 usage at `https://6lab.cisco.com/index.php`.

Even with the increase in IPv6 usage globally across the internet, it's still unlikely that your internal company or home network is using an IPv6 addressing scheme. While there are many advantages to using IPv6 over IPv4, most smaller networks with internal IP ranges still rely on **Network Address Translation (NAT)** with IPv4. Today, possibly as a way of *future-proofing*, network adapters on clients with both IPv6 and IPv4 enabled will use IPv6 as the preferred prefix policy. This can be a concern for networks that do not have DHCPv6 configured or cannot respond to **Stateless Address Auto-Configuration (SLAAC)** requests, making them vulnerable to rogue DHCPv6 and DNS spoofing attacks. If networks aren't configured with DHCPv6, when a client boots up and sends a request for an IPv6 address, an attacker can answer the request and assign an IP address to the client. Once the client has been made aware of the attacker's endpoint, communications can be re-routed to an attacker-controlled DNS, where network packets can be captured and analyzed so that they can steal information or redirect victims to malicious sites. The following diagram shows how a malicious actor can answer an IPv6 multicast request on a network that's not been configured for IPv6:

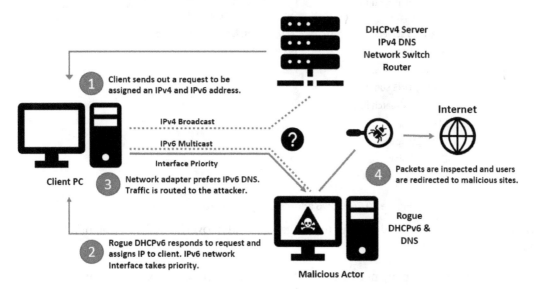

Figure 10.24 – Rogue DHCPv6 and DNS

Using a tool such as **mitm6**, we can see this in action on a network without IPv6 DHCP configured. For more information about the mitm6 tool, visit its public GitHub repository at `https://github.com/dirkjanm/mitm6`.

In the following screenshot, the attacker's endpoint assigned a client PC, `WINDOWS-PRLGT3N` an IPv6 address and pointed a DNS to its endpoint:

```
└─# mitm6 -hw WINDOWS-PRLGT3N.prod.    .com -d prod.    .com --ignore-nofqdn
Starting mitm6 using the following configuration:
Primary adapter: eth0 [00:15:5d:00:02:47]
IPv4 address: 192.168.1.110
IPv6 address: fe80::215:5dff:fe00:247
DNS local search domain: prod.    .com
DNS whitelist: prod.    .com
Hostname whitelist: windows-prlgt3n.prod.    .com
IPv6 address fe80::192:168:1:113 is now assigned to mac=e8:6a:64:25:da:18 host=WINDOWS-PRLGT3N.prod.
168.1.113
```

Figure 10.25 – MiTM attack on IPv6

We can confirm this by running `ipconfig /all` from the command prompt or by opening **Settings | Network & Internet** and choosing the **Ethernet** adapter on the client's PC. In the following screenshot, we can see that the Link-local IPv6 address and IPv6 DNS servers have been assigned by the attacker's rogue DHCP/DNS host:

Link speed (Receive/Transmit):	1000/1000 (Mbps)
IPv6 address:	2600:1700:1e1e:7fd0:2d8c:6efa:c052:d9b5
Link-local IPv6 address:	fe80::192:168:1:113%16
	fe80::2d8c:6efa:c052:d9b5%16
IPv6 DNS servers:	fe80::215:5dff:fe00:247%16 (Unencrypted)
IPv4 address:	192.168.1.113
IPv4 DNS servers:	192.168.1.254 (Unencrypted)
Primary DNS suffix:	
DNS suffix search list:	
Manufacturer:	Intel
Description:	Intel(R) Ethernet Connection (4) I219-LM
Driver version:	12.18.9.8
Physical address (MAC):	E8-6A-64-25-DA-18

Figure 10.26 – IPv6 address assigned by rogue DHCP

There are a few recommended ways to mitigate this often overlooked and potentially damaging attack vector:

- On internal company networks, configure a DHCPv6 server or disable IPv6 on your network appliances to prevent IPv6 multicast requests from being forwarded.

- On clients, block DHCPv6 traffic and ICMPv6 router advertisements in Windows Firewall.

- On clients, reprioritize IPv4 over IPv6.

On a client PC, you can change the prefix policy precedence of a network interface so that it uses IPv4 over IPv6. Let's look at the default prefix policies by opening an admin command prompt and running `netsh interface ipv6 show prefixpolicies`. As shown in the following screenshot, the first prefix, which has a precedence of `50`, is `::1/128`. This is the IPv6 loopback adapter, which is followed by the IPv6 default gateway:

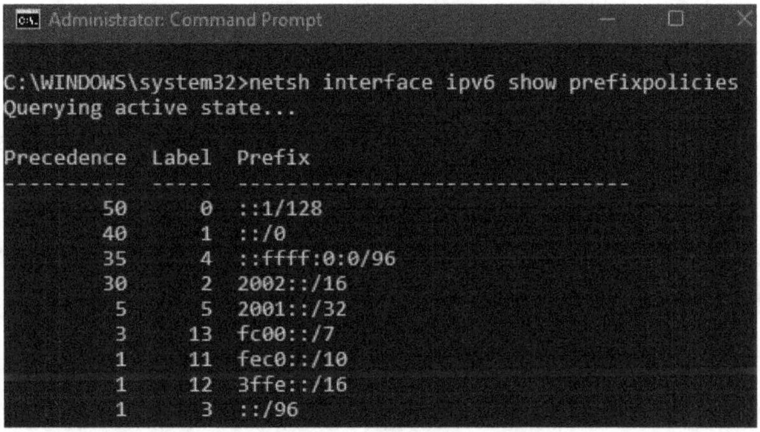
Administrator: Command Prompt
C:\WINDOWS\system32>netsh interface ipv6 show prefixpolicies
Querying active state...
Precedence Label Prefix
---------- ----- --------------------------------
50 0 ::1/128
40 1 ::/0
35 4 ::ffff:0:0/96
30 2 2002::/16
5 5 2001::/32
3 13 fc00::/7
1 11 fec0::/10
1 12 3ffe::/16
1 3 ::/96

Figure 10.27 – IPv6 prefix policies

The IPv4-mapped address block in IPv6 CIDR notation is `::ffff:0:0/96` and has a precedence of `35`. We can run the `netsh interface ipv6 set prefixpolicy ::ffff:0:0/96 55 4` command to change its priority and verify its output, as shown here:

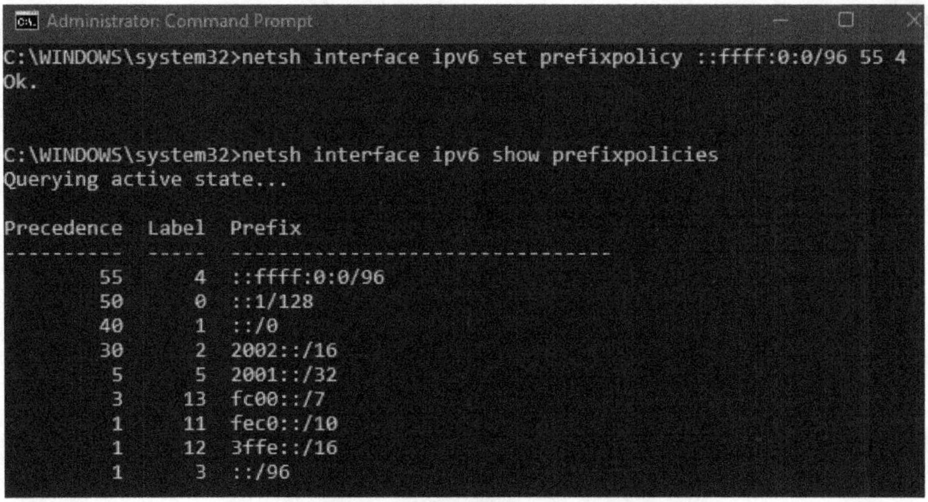
Administrator: Command Prompt
C:\WINDOWS\system32>netsh interface ipv6 set prefixpolicy ::ffff:0:0/96 55 4
Ok.
C:\WINDOWS\system32>netsh interface ipv6 show prefixpolicies
Querying active state...
Precedence Label Prefix
---------- ----- --------------------------------
55 4 ::ffff:0:0/96
50 0 ::1/128
40 1 ::/0
30 2 2002::/16
5 5 2001::/32
3 13 fc00::/7
1 11 fec0::/10
1 12 3ffe::/16
1 3 ::/96

Figure 10.28 – IPv4 priority in the prefix policy

You can confirm this by running `ping localhost`. It should resolve to an IPv4 address of `127.0.0.1` as opposed to `::1` for IPv6.

Next, let's look at how an attacker can use ARP to poison an ARP cache and act as a MiTM.

ARP cache poisoning

ARP is a protocol that's used by computers to discover link-layer addresses, such as a **Media Access Control (MAC)** address on a network. An example of this is when a client PC is trying to create an association of an IP to the MAC address of a local router or internet gateway. To create that mapping, the client will send out a broadcast message and wait for a response from a device that contains the associated IP. The device responds with its MAC address and, as a result, the client updates its local ARP cache with the record. Since ARP has no authentication mechanism, anyone on the local network segment can send a spoofed ARP message and reply to these broadcasts, where the client now receives the MAC address of an attacker's endpoint instead of the local router. This can result in a bi-directional traffic flow that passes through the attacker's machine, creating a MiTM in a technique known as ARP cache poisoning. Communications that pass through the attacker's machine are subjected to being captured and inspected or redirected to try and obtain information. The following diagram represents how ARP cache poisoning works:

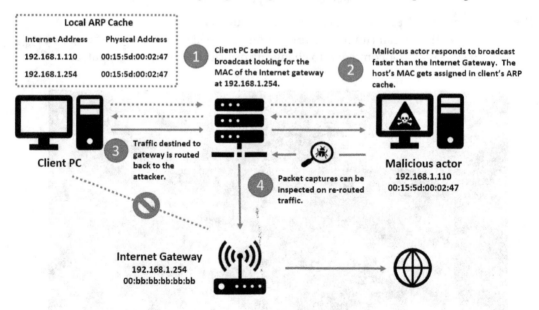

Figure 10.29 – ARP cache poisoning

Using a tool such as **Ettercap**, an attacker can target a victim and gateway to replace the associated MAC address in the ARP cache on the victim's PC by responding faster to broadcast requests. More information about Ettercap can be found at `https://www.ettercap-project.org/`.

Once communications flow through the MiTM endpoint, they are particularly vulnerable if that information is transmitted over a protocol that uses clear text, such as HTTP or FTP. Cleartext values can be extracted from a simple packet capture using a tool such as **Wireshark**. In the following screenshot, we can see that ARP poisoning was successful to the client's PC IP address at `192.168.1.113` and the internet gateway at `192.168.1.254`:

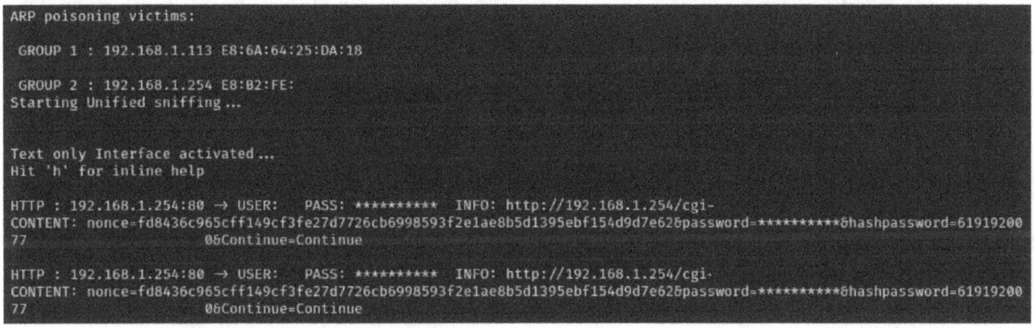

```
ARP poisoning victims:

 GROUP 1 : 192.168.1.113 E8:6A:64:25:DA:18

 GROUP 2 : 192.168.1.254 E8:B2:FE:
Starting Unified sniffing ...


Text only Interface activated ...
Hit 'h' for inline help

HTTP : 192.168.1.254:80 → USER:   PASS: **********  INFO: http://192.168.1.254/cgi-
CONTENT: nonce=fd8436c965cff149cf3fe27d7726cb6998593f2e1ae8b5d1395ebf154d9d7e62&password=**********&hashpassword=61919200
77                06Continue=Continue

HTTP : 192.168.1.254:80 → USER:   PASS: **********  INFO: http://192.168.1.254/cgi-
CONTENT: nonce=fd8436c965cff149cf3fe27d7726cb6998593f2e1ae8b5d1395ebf154d9d7e62&password=**********&hashpassword=61919200
77                06Continue=Continue
```

Figure 10.30 – ARP poisoning

Here, the user visited an HTTP site and entered a password. We can check the ARP cache on a client and confirm that ARP poisoning was successful by opening a command prompt and typing `arp -a` to view the output of the ARP cache. We ran this command twice to show what happened before and after. After using the command again, the physical address changed and is now mapped to the attacker's host at `00-15-5d-00-02-47`:

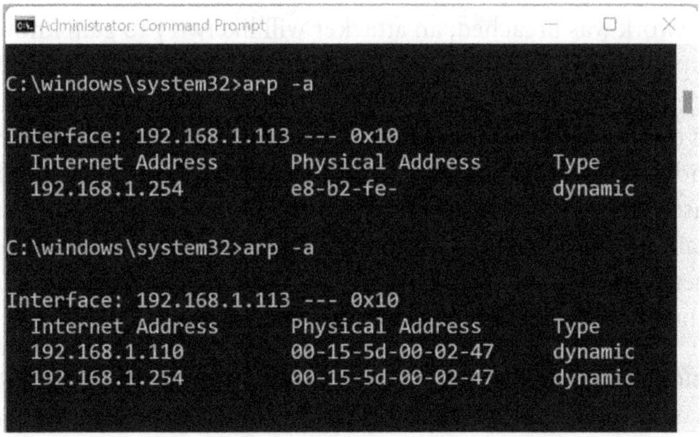

```
Administrator: Command Prompt                          —    □    X

C:\windows\system32>arp -a

Interface: 192.168.1.113 --- 0x10
  Internet Address        Physical Address        Type
  192.168.1.254           e8-b2-fe-               dynamic

C:\windows\system32>arp -a

Interface: 192.168.1.113 --- 0x10
  Internet Address        Physical Address        Type
  192.168.1.110           00-15-5d-00-02-47       dynamic
  192.168.1.254           00-15-5d-00-02-47       dynamic
```

Figure 10.31 – Output of the ARP cache

Unfortunately, ARP is a common communication protocol that's used to translate addresses between the data link and the network layer and is insecure. There are a few recommendations to consider that will help mitigate MiTM through ARP cache:

- For company networks, many network switches support a security feature called **Dynamic ARP inspection** (**DAI**). This helps evaluate ARP messages over the network and prevents **denial-of-service** (**DOS**) attacks through ARP.

- On clients, avoid using services that authenticate using insecure protocols. Services such as HTTP, FTP, and Telnet send passwords in plaintext that can be extracted from a packet capture.

- Use protocols with encryption such as SSL/TLS to ensure that communications are encrypted in transit and cannot be decrypted.

- Having well-defined network segmentation can minimize the number of potential targets as ARP messages do not traverse across subnets.

- Define static ARP tables in your network if possible.

As we mentioned previously, there are many ways to use standard communications protocols that can be exploited once an attacker gets access to your network. Many of these protocols are enabled by default, are standard network functionality, and can present significant security risks without taking the proper precautions. Next, let's learn how to protect against lateral movement and privilege escalation if an attacker can gain access to a system or capture credentials.

Protecting against lateral movement and privilege escalation

Assuming your network was breached, an attacker will likely try to gain knowledge about the network and systems by using discovery and reconnaissance techniques. Many of these techniques require little to no privileges to run once they're inside the network. Some examples of discovery tactics that produce useful information include running network scans to identify potential hosts, running vulnerability scans to find hosts that are not patched against known weaknesses, and gathering information by enumerating cloud services and domain policies. Even the slightest bit of information, as trivial as it may seem, could be used against you to find potential security gaps. A few examples of how these reconnaissance activities could be used to launch an attack are as follows:

- Password and Group Policy discovery information can be used to fine-tune methods that are used in brute-force attacks and reduce the chances of account lockout or identity protection services from being triggered.

- File and directory discovery can be used to identify the location of company data that can be stolen, held for ransom, and used for exploitation.

- System service and process discovery can be leveraged to carefully craft attacks, or used for process injection to avoid detection.

- Account discovery can be used to find additional targets and identify privileged accounts.

The **MITRE ATT&CK** framework provides many additional examples of discovery tactics, including explanations of how they can be used to further gain a foothold. To learn more, go to `https://attack.mitre.org/tactics/TA0007/`.

Many of the security controls we've discussed in this book can help prevent lateral movement and privilege escalation, should one of your endpoints become compromised. No one solution fits all that provides guaranteed protection, so you must have a good understanding of the current attack surface to make the best defense decisions for your company to stay within budget and remain tolerant of your acceptable risk. Let's look at a few best practice principles:

- Protect your identities as well as possible. This includes implementing strong password policies, requiring multi-factor authentication, and going passwordless when possible. For improved protection, enable behavioral and threat monitoring solutions that can audit sign-ins and build a profile of users' activities to alert you when any anomalies occur in an automated way. Use access control policies to restrict access from unapproved devices and require trusted compliant devices through attestation based on zero-trust principles.

- Adhere to the principle of least privilege. Use standard accounts on Windows systems and regularly audit access and remove it unless it can be justified. **Privileged Identity Management** (PIM) and **Privileged Access Management** (PAM) are invaluable in these efforts.

- Strong network designs that include micro-segmentation can help reduce lateral movement by only allowing the necessary systems to talk to each other using only the necessary protocols.

- Enable policies that prevent standard user accounts from enumerating information on the domain or in the cloud. By default, a standard user has a lot of visibility and can be helpful in reconnaissance activities, should the account become compromised. Deploying tools such as Microsoft Defender for Identity can help alert you when anomalies such as reconnaissance occur in Active Directory.

- On clients, enable a solution such as Credential Guard that can block credential-stealing by isolating the **Local Security Authority (LSA)** processes with hypervisor-based isolation. Additionally, **attack surface reduction (ASR)** rules can be enabled to block credential stealing from the LSASS.

Let's look at a few examples of policies that can be enabled to provide some added protection.

Preventing resources from being enumerated

Your infrastructure, resources, and accounts are the jewels of your environment, and they need to remain protected from unauthorized access. As we mentioned previously, if an attacker can get inside your network or compromise an account, one of the first things they will do is gather information. For example, a standard domain user has permission to enumerate all users in your domain simply by running the `net use /domain` command from a computer on your network. This means privileged security groups and accounts can easily be found and targeted. In the following screenshot, the output of the `net group "domain admins" /domain` command returned all the members in the `Domain Admins` security group:

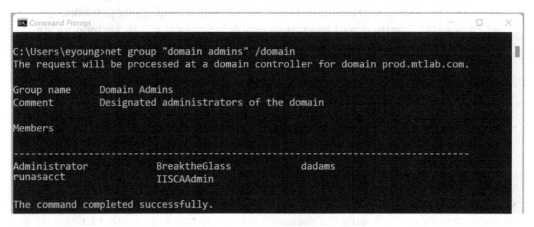

Figure 10.32 – Domain Admins enumeration

Resources in your domain can also be discovered by anonymous or *null* sessions. Using tools such as **enum4linux**, if the domain controllers allow null sessions in SMB, it can be leveraged to discover users, computers, groups, and the password policy. The `net use` command is a very basic example but demonstrates the level of visibility that's available to any domain user account. In addition, security tools such as enum4linux make it easy to gather additional information with a few commands. More information can be found at `https://github.com/CiscoCXSecurity/enum4linux`.

While it may be necessary for this information to be discoverable, it is possible to block access to certain sensitive groups and users by modifying the **Access Control List** (**ACL**) on objects directly, or at the OU and domain level. Applying the **deny read** permission to the ACL effectively stops users from querying these objects. These permissions can be applied to a security group where users can quickly be added and removed. In the following screenshot, you can see where the deny read permissions are configured for a security group named **Deny Read Domain Admins**:

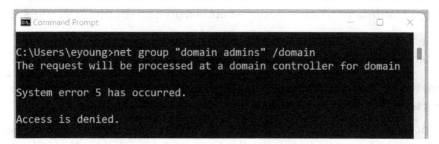

Figure 10.33 – Domain Admins ACL

If a domain user is a member of this group and they run the `net group "domain admins" /domain` command, they will receive an access denied message, as shown in the following screenshot:

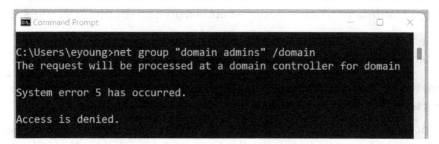

Figure 10.34 – Access denied for enumeration

If all your privileged groups are stored in the same OU, then this permission can be applied to limit what can be discoverable. There is a good blog that goes into more detail if you're interested. It can be found at `https://www.adamcouch.co.uk/disable-domain-user-enumeration/`.

In addition to restricting the information that a domain user can query, we recommend enforcing the policies listed in the CIS Windows Benchmarks in the **Network access** section. A few examples of these policies are as follows:

- **Network access: Do not allow anonymous enumeration of SAM accounts and shares** set to `Enabled`. Enforcing this setting will stop anonymous users from discovering accounts listed in **Security Accounts Manager** (**SAM**) and network shares. SAM is a database on Windows that stores user accounts and security descriptors for local users. Auditing can be enabled to log attempts to access the SAM database.

- **Network access: Named Pipes that can be accessed anonymously** is set to `none`. This will disable null sessions over named pipes and prevent unauthenticated access.

- **Network access: Allow anonymous SID/Name translation** is `disabled`. This will prevent a user with local access from using the known administrator's SID to discover the built-in administrator account if it has been renamed. If a local administrator account is required on client PCs, it is recommended to create a new account and not reuse the built-in account.

- **Enumerate local users on domain-joined computers** is `disabled`. This may help prevent local user accounts on domain-joined computers from being enumerated.

If you host a directory in the cloud such as **Azure Active Directory** (**Azure AD**), it is also susceptible to similar enumeration by standard user account privileges. There are many Azure AD enumeration tools available and once accounts or privileged groups have been discovered, they can be targeted in password spraying attacks. Luckily, Azure has many built-in features to protect you against these attacks, such as security defaults, Conditional Access, and Azure AD Identity Protection. For additional information about security defaults, visit `https://docs.microsoft.com/en-us/azure/active-directory/fundamentals/concept-fundamentals-security-defaults`.

It is possible to prevent account enumeration of the Azure AD tenant for guest and member users by using the following settings:

1. From the Azure portal at `https://portal.azure.com`, open **Azure Active Directory** and select **User Settings**.

2. Restrict access to the Azure AD administration portal by setting it to `Yes`. This will stop non-administrators from using the Azure portal experience. This will not stop them from using PowerShell.

3. Select **Manage external collaboration settings** to view the permission policies for Azure AD B2B guest users.

4. **Guest user access is restricted to properties and memberships of their own directory objects (most restrictive)** must be selected.

By default, directory members have read access to all the users in the directory, which can be disabled using PowerShell. While possible, this is not recommended by Microsoft. To do this, you will need an account with **Global Administrator** permissions and the **MSOnline** PowerShell module. Read access can be disabled by running the `Set-MsolCompanySettings -UsersPermissionToReadOtherUsersEnabled $false` command.

For additional information about the default user access permissions for members and guest users in Azure AD visit `https://docs.microsoft.com/en-us/azure/active-directory/fundamentals/users-default-permissions`.

> **Tip**
> Disabling this could cause other services that rely on user account lookups to break.

Next, let's look at Kerberos authentication and how to provide protection against Kerberos-based attacks in the domain. If you are unable to host services in the cloud to leverage modern authentication capabilities such as FIDO, MFA, and Conditional Access, it's important to protect Kerberos on your domain as it's likely the primary authentication mechanism.

Protecting Kerberos tickets

Kerberos is a ticket-based authentication protocol that allows a client and server to connect through mutual-based authentication to verify each other's identity. It is a more secure protocol than NTLM and protects against eavesdropping and relay-type attacks that are a problem with NTLM authentication. Windows domains make use of Kerberos tickets as their default authentication method. When a client authenticates to a domain, the domain controller will respond with a Kerberos service ticket that the client can send to any requested services as proof they are authenticated. As a result, they are authorized to access the resource. During this authentication process, **ticket-granting tickets (TGT)** are issued through the **Key Distribution Center (KDC)** in Active Directory and are encrypted and signed using the special KRBTGT service account.

This account exists by default and can be accessed by an account with elevated privileges on the domain. Attackers who successfully obtain the KRBTGT account password can leverage it to forge TGTs to impersonate any user and, in effect, access any resource in the domain. This process of forging a TGT through the KRBTGT service account is known as a **golden ticket**. If the KRBTGT account cannot be accessed, attackers can also perform a **pass-the-ticket** (**PTT**) attack using stolen Kerberos tickets and continue to move laterally. PTT attacks steal valid Kerberos tickets through account compromise and with techniques such as credential dumping. Let's look at these attacks in more detail.

Golden ticket and silver ticket attacks

If an attacker can obtain the KRBTGT password hash or *golden ticket*, they can impersonate any account in the domain and abuse the Kerberos authentication process. In a golden ticket attack, the adversary bypasses the issuance of the TGT from the KDC and forges **ticket-granting tickets** (**TGTs**) with a long validity period that will even remain active if a user changes their account password. This process is shown in the following diagram:

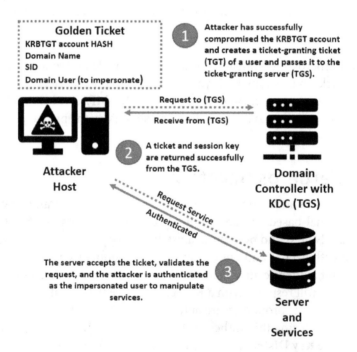

Figure 10.35 – Golden ticket attack

With a forged TGT in hand, a **ticket-granting service** (TGS) ticket can be requested from the authentication server and then sent to other resources to gain access. With the ability to impersonate any user, including a domain admin, the possibilities are almost limited to the creativity of the attacker as to what they can do next.

A **silver ticket** is a similar type of exploit that leverages a falsified Kerberos ticket, but unlike the golden ticket, a falsified TGT ticket is not passed to the KDC. In this attack, a TGS ticket is directly crafted by the attacker leveraging the extracted hash of the compromised service or computer account. From there, the forged TGS ticket can be passed directly to a target server's services (such as HOST, LDAP, WSMAN, RPCSS, and so on) and authenticated without them needing to contact the domain controller for validation. Using one or many forged TGS tickets, an attacker can execute different service types (such as WMI, PowerShell Remoting, and scheduled tasks) on the target system. For example, to connect to PowerShell Remoting, an attacker would pass a silver ticket for the HTTP and WSMAN service to open a shell on the target computer. Even though this temporarily limits the scope of access as the compromised computer account and target service may have limited operations, silver tickets are just as dangerous, not as easy to spot, and can ultimately lead to the same privilege escalation in the long run. A carefully crafted ticket can tell the target service that the computer account is a Domain Admin without fully verifying it against the domain controller. Typically, computer account passwords do not have any password changing requirements, so if the password becomes compromised, it may be valid for an extended period:

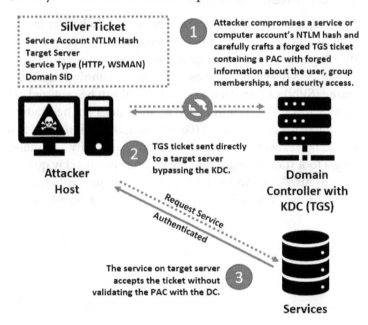

Figure 10.36 – Silver ticket attack

The following are a few recommendations that can help protect Kerberos authentication against these types of attacks:

- Protect privileged accounts, elevated service accounts, administrator accounts, and those that have logon rights to a domain controller. These accounts can be used to dump the KRBTGT account hash using tools such as **Impacket** (`https://github.com/SecureAuthCorp/impacket`) or **Mimikatz** (`https://github.com/gentilkiwi/mimikatz`).

- Enforce strict password requirements on privileged accounts and user-based SPNs to make it difficult for password cracking tools and Kerberoasting.

- Rotate the KRBTB service account password at regular intervals.

- Use endpoint security solutions to alert you if malicious tools and scripts are being used. Monitor for suspicious activity such as service principal creation, TGTs with long lifetimes, and unusual replication. Azure Defender for Identity is a great solution to help monitor against these types of attacks. You could even set up a honeypot account to act as bait and alert you if it's being targeted.

Next, let's look at two additional attacks that can be used to exploit Kerberos authentication.

Kerberoasting and AS-Rep roasting

Kerberoasting is a technique that's used to steal Kerberos TGS tickets by targeting services running with user-based service accounts. In a Kerberoasting attack, any compromised standard domain user can acquire legitimate proof of authentication from the KDC and use it to request the TGS tickets of services. Any service running with a user-based **service principal name** (**SPN**) can return a TGS ticket that can be captured and cracked using password-cracking tools. The compromised account does not need administrative privileges to do this. Unlike host-based SPNs (computer accounts) in silver ticket attacks, user-based SPNs are likely created with weaker passwords, are easily guessable, and set to never expire. Using a tool such as **Impacket**, the attacker can quickly query the entire domain with the compromised user account. In the following screenshot, the TGS ticket hash that was returned for the **MSSQLSvc** service is running as a user-based SPN:

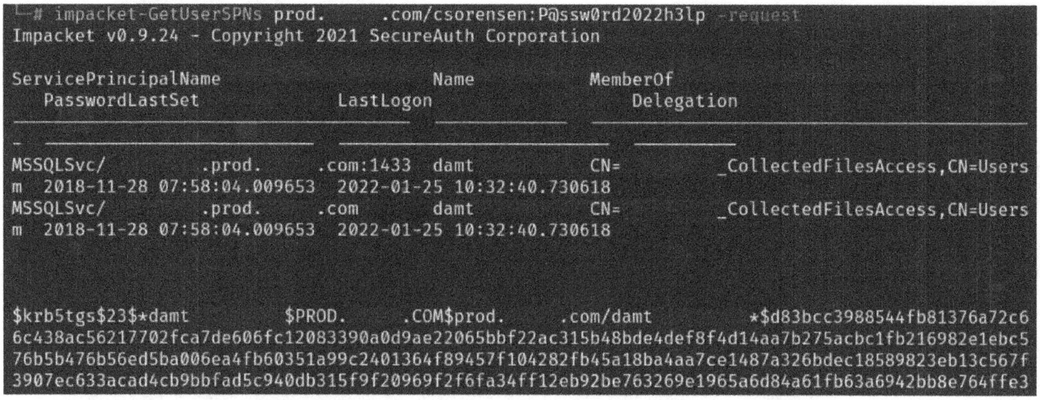

```
└─# impacket-GetUserSPNs prod.      .com/csorensen:P@ssw0rd2022h3lp -request
Impacket v0.9.24 - Copyright 2021 SecureAuth Corporation

ServicePrincipalName                      Name          MemberOf
   PasswordLastSet            LastLogon                  Delegation

-                                         ------        --------
MSSQLSvc/         .prod.      .com:1433   damt          CN=          _CollectedFilesAccess,CN=Users
m  2018-11-28 07:58:04.009653  2022-01-25 10:32:40.730618
MSSQLSvc/         .prod.      .com        damt          CN=          _CollectedFilesAccess,CN=Users
m  2018-11-28 07:58:04.009653  2022-01-25 10:32:40.730618


$krb5tgs$23$*damt         $PROD.      .COM$prod.      .com/damt       *$d83bcc3988544fb81376a72c6
6c438ac56217702fca7de606fc12083390a0d9ae22065bbf22ac315b48bde4def8f4d14aa7b275acbc1fb216982e1ebc5
76b5b476b56ed5ba006ea4fb60351a99c2401364f89457f104282fb45a18ba4aa7ce1487a326bdec18589823eb13c567f
3907ec633acad4cb9bbfad5c940db315f9f20969f2f6fa34ff12eb92be763269e1965a6d84a61fb63a6942bb8e764ffe3
```

Figure 10.37 – Kerberoasting attack

The following are a few ways you can protect against Kerberoasting:

- Enforce strict password policy requirements with long, complex passwords for service accounts.

- Implement a PAM solution that can rotate passwords frequently and automate managing dependencies, such as restarting services or application pools.

If possible, enable AES Kerberos Encryption for service accounts to deter password cracking. By default, Active Directory will use RC4/DES, which is easier to crack offline with tools. AES Kerberos Encryption can be set by opening **Active Directory User and Computers**, opening the service account's properties, clicking the **Accounts** tab, and selecting either **This account supports Kerberos AES 128 bit encryption** or **This account supports Kerberos AES 256 bit encryption**.

Authentication Server Reply (AS-Rep) roasting isn't as prevalent today, but if your domain has poorly configured accounts that do not require Kerberos pre-authentication, it's possible to fall victim to an AS-Rep roasting attack. During the Kerberos authentication process, an encryption key containing the request timestamp that's been encrypted by the user's password hash is sent to the KDC to issue a TGT for use in authorization. The timestamp must match that of the KDC for the ticket to be issued during the reply process. If **Do not require Kerberos preauthentication** is selected as one of the user account attributes, the attacker can send a request to the KDC. They will be issued a ticket due to the lack of password requirements. However, this can later be cracked offline with a password cracking tool. Using a tool such as **Impacket**, the attacker can request the domain to return accounts that do not require pre-authentication and obtain the TGT ticket containing the targeted user's password hash. In the following screenshot, a standard domain user queried the domain for accounts that do not require Kerberos pre-authentication:

Figure 10.38 – AS-Rep roasting attack

This option can be checked in **Active Directory Users and Computers** by opening a user account's properties, clicking on the **Account** tab, and scrolling down to **Account options**. As shown in the following screenshot, the account does not require pre-authentication:

Account options:

- [ ] Use only Kerberos DES encryption types for this account
- [ ] This account supports Kerberos AES 128 bit encryption.
- [ ] This account supports Kerberos AES 256 bit encryption.
- [x] Do not require Kerberos preauthentication

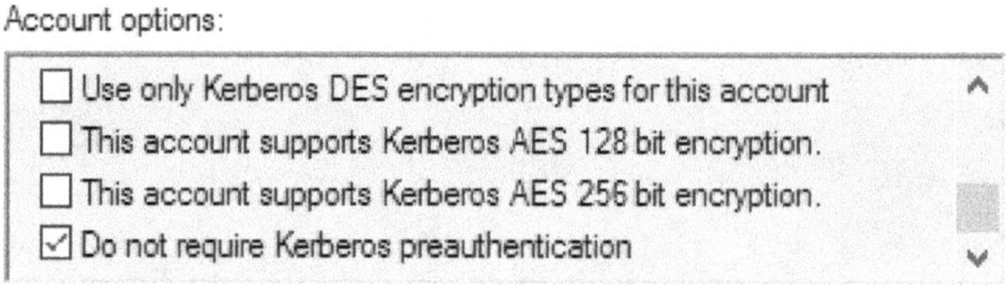

Figure 10.39 – Kerberos pre-authentication

Next, let's look at how an attacker can attempt to retrieve credentials from the host **operating system (OS)** and domain using techniques known as OS credential dumping.

Mitigating OS credential dumping

OS credential dumping is a post-exploitation technique that's used to steal stored credentials from the host OS. In a Windows environment, there are a few common places that attackers target to try and get this information. Successful exploits with credential dumping can affect not only local Windows hosts but put Active Directory domains and the Azure cloud at risk. For example, it's possible to extract a PRT token from an Azure AD registered device and use it to gain access to Azure and bypass security controls such as MFA. Let's look at a few places where OS credential dumping can occur. For reference, we are using the OS Credential Dumping MITRE ATT&CK technique. More information can be found at `https://attack.mitre.org/techniques/T1003/`.

- **Local Security Authority Subsystem Service (LSASS)** is a process that runs on Windows that handles Windows security policy enforcement by verifying account access. Domain credentials, local usernames, passwords, and Azure primary refresh tokens are stored inside the memory of the LSASS process and with the appropriate privileges, they can be extracted from memory.

- **Security Account Manager (SAM)** is a database stored in the Windows Registry that contains local account data. Accounts stored in the SAM are becoming less frequent, especially in fully Azure AD joined environments where the accounts aren't stored here. Nevertheless, if local administrator accounts exist on the system, they can be dumped from the SAM database.

- **Active Directory domain database (AD DS)** contains information about AD objects, including group memberships and password hashes. If the NTDS file can be copied, its contents can be extracted and its hashed passwords can be cracked using password cracking tools.

- **Local Security Authority (LSA)** secrets is the storage component of the local security authority and stores information such as user passwords and system and service account passwords that are used to run services in Windows. These secrets can be extracted, and the encrypted values can be decrypted into readable text.

- Cached domain credentials are stored on Windows systems locally as an alternative authorization mechanism to allow users to sign into their PCs if the domain is unreachable. These are stored in the SAM database and LSA secrets and can be extracted and cracked using password cracking tools.

- **Domain controller replication** (also known as **DCSync**) is a technique where an attacker will attempt to start the DC replication process by impersonating a domain controller. If successful, the replicated data in scope will include information such as AD objects, accounts, and password hashes that can be captured and cracked. This attack can be stealthy as it can be initiated without the attacker logging onto a domain controller directly, making it more difficult to detect.

Let's look at a few recommendations that can help detect and prevent OS credential dumping:

- Enable an **Endpoint Detection and Response** (**EDR**) solution that can detect when suspicious tools and scripts are installed and run. In the following screenshot, Defender for Endpoint detected when a command attempted to exfiltrate the SAM database in the registry, blocked the process, and issued an alert:

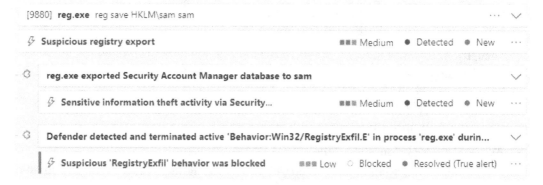

Figure 10.40 – Microsoft 365 Defender for Endpoint alert

- Use a solution such as Windows Defender Credential Guard to prevent exfiltration of the LSA. Credential Guard leverages VBS security to isolate the LSA process that stores secrets (Kerberos, NTLM, and Credential Manager) and can only be accessed by a small set of trusted OS binaries.

- Use **Attack Surface Reduction** (**ASR**) rules to block processes that attempt to steal credentials from Windows LSASS. As we mentioned earlier, LSASS is a process that handles security policy enforcement and can be targeted as a point of exfiltration.

- To protect the Active Directory NTDS database, ensure domain controller backups are stored, encrypted, and in a secure place where access is monitored. Ensure that access to domain controllers is managed with a PAM solution and that administrative activity is monitored.

- To protect against DCSync-style attacks, monitor domain controller logs for replication requests and tightly control the **access control lists (ACLs)** that set permissions for domain controller replication. In the following screenshot, Microsoft Defender for Identity created an alert for a suspected DCSync attack:

Figure 10.41 – DCSync alert from Defender for Identity

Next, let's learn how to prevent user access to the registry to help deter access from standard user accounts.

Preventing user access to the registry

The Windows Registry acts as a database in which settings and information are stored and retrieved by applications and system components. Users should not edit or modify the registry as it can cause incompatibilities and system stability issues. From a security and hardening perspective, the Windows registry can be used to install backdoors and used in evasion and persistence tactics, to name a few. Using Policy, it is possible to lock down the registry to some extent and prevent standard user accounts from opening registry editing tools. Doing this can help act as a deterrent to prevent unauthorized modification through the user's context, but sources with administrative privileges will still have access. You can prevent access to registry editing tools for users in Intune by using **Settings Catalog** and searching for **Prevent access to registry editing tools** in the **Settings picker** area. This setting can be seen in the following screenshot:

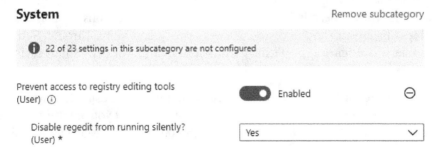

Figure 10.42 – Preventing user registry access

Once the policy has been applied, the next time a user goes to open `regedit.exe`, they will receive the following message:

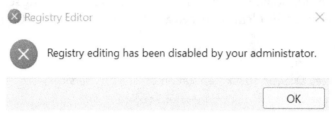

Figure 10.43 – Registry Editor disabled

To configure the same setting with Group Policy, go to **User Configuration | Policies | Administrative Templates | System**. Select **Prevent access to registry editing tools**. This will also help prevent the registry from being modified from the command line in the user context. For more information on hardening the Windows registry, MITRE has provided a list of mitigations that can be used to address known attack techniques for leveraging the registry at `https://attack.mitre.org/mitigations/M1024/`.

Now, let's learn how to protect users' privacy by reviewing Windows privacy settings.

Windows privacy settings

Windows has many great features that provide a personalized and enhanced connected experience for its users. To support this personalization, Windows has permission settings that control what data and device features that applications are allowed to access. A few examples include allowing an application to access the camera, device location, or microphone. Unless controlled by a policy, many of these privacy permissions are allowed by default and could pose a potential privacy risk for some organizations. To view the Windows privacy settings, open **Settings** and choose **Privacy & Security**. Here, you can get an idea of the types of permissions that are available to applications, such as access to speech settings, diagnostics and feedback, activity history, and more. Through **Settings**, you can granularly configure app-specific permissions or allow or deny all for each permission type.

Let's run through a few settings and where we can configure them using Intune. Note that some of these privacy permissions may need to remain enabled if you are using solutions such as Log Analytics or Endpoint Analytics in Microsoft Endpoint Manager to collect telemetry data from the endpoints.

The **Privacy & Security** settings are available in the **Intune Settings** catalog and **Templates**. If the policies don't exist in the UI, they can also be mapped using a custom template if a CSP is available, by pushing a registry key with PowerShell scripts, and so on. Let's look at a few places we can configure these settings as they are hard to find based on the friendly name shown in the Windows **Settings** app. You can search for them using **Settings Picker** in the **Settings Catalog** area:

- **Privacy & Security | General**:

  - Let apps show me personalized ads by using my advertising ID:

    - **Settings Catalog | Disable Advertising ID**

  - Let Windows improve Start and search results by tracking app launches:

    - **Settings Catalog | Turn off user tracking (User)**

  - Show me suggested content in the **Settings** app:

    - **Settings Catalog | Allow Online Tips**

- **Privacy & Security | Speech**:

  - Use your voice for apps using Microsoft's online speech recognition technology:

    - **Settings Catalog | All Input Personalization**

- **Privacy & Security | Inking & typing personalization**:

  - Personal inking and typing dictionary

- **Privacy & Security | Diagnostics & feedback**:

  - Diagnostic Data:

    - **Settings Catalog | Allow Telemetry**

  - Improve inking and typing:

    - **Settings Catalog | Allow Linguistic Data Collection**

  - Tailored experiences:

    - **Settings Catalog | Allow Tailored Experiences with Diagnostic Data (User)**

- Delete diagnostic data:

    - **Settings Catalog | Disable Device Delete**

- **Privacy & Security | Activity history**:

    - Store my activity history on this device:

        - **Settings Catalog | Publish User Activities**

    - Send my activity history to Microsoft:

        - **Settings Catalog | Upload User Activities**

- **Privacy & Security | Search permissions**:

    - SafeSearch

        - **Settings Catalog | Do Not User Web Results**

    - Cloud Content Search:

        - **Settings Catalog | Allow Cloud Search**

We didn't list every setting as some of them don't have mapped CSPs or Group Policy settings. It may be possible to configure them directly with registry keys, but that is outside the scope of this book.

Next, let's look at setting application-specific privacy permissions.

Controlling application privacy permissions

Using Intune, you can configure the access that specific applications have to privacy features. Most of these settings can be found in the **Settings Catalog** area by searching for **Privacy** in **Settings Picker**. For example, in the following screenshot, we have set the **Let Apps Access Camera** policy to `Force deny` and configured a list of allowed apps using **Let Apps Access Camera Force Allow These Apps**:

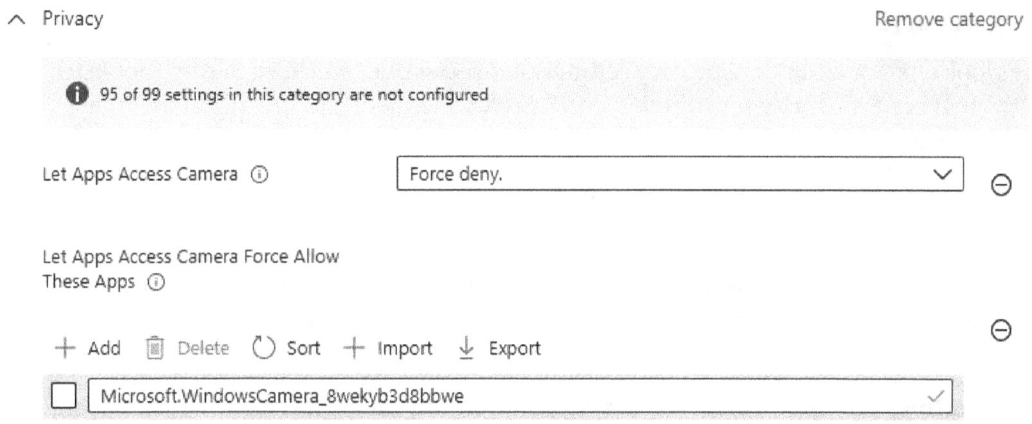

∧ Privacy   Remove category

ⓘ 95 of 99 settings in this category are not configured

Let Apps Access Camera ⓘ | Force deny.   ⌄ | ⊖

Let Apps Access Camera Force Allow
These Apps ⓘ
                                                                                      ⊖
+ Add   🗑 Delete   ↻ Sort   + Import   ↓ Export

☐ | Microsoft.WindowsCamera_8wekyb3d8bbwe   ✓ |

Figure 10.44 – Setting app permissions in Intune

Configuring an application allow list is only supported for Microsoft Store apps at the time
of writing. To do this, you will need to gather the application's **Package Family Name
(PFN)** using the Microsoft Store URL or PowerShell. For example, to find the PFN for the
Camera app using PowerShell, run `Get-AppXPackage *Camera | Select Name,`
`PackageFamilyName`, as shown here:

Figure 10.45 – Windows PackageFamilyName

> **Tip**
> You cannot control camera access to third-party apps selectively. Setting **Let
> Apps Access Camera** to `Force deny` will block third-party apps.

For more information about finding the package family name using PowerShell or
the Microsoft app store, go to `https://docs.microsoft.com/en-us/mem/`
`configmgr/protect/deploy-use/find-a-pfn-for-per-app-vpn`.

Additional privacy settings

Let's look at a few additional privacy settings that you should consider that are not listed in the **Privacy & Security** settings. It's worth evaluating them and determining if they should be disabled on company devices, depending on your privacy controls:

- **Settings Catalog | Allow Game DVR**. Disabling this policy will block Windows Game Recording and Broadcasting.

- **Settings Catalog | Disable Privacy Experience**. Disabling this policy may prevent new users from changing company-managed privacy settings when they log on for the first time.

- **Settings Catalog | Turn off toast notifications on the lock screen (User)**. Enabling this policy will prevent toast notifications from displaying on the lock screen.

- **Settings Catalog | Allow Cortana Above Lock**. Disabling this setting will prevent a user from interacting with Cortana on the lock screen using speech.

- **Settings Catalog | Allow Windows Spotlight (User)**. Disabling this policy will turn off consumer features and Windows tips on the lock screen.

- **Settings Catalog | Allow Advertising**. Disabling this policy will prevent the device from sending out Bluetooth advertisements. We covered additional Bluetooth security settings in *Chapter 4, Networking Fundamentals for Hardening Windows*.

- **Settings Catalog | Allow Location**. Disabling this policy will prevent apps from accessing location services, including Cortana and Windows search.

Summary

In this chapter, we learned how standard computer communications protocols are used as attack vectors in MiTM attacks. We learned how to configure policies to mitigate them and that in a well-administered network, protocols such as LLMNR, NBT-NS, mDNS, and WPAD can be disabled. Next, we talked about relay attacks and covered different security settings that can be used to protect against exploits that target Kerberos authentication, SMB, LDAP, IPv6, and ARP. After that, we covered how an attacker can use discovery tactics to move laterally and escalate privileges. We reviewed different attack techniques, such as golden and silver tickets, that can be used to exploit Kerberos authentication, and covered areas targeted to steal credentials on an OS.

Finally, we reviewed the privacy settings that are listed in the **Privacy & Security** section of the Windows Settings app. We discussed how to control these settings using Intune and listed where to find the relevant policies in the Intune Settings catalog.

In the next chapter, we are going to look at server infrastructure management, including the importance of implementing access management solutions and how to use Azure services to connect and manage Windows servers remotely.

11
Server Infrastructure Management

The data center is constantly evolving, and services traditionally used by hosting servers in physical data centers are now virtualized and using serverless computing models in the cloud. No matter how your infrastructure is deployed or what infrastructure is used, each presents a unique security challenge for an organization. In this chapter, we will provide an overview of the data center and cloud models as they exist today. We will discuss security access strategies for Windows servers as they are relevant to all infrastructure models, to ensure that no one can access Windows without going through the proper access controls. You will learn about the available management tools used for on-premises, hybrid, and cloud deployments, as well as how to leverage Azure services to expand your data center reach to the cloud. Then, we will provide an overview of the Azure services that are used to manage Windows servers, including the Azure portal and **Azure Resource Manager** (**ARM**). It's important to understand the existence of these tools and services so that you have a high-level understanding of each when building out your security program. Depending on the size of your organization, services such as these may require several teams to control access, including physical security, a **security operations center** (**SOC**), and **identity and access management** (**IAM**) teams. All of these play a vital role in ensuring your Windows systems are properly managed and protected.

In this chapter, we will cover the following topics:

- Overview of the data center and the cloud (**infrastructure as a service (IaaS)**, **platform as a service (PaaS)**, and **software as a service (SaaS)**)

- Implementing access management in Windows servers

- Understanding Windows Server management tools

- Using Azure services to manage Windows servers

- Connecting securely to Windows servers remotely

Let's get started!

Technical requirements

Throughout this chapter, we will be referencing different services available in Azure. We've discussed configuring services in Azure already, but if you would like to follow along, you can sign up for a free Azure account for 30 days and get **United States dollars** (**USD**) $200 credit at `https://azure.microsoft.com/en-us/free/`.

To complete the **role-based access control** (**RBAC**) example provided later in this chapter, you will need the following:

- PowerShell (version 5.1 recommended) with the **Azure Active Directory** (**Azure AD**) module installed

- Global Administrator rights to your Azure subscription

- A text viewer to open **JavaScript Object Notation** (**JSON**) files

Let's start by looking at an overview of the different data center types.

Overview of the data center and the cloud (IaaS, PaaS, and SaaS)

Over the years, the data center has changed quite significantly as it relates to the hardware our services run on. Most notably, the operating systems, versions, and virtualization of those services have recently shifted to running fully on cloud-based technologies. In the past, a traditional enterprise data center typically consisted of physical mainframes to store and access information. Data centers during these times were sometimes located on location or at a separate facility under the management of the organization. As the technology evolved, there was a shift from the mainframe to server-based data centers. This is where the Windows Server family became widely adopted and grew in popularity.

Moving beyond standard hardware-based server models is where virtualization technology entered the picture. The ability to run many servers on minimal physical hardware changed the dynamics of the data center significantly. Fast forward to today, and we are in a major shift to cloud computing. Organizations are slowly moving away from the traditional on-premises data center and moving all their workloads into cloud environments. With the cloud data center, organizations can continue to run traditional servers and services, but the overhead of owning and managing physical infrastructure is greatly reduced or eliminated.

Another major change with a shift to the cloud is the elimination of on-site facility management and physical operations. Building and maintaining a data center is an enormous undertaking that is challenging and comes with substantial cost implications when designing for highly available services and **disaster recovery** (**DR**) as part of a **business continuity plan** (**BCP**). Moving to the cloud changes these dynamics significantly. Your cost model changes to a subscription-based model with zero upkeep of the physical hardware and facilities. This allows opportunities for a more robust BCP with the flexibility to spread services across multiple data centers in different locations.

This shift also changes the dynamics of security for the data center. Traditionally, physical security with access controls, locks, badge readers, and security cameras was all needed. This goes away with the cloud, but how do you ensure the cloud provider is protecting the access and controls? How do you ensure your data is safe? These are all valid concerns and change the way we manage security as opposed to the traditional data center perspective.

Next, let's look at the three common data center types.

Types of data center

This section will provide an overview of each of the current scenarios mostly being used today.

On-premises data center

As mentioned previously, an on-premises data center is considered the traditional model. Organizations build out and operate their infrastructure on your business's property or off-site at a separate facility. In this model, you are fully responsible for everything in the physical infrastructure (building, power, cooling, hardware, security, access, and so on) and everything that runs on the hardware.

The following diagram provides an example of a traditional on-premises model:

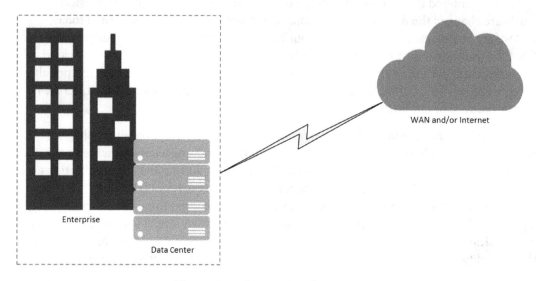

Figure 11.1 – On-premises data center

Next, let's review the cloud data center model.

Cloud data center

As we look further into the cloud model, it is important to understand public and private cloud offerings. A public cloud is where the services are hosted by the provider and the underlying infrastructure is shared with other organizations. Your environment will be logically separated from other organizations, but the underlying hardware, network, and storage are shared with other subscribers on the same service. A private cloud offering is where the services are hosted in a dedicated environment and only your organization runs on the underlying services. Determining the appropriate model will most likely be dictated by your organization's industry and compliance requirements.

> **Tip**
>
> You can find more information on Microsoft Azure's public versus private offerings here:
>
> `https://azure.microsoft.com/en-us/overview/what-are-private-public-hybrid-clouds/`

The following diagram provides an example of a cloud model:

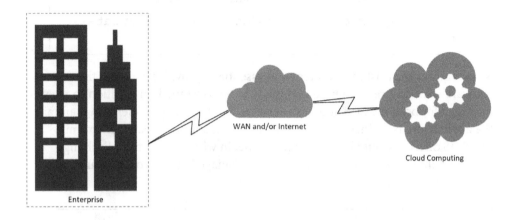

Figure 11.2 – Cloud model

Cloud solutions typically offer three different categories of primary services available for consumption, as outlined here:

- **IaaS** requires the most involvement from your organization and is operated very similarly to a virtualized environment on-premises. The difference is that businesses have no responsibility for physical infrastructure, and the servers, storage, and underlying network fabric are all managed by the hosting provider. You can simply turn on **virtual machines** (**VMs**) and services as needed.

> **Tip**
>
> For additional information about IaaS in Azure, visit this link:
>
> ```
> https://azure.microsoft.com/en-us/overview/what-
> is-iaas/
> ```

- **PaaS** provides you with the needed platform or service, all bundled together from the cloud provider. Typically, with PaaS, the physical infrastructure, operating system, middleware, and other tools to run services are maintained by the hosting provider. For example, in a traditional IaaS Windows environment to host **Internet Information Services** (**IIS**) or a **Structured Query Language** (**SQL**) database, you would need to install these components and roles onto the operating system. With PaaS, you simply subscribe to the service, and you consume it directly. There is no installation or maintenance of any underlying software to run these apps. In Microsoft Azure, IIS is run on Azure App Service, and SQL can be hosted using Azure SQL Database.

> **Tip**
>
> For additional information about PaaS in Azure, visit this link:
>
> `https://azure.microsoft.com/en-us/overview/what-is-paas/`

- **SaaS** requires the least involvement and essentially provides you with the entire software solution preconfigured and ready to be consumed. In addition to what is managed for both the IaaS and PaaS services, the hosting provider also maintains the application itself, including keeping it current and up to date. An example of a SaaS offering would be Exchange Online, in which your entire Exchange environment is hosted, kept up to date, and managed by Microsoft. You simply consume the email services for your organization.

> **Tip**
>
> For additional information about SaaS in Azure, visit this link:
>
> `https://azure.microsoft.com/en-us/overview/what-is-saas/`

Now that we have covered what each of the cloud services is, let's look at some examples within the Microsoft ecosystem. The following diagram provides examples of offerings you can subscribe to within each of the mentioned services:

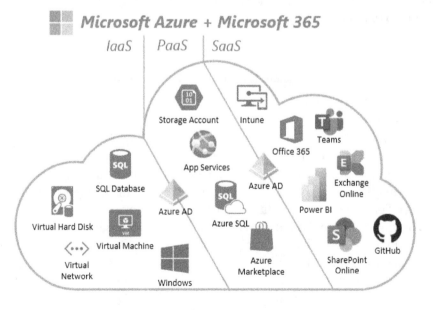

Figure 11.3 – Microsoft IaaS, PaaS, and SaaS examples

Next, we will review the hybrid data center model.

Hybrid data center

The last model we will review is the hybrid model. A hybrid model essentially combines the on-premises model with the cloud model, allowing an organization's on-premises deployment to co-exist with cloud services. This model is most likely going to be preferred for established organizations with existing on-premises data centers simply because they can't easily be moved to a cloud model overnight or due to compliance requirements. What the hybrid model does is allow a pathway from on-premises to the cloud while providing services using both environments.

The following diagram provides an example of a hybrid model:

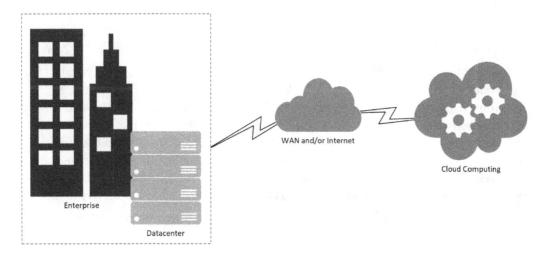

Figure 11.4 – Hybrid model

The focus of this book will be primarily on on-premises, hybrid, and IaaS models as this is where your Windows servers will reside and operate. With the PaaS and SaaS models, the underlying operating system is managed and secured by the **service provider** (**SP**).

Now that we've covered the different models available for operating data centers, let's review access management as it relates to Windows Server. In the next few sections, we will cover securing access to Windows Server and introduce common strategies and security access best practices used by organizations.

Implementing access management in Windows servers

The concept of building a solid access foundation is critical as it relates to server management as part of securing the overall infrastructure. You must consider everything, from the physical access down to protecting unified management consoles that control multiple virtual servers at once. In the next few sections, we will discuss physical access and user access to the infrastructure and the importance of each. Next, we will discuss privileged isolation in AD through a tiered model approach and provide an overview of **privileged access management** (**PAM**) and **privileged identity management** (**PIM**) solutions for **just-in-time** (**JIT**) access. Implementing these tools will provide a robust solution to control server access management. First, let's look at physical access and security.

Physical and user access security

An important factor to consider with the on-premises model is to ensure all physical access to any server location is protected and only accessible to those with approved access. Your Windows servers and other physical infrastructure are typically located in a server room, a closet in your office building, or in a remote facility that is under your ownership. Security access controls should be well enforced, in addition to server hardening. The physical access needs to be locked down to avoid equipment theft and help prevent insider attacks such as installing malware or keyloggers through a **Universal Serial Bus** (**USB**) device. Encryption needs to be enabled on all Windows servers in the event theft does occur to help circumvent any information from being stolen from the data disks.

> **Tip**
> Don't forget that your facility management and physical site access policies are just as important as your **user access management** (**UAM**) to the servers.

As you make your move into the cloud world, your access management changes considerably. Physical access is now the responsibility of the cloud provider. The challenge is this: how do you validate they are protecting your data with the utmost standards? This is where your contracts come into play, but also—more importantly—your due diligence, as it relates to audit requirements and SaaS questionnaires being provided by the cloud provider. We will cover audits in more detail in *Chapter 15, Testing and Auditing*.

The next consideration with the cloud model is the software management plane. Most on-premises deployments by nature allow for management solutions to be isolated from the internet and only accessible from the internal network. With the cloud, the management plane and unified portals are typically internet-accessible, whereby you can access all your resources in one place from anywhere. To help circumvent this risk, very stringent access policies are needed for privileged identities. This includes having standard user accounts separate from privileged accounts and strong access controls including limiting access by safe-listing trusted locations and devices—no exceptions! It's also important to keep a limited number of privileged accounts in your environment. This is where RBAC helps by defining permissions and carving out the scope of who needs to access which resource.

Next, let's review using a tiered model for privileged access.

Using a tiered model for privileged access

A tiered model approach for privileged access in an AD environment is to isolate and build layers of containment between the Windows systems through the directory structure. This is most easily accomplished through AD **organizational units (OUs)** by incorporating a design that is divided into three or more parent containers, commonly labeled as tiers. For example, in a three-tier OU structure, there are tier 0, tier 1, and tier 2 OUs respectively. Tier 0 contains the systems, accounts, and security groups of the highest security concern, such as **domain controllers (DCs)**, Azure AD Connect servers, and identity management systems. The goal of this isolation or *tiered* approach is to prevent escalation across tiers by provisioning access to privileged identities only to the tier they need access to in order to perform an operation. This level of separation provides security boundaries should an account with access from a lower tier get compromised. Its elevation will be restricted to its assigned tier or lower in the model. All access can be controlled through security group assignments and should be audited on a schedule. For example, all privileged accounts that must access a tier 0 server would be in the T0-operators security group, and permissions on the T0 servers, such as membership to the administrator's group, would be managed by a Restricted Groups **Group Policy Object (GPO)**. Let's look at examples of how each of the tiers can be modeled in a three-tiered approach.

Tier 0

Tier 0, the most restricted tier, typically contains a small number of assets and those deemed critical to your infrastructure. Administrators in tier 0 usually have administrative rights to each level tier below using a top-down approach. These assets are of utmost importance and must remain protected and audited for suspicious activity. We would want to limit the use of and accessibility to tier 0 servers and minimize the number of accounts with provisioned access. Access restrictions should apply to resources in tier 0 to ensure only trusted sources can access using expected protocols. Any non-critical function, such as checking email or browsing the internet, should be restricted from running on these servers. Deploy a PAM solution to include password rotation, an approval request flow process, auditing logs, session recording, and even remote **Remote Desktop Protocol (RDP)** or **Secure Shell (SSH)** launchers where credentials are hidden from users.

> Tip
> Take protection a step further by implementing application control policies using Windows Defender Application Control to only allow authorized software to run. You can also limit access to systems by funneling traffic through SSH proxies from a PAM solution.

Tier 1

Tier 1 is your middle tier and contains systems such as business servers, file servers, web application servers, and database servers. Administrators with access to tier 1 will be able to control servers in both tier 1 and tier 2 if using a top-down approach. These servers need to be protected with similar precautions as resources in tier 0. When architecting the organizational structure of the tiered model, create a child OU nested under the tier 1 parent and label them by **business unit (BU)**, application name, or function to create a descriptive structure. This will allow more flexibility for applying Group Policy to define restricted groups that explicitly grant access permissions and allow additional groups to be granted access. Group members will have access to RDP and log on interactively only to resources in their assigned child OU and not all servers in the tier 1 hierarchy. This provides more granularity and organization, especially if access is required by non-**information technology (IT)** staff such as a database administrator. In some scenarios, it may be necessary to restrict non-critical functions such as checking emails or browsing the internet. Certain precautions need to be taken in the event an account becomes compromised. As a best practice, deploy a PAM solution and implement similar features to those in tier 0.

Tier 2

Tier 2 will contain common devices seen in the everyday workplace. This includes end-user workstations, laptops, printers, and virtual desktops. Access to administer resources in this tier is typically assigned to IT support and field IT staff. Child OUs should be created for separation, and allow fine-grained controls if necessary.

Each environment will have unique considerations and the strategy may differ, but the three-tiered approach outlined previously is a great starting point that can be incorporated into your AD structure. The following screenshot shows an example of an OU structure using the tiered model approach. Each parent OU contains a child OU to organize users, computers, and service accounts. This allows for fine-grained policy control using Group Policy:

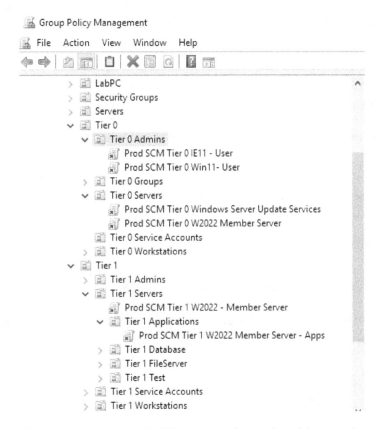

Figure 11.5 – An example OU structure of a tiered model approach

Let's recap by reviewing some important considerations when implementing access management.

Important considerations

In addition to the tiered access model, there are important considerations worth reviewing that can add a substantial layer of security to your Windows systems, as outlined here:

- For RDP and interactive logons, allowed sources should be restricted to a **privileged access workstation** (**PAW**), trusted device, or isolated management environment, preferably requiring a form of passwordless or biometric authentication, or **multi-factor authentication** (**MFA**).

- Network restrictions should be considered for tier 0 access scenarios by restricting RDP connections and other management ports to sources from known **virtual networks** (**VNets**), subnets, and workstations.

- When designing your tiered solution, be mindful and think about built-in security groups with inherited elevated permissions such as Enterprise Admins, Domain Admins, Schema Admins, and Server Operators, to name a few.

- Local accounts can also become a problem if they're not managed properly and are an easy way for someone to create a backdoor without your knowledge. A PAM solution with account discovery can be used to notify security teams when these accounts are added. If you're leveraging local accounts in an AD domain, you can also implement Microsoft's **Local Admin Password Solution** (**LAPS**) to rotate passwords.

- Implement fine-grained password policies and enforce stricter password requirements for administrative accounts.

- Leverage security baselines and Group Policy to further restrict access using restricted groups.

For hybrid cloud or IaaS deployments, PAWs can be deployed using virtualized desktop services such as Azure Virtual Desktop, Cloud PC, or Citrix. Azure also has PaaS offerings available to cater to this scenario, such as the Azure Bastion service. Azure Bastion allows you to securely RDP or SSH directly to VMs over **Secure Sockets Layer** (**SSL**), eliminating the need to expose your servers directly to the internet. This can all be done directly in the Azure portal.

Privileged access strategy

Next, we will look at Microsoft's privileged access strategy, which is now being recommended as a replacement for the **Enhanced Security Admin Environment** (**ESAE**) administrative forest model. We covered ESAE in the first version of this book and, as a reminder, this model requires the setup of a separate AD forest for administrative purposes only. There are no requirements to migrate away from ESAE, but if you are building out a new environment or looking to enhance your security for AD, you will want to follow the privileged access strategy.

The privileged access strategy follows zero-trust principles to provide the best protection for your privileged accounts across all your identity sources throughout your environments. As covered in *Chapter 1, Fundamentals of Windows Security*, zero trust is a foundation that builds in multiple layers of security with your environment, and we should *never trust, always verify*. There are four initiatives to follow with the implementation of this strategy, as outlined here:

1. End-to-end session security.
2. Protect and monitor identity systems.
3. Mitigate lateral traversal.
4. Rapid threat response.

More information on the privileged access strategy can be found in this article:

```
https://docs.microsoft.com/en-us/security/compass/privileged-
access-strategy
```

To help accomplish the privileged access strategy with your traditional on-premises AD deployment, Microsoft advises using the **rapid modernization plan** (**RAMP**). RAMP is built on a roadmap that provides the steps needed to implement the recommended controls. The first section of the roadmap focuses on separate and managed privileged accounts, and is outlined here:

- Ensure you have emergency access accounts in the event of an emergency.
- Enable and implement Azure AD PIM.
- Identity and audit all privileged accounts and limit who needs privileged accounts.
- Ensure separate accounts for on-premises and Azure; ensure no mailbox on administrator accounts.
- Enable and configure Microsoft Defender for Identity.

The second section focuses on how to improve the credential management experience and is outlined here:

- Enable **self-service password reset** (**SSPR**) and enable the combined security information registration experience.

- Require MFA or allow passwordless on all privileged user accounts.

- Block legacy authentication protocols.

- Disable the ability for users to consent to Azure AD applications.

- Enable Azure AD Identity Protection and ensure notifications are set up to review and clean up alerts.

The last section covers administrator workstations' deployment and is outlined here:

- Set up dedicated workstations for privileged users to log in and use with their privileged accounts.

You can learn more about RAMP here:

```
https://docs.microsoft.com/en-us/security/compass/security-
rapid-modernization-plan
```

Setting up the preceding recommendations requires time and investment to implement correctly. It is easy to fall back and apply elevated access to standard users and over-permission accounts to meet deadlines and so on. Don't fall into this trap! Ensure you spend time implementing these recommendations correctly from the ground up and clearly define the account provisioning process.

Next, let's review PAM and its importance as part of access management within the traditional server model.

Understanding privileged account management

Three critical access models for privileged identities that should be considered for controlling access within your server environment are PAM, JIT access, and PIM. All three provide an additional layer of protection around privileged accounts and help ensure they are only available as needed, have an expiration date regarding their usage, and are fully audited and monitored when accessing Windows. Uncontrolled privileged accounts could possibly be your weakest link if not protected properly, and investing in these models should be a requirement. We covered each of these models in detail in *Chapter 5, Identity and Access Management*.

Equally as important as implementing an access management solution is creating a segregation layer between your servers. To do this, organizations architect a tiering model to isolate different groups of servers based on their criticality to the infrastructure. In the next section, you will learn about using a tiered model and what this means for privileged access security.

PAM

PAM is an access model to help secure and monitor your sensitive accounts traditionally focused on the on-premises data center and AD. We covered PAM in detail in *Chapter 5, Identity and Access Management*, and it is an essential tool as part of your Windows Server infrastructure management. It is highly recommended that a PAM tool be implemented in addition to the recommendations provided in the previous sections. For customers who rely heavily on the Microsoft stack, a PAM solution will help fill gaps that aren't covered with native Microsoft products. Some of the limitations with Microsoft in addition to credential management include the ability to automatically rotate passwords, account discovery of your entire directory, management of local and service accounts, a workflow approval process, and the ability to never expose passwords to your administrators.

Although Microsoft does have the ability to deploy PAM within its **Microsoft Identity Manager** (**MIM**) on-premises identity solution, most third-party vendors who specialize in this area will have a richer feature set. An example of a vendor that specializes in PAM is Delinea, with their solution called Secret Server (`https://delinea.com/products/secret-server`). Secret Server can be deployed on-premises or in the cloud and provides a secure secret vault with AD integration. Once up and running, you can set up your secrets (passwords) for your privileged accounts with granular control over how they are accessed and by whom. Access is provided through the Secret Server portal, where users will be able to access and check out secrets they have permission to access. Applying a secret template will determine rules about how the secret is managed. For example, the template in the following screenshot requires the secret to be checked out with a maximum duration set to 12 hours. The password will be changed once the secret is checked back in, and it requires approval before allowing checkout:

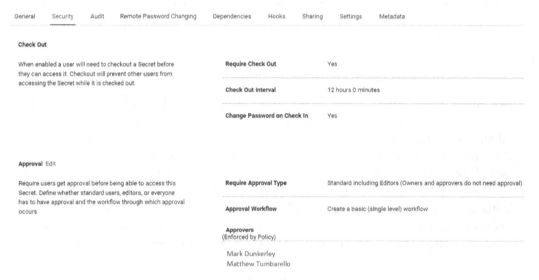

Figure 11.6 – Secret template within Delinea Secret Server

This is a simple example of the benefits of a PAM solution. In addition to controlling account access, you must remember to ensure your PAM solution is protected with multiple layers of security. For example, ensure controls such as **single sign-on** (**SSO**), MFA, Conditional Access, Identity Protection, and audit logging are all enabled.

Now, let's recap and talk about access management best practices.

Access management best practices

Securing access to your environment can be a long and complex journey with many considerations to keep in mind. Security is an ever-evolving space due to the complexity and frequency of cyberattacks. New tools and services are regularly becoming available that help organizations without the resources, funds, or capabilities to simplify the deployment of the solutions needed to protect them. Here is a high-level list of best practices to keep in mind when thinking about the scope of privileged access for your Windows servers and business services:

- Enforcing MFA should be at the top of the list. Require MFA for all cloud-based accounts using Azure MFA or another provider. For on-premises servers, you will need to implement a third-party tool to enforce MFA or restrict access only from a Bastion environment or PAW and apply MFA to access those services in lieu of MFA for servers.

- Deploy a PAM solution.

- Use JIT access to assign permissions dynamically and avoid permanent assignments for your privileged accounts. Helpful services include **Azure Privileged Identity Management (Azure PIM)** coupled with Azure Security Center JIT access.

- Have an effective account provisioning and deprovisioning process. Automate disabling accounts when employees leave the company.

- Constantly audit and monitor privileged accounts in your environment.

- Limit the number of administrators. Always consider job role and function when provisioning administrative accounts and ensure PoLP applies.

- Separate administrative accounts with regular users' accounts. This will help mitigate credential exposure if the administrator's workstation becomes compromised.

- Limit access to email and internet browsing when applicable from privileged systems.

- Enforce strict fine-grained password policies on administrative accounts.

- Limit the amount of emergency "backdoor" accounts and monitor their usage.

- Ensure any changes to the environment go through an approval process by a change advisory board. This can include access to highly sensitive systems.

With technology constantly evolving, so does the need to keep your access model and strategy up to date, especially as more workloads shift to the cloud. Visit Microsoft's *Privileged administration* documentation library to ensure you keep current with the latest recommendations on securing privileged access: `https://docs.microsoft.com/en-us/security/compass/privileged-access-deployment`.

Now that you understand the different recommendations for protecting access to Windows servers, next, we will discuss the tools used to help manage them. Although some are common, it is important to be aware of their utility when building baselines and hardening Windows.

Understanding Windows Server management tools

There are many tools available for Windows Server that are useful for both managing and securing the infrastructure. Management technologies were traditionally developed for on-premises deployments, but now, with cloud-based SaaS offerings and expanded remote work opportunities, it seems the available solutions are growing exponentially. Microsoft offers solutions for enterprise-grade management through its **System Center** suite of tools such as **Operations Manager (SCOM)** and **Configuration Manager (SCCM)**. There are also third-party paid solutions from companies such as ConnectWise, SolarWinds, and CA Technologies, to name a few. Each offers a different feature set depending on your management needs, and with varying price points. In this section, we will review the more common built-in tools available in Windows Server, including Server Manager, Event Viewer, and **Windows Server Update Services (WSUS)** for patch management. Then, we will discuss Windows Admin Center and how it can be used to manage servers and help transition workloads into the Azure cloud.

Introducing Server Manager

Server Manager was introduced in Server 2008. It provides a centralized management plane to view your servers, roles, and services and can support up to 100 remote servers per running instance. The number of available remote servers could vary based on the system performance of the server running it. Server Manager can also be installed on a workstation computer with **Remote Server Administration Tools (RSAT)**. To remotely manage servers, Server Manager remote management must be enabled either through Server Manager, PowerShell, or the command line. This is enabled by default in Windows 2012 and later. Let's review which tasks can be completed through Server Manager, as follows:

- Create, edit, or add custom server groups or pools of servers and clusters.

- Install, uninstall, or make changes to roles and features on both the local and remote servers.

- Open management tools such as Computer Management, Windows PowerShell, Registry Editor, and other **Microsoft Management Console (MMC)** tools.

- Start and stop services, identify events, and collect performance data for analysis.

- Restart servers.

- Export settings to be imported on another system.

To add remote servers to Server Manager from the **Dashboard** view, right-click **All Servers** and choose **Add Servers**. Once servers have been added, all available roles will be visible on the left-side column and can be managed from one centralized point. Creating a server group will create a link on the left column for quick access to different sets of servers for organizational purposes. The following screenshot shows the **MyServers** server group:

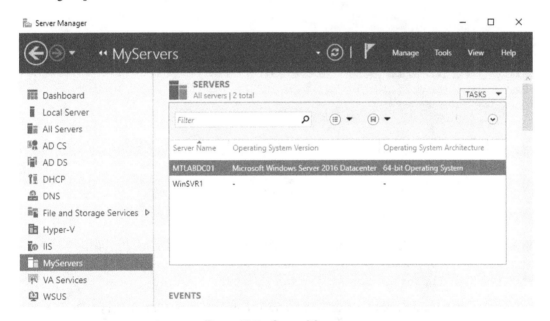

Figure 11.7 – Server Manager

Server Manager is a great place for centralized management to view events, services, and performance from within a single dashboard. Event logs can be viewed and configured from every page except the dashboard. Certain event logs such as application logs are selected by default and include **Critical**, **Error**, and **Warning** severity levels, but they can be customized to fit your needs. The thumbnail alerts can be configured to include other events, and different severities can be included in alert highlighting. Only critical events are highlighted by default.

Using the Best Practices Analyzer tool

The **Best Practices Analyzer** (**BPA**) tool is used to help reduce vulnerabilities by scanning configured roles and comparing your configurations to what experts believe to be best-practice guidelines. BPA can be executed from Server Manager as well as through Windows PowerShell. After the scan completes, the results can show whether a role is compliant with these best-practice recommendations. The summary outlines the problem in detail, as well as lists the impact and resolution steps. This can be seen in the following screenshot:

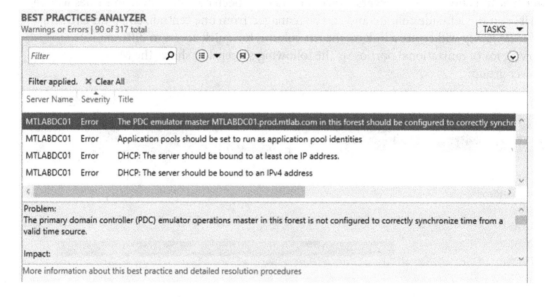

Figure 11.8 – Results from a BPA scan

For more information, please go to `https://docs.microsoft.com/en-us/windows-server/administration/server-manager/run-best-practices-analyzer-scans-and-manage-scan-results`.

Next, we'll look at using Event Viewer and examine common event **identifiers** (**IDs**).

Looking at Event Viewer

Event Viewer is used to view log files from applications, including security and system-related events. Events are categorized by error, warning, and informational events, which can be useful for troubleshooting both security and performance issues. PowerShell can be used in addition to Event Viewer to query for events, including logs from remote computers. To open Event Viewer using Windows Search, type in `event viewer`. To view logs on a remote computer, right-click on **Event Viewer (local)** at the top of the tree and choose **Connect to Another Computer….**

Event Viewer can also be used for automating actions based on certain events. Using **Attach a Task to this Event** action, a basic scheduled task can be created and run based on when an event appears in the log.

> **Tip**
> Event Viewer is very important for monitoring Windows events from a security perspective. Security professionals should pay close attention to sources around login activities, application crashes, network firewall rule changes, clearing of event logs, audit policy changes, and Group Policy changes.

To view security-specific logs, open **Windows Logs** > **Security**. In the following screenshot, event ID `4624` indicates a successful logon. This event ID contains a lot of detailed information about the logon, including the **Logon Type**, account information, and network information about the user who has logged on:

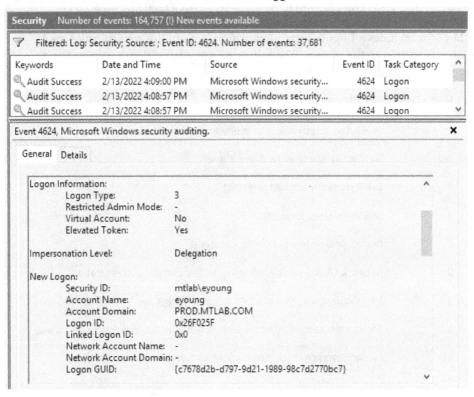

Figure 11.9 – Event 4624: successful logon in Event Viewer

There are common security events to look for under **Windows Logs** > **Security** that could indicate an attacker attempting to access the system. While these event IDs are normal and typically do not indicate attack, in the event a compromise did occur, they are useful in helping build a timeline around the attack and provide additional details during forensic analysis. The event IDs are presented here:

- `Event ID 4625:` Audit Failure Logon

- `Event ID 4624:` Audit Success Logon

- `Event ID 4648:` Login with explicit credentials

- `Event ID 4735:` Security-Enabled local group was changed

- `Event ID 4728, 4732,` and `4756:` Member added to a security group

- `Event ID 4740:` Account Lockout

- `Event ID1102:` Log Clear may indicate an evasive tactic by an attacker

Event ID `4624` includes a **Logon Type** field, which is useful for identifying how an account has logged in to the system. The following screenshot demonstrates the different logon types that are associated with event ID `4624`:

Logon Type	Description
2	Interactive Logon (the user is physically at the device)
3	Network (Logon occurred elsewhere)
4	Batch (typically a scheduled task)
5	Service (system services)
7	Unlock (logged in from the lock screen)
8	Network Clear Text (logon with clear text or basic authentication)
9	New Credentials (typically with run as different user)
10	Remote Interactive (typically with remote desktop or terminal services)
11	Cached Interactive (cached credentials were used)

Figure 11.10 – Logon types for event ID 4624

In addition to monitoring security logs, **Microsoft Defender** logs are useful for detecting evasion techniques such as disabling virus and threat protection settings. They can be found under **Applications and Services Log** > **Microsoft** > **Windows** > **Windows Defender**. Interesting events such as canceling or pausing malware protection scans could indicate a malicious actor or the prevalence of malware or could warrant further investigation.

> **Tip**
>
> Further reading on Microsoft Defender antivirus event IDs can be found at `https://docs.microsoft.com/en-us/microsoft-365/security/defender-endpoint/troubleshoot-microsoft-defender-antivirus?view=o365-worldwide`.

From an operational perspective, especially when a SOC is used for monitoring multiple systems, looking at Event Viewer on servers individually isn't the most effective method. It is recommended to incorporate a **security information and event management** (**SIEM**) solution to better track and analyze event logs. Examples of SIEM solutions include Microsoft's security monitoring tools such as Azure Sentinel, Microsoft Defender for Endpoint, Microsoft Defender for Cloud Apps, and Azure Defender, or third-party tools such as Splunk and QRadar can be used for log repositories and analysis. We'll cover SIEM tools and security monitoring in more detail in *Chapter 13, Security Monitoring and Reporting*, and *Chapter 14, Security Operations*.

Next, let's look at how WSUS can be used to manage Windows updates and patch vulnerabilities in Windows Server.

Using WSUS

WSUS is a solution for managing Windows updates and for keeping security patch-level compliance across your servers. In some instances, maintaining the same patch level across servers is critical for applications to function, and relying on the standalone Windows Update doesn't suffice for this level of control. WSUS allows you to approve updates using a management dashboard and choose when to deploy them. In a simple WSUS architecture, the WSUS downstream server acts as the middleman between the client and the Microsoft Update **content delivery network** (**CDN**). Using a management dashboard, an administrator can download and approve critical updates, security patches, security rollups, service packs, feature packs, Microsoft product updates, and antivirus definition files. Computers connecting to WSUS can then be grouped together and targeted for an update deployment.

Most WSUS implementations require minimal processing power on the host computer, and a single WSUS instance can host upward of 100,000 clients. For environments greater than 100,000 endpoints, multiple WSUS servers can be deployed by using a load-balanced frontend. A single SQL Server database is needed that can be shared by each WSUS instance to allow updates to be controlled for multiple locations and branch offices through a single centralized dashboard.

From a network and firewall perspective on the endpoints, clients connect to WSUS over **Hypertext Transfer Protocol/Transmission Control Protocol (HTTP/TCP)** port 8530. It is recommended to secure communications by deploying a custom SSL for clients to connect HTTPS/TCP over port 8531. IIS is required if you wish to use WSUS in both scenarios.

You can see an overview of the WSUS management console in the following screenshot:

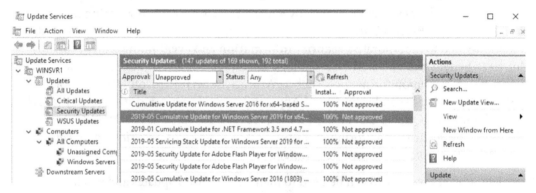

Figure 11.11 – WSUS management console

For an endpoint to receive updates, they must first be approved through the console. To help ease the administrative effort required to manage WSUS, automatic approvals can be configured for updates and antivirus definitions if necessary.

WSUS also supports the deployment of third-party updates for commonly used software vendors such as Adobe. Combining WSUS with a Configuration Manager software update point will allow you to subscribe to a vendors' update catalog that has partnered with Microsoft to approve and deploy third-party software updates. For servers hosted in the Azure cloud, WSUS can be enhanced further by leveraging Update Management in Azure Automation. Update Management can manage updates for Windows and Linux systems hosted both in Azure and on-premises directly through the Azure portal. This service supports Windows Server downward to 2008 **Release 2 (R2)** and will be covered in more detail later in this chapter and in *Chapter 12, Keeping Your Windows Server Secure*.

Here are some helpful links for staying up to date on the latest released security updates:

- Microsoft Security update guide: `https://msrc.microsoft.com/update-guide/en-us`

- Patch Tuesday dashboard: `https://patchtuesdaydashboard.com/`

- Information about **Common Vulnerabilities and Exposures (CVE)**: `https://www.cve.org/About/Overview`

Next, let's review managing servers using Windows Admin Center.

Introducing Windows Admin Center

Windows Admin Center is a browser-based tool that provides an alternative to the classic MMC and is supported on Windows 10/11 and Windows Server 2016 or later. It can support the management of servers down to 2008 R2 but with limited functionality. The **user interface (UI)** frontend is a locally deployed browser-based tool that leverages PowerShell with **Windows Management Instrumentation (WMI)** and supports remote management to remotely manage servers. To support down-level servers (Windows Server 2012 R2 and lower), **Windows Management Framework (WMF)** 5.1 must be installed on those target systems. The official documentation, including installation instructions, can be found at the following link: `https://aka.ms/WindowsAdminCenter`.

> **Tip**
> Windows Admin Center runs in **HyperText Markup Language 5 (HTML 5)** and requires Microsoft Edge or Google Chrome browser to run on Windows Server. It cannot be installed directly on a DC.

Windows Admin Center's toolset for server connections includes Active Directory, DHCP, DNS, Firewall, Remote Desktop, Roles and Features, Scheduled Tasks, and Updates, to name a few. Many more features are available, and features can be extended through extensions that include support for third-party developers.

Windows Admin Center can be used to manage on-premises servers and servers in the cloud. While its core functionality is built around hybrid management, it has no dependency on Azure and is included for free with your Windows Server license. For Azure cloud customers, Windows Admin Center can centrally manage Azure hybrid services and cloud workloads extended to on-premises servers directly from the UI.

Some of the Azure services available through Windows Admin Center include the following:

- Azure Backup

- Azure File Sync

- **Azure Kubernetes Services (AKS)**

- Azure Security Center

- Azure Monitor

- Azure Update Management

- Support for Azure AD authentication

Windows Admin Center can be fully integrated with Azure AD for authentication; it supports MFA and is now available directly from the Azure portal. The following screenshot shows the **Overview** pane of Windows Admin Center using the Microsoft Edge browser:

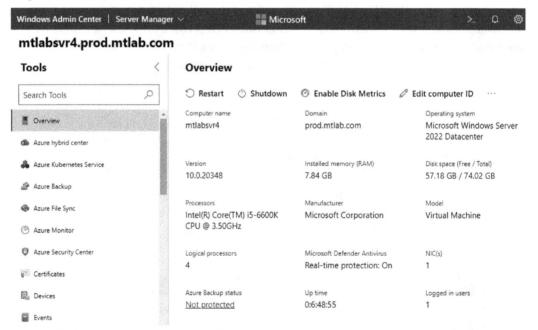

Figure 11.12 – Windows Admin Center in Microsoft Edge

Windows Admin Center can also be used to create and configure standalone Azure VMs. By leveraging Storage Migration Service and Storage Replica, on-premises server workloads including data transfer and VM creation can all be orchestrated from Windows Admin Center. This is helpful for DR, **high availability** (**HA**), or to permanently migrate the server and cut over the workload to the Azure cloud. Additional information about Storage Migration Service in Windows Admin Center can be found at the following link: `https://docs.microsoft.com/en-us/windows-server/storage/storage-migration-service/migrate-data`.

Windows Admin Center can help administrators manage hyper-converged infrastructure workloads. A hyper-converged cluster can be added like any other Windows Server and intuitively managed including cluster monitoring of the underlying resources and management of storage spaces, Hyper-V VMs, and software-defined networks. Windows Admin Center can manage failover clusters too for highly available and fault-tolerant workloads. Nodes can be managed as individual servers or added together and managed as a group in the console. Windows Admin Center deployments that use a gateway server also support a failover cluster architecture to ensure HA of the management console. To view the available deployment options, visit the following link: `https://docs.microsoft.com/en-us/windows-server/manage/windows-admin-center/plan/installation-options`.

Using RBAC, Windows Admin Center has flexibility for securing access to the management console and the underlying resources. It supports authentication using AD, smartcard authentication, or Azure AD with support for Conditional Access and MFA. Using PowerShell **Just Enough Administration** (**JEA**), administrator permissions on the underlying server resources can be limited and temporary. We discuss JEA in more detail later, in *Chapter 12*, *Keeping your Windows Server Secure*.

In this section, we have covered several Windows management tools including Server Manager, Event Viewer, WSUS, and Windows Admin Center. Next, let's look at using Azure services that are useful for managing Windows Server environments.

Using Azure services to manage Windows servers

As discussed in the *Understanding Windows Server management tools* section, Windows Admin Center exposes Azure cloud services that can provide additional benefits for managing Windows servers. It is recommended that you start moving workloads to the cloud not only as part of your digital transformation but also as a security initiative. Azure AD roles and RBAC allow for a fine-grained level of access provisioning over the management plane. In the next section, you will learn about the Azure services that are available for managing Windows servers for both IaaS and on-premises deployments. We will cover the following topics:

- The Azure portal and Marketplace
- ARM
- Implementing RBAC
- Using Azure Backup
- Leveraging **Azure Site Recovery (ASR)**
- Introducing Azure Update Management
- Understanding Azure Arc
- Using Azure Automanage

The Azure portal and Marketplace

The Azure portal is a web UI that's used to manage resources in the Azure cloud. Most management operations today can be performed directly through the Azure portal, but occasionally you may find they require the use of the Azure **command-line interface (CLI)**, Azure PowerShell, or direct calls to the Graph **application programming interface (API)**. Microsoft is doing a great job of consistently adding new operations only supported by the command-line tools directly to the Azure portal interface. To sign in to the Azure portal, follow these steps:

1. Open a browser and navigate to `https://portal.azure.com`.
2. Sign in with your Azure account username and password.
3. If you don't have an Azure account, click **Create one!** to set one up.

To view and access your virtual servers within the Azure portal, simply click on **Virtual Machines**. You will be provided with the **Virtual machines** management page, as illustrated in the following screenshot:

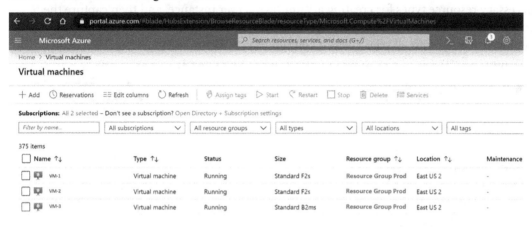

Figure 11.13 – Microsoft Azure portal Virtual machines page

We won't be going into detail about navigating the Azure portal, but we do want to call it out as an important tool for not only managing Windows servers, but also all Azure cloud-based resources. For detailed instructions on customizing its look and feel, as well as creating custom dashboards and setting favorites, you can visit `https://docs.microsoft.com/en-us/azure/azure-portal/azure-portal-overview`.

Using the Azure Marketplace

The Azure Marketplace is a storefront in Azure where you purchase resources and solutions and deploy them directly to your tenant. There are many offerings available, from custom VM images to databases, networking solutions, **internet of things (IoT)**, **development-operations (DevOps)**, and more. It can be accessed from the Azure portal or by going directly to the following link after logging in: `https://azuremarketplace.microsoft.com/en-us/marketplace/`.

For deploying Windows Server, the marketplace has many pre-built images that can be deployed onto a VM in just a few clicks. If you have found a pre-built image that you like, you can use the Azure Marketplace to customize the deployment and specify all the necessary resource types for creating a Windows Server instance, and then capture the customizations in a JSON template. By using this JSON, multiple servers can now be deployed at scale with all your custom configurations. The following screenshot shows a Windows Server 2022 Datacenter core edition as the image type:

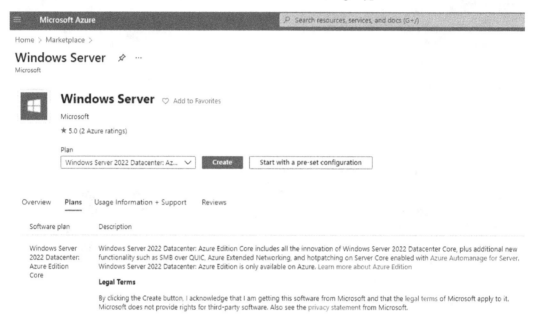

Figure 11.14 – Windows Server purchasing option from the Azure Marketplace in the Azure portal

Using RBAC, resource creation can be locked down to only allow certain users to create resource types that are specified in your deployment configurations.

Next, we'll look at the current management plane for Azure resources, which is called ARM.

ARM

It is important to be aware of ARM if you're working with the Azure cloud to deploy Windows servers and other infrastructure. ARM is defined as a highly resilient management plane for all services that run in Azure. Any controls used that directly affect the management and security of resources are done through ARM. ARM is the underlying plane that is manipulated directly using the Azure portal, Azure PowerShell, Azure CLI, or through APIs and custom tools developed with a **software development kit (SDK)**. Custom templates written in JSON can be created and declaratively deploy resources repeatedly, at scale, and are tracked directly in the Azure portal. More information about creating ARM templates, including the sections that make up the JSON, can be found at `https://docs.microsoft.com/en-us/azure/azure-resource-manager/templates/`.

> **Tip**
>
> Additional information, including key terminology, benefits, and a descriptive understanding of the scope of the management plane, can be found at `https://docs.microsoft.com/en-us/azure/azure-resource-manager/management/overview`.

Next, we'll look at how to lock down access to Azure resources by implementing RBAC. We will also learn how to create a custom role using PowerShell and JSON.

Implementing RBAC

RBAC in Azure is used to authorize access to resources through role assignments. An RBAC role contains a role definition that defines permissions as to what a user can access. Roles are assigned to security principals, which can be users, groups, service principals, or managed identities. When a role is assigned, you can define a scope to determine the set of resources to which the access applies. This can be directly at the subscription level or more granularly to a resource group or resource. Permissions are inherited from parent scopes in a top-down manner, so if a role assignment is set on the resource-group level, all resources nested under that resource group will inherit the permissions defined in the role definition. When assigning permissions to users and groups, RBAC takes an additive model that is slightly different from what we've traditionally seen in an AD environment. For example, if the user is assigned multiple roles, the least restrictive role doesn't take priority and the user will be assigned any additional permissions as outlined in the role definition. Explicit deny assignments must be applied to determine which set of actions is not allowed; otherwise, access is permitted.

Azure has many built-in roles that are ready for assignment. Each role has its own role definition that contains a collection of properties that identify the role and determine permissions to be applied. In addition to basic identification such as role name and description, role definitions include actions that specify what is allowed or denied on the control plane and data plane. These action properties are identified as `Actions`, `NotActions`, `DataActions`, and `NotDataActions`. `Actions` refers to the control plane in which actions are permitted, such as deleting a resource or listing contents of a container in blob storage. `DataActions` refers to the data plane that controls the permissions to the underlying data within the object. An example of a data action is permissions to read data inside a blob in a storage account. Azure offers many built-in RBAC roles that are preconfigured for specific use cases as they pertain to certain resources. If you need to build custom roles, you can export a built-in RBAC role and modify it to fit your needs. For more information regarding the role definition structure, go to `https://docs.microsoft.com/en-us/azure/role-based-access-control/role-definitions`.

Let's look at a basic example of creating a custom role. In the following tutorial, we will copy the `Virtual Machine Contributor` role definition and build a custom role to include an additional `DataAction` to allow administrative login to VMs. To support administrative login, Windows VMs in Azure must support Azure AD authentication. For more information on this, refer to the following link: `https://docs.microsoft.com/en-us/azure/active-directory/devices/howto-vm-sign-in-azure-ad-windows`.

Let's get started. Proceed as follows:

1. Open PowerShell as an administrator and run the following commands to install the Azure module and use the `Connect-AzAccount` cmdlet. We will then use the `Get-AzSubscription` cmdlet and take note of the subscription ID for later:

    ```
    Install-Module -Name Az
    Connect-AzAccount
    Get-AzSubscription | Select ID
    ```

2. The first step is to export the built-in **Virtual Machine Contributor** role as a JSON template. Run the following commands to export the role definition as JSON using the `Get-AzRoleDefinition` cmdlet:

    ```
    Get-AzRoleDefinition -Name "Virtual Machine Contributor"
    | ConvertTo-Json | Out-File "C:\MyFolder\VMContributor.
    json"
    ```

3. Open the `VMContributor.json` file using a code editor of your choice, such as Notepad++, **Visual Studio Code** (**VS Code**), or Notepad. We will need to make some modifications and save the file to be reimported.

4. On *line 2*, change the value of `"Name"` to `Administrative Virtual Machine Contributor`, like so:

   ```
   "Name": "Administrative Virtual Machine Contributor"
   ```

5. On *line 4*, change the value of `"IsCustom"` to `true`, like so:

   ```
   "IsCustom": true,
   ```

6. Modify `"Description"` on *line 5* so that it reads as follows:

   ```
   "Description": "Lets you manage virtual machines, but not
   access to them, and not to the virtual network or storage
   account they are connected to. Allow administrator
   login.",
   ```

   To allow administrative login, we will need to add the following resource provider operations in the `DataActions` property type:

   • `Microsoft.Compute/virtualMachines/login/action`

   • `Microsoft.Compute/virtualMachines/loginAsAdmin/action`

7. Find the `DataActions` property (around *row 55*) and add the two resource operation providers, as in the following example:

   ```
   "DataActions": [
   "Microsoft.Compute/virtualMachines/login/action",
   "Microsoft.Compute/virtualMachines/loginAsAdmin/action",
   ]
   ```

8. Next, we need to populate the `"AssignableScopes"` section. If you exported the built-in role, the `"/"` root value will be prepopulated, but this is only applicable to built-in roles, and we will need to plug in the subscription ID. Assignable scopes can be scoped to multiple subscriptions, management groups, resource groups, or resources. In this example, we will use the subscription ID of our tenant that we saved earlier. Modify the `"AssignableScopes"` section so that it includes the subscription ID, as in the following example:

   ```
   "AssignableScopes": [
   "/subscriptions/8c24xxxa2-xxxx-47xx-3d-8929345de830"
   ]
   ```

9. Finally, save the edited JSON file. We will now import it using the `New-AzRoleDefinition` cmdlet, as follows:

```
New-AzRoleDefinition -InputFile "C:\MyFolder\
VMContributor.Json"
```

If its creation was successful, the custom role will be seen from the **Access control (IAM)** blade in your Azure subscription, as illustrated in the following screenshot:

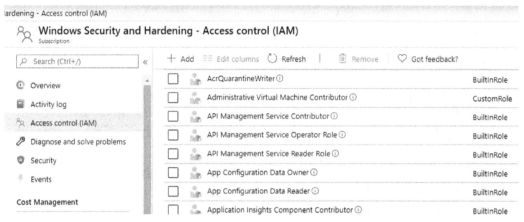

Figure 11.15 – Administrative Virtual Machine Contributor custom role

For more information about the resource provider operations that are available, visit this link: https://docs.microsoft.com/en-us/azure/role-based-access-control/resource-provider-operations.

For more information about JSON, visit this link: https://www.w3schools.com/whatis/whatis_json.asp.

In the next section, we will talk about using Azure Backup for creating backups of our servers.

Using Azure Backup

Having a good backup strategy with clean reliable backups is a critical piece in any business resiliency and DR planning. Part of that strategy should also include planning for recovery from cyberattacks such as ransomware. If there is ever a circumstance where systems needed to be recovered, the inability to restore systems from a backup could have a significant business impact and cost burden. Traditional on-premises backup solutions typically included either network-attached storage, backup servers, or physical tapes. To reduce the risk of environmental disasters destroying backups, IT teams would need to replicate data to another on-premises data center or physically move tapes to another location for safekeeping. Moving to a cloud backup solution could help reduce some of the burden involved with maintaining data backups and help to meet recovery objectives. Azure Backup is a cloud service that can be used to replace an on-premises backup solution or as an automatic storage management tool for hybrid scenarios where data is stored both on-premises and in the cloud. It requires zero infrastructure, has unlimited scaling capabilities, provides application-consistent backups, and data is encrypted both in transit and at rest. Azure Backup offers unlimited data transfer ingress and egress from Azure at no additional cost. There are **locally redundant storage** (**LRS**) and **geo-redundant storage** (**GRS**) options available depending on your HA needs, with no time limits regarding data retention, and up to 9,999 recovery points.

> **Important Note**
>
> For additional information on the Azure Backup service, go to `https://docs.microsoft.com/bs-latn-ba/azure/backup/backup-overview`.

If using a server management tool such as Windows Admin Center, you can take advantage of the built-in management and monitoring tools, all from within the **Backup** dashboard.

Azure Backup's requirements include the following:

- A valid subscription
- Resource group
- Recovery Services vault
- Agent deployment

Azure Backup includes application and crash-consistent backups. Let's look at what these are, as follows:

Application-Consistent Backup: This uses the Windows **Volume Shadow Copy Service (VSS)** to create a backup and capture memory content and pending **input/output (I/O)** operations. When recovering a VM using an application-consistent snapshot, there is no data loss.

Crash-Consistent Backup: This occurs if an Azure VM shuts down during the time of a backup. Only disk data at the time of the backup is captured and recovery doesn't guarantee data consistency.

The following screenshot is of the Azure Recovery Services vault where you can restore a VM from a backup:

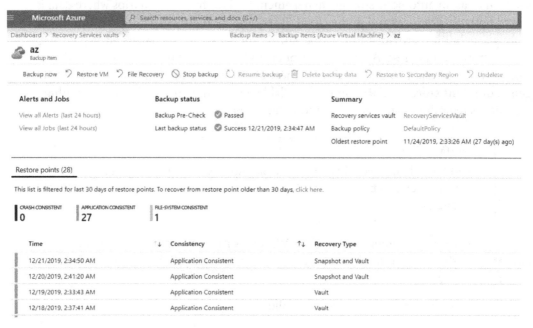

Figure 11.16 – Azure Recovery Services vault

We will now review securing Azure backups.

Securing Azure Backup

In addition to ensuring a good backup strategy is in place for any recovery scenario, it's important the backups are safe and secured from alteration. This includes ensuring backups are encrypted and only the necessary users have access to manage backups.

For on-premises backups, encryption is done using a customer-specified passphrase. Once in transit, data is encrypted using **Advanced Encryption Standard** (**AES**) 256 and sent over **HTTP Secure** (**HTTPS**) to Azure. For Azure VMs, data at rest is encrypted using **Storage Service Encryption** (**SSE**) and protected by HTTPS in transit while never leaving Azure. Azure Backup can also back up VMs that are encrypted using Azure Disk Encryption.

> **Tip**
> For customer-specified passphrases, **DO NOT** lose the encryption passphrase. It is required to restore backups. Microsoft **CANNOT** recover this for you.

Backup data contains highly critical information and needs to be properly secured from unauthorized access. Access can be managed using Azure RBAC. Authentication also takes place through Azure AD where monitoring and auditing are supported. Let's highlight the built-in RBAC roles for Azure Backup, as follows:

- **Backup Contributor** has permissions to create and manage backups but cannot grant access or delete Recovery Services vaults.

- **Backup Operator** has similar permissions to the contributor except for removing backups and changing policies.

- **Backup Reader** can view all backup operations for monitoring purposes only.

For hybrid scenarios, additional security features are available, such as configuring retention periods and notifications for changes, and the ability to create security pins.

Another security feature for backups is soft delete. Soft delete is a feature that retains backups for 14 days after a deletion action and allows recovery without data loss at no additional cost. While this feature is enabled by default, you can permanently delete soft-deleted backup items immediately or disable the feature altogether.

Azure backup also supports a feature known as Resource Guard. Resource Guard helps protect the Recovery Services vault that stores your backups by adding an extra layer of protection for critical operations. When a resource guard is in place, for an Azure backup administrator to perform a critical operation, they must also be granted additional permissions on the resource guard. Critical operations include soft delete, disabling Resource Guard, and modifying backup policies and protections. If the administrator needs to perform an operation, they can request access through a security administrator or using Azure PIM. This model is known as **multi-user authorization** (**MAU**).

> **Tip**
>
> More information about Resource Guard for Azure Backup can be found here:
>
> ```
> https://docs.microsoft.com/en-us/azure/backup/
> multi-user-authorization
> ```

In *Chapter 1, Fundamentals of Windows Security*, we provided an overview of ransomware preparedness. We briefly mentioned backups as an item that needs to be reviewed because intruders are specifically looking to target backups to corrupt or delete them from being used. This can cause severe damage to an organization, so it is critical you spend quality time ensuring your backups are well protected. To expand on what we covered in *Chapter 1*, here are a few recommended best practices for protecting backups on-premises and stored in Azure:

- Maintain multiple copies of backups; for example, the 3-2-1 backup that requires three copies of the backed-up data on two different types of media and one being off-site.

- Maintain a copy of your backups offline and ensure they cannot be accessed by any network of internet connectivity, also known as an airgap backup.

- Encrypt all backups and ensure the private keys are secured.

- Ensure backups are immutable and unable to be altered.

- Ensure both application- and crash-specific backups are being taken, or consult with the business to document the backup.

- Regularly test your backups.

- Review and keep backup procedures up to date.

- Ensure you have a multi-layered access model that includes MFA and PIM or use a PAM solution to keep accounts that can manage backups protected.

- Monitor and alert on backup access.

- Audit all backup activity.

- Enable and enforce a PIM access model to manage backups.

- Use Resource Guard for your Recovery Services vaults.

Next, let's look at using ASR. ASR helps to recover servers and workloads in other regions during DR or as a tool to move systems in Azure.

Leveraging ASR

ASR is a BC and DR solution built into Azure for all types of workloads. The solution covers both Azure and on-premises Windows servers and VMs. ASR consists of two major components, as outlined here:

- **Site Recovery Services**, which is used for the replication of VMs and workloads from a primary to a secondary region in Azure

- **Backup Services**, which uses the Azure Backup service for data backup and recovery

For more information on ASR, visit this link:

```
https://docs.microsoft.com/en-us/azure/site-recovery/site-
recovery-overview
```

When building a BC and DR strategy, most organizations outline **recovery time objectives (RTOs)** and **recovery point objectives (RPOs)** for each business-critical service and application. These objectives determine the amount of time in which services are required to be restored and the point in time at which any incurred data loss is acceptable. Using the ASR service, during a DR scenario, workloads are replicated from the primary to the secondary region based on configurations set around the RTO and RPO strategy. If a regional outage occurs in Azure, you can orchestrate a failover to meet these objectives to restore services. Let's review this in more detail, as follows:

- An **RTO** is the established duration of time for each business process in which services must be restored to meet **service-level agreements (SLAs)**.

- An **RPO** is the maximum amount of data that could be lost during any major outage. For example, if the RPO is 24 hours and the last backup was completed 6 hours ago at the time of an outage, there are 18 hours to restore the data until the business will suffer a significant volume of data loss.

ASR supports building customized recovery plans that allow you to strategically plan which services, VMs, and critical infrastructure fail over and when. Using PowerShell and Azure Automation, many tasks during the failover/recovery process can be automated to reduce the amount of manual effort needed by IT staff. ASR also supports testing the failover process using recovery plans. The following screenshot is of the **Overview** dashboard. Here, you can monitor the site recovery and backup status of resources that are scoped within the ASR service:

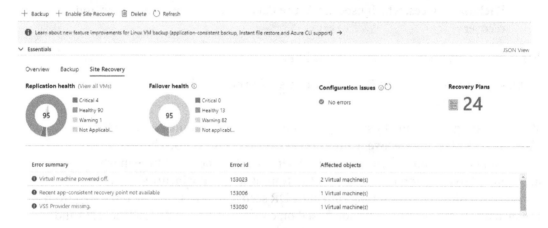

Figure 11.17 – ASR dashboard

> **Tip**
>
> ASR can also be used for migrating on-premises machines to Azure using the same steps that were defined for DR, but without failback! For more information, go to https://docs.microsoft.com/en-us/azure/site-recovery/migrate-tutorial-on-premises-azure.

Next, let's look at keeping Windows servers up to date using Azure Update Management to deploy Windows updates to servers.

Introducing Azure Update Management

As described earlier in the *Using WSUS* section, Update Management is an Azure cloud-based solution for managing system updates for Windows and Linux. Update Management can be used for servers in both the cloud and on-premises.

The requirements to deploy Azure Update Management for Windows are listed as follows:

- An Azure subscription
- Resource group

- Log Analytics workspace
- Azure Automation account
- Deployment of the **Microsoft Monitoring Agent (MMA)**
- Microsoft Update or WSUS configured on your systems

Onboarding servers into Update Management can be done individually on the server directly or using the Azure portal and other methods for bulk enrollment. For step-by-step instructions on how to onboard multiple servers from the Azure portal, visit the following link: `https://docs.microsoft.com/en-us/azure/automation/update-management/enable-from-portal`.

After the servers are onboarded, you can view them in the Azure portal by going to the **Automation Account** selected during the onboarding processing and opening **Update Management**. The following screenshot shows the overall compliance of all your machines onboarded into the solution:

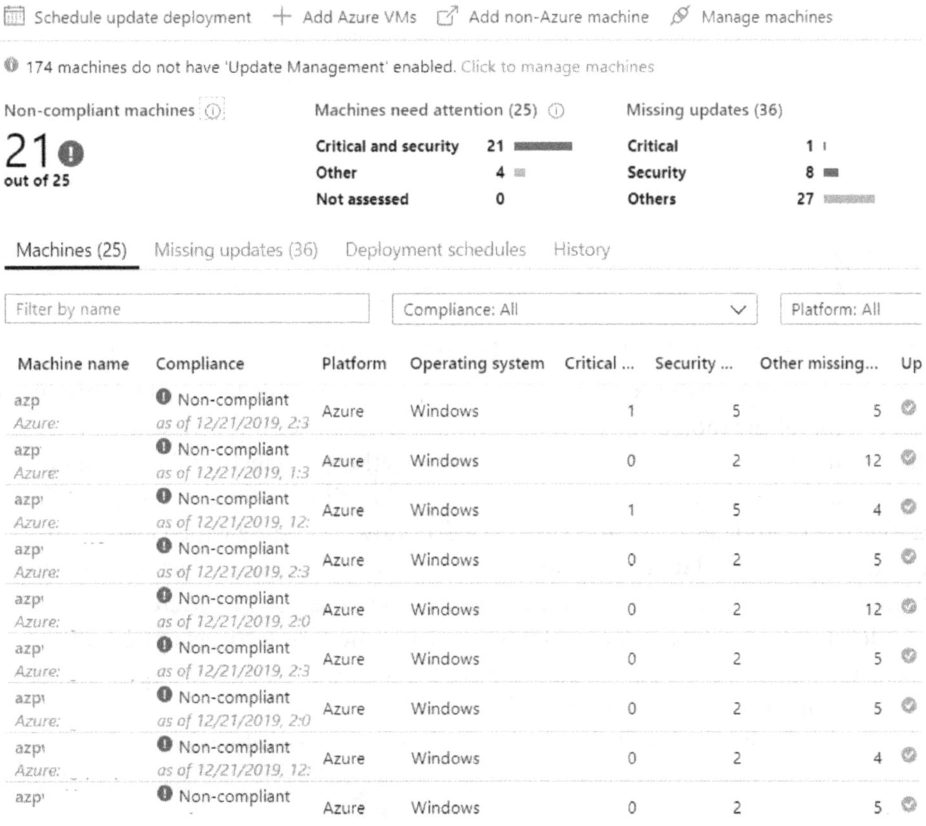

Figure 11.18 – Azure Automation Update Management

The compliance report will be aligned with the latest updates available for Windows Update or approved updates if your deployment is linked to WSUS. To add other machines, click on **Add Azure VMs** at the top of the page. For non-Azure servers, deploy MMA and configure it to use the Update Management Log Analytics workspace during setup. Clicking on **Missing Updates** will provide an overview of all the available updates and a count of machines missing the update. The following screenshot shows all missing updates, and it includes details such as the update name, classification, number of machines missing the update, operating system, and a **knowledge base (KB)** information hyperlink:

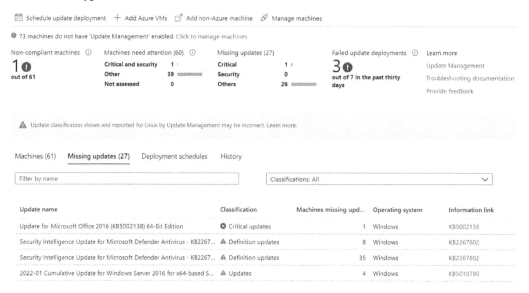

Figure 11.19 – Overview of available updates and machines missing each update

When you're ready to schedule a deployment, click on the **Schedule update deployment** option at the top of the toolbar to configure the deployment settings. By selecting **Groups to update**, you can organize groups of computers together by selecting them manually or by using dynamic queries based on resource groups, locations, and tags. For granularity around the type of updates to deploy, there is a dropdown to scope which classification of update to deploy. The **Update classifications** dropdown will let you select a category of updates (such as **Critical** and **Security**), and you can choose the **Include/exclude updates** option to define specific KBs in the deployment. The **Schedule settings** option allows you to set a one-time deployment or a recurring deployment. Additional options include setting a maintenance window time and reboot behaviors, and the ability to link pre- and post-scripts that should be executed before or after an update deployment run.

The deployment status can be monitored with alerts using Azure Monitor. By configuring **Action groups**, notifications can be sent out via email, **Short Message Service (SMS)**, push notification, voice, and even by triggering a webhook or Azure Automation runbook.

For additional information about the Update Management solution in Azure, go to `https://docs.microsoft.com/en-us/azure/automation/update-management/overview`.

The following Windows operating systems are supported by Azure Automation Update Management:

- Windows Server 2022
- Windows Server 2019
- Windows Server 2016
- Windows Server 2012 and R2
- Windows Server 2008 R2

> Tip
> Windows clients and Windows 2016 Nano Server are not supported at this time.

We will cover how to set up and configure Azure Automation Update Management in more detail in *Chapter 12, Keeping Your Windows Server Secure.*

Next, let's look at how resources outside of Azure can be managed with Azure Arc.

Understanding Azure Arc

Azure Arc is a Microsoft Azure offering that allows you to extend your Azure cloud infrastructure and services management capabilities to your on-premises infrastructure, other public cloud infrastructure, and edge devices. This allows companies to better streamline governance and management by centralizing and unifying the management of resources. The current resource types outside of Azure that can be managed by Azure Arc include the following:

- Windows and Linux servers
- VMs on VMware vSphere and Azure Stack HCI
- SQL servers

- Azure data services
- Kubernetes clusters

Focusing on the management of Windows servers, Azure Arc-enabled servers allow for the management of Windows servers and Windows VMs outside of your Azure environment. This includes the ability to manage Windows servers located on your local corporate network or other cloud providers. To connect your Windows machines that are hosted outside of your Azure environment to Azure, you will need to install the Azure Connected Machines agent on each machine. Then, you can install the Log Analytics agent to monitor the operating system, collect telemetry data, and use Automation runbooks, Update Management, Microsoft Defender for Cloud, and so on. Once your Windows machine is connected to Azure, it can be managed just like any other Azure VM. Some of the actions supported include the following:

- Governance by assigning Azure Policy guest configurations
- Protection using Microsoft Defender for Endpoint, Microsoft Defender for Cloud, and Microsoft Sentinel
- Configurations with Azure Automation, Update Management, Azure Automanage, and VM extensions installed on Arc-enabled servers
- Monitoring using Log Analytics workspaces and VM Insights in Azure Monitor

You can learn more about Azure Arc by visiting the documentation located here:

`https://docs.microsoft.com/en-us/azure/azure-arc/`

Finally, let's look at automating the management of your Windows VMs in Azure with Azure Automanage.

Using Azure Automanage

One of the newer features now available in Azure (currently in preview as of this writing) is Azure Automanage. Azure Automanage provides a fully automated solution for Windows VMs that leverages best-practice recommendations from Microsoft to automatically manage and configure VMs. Automanage can help simplify the management of your VMs by automating tasks, applying security updates and configurations, and detecting and remediating configuration drifts. By leveraging the participating services, Windows VMs can automatically be onboarded to Azure services that support best practices for Azure server management outlined in the **Cloud Adoption Framework for Azure**. For additional information about Azure server management services, visit the following link: `https://docs.microsoft.com/en-us/azure/cloud-adoption-framework/manage/azure-server-management/`.

The following current prerequisites must be met before you can use Azure Automanage on your Windows VM:

- Windows Server 2012/R2 or later.
- VMs must be in a region that supports Azure Automanage.
- RBAC permissions must be configured correctly.
- Sandbox subscriptions and Windows client images are currently not supported.

For more details on the requirements, visit this article:

`https://docs.microsoft.com/en-us/azure/automanage/automanage-virtual-machines#prerequisites`

The following participating services will be auto-onboarded, configured, and monitored for configuration drift using Azure Automanage:

- Machines Insights Monitoring
- Backups
- Microsoft Defender for Cloud
- Microsoft Antimalware
- Update Management
- Change Tracking and Inventory
- Guest configuration
- Boot Diagnostics
- Windows Admin Center
- Azure Automation Account
- Log Analytics Workspace

You can view more details on the services here:

`https://docs.microsoft.com/en-us/azure/automanage/automanage-windows-server#participating-services`

To enable Azure Automanage for your existing Windows servers, follow these steps:

1. Log in to the Azure portal at `https://portal.azure.com`.
2. Search for `Automanage - Azure machine best practices` and select a service to open.

3. Click on **Automanage machines**, then click on **Enable on existing machine**.

4. Select a configuration profile you would like to use. You can click on **View Azure best practices profiles** to view more details on each of the best practices being applied.

5. Click **Select machines** to view all machines and select the ones you would like to apply the profile. Click **Select**, then click **Enable**.

If you would like to create a custom profile for your Windows servers, follow the instructions provided here:

`https://docs.microsoft.com/en-us/azure/automanage/virtual-machines-custom-profile`

There is an overwhelming number of tools and services available in Azure that support the management of both on-premises and cloud deployments of Windows infrastructure. Visit the following link to view a comprehensive list of all Azure services: `https://azure.microsoft.com/en-us/services/`

In this section, we covered many of the Azure services used to manage Windows servers. These services included the Azure portal and Marketplace, ARM, RBAC, Azure Backup, ASR, Azure Update Management, Azure Arc, and Azure Automanage. Next, let's look at how to securely connect to Windows servers remotely.

Connecting securely to Windows servers remotely

Careful consideration should be made when setting up and configuring remote access in your environment. It's important to have a well-defined access model in place, keep access points to a minimum, and limit the availability of these access points from over the internet. Allowing remote access directly over the public internet significantly increases risks and could easily allow an attacker to gain access to your servers. If possible, it's best to eliminate internet availability entirely by properly configuring secure remote access tools and using **virtual private network (VPN)** connections. Let's look in more detail at the tools you can use to enforce a secure remote access policy.

Remote management and support tools

Remote management is a critical task that is needed to support both your end users and the infrastructure within your environment. There are many solutions available to help remotely manage and access your environment in a secure manner. The types of users that will require remote access will include system administrators, business developers, security teams, and IT support staff for accessing end-user devices. Ensure whichever remote support tool is selected is thoroughly reviewed and provides as secure a connection as possible. In remote support scenarios, it's also important that support staff cannot connect to your user's device unattended without the user's proper consent, as this could become a liability issue.

For managing your infrastructure remotely, most system administrators will likely leverage tools such as RDP or PowerShell. Just as with end-user support, there are many options available for remote management with RDP. As an organization, you should define and enforce policies that specify exactly how remote management tasks are handled. Let's look at a few recommendations to consider in your remote access strategy, as follows:

- Never allow direct remote management from the internet.

- Only allow access from a PAW or use a VPN at a minimum.

- If a server must be accessible from the public internet, change the default RDP port from 3389.

- User separate accounts or PAM to manage account access to servers.

- Enforce MFA when applicable.

- Use strong encryption methods for remote connections and limit where connections can originate from. For example, you can lock down access by certain locations or device types.

- Minimize the number of support staff accessing your infrastructure and servers.

- Monitor and audit all access to your infrastructure and servers.

> **Tip**
> To limit interactive logons to servers, you can deploy RSAT to administrative workstations:
>
> ```
> https://docs.microsoft.com/en-us/windows-server/
> remote/remote-server-administration-tools
> ```

Next, let's look at controlling access by using JIT access and Azure Bastion services to provide additional security within your remote access management strategy.

Using Microsoft Defender for Cloud JIT access

As a feature of Microsoft Defender for Cloud, **JIT** allows you to access systems by controlling or gating access at the network layer in Azure. Resources in Azure typically sit behind a firewall or **network security group** (**NSG**) that controls allow/deny rules for network protocols and connectivity. JIT works by creating or modifying NSG and firewall rules programmatically to only allow remote access protocols for a specified time. Ports that you would like to be available are configured through the JIT access configuration. For example, when a user needs JIT access, they issue an access request through Microsoft Defender for Cloud. A workflow approval process is started, and once approved, the solution automatically adds an inbound rule opening the configured ports on the resources' NSG or firewall. The network traffic is allowed for the duration of time specified in the access request, and once the time allotment expires, the NSG rules are returned to the previous deny state. Configuring access to the request workflow can be controlled using a custom RBAC role that can be assigned directly to users or groups. The RBAC role must contain the following actions:

- `Microsoft.Security/locations/jitNetworkAccessPolicies/`
  `initiate/action`

- `Microsoft.Security/locations/jitNetworkAccessPolicies/*/`
  `read`

- `Microsoft.Compute/virtualMachines/read`

- `Microsoft.Network/networkInterfaces/*/read`

We covered creating a custom role earlier in the *Using Azure services to manage Windows servers* section. To read more information about the resource policy providers, visit `https://docs.microsoft.com/en-us/azure/role-based-access-control/resource-provider-operations`.

> **Tip**
> These policies must be scoped to the subscription or resource group in which the VM and network interface reside.

Further information about Microsoft Defender for Cloud JIT access can be found at `https://docs.microsoft.com/en-us/azure/defender-for-cloud/just-in-time-access-usage`.

Let's create a JIT access rule that allows for RDP over port 65001. While it's not recommended to expose servers directly to the internet, this can be useful if a jump box is needed until stronger access controls can be put into place. Earlier, in *Chapter 4, Networking Fundamentals for Hardening Windows*, we provided an example of how to modify the default listening port for RDP to 65001 on a Windows Server. If you don't have Microsoft Defender for Cloud enhanced security and wish to follow the example, you can enable a free 30-day trial and continue with the step-by-step instructions, as follows:

1. Open **Microsoft Defender for Cloud** by searching for it in the Azure portal.

2. Click on **Workload protections** and choose **Just in time VM access** in the **Advanced protection** section.

3. In the **Virtual Machines** section, click on **Not Configured** to view VMs that can be configured for JIT VM access to be applied.

> **Tip**
> The **Unsupported** tab lists VMs that may either not be running or are not protected by an NSG.

4. Select the VM with the modified RDP port. Choose **Enable JIT on 1 VMs**.

5. On the JIT VM access configuration page, default port rules are already created. Let's add a new rule to add port 65001.

    JIT VM access configuration already has default rules for the following ports:

    -   Port 22 is common for SSH.

    -   Port 3389 is the default RDP port.

    -   Port 5985 and 5896 are WinRM HTTP/HTTPS ports.

6. Click on **Add** and enter 65001 for the port number, choose **TCP** for the protocol, and leave **Allowed source IPs** as default. Leave the **Max request time** field at 3 hours. Click on **OK**.

7. Click on **Save**.

Let's look at the new configuration as applied to the NSG. In the Azure portal, navigate to the VM we selected and click on **Networking** under **Settings**. In the following screenshot, notice a new inbound augmented security rule called `SecurityCenter-JITRule` that has all the ports specified in the configuration. By default, the action is set to **Deny**:

Network Interface: wshdc02643   Effective security rules   Topology
Virtual network/subnet: VN-USE2-Prod/SN-Mgmt-USE2-Prod   NIC Public IP: 52.167.213.130   NIC Private IP: **10.240.192.37**   Accelerated networking: **Enabled**

Inbound port rules   Outbound port rules   Application security groups   Load balancing

Network security group NSG-Identity-Prod (attached to subnet: SN-Mgmt-USE2-Prod)   **Add inbound port rule**
Impacts 1 subnets, 0 network interfaces

Priority	Name	Port	Protocol	Source	Destination	Action	
900	⚠ SecurityCenter-JITRule_-2046734702_C0...	22,3389,5985,5986,650...	Any	Any	10.240.192.37	🚫 Deny	...
1000	⚠ IBA_RDP_3389	3389	TCP	Internet	VirtualNetwork	✅ Allow	...
1001	IBA_RDP_65001	65001	TCP	Internet	VirtualNetwork	✅ Allow	...
65000	AllowVnetInBound	Any	Any	VirtualNetwork	VirtualNetwork	✅ Allow	...
65001	AllowAzureLoadBalancerInBound	Any	Any	AzureLoadBalancer	Any	✅ Allow	...
65500	DenyAllInBound	Any	Any	Any	Any	🚫 Deny	...

Figure 11.20 – NSG in Azure

> **Tip**
> Security Center JIT rules default to a priority of 4096, which is the highest integer value that you can set in a custom security rule. If custom rules were created to allow RDP previously, modify the priority of the new Security Center rule to be a lower integer, or delete the previous rules.

Once the JIT access rule has been configured, RDP will effectively be blocked. To request access, follow these steps:

1. Navigate to **Microsoft Defender for Cloud** in the Azure portal.

2. Choose **Just in time VM access** within **Workload protections**.

3. Under the **Configured** tab, select the VM to RDP into, and choose the **Request Access** button.

4. Toggle the rule for port 65001 to **On** and select **My IP** in the **Allow Source IP** column. We can leave the time range at 3 hours.

5. Enter a justification and choose **Open Ports**.

The following screenshot shows the custom port rule we created. During the access request, you can specifically choose the ports you need for connectivity:

Figure 11.21 – JIT VM access

If the appropriate RBAC is applied to request access to a VM, the access request will be approved, as indicated in the **Last access** column in the following screenshot:

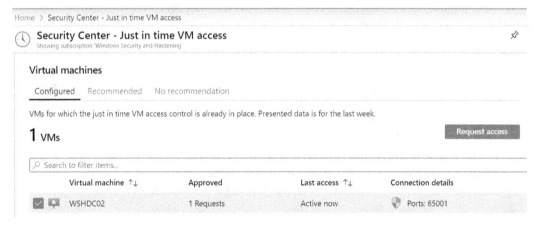

Figure 11.22 – JIT access rule over port 65001 showing as active

The RDP over port `65001` is now active. Browse back to the VM network interface inbound port rules. A new **Allow** rule has been added using your public IP as the source, as seen in the following screenshot:

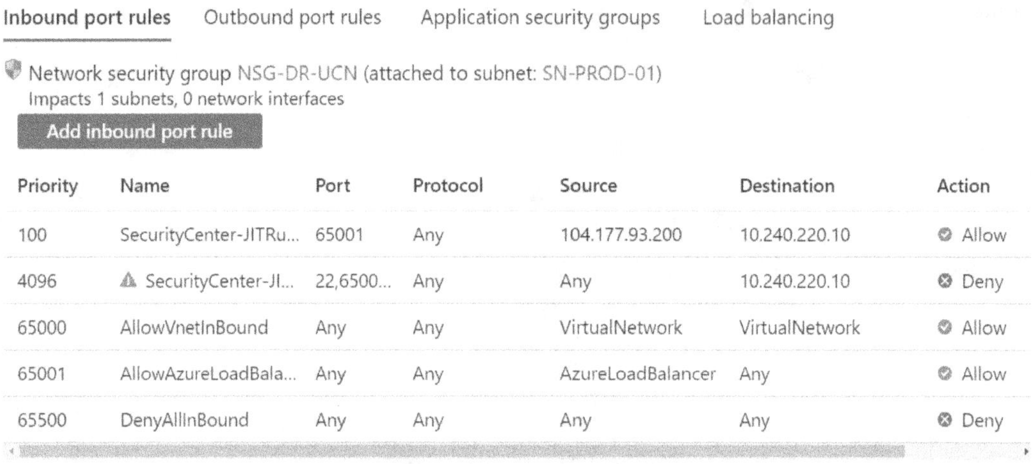

Inbound port rules   Outbound port rules   Application security groups   Load balancing

Network security group NSG-DR-UCN (attached to subnet: SN-PROD-01)
Impacts 1 subnets, 0 network interfaces

Add inbound port rule

Priority	Name	Port	Protocol	Source	Destination	Action
100	SecurityCenter-JITRu...	65001	Any	104.177.93.200	10.240.220.10	✔ Allow
4096	⚠ SecurityCenter-JI...	22,6500...	Any	Any	10.240.220.10	✖ Deny
65000	AllowVnetInBound	Any	Any	VirtualNetwork	VirtualNetwork	✔ Allow
65001	AllowAzureLoadBala...	Any	Any	AzureLoadBalancer	Any	✔ Allow
65500	DenyAllInBound	Any	Any	Any	Any	✖ Deny

Figure 11.23 – Active JIT rule

Following the approval of the JIT rule, the source is configured from your public **Internet Protocol version 4 (IPv4)** address. **Network address translation (NAT)** is automatically handled in Azure to allow for connectivity to the host VM with a private IP address. To connect to this server with RDP, enter the port number at the end of the IP address. In this example, `52.167.213.130:65001` would be used to create a connection.

Next, let's look at using the Azure Bastion service to connect to a VM without having to expose the host directly to the public internet.

Connecting with Azure Bastion

Azure Bastion is a PaaS service in Azure that allows for SSL-secured RDP or SSH connections directly to VMs from the Azure portal. It's a great alternative method for secure remote connections to internal resources without the need for a VPN or jump box.

To configure Azure Bastion, an isolated or *bastion* subnet is deployed directly inside your VNet IP range. This allows all RDP and SSH traffic to remain inside of your private VNet and keeps these protocols from being exposed over the internet. This helps protect your hosts from threat actors looking for targets with active port scanning and significantly reduces the attack surface. Your internal resources can remain on your private IP scheme without the need for a public IP address. To connect to a host, Azure Bastion supports RDP or SSH with HTML 5.

In this example, let's look at configuring Azure Bastion to connect to a host. To deploy the Azure Bastion service, a *bastion* subnet will be created inside your VNet, named `AzureBastionSubnet`. It must contain a minimum **Classless Inter-Domain Routing (CIDR)** range of `/27` to support this service. The subnet must be created first before creating a bastion. Proceed as follows:

1. Log in to the Azure portal at `https://portal.azure.com`.
2. Search for `Bastions` and select **Bastions** under **Services**.
3. Click on **Create Bastion**.
4. Select your **Subscription** type and choose a **Resource group** type. We selected the resource group our VNet was in.
5. Give it a friendly name and choose the region your resources are in.
6. Select your VNet and choose **AzureBastionSubnet**.
7. Create a new IP address and give it a friendly name.
8. Click on **Review + Create**. Then, select **Create**.

> **Tip**
> Azure Bastion does not support customized TCP ports. If you are using the VM configured earlier, the RDP listening port must be changed back to `3389` or you will receive a bad connection error. Additionally, NSG rules will need an allow rule to open access to the default `3389` RDP port, and any JIT access rules need to be accounted for or connections may be blocked.

To connect to a VM with Azure Bastion, follow these steps:

1. Go to the **Virtual Machines** pane inside the Azure portal.
2. Search for your VM and select it to bring it up in the **Overview** pane.
3. Click on **Bastion** under **Operations**.
4. Enter the VM's administrator username and password and click on **Connect**.

You may have to allow popups if your pop-up blocker is enabled. The following screenshot shows the **Connect to virtual machine** menu:

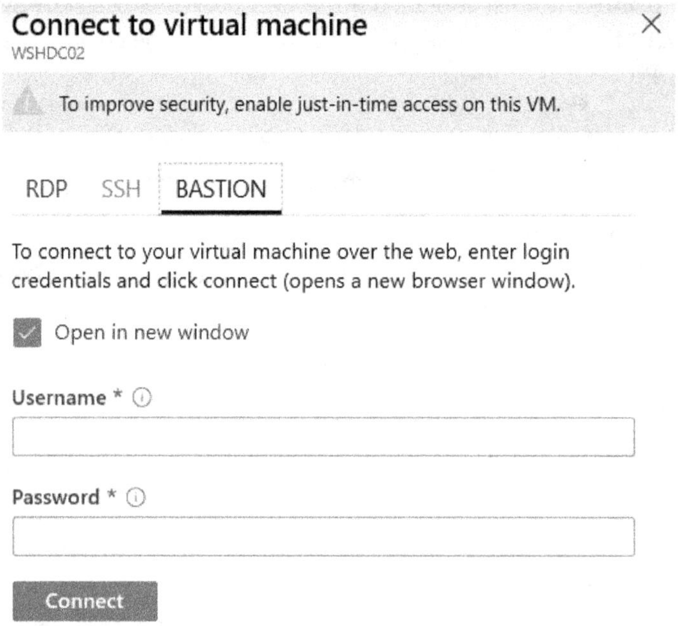

Figure 11.24 – Azure Bastion connection

Select the **BASTION** tab to connect.

Summary

In this chapter, we provided an overview of the traditional on-premises data center, hybrid, and cloud models. Within the cloud model, we covered three primary service offerings known as IaaS, PaaS, and SaaS. Next, we reviewed implementing access management as it relates to both physical and user access to Windows servers and infrastructure. We then covered using the tiered model approach in AD and best practices around implementing the privileged access strategy model.

The following section covered Windows Server management tools. We reviewed local tools such as Server Manager and Event Viewer and discussed deploying Windows Updates using WSUS and managing servers remotely with Windows Admin Center. We then moved on to Azure services for managing Windows servers both on-premises and in the cloud. In this section, we provided details about the Azure portal, using the Marketplace, permissions with RBAC, ARM Azure Backup, Azure Update Management, ASR, Azure ARC, and Azure Automanage. We finished off the chapter with an overview of connecting securely to servers remotely. We hope the acknowledgment of these services has sparked an interest to research further and see how these services can be implemented in your own environment to help increase your overall security posture.

In the next chapter, we focus on keeping your Windows Server secure. This chapter will cover the Windows Server versions, including reviewing the various server roles, configuring Windows updates, protecting servers with Microsoft Defender for Endpoint, and other strategies to harden Windows Server.

12

Keeping Your Windows Server Secure

In this chapter, we will discuss best-practice recommendations for hardening Windows Server. We will review the available Windows Server versions, cover new security features in Windows Server 2022, and discuss the built-in roles and features that add functionality to server deployments. Next, we will cover onboarding Windows Server into Microsoft Defender for Endpoint to enable **endpoint detection and response** (**EDR**) capabilities and deploy Windows Defender security baselines. Then, we will review deploying security updates with **Windows Server Update Services** (**WSUS**) and Azure Automation Update Management. Keeping servers updated with the latest patches and virus definitions could be considered by many the number 1 recommendation for overall hardening. It also requires the most operational overhead, and these two technologies are foundational in that regard.

Next, we will cover the security controls on Windows Server by implementing a security baseline using recommendations from Microsoft and CIS. We will provide general hardening tips and discuss controls for managing accounts and securing the logon and authentication process. We will finish the hardening section by covering disk encryption with Azure Disk Encryption and provide an example of how to apply application control policies with Windows Defender Application Control. Finally, we will finish this chapter by reviewing PowerShell security and **Just Enough Administration (JEA)**.

In this chapter, we will cover the following topics:

- Windows Server versions
- Security roles in Windows Server
- Configuring Windows updates
- Configuring Windows Defender
- Hardening Windows Server
- Application control policies using WDAC
- Implementing PowerShell security

Technical requirements

To follow the examples in this chapter, you will need basic knowledge of PowerShell, access to add server roles and features, and the permissions to modify Group Policy and deploy resources in Azure. The following products and tools will be referenced:

- Windows Server 2022 Core and Desktop versions
- WSUS
- An Azure subscription with contributor rights
- Licensing for Microsoft Defender for Cloud
- PowerShell, including the Azure PowerShell Az module

Windows Server versions

Many servers hosting workloads that contain sensitive data run critical services, and host applications are likely to run in Windows environments. Any security breach inside the server environment has the potential to cause significant damage and loss of data. Windows Server contains many security controls built directly into its operating system, but they may not necessarily be enabled by default.

Just like with Windows clients, there is no one solution or collection of settings that provides a silver bullet in terms of defense for servers. It will require adding security layers that introduce boundaries to make it more difficult for attackers to break through. The number of layers varies by organization and it's up to your company to determine what features to enable that fit your security stature. During the hardening process, enable small subsets of controls at a time, followed by performing rigorous testing and validation to ensure functionalities are not interrupted until your baseline controls are satisfied. As mentioned in *Chapter 1*, *Fundamentals of Windows Security*, organizations such as NIST can offer guidance for making the decisions that are needed that help define your security stature.

The topics in this book mainly focus on features that are available in Windows Server 2016 and later. Many of the features we will cover may be available in earlier versions too, but we recommend reviewing this further if a version is not specifically mentioned. Windows Server follows Microsoft's life cycle policy and currently supports versions of Windows Server 2008 and 2012 hosted in Azure until 2024 and 2026 if extended support is purchased. In addition to the life cycle policy, it is important to be familiar with the release and servicing options. These can be reviewed at `https://docs.microsoft.com/en-us/windows-server/get-started/windows-server-release-info`.

To view the life cycle for a specific Windows Server version, go to `https://docs.microsoft.com/en-us/lifecycle/products/?products=windows`.

> **Tip**
> Starting with Windows Server 2022, the **Long-Term Servicing Channel (LTSC)** is the only release channel available.

Windows Server 2022 has three editions to choose from: **Standard**, **Datacenter**, and **Datacenter: Azure Edition**. The Datacenter versions provide additional features over the Standard version, and each edition has the following installation options available:

- **Desktop Experience** is the traditional deployment of Windows Server and includes a **graphical user interface (GUI)**. This option is only available in the LTSC version.

- **Server Core** has a smaller OS footprint and doesn't include the GUI.

- **Nano Server** was first introduced in Server 2016, has an even smaller footprint than Server Core, and can only be remotely administered. A couple of examples include using Nano Server for **Internet Information Services (IIS)** or a DNS server. This option is only available as a container-based Windows image.

Windows Server now includes a version called Secured-core. A Secured-core server contains certified hardware from equipment manufacturers and OEMs that provide support for the latest hardware-based security features. These hardware and firmware-based security capabilities include System Guard and virtualization-based security. Secured-core servers can be found at Azure Stack HCI or the Window Server catalog at `https://www.windowsservercatalog.com/noresults.aspx?1&bCatID=1333&cpID=0&avc=10&ava=0&avt=0&avq=140&OR=1&PGS=25#/catalog?FeatureSupported=securedCoreServer`.

When choosing an installation option, it's recommended to use the most minimal installation option possible to reduce the footprint of the server. Deploying a server without a GUI and limiting logon types to remote administration using tools, such as PowerShell and Microsoft Management Console, is a significant security improvement and should be implemented as a standard best practice.

Windows Server 2022 includes the following new security highlights:

- Hardware-based security features are available in Secured-core PCs such as UEFI Secure Boot, VBS, and hardware root-of-trust.

- Secured protocols such as HTTPS and TLS 1.3 are enabled by default.

- Support for **DNS-over-HTTPS (DoH)** for secure DNS.

- Supports the latest cryptography for SMB with automatic negotiation for the highest cryptography suite possible. SMBv1 is disabled by default.

- Support for SMB over QUIC (`https://docs.microsoft.com/en-us/windows-server/storage/file-server/smb-over-quic`).

- Support for hotpatching using Azure Automanage in Windows Server 2022 Datacenter: Azure Edition. Hotpatching allows you to install updates on VMs that don't require a reboot after installation to maintain uptime.

The following URL provides a list of all the new features of Windows Server 2022: `https://docs.microsoft.com/en-us/windows-server/get-started/whats-new-in-windows-server-2022`.

Next, will we review Windows Server's roles and list features that have security-specific components to help you secure your environment.

Security roles in Windows Server

Server roles and features in Windows Server help add additional functionality to your Windows deployment. For example, a basic DNS server role allows you to create a catalog of computer name-to-IP mappings for name resolution services on your internal network. As additional roles are installed on a server, new services become enabled, so it's critical that hardening is taken into consideration to ensure vulnerabilities aren't exposed due to misconfigurations. As a best practice, it's recommended not to install any unnecessary roles on servers for the most secure configuration. For example, on a **domain controller** (**DC**), you should not enable additional roles other than Active Directory domain services and DNS services. A DC holds the keys to authentication and hardening will be different than that of a server that hosts a web application and the IIS web server role. Any misconfigurations could provide a hacker with the opportunity to exploit or infiltrate your environment.

Some examples of roles and features in Windows Server 2022 that include security-related features are as follows:

- Active Directory Certificate Services
- Active Directory Domain Services
- Active Directory Federation Services
- Active Directory Rights Management Services
- Device Health Attestation
- Host Guardian Service
- Network Policy and Access Services
- Remote Access
- Remote Desktop Services
- Windows Server Update Services
- BitLocker Drive Encryption
- BitLocker Network Unlock
- Group Policy Management
- **Remote Server Administration Tools** (**RSAT**)
- Windows Defender Features (installed by default)
- Windows PowerShell (installed by default)

The following URL provides a complete list of all Windows Server 2022 roles and features that are available in each edition: `https://docs.microsoft.com/en-us/windows-server/get-started/editions-comparison-windows-server-2022`.

Reducing the Windows Server footprint

Any opportunity where unnecessary roles and features can be reduced, or actions that can be enforced to limit the amount of direct interaction with the server, should be taken. For example, if Windows Server Core has a role for Active Directory domain services, it is recommended to deploy it using Server Core. Once deployed as a Core version, this role can be fully managed through remote tools such as **PowerShell**, **Microsoft Management Console** (**MMC**), and **Windows Admin Center**. This significantly reduces your threat footprint. Most of the roles and features that are included with the **Desktop Experience** are also supported by Server Core.

When you're planning for a new deployment of Windows Server, you can consult the following link to determine if the feature is supported in Server Core: `https://docs.microsoft.com/en-us/windows-server/administration/server-core/server-core-roles-and-services`.

Enabling features on Server Core 2022

Server Core is a lightweight installation option that's available for Windows Server and has fewer roles to choose from but is the most secure option. As we mentioned earlier, with a Sever Core deployment, you can use remote management tools such as PowerShell, MMC, and Windows Admin Center to manage the server without having to log directly into the physical system. The option to install Server Core can be chosen at installation time or by selecting it from the **Plan** dropdown when choosing an image from the Azure Marketplace:

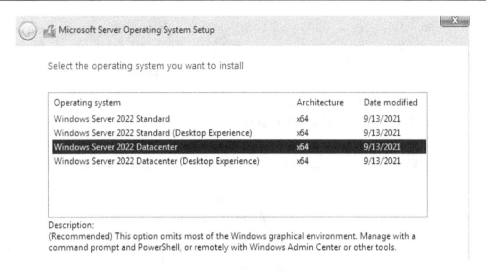

Figure 12.1 – Windows Server 2022 Datacenter

After the installation is complete, you will be asked to configure a local administrator account to log in. Then, you will be brought to the **SConfig** screen, which will present several menu options. Let's look at how to install a role.

Using PowerShell, let's install WSUS on Windows Server 2022 Core edition by following these steps:

1. Deploy Windows Server 2022 Core using the installation wizard or through the Azure Marketplace.

2. After logging into Windows Server Core, choose the **Exit to command line (PowerShell)** option.

3. Type `Install-WindowsFeature UpdateServices` and press *Enter*.

4. The installation will take a few minutes. After it completes, it will look as follows with an **Exit Code** of `Success`:

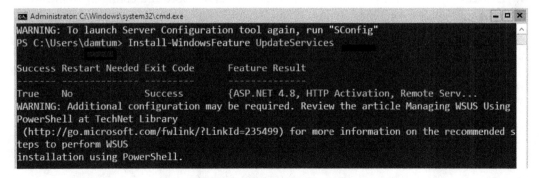

Figure 12.2 – Installing WSUS in Server Core

The WSUS feature is now enabled. We will complete the configuration in the next section.

Next, let's learn how to configure services to manage the deployment of Windows updates to servers, including using WSUS and Azure Update Management.

Configuring Windows updates

One of the top recommendations for hardening Windows Server is to run the latest version of the OS with the latest security patches installed. To be able to install the latest updates, the Windows machine must be able to download and install updates directly from Windows Update or a software update distribution point. Just like Windows clients, you can configure Windows Update for Business policies on servers with Group Policy, but this won't allow for fine-grained control regarding the approval and timing of update deployments. To accommodate this scenario, WSUS has been the standard method for deploying Windows updates for on-premises deployments.

WSUS allows companies to effectively manage, distribute, and orchestrate update installations to clients and servers to maintain security compliance and patch-level continuity within the environment. A WSUS deployment is a great method for deploying updates, but it does have prerequisites that require deploying infrastructure to support its functionality. To get around these requirements, Microsoft offers an alternative solution for Azure customers known as Azure Automation Update Management. Using Update Management, IT can manage both Windows and Linux updates for cloud and on-premises resources without the need for a WSUS implementation.

We provided an overview of WSUS and Azure Automation Update Management in *Chapter 11*, *Server Infrastructure Management*. Additionally, if you don't want to allow servers to have direct connectivity to Windows Update services, you can configure a WSUS instance in a DMZ and securely distribute update packages on your internal network in Azure.

Now, let's learn how to configure WSUS on Windows Server Core and configure Windows Update Group Policy to onboard clients.

Implementing WSUS

Earlier, in the *Enabling features on Server Core 2022* section, we installed the WSUS role within Windows Server Core. Before we can access the management console, there are a few post-installation tasks to complete. Let's use PowerShell to complete these tasks:

1. Log back into your Windows Core WSUS server and choose the **Exit to command line (PowerShell)** option from the **Sconfig** menu.

2. Enter each of the following commands one at a time to complete the setup process:

    ```
    New-Item -Path "c:\" -Name "WSUS" -ItemType "directory"
    CD "C:\Program Files\Update Services\Tools"
    .\wsusutil.exe postinstall CONTENT_DIR=C:\ "WSUS"
    ```

After the post-installation activities are complete, we can install RSAT on a Server Desktop edition or Windows client to manage WSUS remotely.

Let's install them on a Windows 2022 Desktop edition by completing the following steps:

1. Log onto a Windows Server that's used for management and open **Server Manager**.

2. Click on **Manage**, then **Add Roles and Features**.

3. Click **Next** (leave everything with the default settings) until you get to **Features**.

4. Scroll down to **Remote Server Administration Tools** and expand the menu, expand the **Role Administration Tools** menu, and then select **Windows Server Updates Services Tools**.

5. Click **Next**, then **Install**. Click **Close** once the installation is complete.

Now, we can use the MMC to connect to the Windows Server Core WSUS server and configure it further:

1. Search for **Windows Server Update Services** and open it.

2. Click **Connect to Server** in the **Actions** menu. The following dialog box will appear:

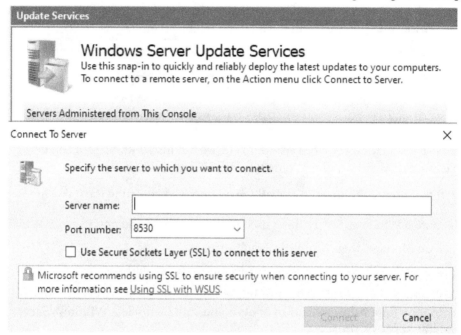

Figure 12.3 – Connecting to the WSUS server

3. Enter the *WSUS Server name* or *IP* that was configured, leave the port as 8530 if SSL is not configured, and then click **Connect**.

> **Tip**
> The following guide provides instructions on how to configure and enable SSL for your WSUS environment: https://docs.microsoft.com/en-us/windows-server/administration/windows-server-update-services/deploy/2-configure-wsus#25-secure-wsus-with-the-secure-sockets-layer-protocol.

4. In the WSUS Configuration Wizard, click **Next**, then **Next**. Select **Synchronize from Microsoft Update** and then click **Next**.

5. If you're using a proxy, select **Use a Proxy** and then enter the required information. If not, leave this *unchecked* and click **Next**. Then, click **Start Connecting** (this step may take a while to complete).

6. Once complete, click **Next**, select the languages needed, and click **Next**.

7. Select all the products you want to scope into WSUS management and click **Next**.

8. Select **Classification of Updates**. At a minimum, select **Critical Updates** and **Security Updates**. Then, click **Next**.

9. Select **Synchronize automatically** and schedule your *first sync (Outside of Business hours)*. Select how many times you wish to sync per day and click **Next**.

10. Select **Begin initial synchronization** and click **Next**. Click **Finish** to view the management console. You should see an overview similar to the following:

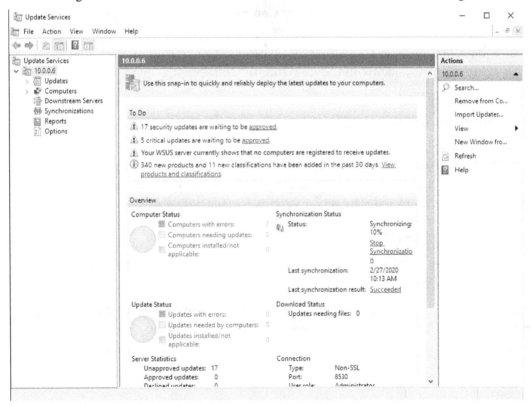

Figure 12.4 – WSUS management console

Additional configurations, including modifying those chosen during setup, can be viewed by clicking on **Options** in the left management pane under **Update Services**.

A few additional configurations are needed before we can connect clients to WSUS. First, we must create a computer group that can be used to test updates. Computer groups allow you to organize groups of clients and servers to schedule update deployments. Navigate to and expand the **Computers** menu, right-click **All Computers**, and then click **Add Computer Group....** Name your group and then click **Add**.

Next, we will configure WSUS so that it uses Group Policy to efficiently manage the auto-configuration of Windows servers. To do this, choose **Options** within the management console and click on **Computers**. Change the option to **Use Group Policy or registry settings on computers** and click **OK**.

Finally, we are ready to configure Group Policy so that it instructs your Windows servers where to connect for Windows updates:

1. Open the **Group Policy Management Console** and create a new GPO in an OU where there are server computer objects.

2. Browse to **Computer Configuration | Policies | Administrative Templates | Windows Components** and select **Windows Updates**.

3. Configure the following settings:

   - Set **Configure Automatic Updates** to `Enabled`. Choose the options that fit your preferences.

   - Set **Specify intranet Microsoft update service location** to `Enabled`. Add your WSUS server to the **Set the intranet update service for detecting updates** and **Set the intranet statistics server** settings. Use the `http://servername.domain.com:8530` format for HTTP or `https://severname.domain.com:8531` for HTTPS if you're using an SSL.

   - Set **Enable client-side targeting** to `Enabled`. Enter the group name that you created within the WSUS management console for testing.

The following screenshot shows the three settings you need to configure within the GPO to force WSUS to be the update endpoint for your Windows servers:

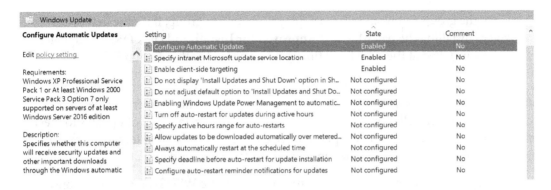

Figure 12.5 – Group Policy configuration

Now that the servers have been configured to connect to the WSUS server, the last step is to approve and push the updates to the clients:

1. From the WSUS management console, browse to **Updates** and click **Updates needed by computers** on the main screen. Select all the available updates that need to be deployed, right-click, and then click **Approve** to open the **Approve Updates** dialog box, as shown in the following screenshot:

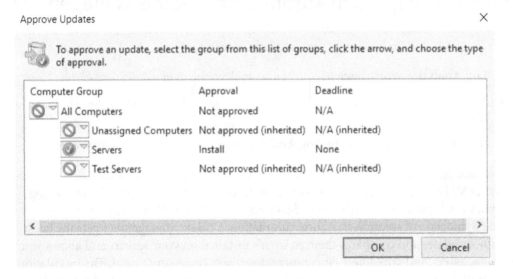

Figure 12.6 – Approve Updates

2. Click the group you would like to deploy to (your test group, if it's the first time you're installing the patch), select **Approved for Install**, and click **OK**. The updates will be installed at the scheduled time you configured within Group Policy.

To help reduce the administrative overhead with approving updates, you can configure automatic approval rules in WSUS. Each computer group can have different rules and include settings for installation deadlines that will force the installation of updates to meet compliance requirements. Update types that are good candidates for automatic approval rules include critical security updates and antivirus definition updates.

To set up automatic approvals, browse to **Options** within the WSUS management console and click on **Automatic Approvals** to set up the rules.

Next, we will review Azure Automation Update Management. Update Management is a cloud-based solution that includes support for Linux and Windows servers, both on-premises and in Azure. In addition, Update Management is provided as a service within Azure and can eliminate the need for on-premises resources to support a WSUS deployment. Currently, Update Management in Azure does not support Windows clients.

Implementing Azure Automation Update Management

To use Update Management in Azure, you will need permissions to create the following resources in addition to having an active Azure subscription:

-   **Log Analytics Workspace** to collect and analyze the data

-   **Azure Automation Account** to be able to configure Azure Automation Update Management

-   **Resource Group** to store the Log Analytics workspace and Automation account

Update Management works by installing an agent on a server known as the Log Analytics VM extension. The extension is configured with the workspace ID of the Log Analytics workspace to send telemetry data into Azure. This is then analyzed by the Update Management solution in Azure Automation. Update Management gets the latest Windows updates and compares them to what's installed on your servers and allows you to review, assess, and schedule a deployment directly in the Azure portal. The installation orchestration of Windows updates is handled by the Azure Automation Hybrid Runbook Worker role. This worker allows automation runbooks to be executed directly against local resources on the target server.

For the server to be able to download available Windows updates, it must have connectivity to Microsoft update services, a WSUS server, or a similar repository. Onboarding virtual machines into Update Management is handled directly through the Azure portal; otherwise, the agent can be downloaded and deployed using an offline application installation methodology. To allow servers to send data into Log Analytics, there are a few additional networking requirements. Details can be found at `https://docs.microsoft.com/en-us/azure/automation/automation-network-configuration`.

First, let's create the Automation account and Log Analytics workspace:

1. Log into `https://portal.azure.com` and search for `Automation Accounts`.

2. Click **Add** and enter a **Name, Subscription, Resource group**, and **Location**.

3. Choose **Yes** to create an **Azure Run As account** and click **Create**, as shown in the following screenshot:

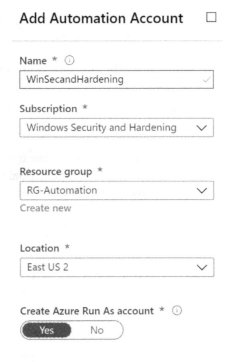

Figure 12.7 – Add Automation Account

4. Once the Automation account has been created, click on it to view the overview.

5. Click on **Update Management** within the **Log Analytics Workspace** section, click the drop-down menu, select **Create New Workspace**, and click **Enable**.

6. Once complete, choose **Update Management** to get to the overview pane, as shown here:

Figure 12.8 – Azure Update Management console

Now that the Update Management solution has been configured, we can onboard virtual machines using the following methods in the Azure portal:

- From within the VM settings in Azure, click on **Update Management** and then click **Enable**.

- From within the Update Management console, click on **Add Azure VMs**. Select the VMs you would like to enable and then click **Enable**.

Once VMs have been enabled, it can take up to 6 hours for the telemetry data to become available in Log Analytics. Once the data has populated, the update data will be parsed in the Update Management solution, and you will see the overview dashboard populated with data, as shown here:

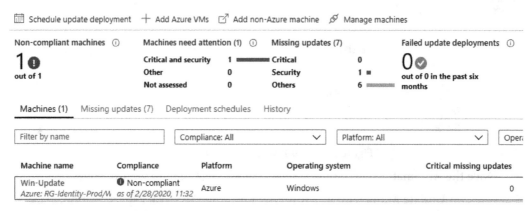

Figure 12.9 – Azure Update Management console

> **Tip**
> The following article provides troubleshooting advice for Update Management:
> `https://docs.microsoft.com/en-us/azure/automation/`
> `troubleshoot/update-management.`

To add non-Azure VMs, you will need to manually install the agent on the machines and configure the workspace ID and primary key of the Log Analytics workspace. The agent can be deployed using any enterprise deployment tool such as Configuration Manager or Group Policy and can be found by following these steps:

1. Open the Log Analytics workspace that's connected to your Automation account.

2. In the **Overview** section, click **Agents Management** under **Settings**.

3. Click on **Windows servers** and download the 32 or 64-bit Windows agent. Make note of the **Workspace ID**, **Primary key**, and **Secondary key** details.

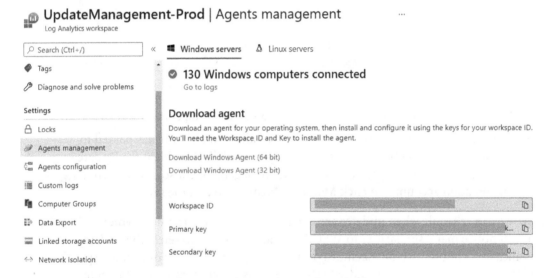

Figure 12.10 – Log Analytics – Download agent

4.   Log on to the Windows server, install the downloaded agent, and configure the workspace ID and key following the onscreen instructions. The Microsoft Monitoring Agent applet is available through the control panel, and you can view the status of the connection or add additional workspaces, as shown here:

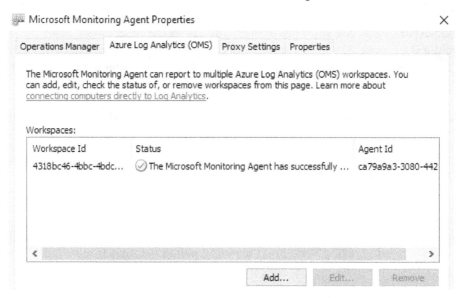

Figure 12.11 – Microsoft Monitoring Agent Properties

5.   Once the workspace ID has been configured and enough time has passed to allow for data collection, an additional step is needed to enable it for Update Management. Browse back to the Update Management console within your Automation account and click **Manage Machines** to add the server.

Now, we can schedule when updates are deployed by configuring a deployment schedule. You will notice many similarities to WSUS, including options to select the update classification, choose specific machines or machine groups, include or exclude specific updates, configure scheduling, and set reboot options. To configure a deployment schedule, follow these steps:

1.   Open Update Management from **Automation Accounts** and click on **Schedule update deployment**.

2.   Complete the setup by providing the required information to create your new update deployment. Some of the available settings can be seen in the following screenshot:

New update deployment
Update Deployment

Name * ⓘ

Critical Updates Only   ✓

Operating system

(**Windows**   Linux)

Schedule Settings   ⋯

* Items to update

Add groups and/or machines to update

Groups to update Click to Configure	>
Machines to update Click to Configure	>

Update classifications

Critical updates   ⌄

Starts * ⓘ

| 02/23/2022 🗓 | 3:00 AM |

Time zone

United States - Eastern Time   ⌄

Recurrence

(**Once**   Recurring)

Figure 12.12 – New update deployment

3. Click **Create** once you're finished.

The deployment schedule is now active, and you can use the Update Management console to track the overall compliance of the deployment. In the following screenshot, **Non-compliant machines** is **0**. This means that all the latest available updates that have been scoped in the deployment schedule have been deployed:

Figure 12.13 – Azure Update Management console with a compliant server

---

Tip

When choosing **Update Classifications** in your deployment schedule, don't forget to include **Definition Updates** if you're using Microsoft Defender Antivirus.

Update Management still relies on client-specific policies for automatic update behavior. For additional information about configuring update settings on servers connected to Azure Automation Update Management, visit `https://docs.microsoft.com/ en-us/azure/automation/update-management/configure-wuagent`.

Alert rules can be configured in Azure Monitor to help track the status of scheduled deployments in Update Management. To learn more about the available configurations, we recommend reviewing the details at `https://docs.microsoft.com/en-us/ azure/automation/update-management/overview`.

Keeping servers updated plays a critical role in any well-rounded security program. Today, many IT organizations have adopted minimum security standards that require systems to have the latest critical and security updates within 30 days of release to maintain compliance. It's necessary to have a solution in place that can effectively deliver and orchestrate update installations.

Next, let's look at onboarding servers to Microsoft Defender for Endpoint to take advantage of Microsoft's EDR capabilities.

Configuring Windows Defender

Microsoft Defender Antivirus or Windows Defender is the available antivirus solution for Windows Server. It is enabled by default, but the service can be toggled on or off by adding or removing the server role in Server Manager, or with PowerShell. Windows Server supports other third-party antivirus solutions and Defender is designed to run congruently alongside them if you enable passive mode or disable the feature. In Windows Server 2022, you can use the Windows Security app to configure features of Windows Defender such as **Virus & threat protection** settings and other advanced settings. To enhance server protection and enable EDR capabilities, Defender Antivirus for servers and other third parties is designed to work with Microsoft Defender for Endpoint. To view a list of benefits for connecting to Defender for Endpoint, visit `https://docs. microsoft.com/en-us/microsoft-365/security/defender-endpoint/ why-use-microsoft-defender-antivirus?view=o365-worldwide`.

Defender for Endpoint was initially designed to support Windows clients but now includes support for Windows Server if you follow the onboarding process. Once the servers have been connected, the Microsoft 365 Defender portal can be used as a unified console to investigate threats and provide real-time protection and remediation capabilities across all onboarded endpoints. The following Windows Server versions are currently supported:

- Windows Server 2008 R2 SP1
- Windows Server 2012 R2
- Windows Server 2016
- Windows Server Semi-Annual Channel
- Windows Server 2019 and later
- Windows Server 2019 Core
- Windows Server 2022

Let's review a few methods for onboarding servers in bulk.

Connecting to Microsoft Defender for Endpoint

Onboarding Window Server requires both an installation and onboarding package, depending on the version. Any of the following methods can be used to deploy these components:

- Local script or Group Policy
- Microsoft Configuration Manager
- Microsoft Defender for Cloud

These components can be installed independently from one another or using the unified solution package from Microsoft. If you have previously onboarded Windows Server 2016 and below to Defender for Endpoint, it is recommended to migrate to the unified solution package. This will enable support for new advanced protection features such as blocking potentially unwanted apps, network protection, ASR rules, and tamper protection.

Onboarding with Microsoft Defender for Cloud

Onboarding servers can be completed automatically if you are using server protection in Microsoft Defender for Cloud. Auto-provisioning capabilities will automatically onboard servers to Defender for Endpoint, without requiring any additional steps. Defender for Cloud is designed to work congruently with Defender for Endpoint, adds additional vulnerability detection capabilities, and allows you to monitor Defender for Endpoint signals directly in the console in Azure. There are additional costs to using Defender for Cloud, depending on the plan you choose, and any data consumption costs for using Log Analytics to store telemetry data. The following screenshot shows the **Auto provisioning** feature from the Azure portal, which automatically onboards servers by deploying the Log Analytics agent for Azure VMs:

Figure 12.14 – Microsoft Defender for Cloud – Auto provisioning

For additional pricing information for Microsoft Defender for Cloud, visit `https://azure.microsoft.com/en-us/pricing/details/defender-for-cloud/`.

For Windows Server 2019 and above, only the onboarding package is required for Defender for Endpoint. Let's look at how to onboard Windows Server to Defender for Endpoint by using Group Policy.

Onboarding with Group Policy

In this example, we will use Group Policy to onboard Windows Servers 2019 and later. If you are looking to onboard Windows Server 2016 and earlier, visit the following link for detailed instructions: `https://docs.microsoft.com/en-us/ microsoft-365/security/defender-endpoint/configure-server- endpoints?view=o365-worldwide#windows-server-2012-r2-and- windows-server-2016`.

You will need access to download the onboarding package from the Microsoft 365 Defender portal and permission to modify Group Policy. A network share with a UNC path that allows read-only access for computer accounts is required to be able to deploy the package. Follow these steps:

1. Sign into the **Microsoft 365 Defender** portal at `https://security. microsoft.com`.

2. Choose **Settings** and select **Endpoints**. Click on **Onboarding** in the **Device management** section.

3. Choose **Windows Server 1803, 2019, and 2022** for the operating system type and select **Group Policy** from the **Deployment Method** dropdown. Click on **Download onboarding package**.

   Extract the contents of the onboarding package to a directory on your network share. The extracted contents should look like the image below.

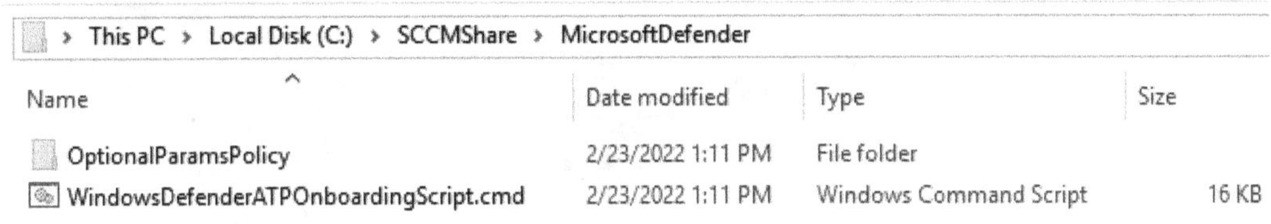

Name	Date modified	Type	Size
OptionalParamsPolicy	2/23/2022 1:11 PM	File folder	
WindowsDefenderATPOnboardingScript.cmd	2/23/2022 1:11 PM	Windows Command Script	16 KB

> This PC > Local Disk (C:) > SCCMShare > MicrosoftDefender

Figure 12.15 – Windows Defender for Endpoint onboarding package

4. Open **Group Policy Management Editor** and create a new policy. Link it to an OU where your target Windows Server objects reside. Give it a friendly name, such as `Windows Server - Defender for Endpoint`.

5. Go to **Computer Configuration | Preferences | Control Panel Settings** and right-click **Scheduled Tasks**. Select **New** and choose **Immediate Task (At least Windows 7)**. Give it a friendly name, such as `Defender for Endpoint Onboarding`.

6. Under **Security Options**, select **Change User or Group** and type in `SYSTEM`. Click **Check Names** and choose **OK** to select **NT AUTHORITY\System**.

7. Select the radio button next to **Run whether user is logged on or not**.

8. Choose **Run with highest privileges**.

    The **New Task** window should look as follows:

Figure 12.16 – Creating a scheduled task with Group Policy

9. Click the **Actions** tab, click **New**, and choose **Start a program** under **Action**.

10. Under **Program/script**, enter the UNC path of the
    `WindowsDefenderATPOnboardingScript.cmd` file from the extracted
    contents earlier using the fully qualified domain name. Click **OK** and then **OK**
    again to commit the change.

    The **New Action** settings should look as follows:

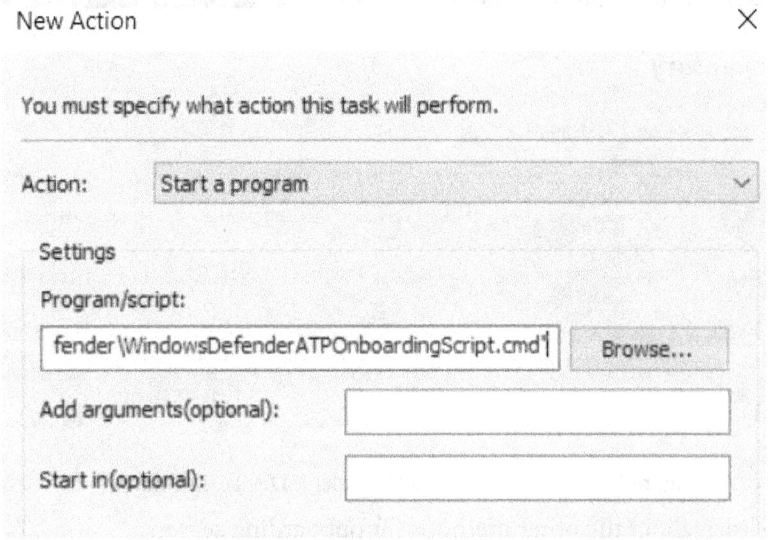

Figure 12.17 – Configuring a new action for a scheduled task

The scheduled tasks should now be visible in the GPO in the **Scheduled Tasks** section.

11. Close any open Group Policy windows.

Target servers should begin to receive policies within 90 minutes if a policy update isn't forced. Typically, machines onboard within 24 hours of executing the file, but it can take several days for telemetry data to flow in Defender for Endpoint. To check the machine's status, go to the Microsoft 365 Defender portal at `https://security.microsoft.com`. In the search bar, enter the hostname of a target server to search for results. You can also click on **Device Inventory** to view the list of discovered endpoints, as shown here:

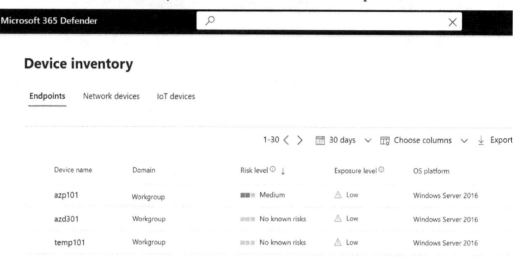

Figure 12.18 – Microsoft 365 Defender – Device inventory

If you want to learn about the other methods for onboarding servers, clients, and VDI hosts, visit `https://docs.microsoft.com/en-us/microsoft-365/security/defender-endpoint/configure-server-endpoints?view=o365-worldwide`.

Next, let's look at configuring a Windows Defender security baseline using Microsoft Security Baseline recommendations.

Windows Defender security baseline

A Windows Defender security baseline is available in the latest version of Microsoft Security Baseline for Windows Server 2022. These controls focus specifically on configuring features of Windows Defender Antivirus. When using Policy Analyzer to analyze the Microsoft recommendations against an out-of-the-box Windows Server 2022 installation, you will see that 24 settings are not configured by default. This includes settings for real-time protection, PUA protection, network protection, cloud-delivered protection, and ASR rules.

The baseline recommendations can be applied by importing the reference GPO into a new or existing GPO that's been scoped to an OU where our target servers reside. The process to do this is the same as what we explained in *Chapter 8, Keeping Your Windows Client Secure*, in the *Deploying Windows Security Baselines* section.

The Windows Defender GUI in Server 2022 can be found by opening the **Windows Security** app. Here, we can view all the enabled features and configure additional settings such as exclusions if needed. If the security baseline was deployed using Group Policy, we can validate the settings that are being managed by clicking on **Virus & threat protection | Manage settings** under **Virus & threat protection settings**. Policies that are managed will be grayed out and a message stating **This setting is managed by your administrator** will be shown, as shown here:

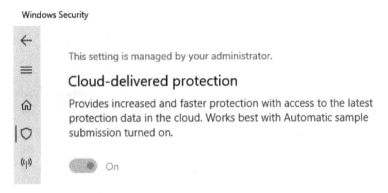

Figure 12.19 – Setting controlled by Group Policy

Windows Defender can also be queried with PowerShell to view and configure additional features. The main PowerShell cmdlets to get information about Defender Antivirus are as follows:

- `Get-MpPreference` will return information about the configured Defender Antivirus settings, including scanning behavior and the status of enabled features.

- `Get-MpComputerStatus` will return details about the antimalware version, signature definition date, and scanning details.

Microsoft Security Baselines provides a good set of configurations, but it is recommended to also enable Tamper Protection for Windows Server 2012 R2 and above. Tamper Protection helps prevent unauthorized modification of Defender security features through the registry, Group Policy, and PowerShell. At the time of writing, the feature is not available in Group Policy, but it can be enabled using Microsoft Endpoint Configuration Manager, through the Microsoft 365 Defender portal, or individually on the server. If you use the Defender portal, this setting will be applied to all onboarded devices in the tenant. It can be enabled by going to `https://security.microsoft.com` and selecting **Settings | Endpoints | Advanced features | Tamper protection**. To enable it on an individual device, open the **Windows Security** app, choose **Virus & threat protection | Manage settings** (under **Virus & threat protection settings**), and toggle **Tamper Protection** to On, as shown in the following screenshot:

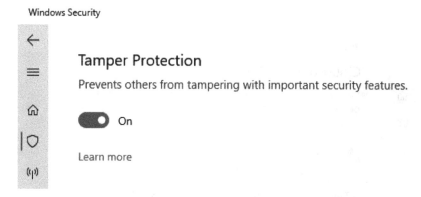

Figure 12.20 – Tamper Protection in Windows Server 2022

Windows Server versions 2016 and lower won't have this setting in the Window Security app, so you must validate that it's enabled by opening PowerShell as an administrator and typing `Get-MpComputerStatus | Select-Object IsTamperProtected, RealTimeProtectionEnabled`. A value of `true` means it's enabled.

> **Tip**
> To use Tamper Protection on Windows Server 2012 R2, you must onboard devices using the modern unified solution package.

In addition to the Microsoft security baselines, a few additional policies are recommended in the CIS Level 1 benchmarks for Windows member servers. To configure them using Group Policy, navigate to **Computer Configuration | Administrative Templates | Windows Components | Microsoft Defender Antivirus**. Do the following:

- Set **Turn off Microsoft Defender Antivirus** to Disabled .

- Set **Turn on behavior monitoring** to Enabled. This can be found under **Microsoft Defender Antivirus | Real-Time Protection**.

- Set **Configure local setting override for reporting to Microsoft MAPS** to Disabled. This can be found under **Microsoft Defender Antivirus | MAPS**.

- Set **Turn on e-mail scanning** to **Enabled**.

Next, let's look at implementing baseline controls to harden Windows Server.

Hardening Windows Server

Hardening Windows Server with the appropriate security controls plays a key role in maintaining a good security posture. Many security controls are built directly into Windows Server, but they may not necessarily be enabled by default. Security teams must do their due diligence to find the appropriate balance of usability to risk when determining what controls to enable. To help with the hardening process and as a fundamental first step, we can implement a preconfigured security baseline designed for Windows Server. In the next section, we will cover implementing these preconfigured recommendations from Microsoft and CIS.

Implementing a security baseline

From a program level, a security baseline is a foundation that consists of clearly defined policies, standards, procedures, and guidelines. They are more than just a set of configurations that apply to devices. In *Chapter 2, Building a Baseline*, we covered these concepts and discussed their importance, as well as providing an overview of the different frameworks seen today. We also introduced benchmarks from the **Center for Internet Security** (**CIS**) and Microsoft as a guideline for building the specific configurations that apply to your Windows servers.

First, let's look at a comparison of these recommendations against the default server settings and review the results. Then, we will use the preconfigured GPO from Microsoft and apply it to a group of target systems. The PDF version of the CIS benchmark for Windows Server 2022 can be downloaded from `https://www.cisecurity.org/cis-benchmarks/`.

The latest baselines from Microsoft Security and Compliance Toolkit can be downloaded from `https://www.microsoft.com/en-us/download/details.aspx?id=55319`.

Using Policy Analyzer, we can compare the preconfigured GPOs from Microsoft and the CIS build kits. Looking first at Microsoft Security Baseline for Windows Server 2022, we analyzed three different policy sets against the effective state of the system. The **Total** column contains the total number of policies analyzed. The **Differences** column shows values that haven't been configured by default in Windows Server or that conflict with the baseline. The **Score** column shows the total percentage of compliance with the recommendations. This analysis can be seen in the following table for the Member Server, Defender Antivirus, and Credential Guard policies:

Windows Security Baseline Server 2022	Total	Differences	Score
MSFT Windows Server 2022 – Member Server	140	95	32%
MSFT Windows Server 2022 – Defender Antivirus	24	24	0%
MSFT Windows Server 2022 - Credential Guard	6	6	0%

Figure 12.21 – Microsoft Security Baseline analysis

Next, we used the CIS Microsoft Windows Server 2022 Build Kit and compared the Level 1 and Level 2 benchmarks using Policy Analyzer. In the CIS build kits, the Level 2 reference GPO only contains additions to the Level 1 to avoid duplication. In the following table, we can see the comparison to the default Windows settings using the same method that was used previously:

CIS Microsoft Windows Server 2022 Benchmark	Total	Differences	Score
Level 1 – Member Server	366	269	27%
Level 2 – Member Server (extended L1)	79	75	0.5%

Figure 12.22 – CIS Windows Server 2022 Benchmark analysis

It's worth mentioning that the CIS build kit also includes an additional GPO for Windows services. It includes a recommendation to disable the **Print Spooler** service in response to CVE-2021-34527, also known as the **PrintNightmare** vulnerability.

CIS build kits are included as part of a SecureSuite membership. More information about them can be found at `https://www.cisecurity.org/cis-securesuite/cis-securesuite-build-kit-content/build-kits-faq`.

Now that the analysis is complete, we can implement the Microsoft recommendations using the preconfigured Group Policy. Then, we can run an analysis to compare them against the CIS benchmarks. To import the Member Server baseline, follow these steps:

1. Open **Group Policy Management Editor**. Create a GPO that is linked to an OU that contains member servers. Give it a friendly name, such as `MSFT Windows Server 2022 - Member Server`.

2. Click on the **Group Policy Objects** folder, find the GPO, then right-click and choose **Import Settings**.

3. Click **Next** twice and choose **Browse**. Navigate to the GPO folder of the extracted ZIP file and click **OK**. Then, click **Next**.

> **Tip**
> The Windows Server 2022 Security Baseline can be downloaded from `https://www.microsoft.com/en-us/download/details.aspx?id=55319`.

4. Select the **MSFT Windows Server 2022 – Member Server** GPO and click **Next**. Click **Next** again. Click **Next** on the **Migrating References** page to copy them identically from the source. Then, click **Finish**.

5. Click **OK** to finish importing the policy settings.

Once the GPO has been imported, we must force a policy update and restart a target server before running the analysis. The following table shows the comparison of Microsoft Security Baseline to the CIS benchmarks:

MSFT Windows Server 2022 – Member Server	Total	Differences	Score
CIS Level 1 – Member Server	366	179	51%
CIS Level 2 – Member Server (extended L1)	79	74	0.6%

Figure 12.23 – Microsoft Security Baseline compared to CIS

This is not inferring that Microsoft Security Baseline doesn't offer great protection. We wanted to demonstrate the differences between the two baselines by running this analysis.

Now that a baseline has been applied, it's important to monitor settings to watch for configuration drift and report on compliance when baseline recommendations are updated. This can be done using tools such as Configuration Manager or by deploying assessment tools like CIS CAT Pro Dashboard or other vulnerability scanning solutions like Nessus Tenable, that support auditing policy configurations. Azure Log Analytics also has a solution called **Change Tracking and Inventory** that can monitor changes to software, Windows services, and registry entries. In *Chapter 8*, *Keeping Your Windows Client Secure*, we covered importing a GPO to a Configuration Baseline in the *Converting a GPO into a Configuration Baseline* section. Following this guide, you can deploy a baseline using Configuration Manager in monitor mode and report on non-compliance against your GPO settings.

In addition to Group Policy enforcement, Azure Automation State Configuration can be used as a configuration management solution to monitor and remediate any configuration drifts. Azure Automation State Configuration is built on **PowerShell Desired State Configuration** (**DSC**) technology and is useful for maintaining consistent base configurations when deploying new resources. It also helps ensure systems remain in the state outlined in the configuration file. For additional information about Azure Automation State Configuration, visit `https://docs.microsoft.com/en-us/azure/automation/automation-dsc-overview`.

Next, let's look at general best practices for hardening Windows Server.

Hardening tips for Windows Server

Security baselines are a fundamental step in your overall hardening efforts. In addition to well-defined controls, any opportunity to reduce unnecessary roles, services, and unintentional privileges will greatly help reduce the overall attack surface. At the network level, carefully monitor communications (both ingress and egress) and only allow the required ports and protocols to support application functionality.

Let's look at a few additional general security recommendations for Windows Server and domain controllers:

- Domain controllers should be updated with priority and run the latest Windows versions if possible. They hold the keys to authentication and authorization in Active Directory environments.

- DCs should only run **Active Directory Domain Services** (**AD DS**) and DNS server roles.

- Use a Server Core installation for DCs and purchase hardware that supports Secure-Core if you're hosting physical servers on-premises.

- Block the DC from connecting directly to the internet. Only allow exclusions for WSUS or Windows Update services if possible.

- Admins should only interact with DCs from known workstations, jump servers, or Bastion environments. This includes remoting tools such as PowerShell, Windows Admin Center, and MMCs.

In addition to the recommendations for DCs, the following tips also apply to member servers:

- Enable the Credential Guard, Device Guard, and Code Integrity policies for application control.

- Only allow signed scripts to be executed.

- Restrict hosts that are allowed to perform administrative actions. Use safelists by IPs and only allow access from a PAW or other trusted device.

- Use PowerShell Just Enough Administration for remote administration with PowerShell. This will only allow access to certain functions within PowerShell and help to remove any unintentional privileges.

- For remote management over WinRM, enable WinRM over HTTPS to encrypt data in transit using certificates.

- Avoid creating privileged accounts that have permission to perform multiple administrative functions when possible.

Next, let's review how to configure account control settings on Windows Server. Many of these best practice recommendations will already be applied if you imported one of the security baselines from CIS or Microsoft.

Account controls for Windows Server

Account controls in Windows Server include settings that help define policies such as configuring password requirements, account lockout thresholds, and security options around local accounts on the system. First, let's look at some of the controls for configuring local accounts.

Configuring accounts

Implementing policies around the types of accounts that have access to servers is part of any good security and access strategy. In domain environments, administrative access should be controlled using domain-based Kerberos authentication and not with the use of local accounts. Managing local accounts on systems and scheduling password changes can be challenging and may require third-party tools to increase management efficiency. Controlling the identity is a key component of the overall security framework, and we want to ensure that local accounts are locked down and protected.

Using Group Policy, go to **Computer Configuration | Windows Settings | Security Settings | Local Policies | Security Options** to find account policies. The following table includes some recommended policies you should consider enabling for local accounts:

Policy Title	Recommendation
Accounts: Administrator account status	**Disabled**. Be careful if disabling the account. If the member server loses connectivity to the domain, you may need to boot into safe mode to recover the server.
Accounts: Block Microsoft Accounts	**Users can't add or log on with Microsoft Accounts**.
Accounts: Guest account status	**Disabled**
Accounts: Limit local account use of blank passwords to console logon only	**Enabled**
Accounts: Rename administrator account	**Choose another name and change description**. This does not change the SID of the account. If a local account is required, it is recommended to re-create it.
Accounts: Rename guest account	**Choose another name and change description**. The Guest account should be disabled by default.

Figure 12.24 – Accounts policy settings

Next, let's look at configuring account policies specifically in domain environments.

Configuring account policies

Setting account policies for domain accounts includes setting password requirements and lockout settings. They are commonly applied to all accounts at the domain level using the **Default Domain Policy** GPO. If any account policies are created in a child OU, they will only apply to local accounts on the member servers in scope unless you're using a **Password Settings Object** (**PSO**) from Active Directory Administrative Center. The Default Domain Policy is configured with predefined settings once Active Directory Domain Services is enabled and can be found nested directly under the root of the domain. Let's look at modifying some of the default values.

To find the policy settings in GPMC, go to **Computer Configuration | Windows Settings | Security Settings | Password Policy**. The following screenshot shows the best practice recommendations from the predefined Default Domain Policy:

Policy	Policy Setting
Enforce password history	24 passwords remembered
Maximum password age	365 days
Minimum password age	1 days
Minimum password length	14 characters
Minimum password length audit	Not Defined
Password must meet complexity requirements	Enabled
Relax minimum password length limits	Enabled
Store passwords using reversible encryption	Disabled

Figure 12.25 – Default Domain Policy – Password Policy

In the preceding screenshot, we can see the following:

- **Maximum password age** is set to 365 days.

- **Minimum password length** is set to 14 characters.

- **Relax minimum password length limits** is set to Enabled. This policy was introduced in Windows 10 and Server 2022 and allows the minimum password length to be increased beyond the limit of 14.

> **Tip**
>
> Based on recent research from NIST, Microsoft, and others, the latest password recommendations are to remove periodic password change requirements and to only change a password in case of potential threat or compromise.

Next, let's look at recommendations for **Account Lockout Policy**. To find the policy settings in GPMC, go to **Computer Configuration | Windows Settings | Security Settings | Account Lockout Policy**. The policy settings can be seen in the following screenshot:

Policy	Policy Setting
Account lockout duration	15 minutes
Account lockout threshold	5 invalid logon attempts
Reset account lockout counter after	15 minutes

Figure 12.26 – Account Lockout Policy

In the preceding screenshot, we can see that the following policies have been configured:

- **Account lockout duration** is set to **15 minutes**.

  Leaving this setting as 0 will require an administrator to manually unlock an account every time it's in a locked state. This can lead to increased calls to the support desk. Setting a duration can also help mitigate a potential **denial of service (DoS)** attack.

- **Account lockout threshold** is set to **5 invalid logon attempts**.

  Setting a lockout threshold helps to prevent brute-force attacks by triggering an account lockout. The account will remain locked out for the duration set in the first policy. Be mindful that if all accounts are subjected to an account lockout threshold, an attacker could effectively lock out all the accounts in a DoS-type attack. Use fine-grained password policies to change these settings on a subset of privileged accounts if needed.

- **Reset account lockout counter after** is set to **15 minutes**.

  This setting determines how long before the lockout threshold resets. Keep this setting aligned with the lockout duration. The default settings could be vulnerable to DoS attacks and can lock out users in your environment permanently until they're manually unlocked by an administrator.

For more granular control over account policies, configure a Default Domain Policy and supplement it with **fine-grained password policies** or **Password Settings Objects (PSOs)**.

Implementing fine-grained password policies

Fine-grained password policies and PSOs are configured separately from Group Policy account policies. Some use cases include setting stronger password requirements for administrator and service accounts or changing the account lockout thresholds for DoS mitigations. Fine-grained password policies are configured in Active Directory Administrative Center and can be applied directly to specific users or security groups. To learn how to implement a fine-grained password policy, go to *Chapter 5, Identity and Access Management*, and find the *Securing your passwords* section.

As a recap, we can configure a PSO by opening **Active Directory Administrative Center** and changing to the **Tree** view. Expand the domain, go to **System container**, and find **Password Settings Container**. Click **New**, then select **Password Settings** under **Tasks** to create a new policy. The following screenshot shows a password policy targeted at the **Domain Admin** security group:

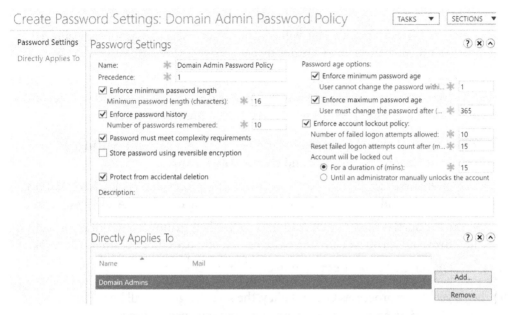

Figure 12.27 – Fine-grained password policy in Active Directory

Next, we'll look at controlling user rights assignment in Windows Server.

Controlling user rights assignment

User Rights Assignment policies define the ways a user or service can log on and interact with a server locally. Examples of rights assignments include who can log on locally or through Remote Desktop Services, who can join workstations to the domain, and blocking access to run batch jobs or services locally on the system. User Rights Assignment policies can be found under **Computer Configuration** > **Windows Settings** > **Security Settings** > **Local Policies** > **User Rights Assignment**. The CIS benchmark includes 48 combined recommendations across Level 1 and Level 2 for configuring User Rights Assignment policies in Windows Server 2022. A few examples of additional recommendations from CIS to complement the Microsoft Security Baseline can be seen in the following table:

Policy	Policy Setting
Adjust memory quotas for a process	Administrators, LOCAL SERVICE, NETWORK SERVICE
Allow log on through Remote Desktop Services	Administrators, Remote Desktop Users
Change the system time	Administrators, LOCAL SERVICE
Change the time zone	Administrators, LOCAL SERVICE
Deny log on as a batch job	Guests
Deny log on as a service	Guests
Deny log on locally	Guests
Deny log on through Remote Desktop Services	Guests, Local Account
Shut down the system	Administrators

Figure 12.28 – Recommended User Rights Assignment settings

We can further control access to built-in local groups by governing the memberships using a Restricted Groups policy. This will control which users and groups are allowed to be members of the built-in security groups on each server.

Let's look at creating a Restricted Groups policy using GPO. In this example, we will add a security group named T1-Operators to the local **Administrators** group on a server by modifying our existing **MSFT Windows Server 2022 – Member Server** GPO:

1. Go to **Computer Configuration | Windows Settings | Security Settings | Restricted Groups**. Right-click and choose **Add Group**.

2. Click on **Browse**, search for the T1-Operators group, and click **Check Names**. Once you've found the group, click **OK**. Click **OK** again on the **Add Group** menu.

3. On the **Properties** menu, click **Add** next to the **This group is a member of** section. Click **Browse**, type in Administrators, and choose **Check Names**. Click **Ok**.

4. Click **OK** on the **Properties** menu to commit the changes.

After the target server processes Group Policy, the security group will be in the BUILTIN\Administrators group. The following screenshot shows the security group listed in the Restricted Groups policy in the Group Policy Management MMC:

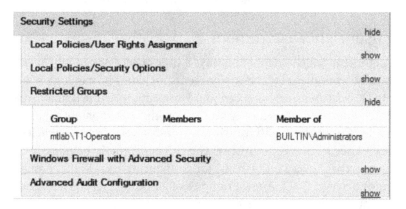

Figure 12.29 – Restricted Groups policy

> **Tip**
> If a user is added locally to a built-in group on the server and they are not a member of the domain group or added through the Restricted Groups policy, they will be removed on the next policy sync.

Next, let's look at User Account Control settings.

User Account Control

User Account Control (UAC) helps to control the context in which an application can run. Its protection dictates that applications will be opened in standard user context unless otherwise specified or invoked with the aid of an elevated parent process. This helps prevent unauthorized apps from automatically installing and prevents users from making changes to protected system settings. It also keeps permission continuity if a parent process running as a standard user invokes a child process. The child process will need to be explicitly elevated by admin consent. The familiar UAC experience can be seen in the following screenshot, which shows the consent prompt that appears when an application requires an administrator's access token:

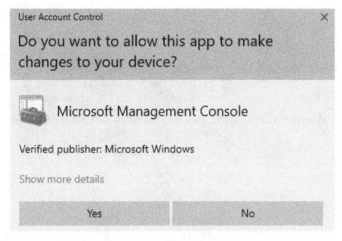

Figure 12.30 – UAC in Windows Server 2022

By design, when an administrator logs into a system, both a standard and an administrator access token are granted to the user object. When the administrator runs an application, it will attempt to use the standard user access token until a higher integrity level is required and will prompt for consent for permissions.

To protect the UAC consent process flow, it is recommended to secure UAC prompts using the secure desktop. This is the default Windows behavior and can be noticed by the dimming of the desktop during the elevation attempt. The secure desktop helps ensure that only Windows processes can interact with the consent flow and prevents the risk of an attacker stealing credentials by using a fraudulent UAC prompt. The following table lists a few recommendations for UAC policies:

Policy Title	Recommendation
Admin Approval Mode for the Built-in Administrator account	**Enabled**. This will ensure UAC will prompt for consent whenever privileges are escalated.
Behavior of the elevation prompt for administrators in Admin Approval Mode	**Prompt for consent on the secure desktop**. Choosing prompt for credentials could allow a malicious app fake a UAC prompt and steal credentials.
Behavior of the elevation prompt for standard users	**Automatically deny elevation requests**. This will block standard users from being prompted and require they choose "Run as Administrator." The block message is configurable to avoid calls to the service desk.
Detect application installations and prompt for elevation	**Enabled**. This will ensure applications will prompt for credentials to continue.
Run all administrators in Admin Approval Mode	**Enabled**. This is the default behavior.

Figure 12.31 – User Account Control policies

The location of User Account Control policies in Group Policy can be found by going to **Computer Configuration | Windows Settings | Security Settings | Local Policies | Security Options | User Account Control**. For detailed information about UAC security policy settings, visit `https://docs.microsoft.com/en-us/windows/security/identity-protection/user-account-control/user-account-control-security-policy-settings`.

Next, let's review protecting domain accounts by using the **Protected Users** security group in Active Directory.

Protected Users

The Protected Users security group helps safeguard domain accounts by reducing the exposure of credential caching that occurs in an Active Directory environment. It also limits the authentication and encryption types that are accepted for these accounts. Once a user object has been placed into this group, the protections that apply are non-configurable and if any authentication issues occur, the user will need to be removed from the group. Due to some limitations with protected users, it is not recommended to use this group for user-based SPNs or computer accounts as it could result in authentication errors. Let's review some of the protections that are available to members:

- User credentials will not be cached in plain text when using Windows Digest, NTLM, Kerberos, or CredSSP.

- Offline sign-in will not work.

- NTLM authentication does not work when authenticating with a Domain Controller.

- Kerberos will no longer use DES or RC4 encryption.

- Kerberos **Ticket Granting Tickets (TGTs)** cannot be renewed after their initial 4-hour lifetime.

In domain environments, it is recommended to deprecate NTLM authentication as soon as possible in favor of Kerberos. Using the Protected Users group is a good way to start testing applications for Kerberos support. To get the full protection benefits, clients and hosts must run, at a minimum, Windows 8.1 and Windows Server 2012 R2 or later. For additional information about configuring protected users, visit `https://docs.microsoft.com/en-us/windows-server/identity/ad-ds/manage/how-to-configure-protected-accounts`.

Next, let's look at ways to secure the logon and authentication process.

Securing the logon and authentication process

To log in and authenticate with a server, most administrators use remote administration tools such as Remote Desktop Services, PowerShell, or Microsoft Management consoles. We've mentioned the importance of limiting the locations administrators can log in from, but there are additional controls that can be enabled further to protect the logon process. Some of these controls include configuring interactive logon settings, configuring the allowed authentication protocols, and using authentication policies and silos. Each control will add another boundary to your layer of defenses and help reduce the attack surface. First, let's look at configuring interactive logon authentication settings.

Configuring interactive login

Interactive login is when a user logs into a server either physically or remotely with terminal services. In an interactive logon, users typically enter a password or supply credentials through a PIN, smart card, or biometrics. Through a remote interactive logon, the user connects using Windows **Remote Desktop Services (RDS)**, also known as **Terminal Services**, using **Remote Desktop Client (RDC)** software. The RDC client that's built into Windows is often referred to as **MSTSC**, which references the name of the application located in `%windir%\System32`. Interactive logon policies help secure the logon process for these logon types. CIS includes nine recommendations in the L1 profile, and they can be configured in Group Policy under **Computer Configuration | Windows Settings | Security Settings | Local Policies | Security Options**. Let's review some of the recommendations:

Policy Title	Recommendation
Do not require CTRL+ALT+DEL	**Disabled.** Helps to protect against malicious programs that attempt to fake a login screen to steal credentials.
Don't display last signed-in	**Enabled.** Hiding the last logged on user will prevent an attacker from identifying a potential target for brute force or password cracking.
Machine inactivity limit	**900 seconds or fewer.** This will protect a user who forgets to lock their session based on the timeout value.
Message title/text for users attempting to log in	**Enabled.** This will act as a deterrent for attackers and raise awareness to users about potential consequences for misusing company assets.
Require Domain Controller Authentication to unlock workstation	**Enabled.** Protects against the risks of cached credentials by requiring domain authentication.
Smart card removal behavior	**Lock Workstation.** If a user leaves their workstation and removes the smart card, the session will automatically lock instead of remaining logged in.

Figure 12.32 – Interactive logon policies

For example, configuring the message title and text policy for users attempting to log in can help raise awareness about the potential consequences of misusing company assets. This messaging can help educate users to follow security best practices and ensure they lock or log out of their sessions. The following screenshot shows an example of what this looks like when connecting with RDS in Windows Server 2022:

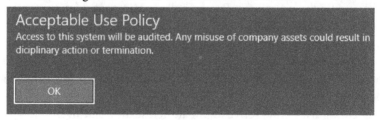

Figure 12.33 – Interactive logon warning

> **Tip**
>
> If you're configuring RDS, make sure to allow connections through Windows Firewall and enable RDS through Group Policy or locally.

Next, we'll look at configuring timeout limits for RDS. This will increase security by disconnecting idle user sessions and free up system resources.

Setting RDP session time limits

RDP session time limits can be set to limit the amount of time that idle users remain logged into a session. Some of the advantages include preventing password lockouts, protecting inactive sessions, and freeing up system resources. If a **Privileged Access Management** (**PAM**) solution is configured with password rotation, allowing idle sessions could result in increased account lockouts. You can also avoid configuring RDS licensing by limiting the number of concurrent sessions through automatic logout. RDP session time limits can be configured using Group Policy by going to **Computer Configuration | Policies | Administrative Templates | Windows Component | Remote Desktop Services | Remote Desktop Session Host | Session Time Limits**. The recommendations shown in the following table include inactive and idle time limits:

Policy Title	Suggestion
Set time limit for active but idle Remote Desktop Services sessions	**Enabled** **Idle session limit: 15 minutes**
Set time limit for disconnected sessions	**Enabled.** **End a disconnected session: 1 minute**

Figure 12.34 – Session time limits for Remote Desktop Protocol

To enforce automatic logoff, set the **End session when time limits are reached** policy to Enabled. Be mindful when configuring automatic logging off as users may complain if they run long batch processes using RDS.

Next, we'll look at configuring policies that protect information exposure on the logon screen.

Securing the logon screen

Securing the logon screen can help reduce the risk of exposing potential attack targets by setting policies that limit what's visible at Windows logon. Configuring these settings can stop the enumeration of users, hide account details, and limit interaction with network connections. CIS has 7 recommendations in the L1 profile regarding logon screen policies. The following table shows the additional recommendations not enforced in the Microsoft baseline.

Policy Title	Recommendation
Block user from showing account details on sign-in	**Enabled**. Hides the last logged on user to prevent an attacker from identifying a potential target.
Do not display network selection UI	**Enabled**. Helps prevent a user from disconnecting the PC from the network without logging into Windows.
Do not enumerate connected users on domain-joined computers	**Enabled**. Prevents enumeration of domain joined users.
Turn off app notifications on the lock screen	**Enabled**. App notifications could contain sensitive information that shouldn't be displayed on the lock screen.
Turn off picture password sign-in	**Enabled**. Picture passwords are not complex logon methods.
Turn on convenience PIN sign-in	**Disabled**. PINs are typically less complex than passwords.

Figure 12.35 – Recommended login processes

> **Tip**
> If picture password sign-in is enabled, the domain password will be cached in the system vault and subject to extraction.

Many of these settings are already disabled by default in Windows, so it's best to enforce them to ensure a user with administrative access does not modify the default logon settings.

Next, we will review authentication silos and how they can be used to restrict account access and the scope of what an account has rights to do on a privileged system.

Using authentication policies and silos

Authentication silos are containers inside AD DS that contain sensitive users, computers, and service account objects. The objects inside this container are assigned authentication policies that provide enhanced protection by creating conditions that must be met to allow access to a system. An authentication silo, like a security group, is another way to organize privileged accounts. Policies can be applied to silos or directly to the objects inside the silo. If the conditions of the policies are not satisfied during the authentication process, access is denied. Authentication silos are commonly used to complement the Protected Users group we mentioned earlier.

When an authentication policy is assigned to an object, a flag is assigned to the account to notify the domain controller that additional security checks are required during the authentication request to verify access. These policies contain restrictions that are placed on the account and include configuring a short TGT lifetime, restricting which devices can request a TGT, and access control conditions or claims that require the authentication request to come from a specific computer to receive a ticket. To support this functionality, both the client host and the domain controller must be able to support **Kerberos Armoring**. Kerberos Armoring or **Flexible Authentication Secure Tunneling (FAST)** provides a secure channel between the client and the **Key Distribution Center (KDC)** to protect pre-authentication data during a Kerberos ticket request. It also denies any downgrade or fallback attempts to use NTLM authentication as Kerberos becomes the only acceptable method.

To create and assign users to authentication policies and silos, you can use Active Directory Administrative Center. To support the use of user claims with **Dynamic Access Control (DAC)**, enable the **Kerberos client support for claims, compound authentication, and Kerberos armoring** policy located in **Computer Configuration | Administrative Templates | System | Kerberos**.

An authentication policy must exist before a silo can be created. To do this, open ADAC, click on **Authentication Policies**, and choose **New** to create a new policy. In the following screenshot, you can see the **User Sign On** properties specifying a TGT lifetime and an access control condition so that the TGT request can only come from the specified computer account:

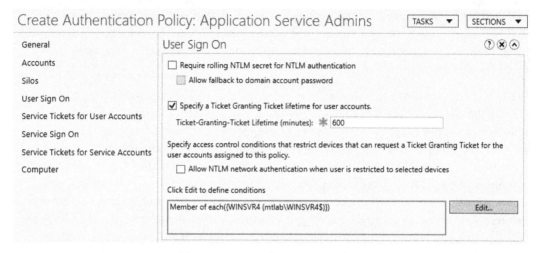

Figure 12.36 – Authentication Policy

Once a policy has been created, a user can be assigned to a policy or silo directly in Active Directory Administrative Center, as shown in the following screenshot:

Figure 12.37 – Assigning a policy to a user in ADAC

We recommend reviewing the Microsoft docs in detail to understand all the pre-requisites that are required to use authentication policies and silos, including troubleshooting details. Additional information can be found at `https://docs.microsoft.com/en-us/windows-server/identity/ad-ds/manage/how-to-configure-protected-accounts#BKMK_CreateAuthNPolicies`.

Next, let's discuss disabling NTLM authentication in favor of Kerberos in a domain environment for a more secure authentication protocol.

Disabling NTLM authentication

NTLM authentication is a challenge and response protocol family that contains multiple security protocols that are used in the authentication process. These protocols include LAN Manager, NTLMv1, NTLMv2, and the NTLM2 session protocols. NTLM authentication is still supported today and is enabled by default in Windows Server. It is recommended to reduce or eliminate the use of NTLM authentication in favor of Kerberos wherever possible as it is vulnerable to MiTM and pass-the-hash types of attacks. NTLM hashed passwords can also be easily cracked if password policies do not enforce lengthy and complex password requirements. In *Chapter 10, Mitigating Common Attack Vectors*, we demonstrated how an attacker can use an Adversary-in-the-Middle attack to capture NTLM hashes and use unsecure SMB and LDAP protocols in an NTLM relay attack to authenticate with other systems.

To deprecate NTLM usage in your environment, you must do the following:

1. Understand how the NTLM protocols work in Windows environments and what systems are using them.
2. Assess their usage by enabling auditing and review logs.
3. Restrict their use through Group Policy.

We covered how to do this in the *Enabling Kerberos authentication* section of *Chapter 10, Mitigating Common Attack Vectors*, in detail, but we recommend reviewing the official documentation from Microsoft to fully understand the effects of disabling NTLM. You can find detailed information at https://docs.microsoft.com/en-us/previous-versions/windows/it-pro/windows-server-2008-R2-and-2008/dd560653(v=ws.10).

If you need to add exceptions and allow NTLM, the following policies can act as a safe list to allow servers to be exempt from enforcement until Kerberos can be supported:

- **Network Security: Restrict NTLM: Add server exceptions for NTLM authentication in this domain** and add any servers for which an exception must be listed.

- **Network Security: Restrict NTLM: NTLM authentication in this domain** must be set to **Deny for domain servers** or **Deny all**. Setting the policy to **Deny for domain servers** will allow those listed in the exemption list.

The Protected Users group is a great way to begin testing application support for Kerberos authentication, as it will be enforced for all the members in the group and deny any downgrade attempts to NTLM.

Next, let's review how to harden the SMB protocol by enabling signing requirements.

Configuring SMB signing requirements

Server Message Block (SMB) is a common protocol that's used for transferring files and facilitating **Microsoft Remote Procedure Calls (MSRPC)** in Windows. It is most notably used for file-sharing capabilities on workstations and servers. Today, there are three versions of SMB known as SMB1, SMB2, and SMB3. In a modern environment, SMB1 should be disabled due to high-security risks and because it is missing many of the protections that were added by the later versions. Without the proper security controls in place, clients attempting a connection over SMB will negotiate any way possible and can be tricked into downgrading the version, thus becoming susceptible to attacks. Additionally, without requiring SMB signing, newer versions of SMB are also vulnerable to MiTM-type attacks. In *Chapter 10, Mitigating Common Attack Vectors*, we covered how an attacker can scan a domain to identify targets that don't require signing and still have SMB1 enabled. We also reviewed how to enable SMB signing in the *Requiring SMB signing* section. We recommended these steps to help secure SMB connections:

- Enforce SMB signing requirements for clients and servers for all three versions of SMB.

- Disable SMBv1 clients whenever possible.

Starting with Windows Server 2019, SMB1 (both client and server) is disabled by default. To check the status of SMB1 on a Windows Server instance, run `Get-WindowsOptionalFeature -Online -FeatureName SMB1Protocol` using an administrative PowerShell window. In the following screenshot, you can see the SMB1's **State** is `Enabled` on a Windows Server 2016 domain controller:

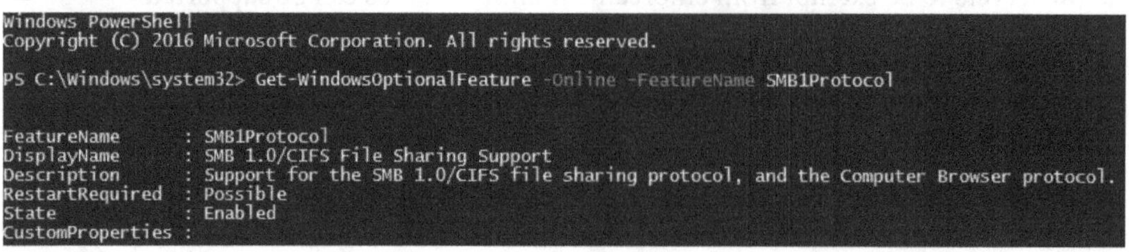

```
Administrator: Windows PowerShell

Windows PowerShell
Copyright (C) 2016 Microsoft Corporation. All rights reserved.

PS C:\Windows\system32> Get-WindowsOptionalFeature -Online -FeatureName SMB1Protocol

FeatureName      : SMB1Protocol
DisplayName      : SMB 1.0/CIFS File Sharing Support
Description      : Support for the SMB 1.0/CIFS file sharing protocol, and the Computer Browser protocol.
RestartRequired  : Possible
State            : Enabled
CustomProperties :
```

Figure 12.38 – SMB1's State

To disable SMB1 on a server, three registry settings need to be set to disable the service and remove SMB1 from the dependency chain of other SMB services:

1. To do this, deploy a Group Policy that configures the following registry key:

   ```
   HKLM\SYSTEM\CurrentControlSet\Services\LanmanServer\
   Parameters
   ```

   The following entry should be configured:

   - **Entry:** SMB1
   - **REG_DWORD:** 0

2. Next, disable the SMB1 service by setting the following registry keys:

   ```
   HKLM\SYSTEM\CurrentControlSet\Services\mrxsmb10
   ```

   The following entry should be set:

   - **Entry:** Start
   - **REG_DWORD:** 4

3. Finally, remove SMB1 from the dependency chain of other SMB services by modifying the following registry key:

   ```
   HKLM\SYSTEM\CurrentControlSet\Services\LanmanWorkstation
   ```

   The following entry should be modified:

   - **Entry:** DependOnService
   - **REG_MULTI_SZ:** "Browser","MRxSMB20","NSI"

Failure to update the dependency chain will break SMB once the service is stopped. To query the enabled protocols for the client, open an administrative Command Prompt and run `sc.exe qc lanmanworkstation`. In the following screenshot, the `MRxSMB10` service is not listed as a dependency on Window Server 2022:

```
Administrator: Command Prompt

Microsoft Windows [Version 10.0.20348.524]
(c) Microsoft Corporation. All rights reserved.

C:\Windows\system32>sc.exe qc lanmanworkstation
[SC] QueryServiceConfig SUCCESS

SERVICE_NAME: lanmanworkstation
        TYPE               : 20  WIN32_SHARE_PROCESS
        START_TYPE         : 2   AUTO_START
        ERROR_CONTROL      : 1   NORMAL
        BINARY_PATH_NAME   : C:\Windows\System32\svchost.exe -k NetworkService -p
        LOAD_ORDER_GROUP   : NetworkProvider
        TAG                : 0
        DISPLAY_NAME       : Workstation
        DEPENDENCIES       : Bowser
                           : MRxSmb20
                           : NSI
        SERVICE_START_NAME : NT AUTHORITY\NetworkService
```

Figure 12.39 – Status of the SMB client on Windows Sever

For detailed information about detecting and disabling SMB, go to `http://docs.microsoft.com/en-us/windows-server/storage/file-server/troubleshoot/detect-enable-and-disable-smbv1-v2-v3`.

Next, let's review securing LDAP by enabling LDAP signing and channel binding.

Configuring LDAP signing and channel binding

Lightweight Directory Access Protocol (LDAP) is an application protocol that can be used to query directory services such as Active Directory. This includes looking up data such as usernames, passwords, and computer objects. LDAP is also used in authentication and to perform administrative actions such as resetting passwords and modifying security groups, which makes it a flexible and powerful protocol. To help secure LDAP, we can enable LDAP signing and channel binding to secure the default configuration in Active Directory and reduce the risk of MiTM-style attacks. In *Chapter 10, Mitigating Common Attack Vectors*, we discussed using LDAPS and how to enforce security controls using Group Policy. It's important to understand the potential effect of enabling these controls for applications that do not support signing and channel binding. To harden these configurations, Microsoft recommends performing the following steps, as outlined in the security advisory ADV190023:

1. Install the latest available Windows security updates to enable logging and policies for LDAP signing and channel binding events.

2. Set the LDAP events with diagnostic logging to capture LDAP signing and channel binding failure events.

3. Identify any sources of failures and investigate them. Plan for remediation activities.

4. Enable LDAP signing and channel binding on supported devices.

You can review how to enforce these policies in the *Enabling LDAP signing and LDAP channel binding* section of *Chapter 10, Mitigating Common Attack Vectors*. We encourage you to read the advisory and detailed enablement steps by Microsoft at `https://support.microsoft.com/en-us/topic/2020-ldap-channel-binding-and-ldap-signing-requirements-for-windows-kb4520412-ef185fb8-00f7-167d-744c-f299a66fc00a`.

Next, we'll look at adding another security boundary by using a domain isolation policy design.

Using domain isolation policies

The concept of domain isolation is to build barriers between trusted and non-trusted devices both inside and outside of your organization and its network. One of the core design principles ensures that each network connection is authenticated using either an IPsec or certificate-based connection unless it has been explicitly exempted. This design helps protect systems from unsolicited inbound network traffic and non-managed, untrusted devices. To implement a domain isolation policy, the environment relies on Active Directory, Group Policy, and connection security rules in Windows Firewall. Source systems are allowed to securely authenticate from outside of the domain when you plan for a certificate-based isolation design in the domain isolation policy. A properly configured domain isolation policy looks to accomplish the following:

- Protect devices in domain isolation from untrusted network traffic.

- Restrict access only to devices you trust.

- Add additional layers of encryption when accessing sensitive systems.

There are a few design considerations when planning the architecture, including overall network perimeter design, domain isolation zones, boundary zones, encryption zones, and any exemptions to connection security rules. In the domain isolation zone, Windows devices joined to Active Directory rely on Kerberos authentication to satisfy the security requirements. They typically do not allow inbound connections from untrusted sources.

To create isolation inside the domain, connection security rules are used to define the boundaries of communications inside of Active Directory, instead of trusting the entire domain. In the boundary zone, devices are trusted but configured to accept inbound connections from both trusted and untrusted sources. In the encryption zone, trusted Windows and non-domain-joined systems are configured to communicate with each other using IPsec cryptographic certificates created by an internal certificate authority or a third-party provider. For enhanced protection on domain-joined systems, connection security rules can then be configured to accept both Kerberos and certificate-based connections from within the encryption zone.

In terms of **Advanced Authentication Methods**, Windows Defender Firewall with Advanced Security is configured to use Kerberos first, followed by a certificate issued from the internal certificate authority, as shown here:

Figure 12.40 – Connection methods in Windows Defender Firewall

An example of how a domain isolation policy can be architected is shown in the following diagram:

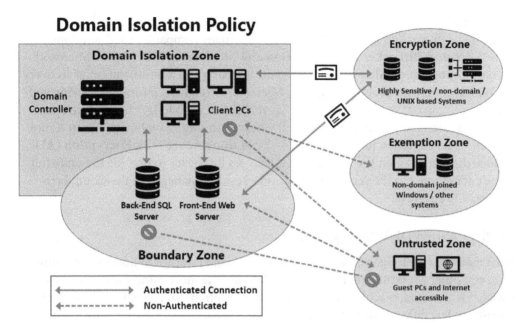

Figure 12.41 – Domain isolation policy architecture

By default, network traffic between zones is not encrypted, making it susceptible to eavesdropping. Enabling IPSec connection security rules will ensure that all inbound and outbound traffic is encrypted.

For additional information on planning and designing domain isolation policies, visit `https://docs.microsoft.com/en-us/windows/security/threat-protection/windows-firewall/domain-isolation-policy-design`.

Next, let's look at BitLocker encryption on servers and review enabling Azure Disk Encryption for virtual machines running in Azure.

Enabling Disk Encryption to prevent data theft

Enabling Azure Disk Encryption on Windows clients and servers can help protect against data being stolen from lost or stolen disk drives. BitLocker is the primary full volume data encryption technology that is included in Windows and can be enabled on Windows Server 2008 and later. The encryption uses the AES encryption algorithm and data can be recovered using a generated recovery key or by placing the drive back into the system with the original components that were used to initially encrypt the drive. In Windows Server, BitLocker can be enabled by adding the feature using Server Manager. After the installation is complete, the server will be enabled to encrypt its drives and includes BitLocker remote administration tools.

Using Group Policy, BitLocker can be deployed to systems and configured to store the recovery key information in AD DS. Until recently, many companies used solutions such as **Microsoft BitLocker Administration and Monitoring (MBAM)** or Microsoft Endpoint Configuration Manager to handle the indexing and administration of recovery keys. With the increase in cloud adoption, Microsoft strongly recommends moving your BitLocker administration workloads into Azure. MBAM is no longer being developed but includes extended support until 2024. If you are running virtual machines in Azure, Microsoft offers an option to encrypt server drives using **Azure Disk Encryption (ADE)** to reduce the workload on the IaaS server. ADE uses BitLocker to provide the same full volume encryption, but the service and recovery keys are managed in the cloud. Let's review how to enable it.

Enabling Azure Disk Encryption

Virtual machines running in Azure need to have attached storage to host the operating system and other application data. Managed disks are the most common storage type seen associated with virtual machine resources. Disk encryption can be enabled for managed disks using a few different methodologies:

- **Client-side encryption** is where all encryption and decryption operations are handled by services outside of the Azure resource provider, either by a customer-managed application or from inside a customer data center. Azure can be used to host the application and handle storage and encryption operations, but it does not have access to decrypt the data itself.

- **Server-side encryption** uses Azure resource providers to handle all the encryption operations, including key management and storage. Depending on the security and compliance requirements, you can bring an encryption key, known as a customer-managed key, and choose whether to store recovery keys in Azure or your repository elsewhere.

When it comes to a full end-to-end solution for all disks attached to a VM, ADE is the available solution from Microsoft to encrypt OS, data disks, and temporary volumes. With ADE, disk encryption keys are stored in Azure Key Vault with the option to provide a customer-managed key. For more information about ADE, visit the FAQ page at `https://docs.microsoft.com/en-us/azure/virtual-machines/windows/disk-encryption-faq`.

Let's learn how to enable ADE on managed disks attached to a virtual machine running Windows Server in Azure. To follow along with these steps, you will need a VM that's been deployed with an attached OS disk running Windows Server. You will also need permissions to create resources in a resource group, including an Azure Key Vault. In this example, we will use the Azure portal to achieve this in just a few steps:

1. Log in to the Azure portal at `https://portal.azure.com`.

2. Select a virtual machine to view the overview page and choose **Disks** on the left-hand side under **Settings**.

3. Click on **Additional Settings** on the top bar to view the encryption options.

4. Under **Encryption Settings**, click the **Disks to encrypt** dropdown and choose **OS and data disks**.

5. Under **Key Vault**, choose **Create new**.

6. Leave the resource group set with its default selection and enter a **Key vault name**. Click on **Next: Access policy**.

> **Tip**
>
> The Azure Key Vault that stores the encryption keys must exist in the same subscription and geographic region as the disk it's encrypting.

7. Select the **Azure Disk Encryption for volume encryption** checkbox.

8. Click **Review + create**. Review the details and click **Create**.

9. Back on the **Disk settings** page, leave the **Key** field blank and click **Save**.

10. After the settings update, the server will reboot and enable encryption. During the deployment, the `AzureDiskEncryption` extension will be installed on the server.

After the deployment finishes, we can log back into our server and validate that BitLocker is enabled. If we open **Server Manager** and go to **Add roles and features**, we will see that the **BitLocker Drive Encryption** feature is enabled. The OS disk will also have a padlock over it, indicating it's encrypted, or that encryption is in progress. Next, we can open the new Azure Key vault we created and click on **Secrets** under **Settings**. In the following screenshot, the **BitLocker Encryption Keys** (**BEKs**) have been uploaded for drive recovery from our encrypted disk:

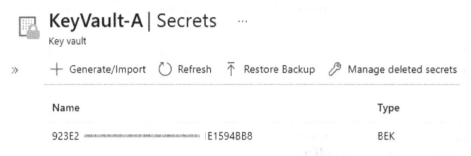

Figure 12.42 – Azure Key Vault – disk encryption

ADE can be enabled at scale using Azure CLI, PowerShell, or Azure Resource Manager templates instead of the Azure portal. If you're tracking vulnerabilities in Microsoft Defender for Cloud, ADE also satisfies the **Enable encryption at rest** control in the available recommendations.

Now that we have covered enabling ADE, let's review deploying a code integrity policy using WDAC. WDAC allows admins to create application controls that significantly reduce the attack surface by blocking untrusted processes.

Deploying application control policies using WDAC

WDAC policies allow you to configure zero-trust application control principles that determine which applications and drivers are allowed to run on Windows systems. A WDAC policy, previously referred to as a **Configurable Code Integrity** (**CCI**), contains policy rules and policy rule options that apply entirely to the system, irrelevant of the user who is logged in. This is important to note, as user-based targeting is currently not supported. In Windows Server 2022, WDAC can support up to 32 active policies at once. Policy rules can be defined based on the following attribute principles:

- Attributes of a code-signing certificate

- Attributes of app binaries that come from metadata, such as the original filename, version, or hash

- The app's reputation, as defined by **Microsoft Intelligent Security Graph (ISG)**
- The process that launched the app or the path where the process or file is located
- Managed packaged apps or universal Windows apps
- COM object configurability
- Application plugins, add-ins, and modules

To deploy a WDAC policy for Windows systems, you can use Group Policy, Configuration Manager, Intune, or PowerShell. It's strongly recommended to carefully design, plan, audit, and test before deploying WDAC in a production environment. An improper policy configuration could cause systems not to boot. During the initial design and testing phases, WDAC supports an audit-only rule option that enables event logging. These logs should be carefully analyzed to understand all the effects of the policy before enforcing it. The logs can be found in Event Viewer and are sent to Microsoft Defender for Endpoint if the server is onboarded.

> **Tip**
> WDAC policies can be highly complex. We strongly encourage you to read the Microsoft policy design guide and build a strategy that fits your organization's needs: `https://docs.microsoft.com/en-us/windows/security/threat-protection/windows-defender-application-control/windows-defender-application-control-design-guide`.

For companies with customized line-of-business applications, Microsoft recommends signing the application with a code signing certificate for the greatest compatibility. This helps ease how the policies are administrated. If code signing is not an option, for the most secure configuration, you can safelist the application using the hash value from the app's binary metadata. Just note that relying on the hash will require the WDAC policy to be updated every time application updates are installed as the hash value will change. To create a signing certificate, you can do the following:

- Use an internal digital certificate or **public key infrastructure (PKI)**.
- Use a third-party signing authority such as Entrust, DigiCert, or GlobalSign.
- Use the Windows Defender Device Guard signing portal through Microsoft Store for Business.

Sometimes, multiple policies will need to be created, depending on the system workload type. For example, in a managed device scenario such as an Intune-joined Windows client, you may have more leniency with the policy rule options by allowing apps with a known good reputation (enabling the **Intelligent Security Graph** option) or enabling the managed installer option. Windows servers may have fixed workloads, and the rules can be stricter as the level of variability regarding the apps and binaries could be a smaller sample set. The following link includes tips for the fixed-workload scenario: `https://docs.microsoft.com/en-us/windows/security/threat-protection/windows-defender-application-control/create-initial-default-policy`.

To help with policy creation, Microsoft has released a tool known as the WDAC Wizard that can be used to create a baseline and supplemental policies with a friendly user interface. A reference to the tool can be found at `https://webapp-wdac-wizard.azurewebsites.net/`

Let's look at creating a WDAC policy using PowerShell. To build the policy, we will use a reference system to act as the starting point. After the policy has been built and tested on the reference system, we can expand the deployment to other systems in audit mode for analysis and append new rules to accommodate additional apps and binaries. Once we are comfortable that the event logs are not showing errors that will interrupt expected application functionality, we can remove the audit mode rule option to ensure the policy is enforced.

First, let's learn how to build a reference policy from a clean Windows Server installation. Windows includes a directory that has several base templates to choose from. We will use the `DefaultWindows_audit.xml` template, which has the rule option for audit mode enabled, by following these steps:

1. In File Explorer, create a new directory on the root of the OS drive called **WDAC**.

2. Go to `C:\Windows\schemas\CodeIntegrity\ExamplePolicies`. Copy the `DefaultWindows_Audit.xml` policy and paste it into the `C:\WDAC` directory. Rename it something friendly, such as `WSHMemberSVR_CIPolicy.xml`.

3. Open **PowerShell ISE** as an administrator. First, let's define a variable for the policy file called `$PolicyFile`, as follows:

```
$PolicyFile = "C:\WDAC\WSHMemberSVR_CIPolicy.xml"
```

4. For version control, we need to assign a unique ID, name, and version using the `Set-CIPolicyIDInfo` and `Set-CIPolicyVersion` cmdlets. Enter the following in the script pane and execute the selection:

```
$PolicyFile = "C:\WDAC\WSHMemberSVR_CIPolicy.xml"
Set-CIPolicyIDInfo -FilePath $PolicyFile -PolicyName
"WSHMemberSVR_CIPolicy"
Set-CIPolicyVersion -FilePath $PolicyFIle -Version
"1.0.0"
```

Open the XML file using Notepad or another text editor. The version number you set previously will be under the `<VersionEx>` element. The policy name will be under the `<PolicyInfo>` element toward the bottom of the file. Toward the top of the XML is where the rules are defined.

Information about the WDAC policy rules options can be found at `https://docs.microsoft.com/en-us/windows/security/threat-protection/windows-defender-application-control/select-types-of-rules-to-create`.

Let's set a few additional rules to enable Microsoft Intelligent Security Graph and its managed installer options:

1. To configure the policy rule options, we will use the `Set-RuleOption` cmdlet and enable options `13` and `14`. We can add the following to the script pane and execute the selection:

```
Set-RuleOption -FilePath $PolicyFile -Option 13
Set-RuleOption -FilePath $PolicyFile -Option 14
```

`Option 13` is used to allow applications that have been installed by a managed installer. `Option 14` is used to allow apps with a known *good* reputation by attesting them against Microsoft Intelligent Security Graph. Additional information on the managed installer can be found at `https://docs.microsoft.com/en-us/windows/security/threat-protection/windows-defender-application-control/configure-authorized-apps-deployed-with-a-managed-installer`.

2. Next, convert the policy file into a binary file format using the `ConvertFrom-CIPolicy` cmdlet.

```
ConvertFrom-CIPolicy $PolicyFile "C:\WDAC\WSHMemberSVR_CIPolicy.bin"
```

To deploy the policy, the `.bin` file can be placed on a network share that servers can access. In this example, we are building a reference WDAC policy and are not ready for a deployment, so we will leave it in the exported path.

3. Open **Local Group Policy Editor**, go to **Administrative Templates | System | Device Guard**, and select the **Deploy Windows Defender Application Control** policy.

4. Select **Enabled** and enter the `C:\WDAC\WSHMemberSVR_CIPolicy.bin` path. Click **OK**. Then, restart the server.

To confirm that the policy has been applied, open **Event Viewer** and go to **Applications and Services | Microsoft | Windows | Device Guard | Operational**. Event ID `7010` will appear if the policy was successfully processed, as shown in the following screenshot. Restart the server for the policy to take effect:

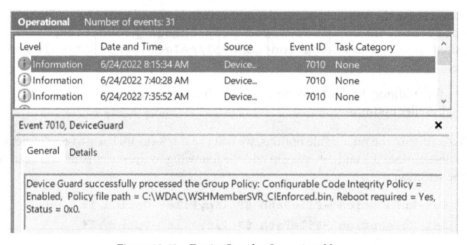

Figure 12.43 – DeviceGuard – Operational logs

To view the audit events that were generated by the WDAC policies, go to **CodeIntegrity | Operational** under **Applications and Services**. Here, we can review any processes that tried to execute and whether they violated the WDAC policy. In the following screenshot, the warning level shows that the file would have violated the policy:

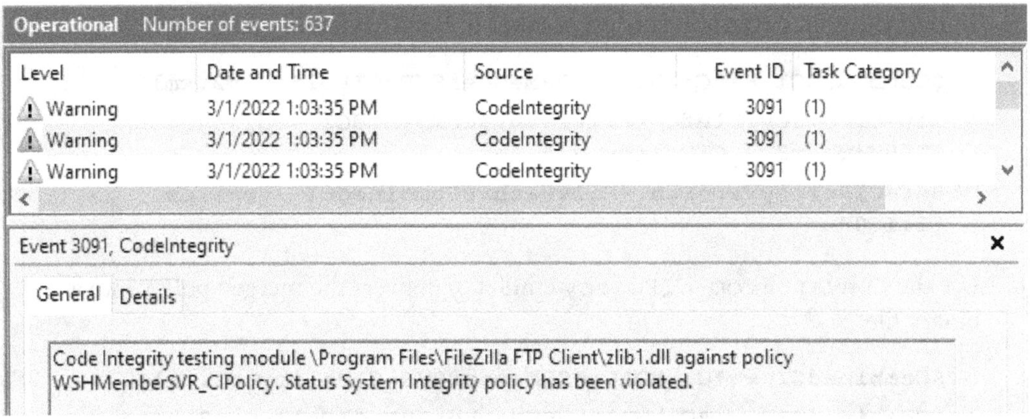

Figure 12.44 – CodeIntegrity – Operational logs

After enough time has passed for a good data sample to be captured, the event logs will be ready to be reviewed. Using PowerShell, we can export the audit entries and merge them with the template policy. To do this, we will use the New-CIPolicy cmdlet by following these steps:

1. Run the following command to export the audit logs, set the policy level, and create a new CI policy:

   ```
   New-CIPolicy -Audit -Level Publisher -Fallback Hash
   -FilePath "C:\WDAC\EventLog_Audit.xml" -UserPEs
   ```

   Using the Publisher level will allow all signed software from the application's publisher based on the certificate trust chain. Using a fallback of Hash will allow any unsigned applications that have been identified in the audit logs to be safely listed by their hash metadata. To review the list of file rule levels, go to https://docs.microsoft.com/en-us/windows/security/threat-protection/windows-defender-application-control/select-types-of-rules-to-create#windows-defender-application-control-file-rule-levels.

2. Next, we will set a few variables to specify the policy paths and run the Merge-CIPolicy cmdlet to combine both the original and audit log policies:

   ```
   $BasePolicy = "C:\WDAC\WSHMemberSVR_CIPolicy.XML"
   $AuditCI = "C:\WDAC\EventLog_Audit.xml"
   $CombinedCI = "C:\WDAC\WSHMemberSVR_CIPolicy_v2.xml"
   Merge-CIPolicy -PolicyPaths $BasePolicy,$AuditCI
   -OutputFilePath $CombinedCI
   ```

3. Update the policy rule and version number in the XML file:

    ```
    $CombinedCI = "C:\WDAC\WSHMemberSVR_CIPolicy_v2.xml"
    Set-CIPolicyIDInfo -FilePath $CombinedCI -PolicyName
    "WSHMemberSVR_CIPolicy_v2"
    Set-CIPolicyVersion -FilePath $CombinedCI -Version
    "1.1.0"
    ```

4. Use the `ConvertFrom-CIPolicy` cmdlet to convert the merged policy into a binary file:

    ```
    $CombinedCI = "C:\WDAC\WSHMemberSVR_CIPolicy_v2.xml"
    $CombinedBIN = "C:\WDAC\WSHMemberSVR_CIPolicy_v2.bin"
    ConvertFrom-CIPolicy $CombinedCI "C:\WDAC\WSHMemberSVR_
    CIPolicy_v2.bin"
    ```

5. If you wish to audit the new policy before enforcing it, update the local Group Policy setting to the new `.bin` file. Restart the server.

    Depending on how strict you need the policy to be, you can use the `FilePublisher` level with a fallback level of `Hash` and scan the entire OS drive. To do this, run the following cmdlet (assuming the OS disk is drive `C`):

    ```
    $OSPolicyCI = "C:\WDAC\OSDrive_CI.xml"
    New-CIPolicy -ScanPath "C:\" -Level FilePublisher
    -Fallback Hash -UserPEs -FilePath $OSPolicyCI
    ```

    > **Tip**
    >
    > Running the `New-CIPolicy` cmdlet on the OS drive to complete the scan will take some time.

If you decide to use `-ScanPath` to create a new CI policy, be sure to merge the policy rules and increment the version by following *Steps 2* to *4*.

Before enforcing the policy, it's recommended to set rule options 9, **Advanced Boot Options Menu**, and 10, **Boot Audit on Failure**. This way, if the WDAC policy blocks a driver during startup, Windows will continue to boot.

To do this, follow the steps below.

1. To enable the **Advanced Boot Options Menu** and **Boot Audit on Failure** rule options, run the following commands:

    ```
    $CIPolicyPath = "C:\WDAC\WSHMemberSVR_CiPolicy_v2.xml"
    Set-RuleOption -FilePath $CIPolicyPath -Option 9
    Set-RuleOption -FilePath $CIPolicyPath -Option 10
    ```

2. Let's make a copy of the policy for a backup and append `Enforced` to the end of the filename by running the following command. This will help us differentiate between the XML files:

    ```
    Copy $CIPolicyPath "C:\WDAC\WSHMemberSVR_CIEnforced.xml"
    ```

3. To set the policy to enforced mode, we will delete rule option 3, which enables auditing, by using the following commands:

    ```
    $EnforcedCI = "C:\WDAC\WSHMemberSVR_CIEnforced.xml"
    Set-RuleOption -FilePath $EnforcedCI -Option 3 -Delete
    ```

4. Next, let's update the ID information and policy version using the following commands:

    ```
    $EnforcedCI = "C:\WDAC\WSHMemberSVR_CIEnforced.xml"
    Set-CIPolicyIDInfo -FilePath $EnforcedCI -PolicyName
    "WSHMemberSVR_CI_Enforced"
    Set-CIPolicyVersion -FilePath $EnforcedCI -Version
    "1.2.0"
    ```

5. Convert the policy into binary format by running the following command:

    ```
    ConvertFrom-CIPolicy $EnforcedCI "C:\WDAC\
    WSHMemberServer_CIEnforced.bin"
    ```

The WDAC policy is now ready to be tested in enforced mode. After updating Group Policy and restarting the server, we can open the **DeviceGuard | Operational** log under **Applications and Services** to confirm it's enforced. In the following screenshot, you can see that Event ID 7010 will show that **Configurable Code Integrity policy = Enabled**:

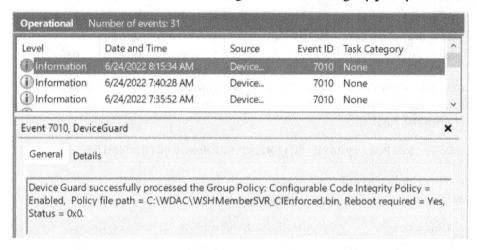

Figure 10.45 – Device Guard – Operational logs in Event Viewer

> **Tip**
> After time has passed and you're comfortable that applications are functioning correctly, rule options 9 and 10 can be removed to add additional protection.

Earlier, we enabled rule option 14 for **Intelligent Security Graph Authorization**. This option allows file hash and signing information to be sent to the Microsoft **Intelligent Security Graph (ISG)** for analysis. Even if the app or binary is not explicitly listed in your WDAC policy, if it passes the integrity check by ISG, it will be allowed to run. ISG is the same technology that's used for SmartScreen and the PUA protection features of Microsoft Defender. If WDAC policies are deployed using MDM or Configuration Manager, the necessary services will automatically be started. Otherwise, for local testing, you can run the `appidtel start` command in an administrative command prompt to start the necessary components. For additional information about confirming application reputation using the ISG, visit https://docs.microsoft.com/en-us/windows/security/threat-protection/windows-defender-application-control/use-windows-defender-application-control-with-intelligent-security-graph.

When attempting to launch an application that is not safely listed in WDAC, the users will be presented with a popup that looks similar to the following:

Figure 12.46 – Application blocked by WDAC

The **AppLocker** logs under **Applications and Services** can also show us whether applications and scripts have been blocked because of the WDAC policy. Event ID 8029, as shown in the following screenshot, shows a PowerShell script that was not signed and was blocked by the WDAC policy:

MSI and Script Number of events: 471				
Level	Date and Time	Source	Event ID	Task Categor
ⓘ Information	3/2/2022 6:05:57 AM	AppLocker	8038	None
⬤ Error	3/2/2022 6:05:57 AM	AppLocker	8029	None
ⓘ Information	3/2/2022 6:05:57 AM	AppLocker	8038	None
⬤ Error	3/2/2022 6:05:57 AM	AppLocker	8029	None

Event 8029, AppLocker   ✕

General Details

C:\WriteText.ps1 was prevented from running due to Config CI policy.

Figure 12.47 – AppLocker logs in Event Viewer

For more information regarding the full list of rules, system requirements, and tips for deployment planning, visit `https://docs.microsoft.com/en-us/windows/security/threat-protection/windows-defender-application-control/windows-defender-application-control`.

Next, let's look at how we can secure PowerShell by enabling logging, discussing language modes, and implementing just-enough administration.

Implementing PowerShell security

PowerShell has become an invaluable command-line interface in the administrator's toolbox, with uses ranging from executing remote management tasks to fully automating processes. It has deep integration with Windows and can be used to manipulate most aspects of the OS, including **Windows Management Instrumentation** (**WMI**) and other security and hardware-based features. As a result, PowerShell can be a viable attack tool. Due to its flexibility and general trust in Windows as a safe utility, PowerShell can be exploited as a **living-off-the-land binary** (**LOLBin**) and used for malicious intent, such as downloading payloads and executing code.

This can be a security concern because the integration with Windows allows for defense evasion and makes it difficult to alert users about suspicious commands. While it's not recommended to disable PowerShell completely, we can make a few setting changes to help secure its operations. The first step is to configure PowerShell logging to help aid in forensic analysis, should an incident occur.

Configuring PowerShell logging

PowerShell logging enables audit logs to be stored in Event Viewer for auditing purposes. This can be valuable for forensic investigations where PowerShell was used to help trace the activities that occurred during an incident. Some of the available logging options are as follows:

- **PowerShell Transcription** allows Windows to capture the input and output of PowerShell commands into text-based transcripts. PowerShell Transcription can be enabled through the following Group Policy or registry setting:

  ```
  Computer Configuration / Policies / Administrative
  Templates / Windows Components / Windows PowerShell /
  Turn on PowerShell Transcription
  ```

  - Enable the policy and specify an output directory. `Documents` is the default output directory if nothing is specified.

  - `HKEY_LOCAL_MACHINE\SOFTWARE\Policies\Microsoft\Windows\PowerShell\Transcription`:

    - **EnableTranscription DWORD** equals `1`

    - **OutputDirectory REG_SZ** equals `PATH TO DIRECTORY`

> **Tip**
>
> CIS recommends that PowerShell Transcription is set to `Disabled`. Enabling the policy risks passwords being stored in plain text in the transcript files, but you trade off on auditing PowerShell events.

- **PowerShell Script Block Logging** enables the logging of all PowerShell script blocks in Event Viewer. PowerShell Script Block Logging can be enabled through Group Policy or the registry:

  ```
  Computer Configuration / Policies / Administrative
  Templates / Windows Components / Windows PowerShell /
  Turn on PowerShell Script Block Logging
  ```

  - Enable the policy and select **Log script block invocation start/stop events** for additional logging.
  - `HKEY_LOCAL_MACHINE\SOFTWARE\Policies\Microsoft\Windows\PowerShell\ScriptBlockLogging`:

    - **EnableScriptBlockLogging REG_DWORD** equals 1
    - **EnableScriptBlockInvocationLogging REG_DWORD** equals 1

> **Tip**
>
> CIS recommends not enabling script block invocation logging as it can generate a high volume of event logs.

To view the PowerShell logs in Event Viewer, go to **Applications and Services Logs | Microsoft | Windows | PowerShell | Operational**. If Window Server has been configured to send telemetry data to Log Analytics, PowerShell logs can be collected by enabling the **Microsoft-Windows-PowerShell/Operational** logs under **Agents configuration | Windows event logs** from the Log Analytics workspace. This will allow you to configure alerts based on the events or forward them to a **Security Information and Event Management (SIEM)** solution such as Azure Sentinel for monitoring. Defender for Endpoint also does a great job of alerting users about malicious PowerShell commands that have been executed on the system if the server is onboarded with real-time protection enabled.

> **Tip**
>
> Consider expanding the Windows Event Log size if you're enabling script block logging to avoid logs being overwritten too quickly.

Next, let's explore how to use PowerShell constrained language mode to restrict the available commands.

Enabling PowerShell constrained language mode

PowerShell constrained language mode is used to restrict access to commands that can invoke Windows APIs. Some of the default restrictions of constrained language mode include creating instances of most COM objects, using Add-Type to load arbitrary C# code or Win32 APIs, and some .NET methods. To view the list of allowed types, along with a detailed definition of constrained language mode, open PowerShell and run `Get-Help about_ConstrainedLanguage`.

PowerShell allows you to enforce constrained language mode through **User Mode Code Integrity** (**UMCI**) policies using **Windows Defender Application Control** (**WDAC**). This will ensure scripts can only execute in constrained language mode unless your code-signing authority is added to the WDAC policy, allowing them to run in full language mode. Alternatively, constrained language mode can be configured by creating an environment variable via a GPO and setting __PSLockdownPolicy to a value of 4. However, this can easily be circumvented if the user has the right to override these settings.

To check the current language mode of the execution context in PowerShell, run the following command from a PowerShell window:

```
$ExecutionContext.SessionState.LanguageMode
```

The following current language modes are supported in PowerShell:

- FullLanguage
- ConstrainedLanguage
- RestrictedLanguage
- NoLanguage

To read more about the different types of language modes in PowerShell, visit `https://docs.microsoft.com/en-us/powershell/module/microsoft.powershell.core/about/about_language_modes?view=powershell-7.2`.

Next, let's look at enforcing PowerShell script execution.

PowerShell script execution

PowerShell execution policies determine which types of scripts are allowed to run on a system and help prevent unintentional script execution. Script execution policies should not solely be relied upon for security purposes, as they can easily be circumvented by typing the contents of a script manually into the command line. A few examples of script execution policies are as follows:

- `Restricted` is the default for Windows workstations and allows individual commands but prevents scripts from running.

- `Bypass` is where scripts and commands run freely without warning.

- `RemoteSigned` requires scripts to be signed by a trusted publisher if they've been downloaded from the internet.

- `AllSigned` requires all scripts to be signed by a trusted publisher, including ones created locally. PowerShell will prompt you before allowing the script to run from unknown publishers.

To learn more about script execution policies in PowerShell, type in `Get-Help about_execution_policies`. It is recommended that you keep the default setting of `Restricted` or use `RemoteSigned`. PowerShell script execution policies can be configured for both computer and user scopes. To configure the machine policy with Group Policy, go to the following locations:

- **Computer Configuration | Administrative Templates | Windows Components | Windows PowerShell | Turn on Script Execution**:

  - Choose **Allow local scripts and remote signed scripts** to set the `MachinePolicy` scope to `RemoteSigned`.

To view the current execution policy list by scope, from an administrative PowerShell window, type `get-executionpolicy -list`. The following screenshot shows five different scopes. If all the scopes are set to `Undefined` except for 1, then that is the currently effective policy. The computer scope takes precedence over the user scope when configured with Group Policy:

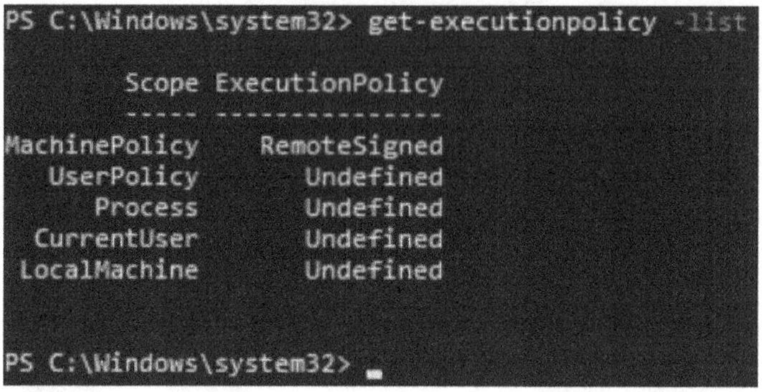

```
PS C:\Windows\system32> get-executionpolicy -list

        Scope ExecutionPolicy
        ----- ---------------
MachinePolicy      RemoteSigned
   UserPolicy         Undefined
      Process         Undefined
  CurrentUser         Undefined
 LocalMachine         Undefined


PS C:\Windows\system32>
```

Figure 12.48 – Execution policy scope list

Next, let's review enabling **Just Enough Administration (JEA)** to control the available cmdlets administrators can run and reduce the scope of unintentional privileges on systems through PowerShell remoting.

JEA

PowerShell JEA helps reduce the available attack surface on Windows servers by limiting the scope of an administrator's capabilities using principle of least privilege concepts through PowerShell remoting. JEA enables admins to run defined sets of cmdlets and functions through remote PowerShell without providing access to log on locally or the need to provide administrative rights to the underlying system.

For example, suppose a member server contains multiple server roles and that multiple IT admins manage each role. Traditionally, each admin would be provided full administrative rights to the system and, in effect, administer the roles for their job function. By specifying roles capabilities in JEA, admins can access the JEA endpoint, execute their assigned cmdlets, and don't require unintentional access that grants them permissions to other roles or the OS and filesystem. If the cmdlets require administrative privileges, JEA can be configured with a temporary privileged virtual account to obtain the correct access tokens. This also allows admins to connect to sensitive servers with non-admin privileges and perform their tasks securely.

JEA can be implemented by following these high-level steps:

1. Validate that the server meets the minimum prerequisite requirements and allows for PSRemoting. PowerShell script block and module logging can be enabled to track all actions that are taken by JEA users.

   To view the list of prerequisites, go to `https://docs.microsoft.com/en-us/powershell/scripting/learn/remoting/jea/prerequisites?view=powershell-7.2`.

2. Configure the necessary role capabilities. **Role capabilities** define what someone can do in a JEA session by defining the available cmdlets, functions, scripts, and providers they can access from inside the PowerShell session.

3. Create the identity that JEA will use to act as the account broker between the IT admin and the underlying system. This can be in the form of a local virtual account or a group-managed service account and is configured in a session configuration file.

4. Define role definitions to determine the mapping of users to their roles.

5. Create the session configuration file and register it with the JEA endpoint. This file applies the role definitions, assigns the JEA identities, and determines who has access to the endpoint.

To enhance JEA, conditional access rules are supported in the session configuration for **just-in-time** (JIT) access management. For example, you can map a security group to a role definition and require an IT admin to request access using a PIM or PAM solution to become a member of the security group. In the following diagram, you can see how JEA administration works. In this example, the JEA user only has access to PowerShell cmdlets that can manage DNS services:

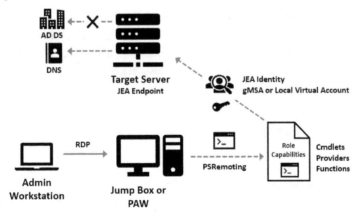

Figure 12.49 – Just Enough Administration

To review all the prerequisites and requirements for enabling JEA, visit the official Microsoft documentation at `https://docs.microsoft.com/en-us/powershell/scripting/learn/remoting/jea/overview?view=powershell-7.2`.

We have just provided an overview of several methods that can be used to secure your PowerShell sessions. Ensure you understand how to configure PowerShell logging, how to enable PowerShell constrained language mode, how to enable PowerShell execution policies, and how to correctly implement JEA.

Summary

In this chapter, we covered securing Windows Server. First, we reviewed different installation options for Windows Server and looked at new security enhancements for Windows Server 2022. In the next section, we discussed different server roles and features and highlighted specific roles that could be used as part of your security strategy. Then, we learned how to install WSUS on a Windows Server Core installation.

In the next section, we covered managing Windows updates with WSUS and Azure Automation Update Management. We reviewed extending Microsoft Defender for Endpoint capabilities to Windows Server and the methods available to onboard them using Group Policy or Azure Defender. After, we discussed hardening Windows Server and walked through implementing a baseline. We analyzed both CIS benchmarks and Microsoft Security Baseline using Policy Analyzer and reviewed controls that protect user accounts and secure the login process. We finished this section by discussing encryption using Azure Disk Encryption and walked through deploying a Windows Defender Application Control policy. Finally, we reviewed PowerShell security and how JEA can be used to limit administrative access to servers.

In the next chapter, we will cover security monitoring and reporting. We will review onboarding Windows clients to Defender for Endpoint with Intune, monitoring services in Azure with Log Analytics and Azure Monitor, and reviewing onboarding servers to Microsoft Defender for Cloud for vulnerability assessments.

Part 3: Protecting, Detecting, and Responding for Windows Environments

This concluding section will focus on proactive methods of protecting Windows endpoints, along with giving an insight into **Security Operations** (**SecOps**) for Windows-based environments. We will begin the section with a detailed review of the monitoring options available for Windows environments and a study of the various reporting capabilities. You will then learn about SecOps, with a primary focus on the cloud-based Microsoft technologies used to continually monitor Windows systems. After that, we will have an overview of security testing and auditing to ensure no gaps exist within your Windows environment and the configuration of endpoints. We will finish the section with our thoughts about the top security recommendations you should focus on as part of your Windows security program.

This part of the book comprises the following chapters:

13
Security Monitoring and Reporting

In this chapter, we will review the methods that are used for reporting and monitoring endpoints within the environment. Throughout this book, we've provided recommendations that help secure and harden Windows environments by enforcing baselines and security controls. After these controls are in place, proper monitoring should be established to ensure application availability, storage alerting, track assets and inventory, and identify weaknesses and vulnerabilities in the underlying software and operating systems. Many products are available that can help provide these insights, so let's review where they can be found based on the solutions we've discussed throughout this book.

First, we will review the features of **Microsoft Defender for Endpoint** (**MDE**). We will cover the MDE dashboards that are used to report on threats, inventory software, and identify weaknesses in the software and configurations on your endpoints. We will cover onboarding Windows clients into MDE, configuring the data connector to Intune, and assigning device-based compliance policies based on the MDE machine risk score. After that, we will discuss collecting telemetry data in Log Analytics workspaces and using gallery solutions and Azure Workbooks to enhance the data to build reporting dashboards.

Later in this chapter, we will provide an overview of enabling vulnerability management with Microsoft Defender for Cloud. We will learn about the services it provides and the methods available to onboard Windows servers. Next, we will review the reporting capabilities in **Microsoft Endpoint Manager** (**MEM**). We will review the different reporting types available in MEM and cover how to collect MDM diagnostic logs both remotely and from the clients. Finally, we will provide an overview of monitoring the application health and update status of Office apps using the Microsoft 365 Apps admin center.

In this chapter, we will cover the following topics:

- MDE features
- Onboarding Windows clients into MDE
- Collecting telemetry with Azure Monitor Logs
- Monitoring with Azure Monitor and activity logs
- Overview of Microsoft Defender for Cloud
- Reporting in MEM
- Monitoring the health and update status of Office apps

Technical requirements

To complete this chapter, we recommend the following minimum requirements:

- An Azure subscription and tenant
- Microsoft Defender for Endpoint licensing
- Microsoft Endpoint Manager licensing
- Microsoft 365 Apps admin center
- An Azure Log Analytics workspace and automation account
- Microsoft Defender for Cloud

Let's start by reviewing the features in Microsoft Defender for Endpoint and the dashboards that can help monitor and protect your endpoints.

MDE features

MDE is part of the Microsoft 365 Defender suite of security products. It provides full **Endpoint Detection and Response (EDR)** capabilities with a focus on protecting connected endpoints. When combined, the M365 Defender suite protects four domains (application, identity, endpoint, and data), allowing threat signals to be viewed holistically by adding context that paints a clearer picture of the security threat. For example, **Microsoft Defender for Office (MDO)** can alert when a user clicks on a phishing link in an email. This threat can then be correlated and investigated in MDE by looking at the device event timeline, using advanced hunting, or reviewing any relevant alerts to understand the potential impact. Security teams can easily add any malicious domains that have been identified to the IOC list, thus blocking connections from the endpoints and automatically purge emails targeting other users using MDO. By analyzing near-real-time telemetry data, MDE can further correlate similar activity seen across multiple endpoints and group alerts into an incident, should the phishing email get delivered and clicked on by multiple users. Through automation and remediation, connections from the endpoint can automatically be blocked and the alerts can be resolved without the need for human intervention. The built-in MDE dashboards can easily be incorporated into any **Security Operations Center (SOC)** workflow or connected to other SIEMs through API connectors.

The capabilities of the MDE solution can be broken down into the following areas:

- **Threat and vulnerability management (TVM)** discovers vulnerabilities and misconfigurations across your endpoints and prioritizes recommendations based on analysis from the Microsoft Intelligent Security Graph. TVM provides real-time detection and response insights and determines the overall device exposure score of your organization.

- **Attack surface reduction (ASR)** helps reduce the potential attack surface by using controls that leverage hardware-based security and advanced software protection features such as application control, exploit protection, network protection, and controlled folder access.

- **Next-generation antivirus** brings machine learning, analytics, and threat-resistance research from the cloud to Defender Antivirus. Next-generation features offer near-real-time protection against new and emerging threats and can detect and block attacks automatically.

- **EDR** refers to the use of sensors to detect near-real-time events and create alerts based on suspicious activities. EDR correlates alerts across endpoints into incidents that can be tracked and actioned by the SOC or with automation. Through advanced features, security teams can build custom threat detection queries, track progress with remediation requests, and submit files for deep analysis by using a secure sandboxed environment.

- **Automated investigation and remediation** allows MDE to use automation based on insights from common alerts and known attack techniques to auto-remediate alerts without requiring the security team's interaction.

- **Microsoft Threat Experts** is an additional managed service offering that allows security teams to consult with a Microsoft security expert directly from the MDE portal.

- **Management and APIs** allow you to manage security-related features on devices in Intune through Endpoint Security. API connectors allow SIEM solutions to be integrated using the Defender for Endpoint APIs.

Let's review the different dashboards in more detail and see how they can be used for monitoring. To sign into the Microsoft 365 Defender portal, go to `https://security.microsoft.com`. Most of the dashboards can be found under **Endpoints** in the left navigation pane.

The Threat analytics dashboard

The **Threat analytics** dashboard provides an overview of any current and emerging threats and analyzes how devices in your organization could potentially be impacted. Within each threat, you can view if there are active alerts that would indicate an attack, the total number of affected assets, and any misconfigurations that could leave devices vulnerable. Each threat provides an executive summary report from Microsoft threat intelligence and lists any exposures and mitigations that can be correlated from Defender security products. The following screenshot shows the **Threat analytics** dashboard showing the latest and high-impact threats:

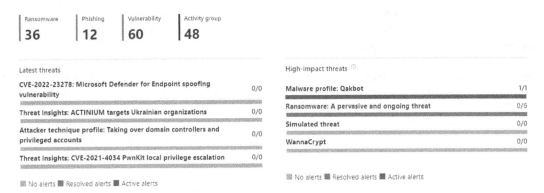

Figure 13.1 – The Threat analytics dashboard in MDATP

More information on **Threat analytics** can be found at `https://docs.microsoft.com/en-us/microsoft-365/security/defender-endpoint/threat-analytics?view=o365-worldwide`.

Next, let's review the TVM dashboard.

The TVM dashboard

The TVM dashboard displays an analysis of the collected EDR data and calculates exposure and configuration scores based on recommendations from Microsoft security best practices. The lower the exposure score, the less vulnerable your devices are to security threats. Selecting **Show more** under **Top security recommendations** will provide a list of actionable security recommendations to help secure devices and improve your devices' **Secure Score**. The higher the secure score, the better that devices are protected against security threats. The TVM dashboard also includes a **Device exposure distribution** pie chart to help you quickly view devices by their associated machine risk level. Risk levels are broken down into high, medium, low, and informational. The **Threat & Vulnerability Management** dashboard can be found under **Endpoints**, on the Microsoft 365 Defender portal. The following screenshot shows the overall exposure score of your devices, top security recommendations, and threat awareness pane analyzing devices exposed to a high-impact threat:

Threat & Vulnerability Management dashboard

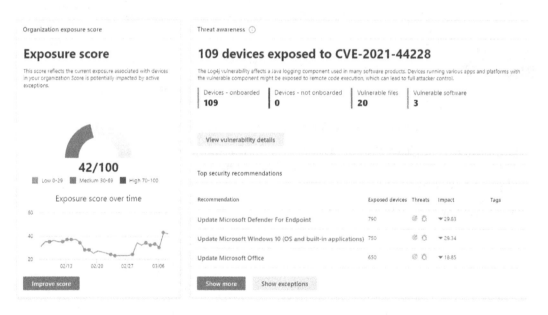

Figure 13.2 – Threat & Vulnerability Management dashboard

The following screenshot of the TVM dashboard shows **Microsoft Secure Score for Devices** and the **Device exposure distribution** pie chart, where you can quickly view devices based on their machine risk level:

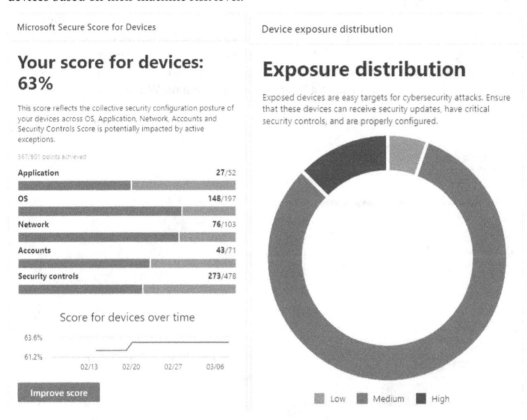

Figure 13.3 – Secure Score and Exposure distribution

To learn more about the **Threat & Vulnerability Management** dashboard, visit https://docs.microsoft.com/en-us/microsoft-365/security/ defender-endpoint/tvm-dashboard-insights?view=o365-worldwide.

Next, let's look at viewing the MDE Device Inventory dashboard.

Device Inventory dashboard

The **Device Inventory** dashboard lists all devices that have been onboarded into MDE and those discovered through device discovery and network assessment scans. Device discovery leverages active onboarded devices to find unmanaged devices on the same network. There are many filtering options available for sorting the device inventory list, including risk and exposure level, OS platform, Windows version, sensor health, onboarding status, and antivirus status. The following screenshot shows the **Device Inventory** dashboard, which includes a filter for **OS platform** on Windows 11 devices:

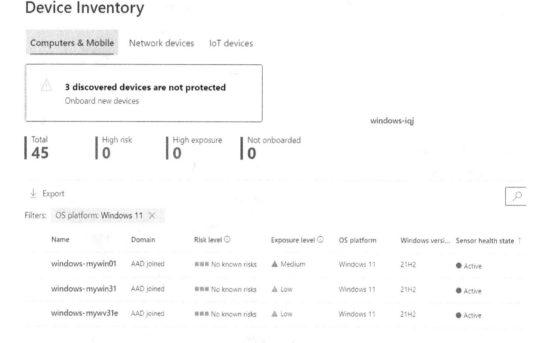

Figure 13.4 – Device Inventory

Device Inventory can be found under **Endpoints** in the Microsoft 365 Defender portal. To learn more about **Device Inventory**, visit `https://docs.microsoft.com/en-us/microsoft-365/security/defender-endpoint/machines-view-overview?view=o365-worldwide`.

Next, let's review the Device health and compliance report.

Device health and compliance

The **Device health and compliance** report allows you to visually monitor the health state of your devices and track trends based on sensor activity, antivirus status, and OS platforms and versions. The following screenshot shows **Health state** information about sensor data activity across all endpoints onboarded into MDE:

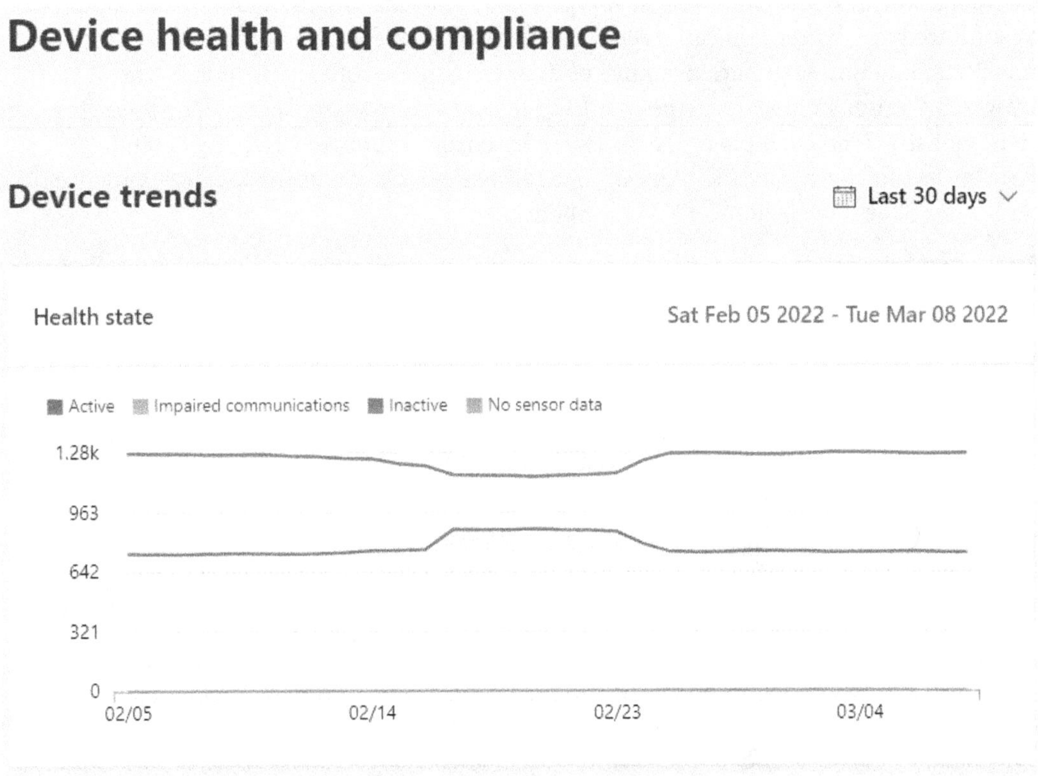

Device health and compliance

Device trends

📅 Last 30 days ∨

Health state   Sat Feb 05 2022 - Tue Mar 08 2022

■ Active   ▨ Impaired communications   ■ Inactive   ▨ No sensor data

1.28k	
963	
642	
321	
0	

02/05   02/14   02/23   03/04

Figure 13.5 – Device health and compliance

To view the **Device health and compliance** report, from the Microsoft 365 Defender portal, choose **Reports**, and click **Device health and compliance**.

For more information, visit `https://docs.microsoft.com/en-us/microsoft-365/security/defender-endpoint/machine-reports?view=o365-worldwide`.

Next, let's look at the **Software inventory** report.

Software inventory report

MDE includes an inventory report of software that has been identified on onboarded devices. The report includes details identifying the software by name and version, as well as reports on any detected vulnerabilities or known weaknesses. For any identified weakness, you can quickly view the count of exposed devices and export a list if needed. Selecting a software title from the list will pull up additional details, including a link to the software page, where you can track security recommendations, version distributions, known vulnerabilities, and an inventory of devices with the software installed. The **Software inventory** report leverages the EDR data as its discovery source. The following screenshot shows an example of the **Software inventory** report in MDE. By default, it's sorted by the software with the highest impact score, which affects the **Device exposure** and **Secure Scores** parts of the TVM dashboard:

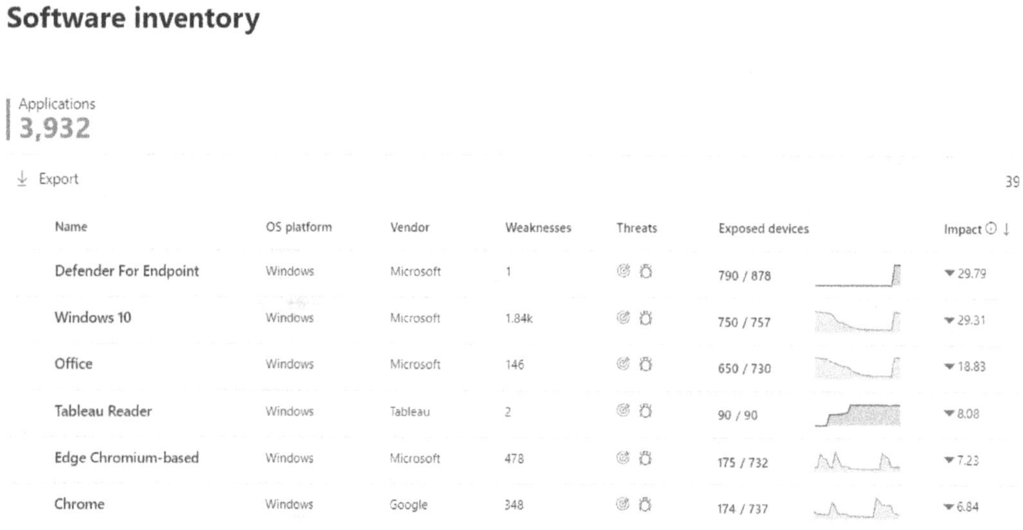

Software inventory

Applications
3,932

↓ Export   39

Name	OS platform	Vendor	Weaknesses	Threats	Exposed devices		Impact ⊙ ↓
Defender For Endpoint	Windows	Microsoft	1		790 / 878		▼ 29.79
Windows 10	Windows	Microsoft	1.84k		750 / 757		▼ 29.31
Office	Windows	Microsoft	146		650 / 730		▼ 18.83
Tableau Reader	Windows	Tableau	2		90 / 90		▼ 8.08
Edge Chromium-based	Windows	Microsoft	478		175 / 732		▼ 7.23
Chrome	Windows	Google	348		174 / 737		▼ 6.84

Figure 13.6 – Software inventory

For more information about how the **Software inventory** report works, visit `https://docs.microsoft.com/en-us/microsoft-365/security/defender-endpoint/tvm-software-inventory?view=o365-worldwide`.

Next, let's look at the **Security recommendations** dashboard.

Security recommendations

The **Security recommendations** dashboard helps by prioritizing security misconfigurations by assigning an impact score and provides actionable steps to remediate the weakness. By default, the recommendations are sorted by highest to lowest impact rating, and resolving them will help increase your secure score and reduce the overall exposure score. Clicking each recommendation offers more details, including a general overview with a short description, any associated CVEs, and a list of exposed devices:

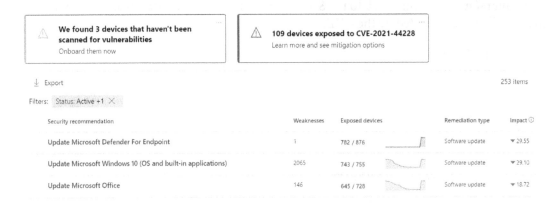

Figure 13.7 – Security recommendations

From this dashboard, you can use filters to sort by remediation types to help narrow down the list of recommendations. For example, by selecting **Configuration change**, you can view recommendations that advise implementing additional security controls that can help harden your endpoints.

Security recommendations can be found in the M365 Defender portal under **Endpoints** by clicking on **Recommendations**. The list of top security recommendations is also listed in the TVM dashboard.

Identifying weakness

The **Weaknesses** dashboard provides an overview of the latest emerging threats and vulnerabilities known to Microsoft security intelligence. It's another method of gaining insights into active threat types and assessing the exposure risk to your devices based on the weaknesses these vulnerabilities use to achieve the exploit. The dashboard provides the CVE name, a severity rating, CVSS score, the impacted software, threat published date, and the number of exposed devices. Clicking on each CVE will provide a short description of the vulnerability and additional threat insights, such as if a publicly known exploit is available. From the CVE overview, clicking on **Go to related security recommendation** will link back to the **Security recommendations** dashboard and filter the list of software affected by the CVE:

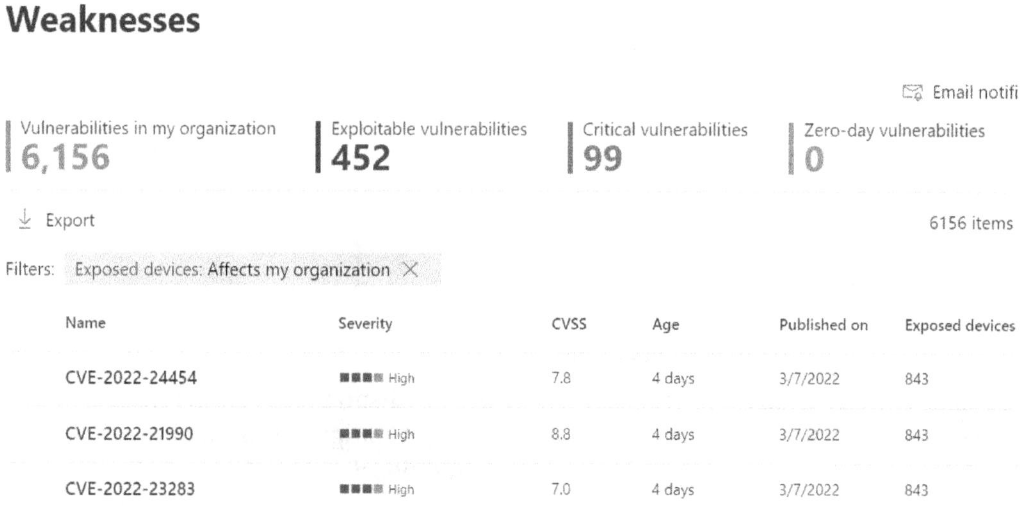

Weaknesses

| | | | | ⊠ Email notifi |
| | | | | |

Vulnerabilities in my organization	Exploitable vulnerabilities	Critical vulnerabilities	Zero-day vulnerabilities
6,156	452	99	0

↓ Export   6156 items

Filters: Exposed devices: Affects my organization ✕

Name	Severity	CVSS	Age	Published on	Exposed devices
CVE-2022-24454	▪▪▪▪▪ High	7.8	4 days	3/7/2022	843
CVE-2022-21990	▪▪▪▪▪ High	8.8	4 days	3/7/2022	843
CVE-2022-23283	▪▪▪▪▪ High	7.0	4 days	3/7/2022	843

Figure 13.8 – The Weaknesses dashboard

The filters in the **Weaknesses** dashboard can help narrow down results that are important to focus on first. For example, you can show only critical severity threats, weaknesses that have open active alerts in MDE, or exploits that are known to be publicly available and verified in the wild. The **Weaknesses** dashboard can be found under the **Endpoints** section on the M365 Defender portal home page.

Reviewing advanced features

MDE is frequently adding new features to enhance the protection capabilities of the product. Often, they are not enabled by default once generally available. These features include configuring connections to other third-party security solutions and configuring service connections to enable telemetry data in other Microsoft products. We recommended reviewing the **Advanced features** area of MDE regularly to check for newly added functionality that could help enhance the EDR protection capabilities. To view the options in the Microsoft 365 Defender portal, go to `https://security.microsoft.com`, choose **Settings**, click on **Endpoints**, and select **Advanced features** under **General**.

To learn more about **Advanced features**, visit `https://docs.microsoft.com/en-us/microsoft-365/security/defender-endpoint/advanced-features?view=o365-worldwide`.

Next, let's look at configuring API connectors for programmatic access to MDE.

Configuring API connectors

MDE allows access to its data and control management actions by connecting partner applications through programmatic APIs. This flexibility allows other security solutions to analyze and aggregate MDE security event data, help orchestrate and automate actions based on threats, and cross-analyze telemetry for enhanced threat intelligence not limited to Microsoft. To view the list of approved partner applications, including links to the integration documentation, log in to the Microsoft 365 Defender portal and click on **Partner applications** under **Partners and APIs** within the **Endpoints** section. A few of the available providers are as follows:

- Splunk
- HP ArcSight
- IBM QRadar
- Demisto
- ServiceNow
- BETTER Mobile
- Palo Alto Networks

Access to the APIs is configured securely through Azure AD app registration using the OAuth 2.0 authorization code flow and is granted consent with both user and application context, depending on the API endpoints you wish to make available. After the connection has been configured with a partner application, you can view the connectivity status from the Microsoft 365 Defender portal by clicking on **Partners and APIs** under **Endpoints** and then clicking on **Connected applications**. The following screenshot shows a list of connected applications and request trends over the last 30 days:

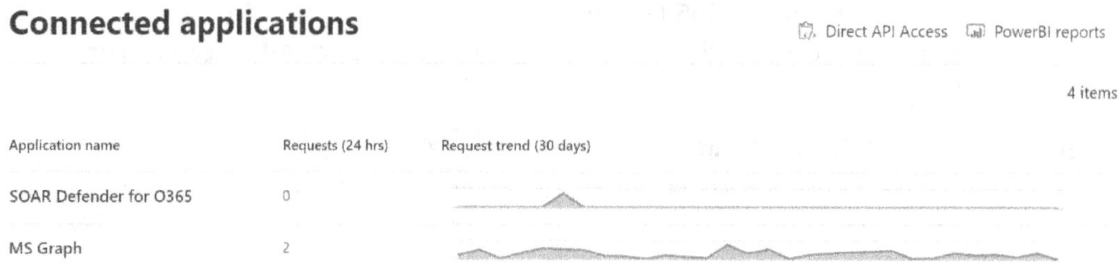

Connected applications   Direct API Access   PowerBI reports

4 items

Application name	Requests (24 hrs)	Request trend (30 days)
SOAR Defender for O365	0	
MS Graph	2	

Figure 13.9 – Connected applications in MDE

For additional information about the API schema for Defender for Endpoint, visit `https://docs.microsoft.com/en-us/microsoft-365/security/defender-endpoint/exposed-apis-list?view=o365-worldwide`.

Next, we will discuss onboarding Windows clients into MDE using Intune.

Onboarding Windows clients into MDE

To get value out of MDE and configure many of its advanced protection capabilities, clients must be onboarded to an instance of MDE. There are several ways to achieve this, and they are determined by how your devices are managed and the architecture of your network. Most methods to onboard clients require deploying an onboarding package that can be distributed using any of the following methods:

- Local script or Group Policy
- Microsoft Endpoint Configuration Manager
- Mobile Device Management (Intune or other MDM providers)
- VDI onboarding script for non-persistent devices

The onboarding package can be obtained in the Microsoft 365 Defender portal by logging into `https://security.microsoft.com` and choosing **Settings | Endpoints | Onboarding** under the **Device management** section. However, if you're using Intune and a cloud-native architecture, clients can be onboarded automatically by enabling **Microsoft Intune Connection** from MDE and assigning an **Endpoint detection and response policy** in Intune **Endpoint Security**. To review the available MDE onboarding deployment strategies, and to determine what may work best for your architecture, review the following PDF:

`https://download.microsoft.com/download/5/6/0/5609001f-b8ae-412f-89eb-643976f6b79c/mde-deployment-strategy.pdf`

In the next example, we will be using the Intune method based on a cloud-native architecture. For details about using other methods, visit the official Microsoft docs page at `https://docs.microsoft.com/en-us/microsoft-365/security/defender-endpoint/onboard-configure?view=o365-worldwide`.

Let's start by configuring the Microsoft Intune connection to establish a service connection from MDE to MEM.

Configuring the Microsoft Intune connection

The Microsoft Intune connection allows you to share device information between MDE and MEM, enables enhanced compliance policy enforcement, and helps power the device's secure score in MDE. Enabling the connector is the first step to automating onboarding for Windows clients. Once the services begin to sync, assigning an EDR policy using Endpoint Security will automatically onboard new clients into MDE without the need to deploy the onboarding package. To enable the Microsoft Intune connection, follow these steps:

1. Log in to the **Microsoft 365 Defender** portal at `https://security.microsoft.com`.

2. Go to **Settings**, click on **Endpoints**, and choose **Advanced features**. Switch the slider to **On** next to **Microsoft Intune connection** and choose **Save Preferences**.

3. Log in to Microsoft Endpoint Manager at `https://endpoint.microsoft.com`.

4. Choose **Endpoint Security** and select **Microsoft Defender for Endpoint** under the **Setup** menu.

5. Turn **Connect Windows devices to Microsoft Defender for Endpoint** to the **On** setting. Click on **Save**.

Synchronization between the two services occurs at least once every 24 hours. The connection status and last synchronized date can be monitored in the MDE setup, as shown in the following screenshot:

Figure 13.10 – Microsoft Intune connection

Once the service connection has been established, Intune automatically becomes aware of the onboarding package, and an EDR profile can be assigned to automatically onboard Intune-managed clients.

Creating an EDR policy

An EDR policy can be created by using the **Endpoint detection and response** template type in **Endpoint Security**. This method also works for MDM-managed Intune devices and with tenant-attached devices if you are using co-management with Configuration Manager. To deploy an EDR policy, follow these steps:

1. Sign into Microsoft Endpoint Manager at `https://endpoint.microsoft.com`.

2. Click on **Endpoint Security** and choose **Endpoint detection and response**. Click on **Create Policy**.

3. Choose **Windows 10 and later** for **Platform** type. Select **Endpoint detection and response** for **Profile** and click **Create**.

4. Give the policy a name and description and click **Next**.

5. On the **Settings** page, keep **Block sample sharing for all files** set to Not Configured and set **Expedite telemetry reporting frequency** to Yes.

6. Click **Next** to assign a scope tag if needed, and **Next** again to assign the policy to groups of users or devices. Click **Next** to **Review + Create** and finally **Create** to finish building the policy.

> **Tip**
> If the policy is assigned to groups of users, the policy will not apply until a user signs in, which may cause a slight delay before devices appear in MDE.

Windows clients will typically begin showing up in the portal within 24 to 48 hours after receiving the policy. To confirm that endpoints have been onboarded and are healthy, go to the Microsoft 365 Defender portal at `https://security.microsoft.com` and choose **Device inventory** under **Endpoints**. The endpoint's sensor health state should be active and its onboarding status should show as onboarded.

Next, let's look at enhanced policy enforcement by using the MDE machine risk score as a condition in a device-based compliance policy.

Creating a machine risk compliance policy

Adding a compliance policy condition to evaluate the MDE machine risk score adds zero-trust protection capabilities to your access controls. For example, any client that has a machine risk score of *high* will be immediately marked as a non-compliant device. Using Azure AD Conditional Access, policies can be configured to block access to certain cloud apps, such as Exchange Online, or require the use of multi-factor authentication for extra verification. To create a compliance policy based on the MDE machine risk score, follow these steps:

1. Log in to MEM, go to **Devices**, and choose **Compliance Policy** under **Policy**. Then, click on **Create Policy**.

2. Choose **Windows 10 and Later** for **Platform** and click **Create**.

3. Enter a friendly name, such as `Windows Device Security Compliance`, and give it a description. Click **Next**.

4. Under **Compliance settings**, choose **Microsoft Defender for Endpoint**. Click on the dropdown and choose **Medium** for **Require the device to be at or under the machine risk score**. Click **Next** and leave **Mark device noncompliant** set to **Immediately**. Click **Next** and assign the policy to a user or device group. Click **Next** to review the settings and click **Create** to finish.

To track the progress of the compliance policy, choose **Device status** under **Monitor** from within the compliance policy overview. We can also confirm the status of the evaluation directly on a device in Intune. In the following screenshot, we can see that the device is **Compliant** based on the machine risk score evaluation:

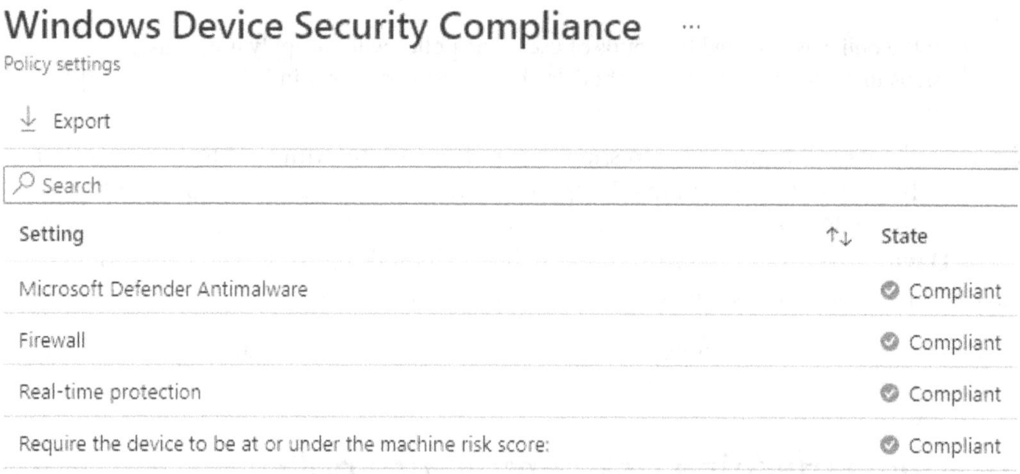

Windows Device Security Compliance ···

Policy settings

↓ Export

Setting	↑↓	State
Microsoft Defender Antimalware		✅ Compliant
Firewall		✅ Compliant
Real-time protection		✅ Compliant
Require the device to be at or under the machine risk score:		✅ Compliant

Figure 13.11 – Compliance policy evaluation on a device

In the preceding screenshot, the **Windows Device Security Compliance** policy also included checks to ensure that the **Microsoft Defender Antimalware** service, **Firewall**, and **Real-time protection** are all enabled. We provided an overview of recommended device-based compliance policies in *Chapter 6, Administration and Policy Management*, which should be considered as part of implementing a zero-trust access strategy.

MDE's monitoring and reporting features greatly enhance the built-in protection capabilities provided by Windows Defender. By analyzing and aggregating telemetry data from endpoints into easy-to-read dashboards, security teams can quickly analyze threats and prioritize top security tasks. Taking a proactive approach, in addition to normal reactionary measures, will greatly enhance your organization's cyber defenses. We encourage incorporating MDE's threat and vulnerability reporting into the normal operating procedures of the security program to help remediate the known risks and threats.

Next, let's learn how to monitor Windows systems by configuring data collection using Azure Monitor Logs in Azure.

Collecting telemetry with Azure Monitor Logs

Azure Monitor Logs, also known as Azure Log Analytics, is a data collection repository and analysis tool that supports Windows systems and many other resources and services running in Azure. The log data that is collected is stored in a repository known as a Log Analytics workspace, where it can be parsed and analyzed using the **Kusto Query Language (KQL)**. Log Analytics workspaces are used to power dashboards in Azure Workbooks, feed Azure Monitor to aggregate performance data, and provide many other telemetry-driven solutions. Log Analytics workspaces are also used to source data for analysis in security solutions such as Defender for Cloud, Azure Sentinel, and other SIEMs through API connections. A few examples of the data that can be sent into Log Analytics from Windows systems are as follows:

- Windows event logs such as application, system, and security event logs
- Windows performance counters such as memory and processor usage, as well as disk space

This wide dataset allows for a lot of flexibility in building reporting dashboards and configuring alerts.

To get started with Azure Monitor Logs, you will need to deploy a Log Analytics workspace in Azure by following these steps:

1. Log in to `https://portal.azure.com` and search for `Log Analytics workspaces`.
2. Click **Create** and enter details for **Subscription** and **Resource Group**. Enter a name for the workspace and choose a region under **Instance details**.
3. Click **Review + Create** to complete the setup.

Once the workspace has been configured, telemetry data is collected from Windows servers by installing the **Microsoft Monitoring Agent** (**MMA**). For Windows clients, a device configuration profile is configured with a commercial ID of the workspace and deployed to managed devices.

> **Tip**
> Azure Monitor agent will be replacing MMA in the future. If you already have MMA deployed, you can visit the following link to learn about the migration path to AMA: `https://docs.microsoft.com/en-us/azure/azure-monitor/agents/azure-monitor-agent-migration`.

Now, let's look at the methods for onboarding Windows servers into Log Analytics.

Onboarding Windows Servers to Log Analytics

To connect Windows Servers to Log Analytics, the Log Analytics agent or MMA can be installed directly onto the server, or through an extension on Azure virtual machines. Although the Log Analytics agent is on a deprecation path for August 31, 2024, it remains fully supported, but it's recommended to plan a migration to the **Azure Monitor Agent** (**AMA**) in the future. Microsoft is frequently adding new supported services to the AMA to bring it as close to feature parity with the MMA and make it compatible with many solutions that rely on MMA today. AMA can also coexist with other agents if you're considering starting the migration path. To find more information about the supported services and features of the Azure Monitor agents, visit `https://docs.microsoft.com/en-us/azure/azure-monitor/agents/agents-overview`.

The following methods can be used to install the Log Analytics agent on Windows Server:

- Auto-provision of Azure VMs through Defender for Cloud
- Directly on a VM from the Azure portal
- Software deployment tools such as Configuration Manager for Azure Arc-enabled or on-premises servers
- **Azure Resource Manager** (**ARM**) templates
- Manually with the UI or through the command line

Most of the installation types will require the workspace ID and key of the Log Analytics workspace to complete the onboarding setup. These can be found by going to the workspace and choosing **Agents management** under **Settings**. The Windows agent is also located here for use with the manual installation method. For detailed instructions on the available deployment methods, visit `https://docs.microsoft.com/en-us/azure/azure-monitor/agents/agent-windows`.

In addition to deploying the agent, data collectors will need to be configured to determine the type of data that is being sent into the workspace. These collectors can be configured automatically by installing solutions available in the Azure Marketplace and by enabling them manually by going to **Agents configuration** under **Settings** in the Log Analytics workspace. To keep track of the overall health and responsiveness of the deployed agents, we recommended installing the **Azure Log Analytics Agent Health** solution from the marketplace directly onto the workspace. This can monitor the number of agents that have been deployed, their geolocation, and the overall count of unresponsive agents, should an issue occur.

If you decide to deploy Azure Monitor Agent instead of MMA, this can be accomplished interactively or at scale using the same methods outlined in the preceding bulleted list. To support on-premises servers, the Azure Connected Machine agent must also be installed, in addition to AMA. For additional information about the Azure Connected Machine Agent, visit `https://docs.microsoft.com/en-us/azure/azure-arc/servers/agent-overview#installation-and-configuration`.

Once you have decided on the deployment method, data collection can be configured using **Data Collection Rules (DCRs)** in Azure Monitor. This is a fundamental change compared to the Log Analytics agent, as this was previously handled using the **Agent Configuration** settings directly in the Log Analytics workspace. DCR rules include similar datasets, including Windows event logs and performance counters, and can be configured through the Azure Monitor UI, REST API, or with the Azure CLI or PowerShell.

For more information about installing and managing servers using AMA, visit `https://docs.microsoft.com/en-us/azure/azure-monitor/agents/azure-monitor-agent-overview?tabs=PowerShellWindows`.

Next, let's look at configuring Windows clients to send telemetry data into Log Analytics for use with the Update Compliance solution.

Onboarding Windows clients to Log Analytics

Window clients can be configured to send telemetry data into Log Analytics to support solutions such as Update Compliance to monitor and track Windows Update deployments. In addition to collecting data from clients, Intune diagnostic data such as auditing, operational logs, and compliance data can also be sent into Log Analytics as additional datasets. Clients that meet the prerequisites are already compatible and can send data into Log Analytics without requiring an additional agent. All that is needed are telemetry policies that have been configured with MDM or Group Policy to tell the client where to send the data and what it's allowed to send. When the Update Compliance solution is installed, the data collection set is automatically configured and sent to the workspace. It's recommended to review the prerequisites before deploying policies to ensure your devices are supported. You can find these at `https://docs.microsoft.com/en-us/windows/deployment/update/update-compliance-get-started#update-compliance-prerequisites`.

First, we will create a device configuration profile in Intune to configure telemetry settings for the clients:

1. Log in to Microsoft Endpoint Manager at `https://endpoint.microsoft.com`.

2. Click **Devices** and choose **Configuration Profile**. Click **Create profile**.

3. Choose **Windows 10 and later** under **Platform** and **Settings catalog** under **Profile type**.

4. Give the profile a name, such as `Windows Data Collection and Telemetry`, and a description. Click **Next**.

5. In **Configuration settings**, click on **Add settings** to open the **Settings picker** area.

6. Search for **System** and select it. Add and configure the following policies:

    -   **Allow device name to be sent in Windows diagnostic data** to Allowed

    -   **Allow Telemetry** to Full

    -   **Allow Update Compliance Processing** to Enabled

    -   **Configure Telemetry Opt in Change Notification** to Disable Telemetry change notifications

    -   **Configure Telemetry Opt in Settings Ux** to Disable Telemetry opt-in Setting

7. Click **Next** and add user or device groups to the assignment. Click **Next**, then **Next** again, and finally **Create** to finish the policy.

To finish the setup, the commercial ID of the Log Analytics workspace is still needed so that the clients know where to send the telemetry data. This will be covered in the *Update compliance* section. Now that we have reviewed methods to onboard Windows servers and clients into Log Analytics, let's look at using monitoring solutions and Azure Workbooks to aggregate the data to build dashboards and create reports.

Monitoring solutions and Azure Workbooks

Monitoring solutions can be added from the Azure marketplace to a Log Analytics workspace to help configure the data collectors, where it is aggregated in the workspace and parsed into pre-built visualizations. Solutions are typically free to install but enabling them could incur data consumption and storage costs, depending on the workspace tier. To view any currently installed solutions in your Azure subscription, log in to the Azure portal and search for **Solutions**. To add a new solution, click **Create** to browse through the Azure Marketplace. In the following screenshot, the `updatemanagement` solution has been added to a Log Analytics workspace to monitor security updates for servers:

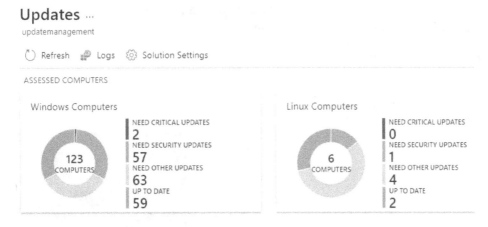

Figure 13.12 – The updatemanagement solution

Let's look at a few solutions that are applicable for Windows servers and clients as there are many options available in the marketplace.

Update Compliance

The **Update Compliance** solution in Azure is specifically designed to track Windows servicing statistics for Windows clients. Using the collected dataset, it parses the data to provide update deployment reports, configuration details for **Windows Update for Business (WUfB)** settings, and statistics around Delivery Optimization usage and settings. **Update Compliance** can analyze and report on the following information:

- Feature and quality update installation status for Windows servicing
- Update issues during feature and quality updates
- WUfB configurations such as feature and update deferral windows
- Delivery Optimization stats such as content distribution and bandwidth utilization for updates and apps
- Windows 11 readiness report

The following screenshot shows sample data that's been analyzed by the **WaaSUpdateInsights** solution overview page:

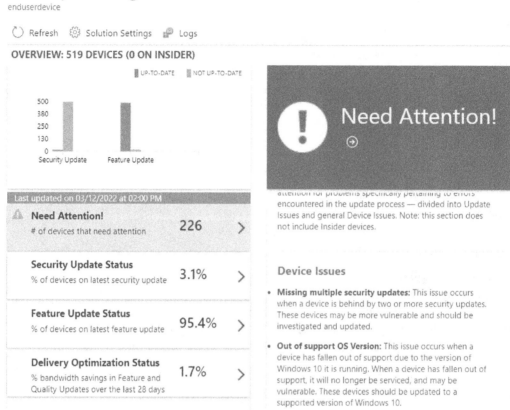

Figure 13.13 – Update Compliance solution in Log Analytics

Clicking each informational blade in the overview will provide greater details, along with a deeper analysis of the status of your devices. The Update Compliance (**WaasUpdateInsights**) solution can be installed on an existing Log Analytics workspace by following these steps:

1. Log in to the Azure portal at `https://portal.azure.com`.

2. Search for **Solutions** and click on **Create**. In the **Marketplace** area, search for **Update Compliance**. Click the solution and click **Create**.

3. Choose the subscription and resource group, and select an Azure Log Analytics workspace. Click **Next: Review + create**. Then, click **Create** to finish.

Now that the deployment has finished, let's gather the commercial ID from the solution settings by going to **Log Analytics workspace | Solutions | WaasUpdateInsights** and choosing **Update Compliance Settings**. Next, let's update the **Device Configuration** profile that we created earlier in the *Onboarding Windows clients to Log Analytics* section with the commercial ID by following these steps:

1. Go back to Microsoft Endpoint Manager at `https://endpoint.microsoft.com`.

2. Click on **Devices**, choose **Configuration profiles**, and open the Windows Data Collection and Telemetry profile we created earlier.

3. Click **Edit** under **Configuration settings**. Click **Add settings** to open the **Settings picker** area. Search for `Commercial ID`.

4. Click on **Administrative Template\Windows Components\Data Collection and Preview Builds**. Then, click on **Configure the Commercial ID** to add it to the **Configuration settings** area.

5. Change the setting to **Enabled** and paste the Commercial ID from the **Update Compliance solution** settings. Click **Review + Save** to commit the changes.

Once the profile has been updated, devices will start to send data into the Log Analytics workspace and data will appear in **Update Compliance**. Generally, devices that are active and frequently connected to the internet will begin sending telemetry data within 72 hours.

Next, let's look at enabling a solution for tracking configuration changes on Windows servers called **Change Tracking and Inventory**.

Enabling Change Tracking and Inventory

The **Change Tracking and Inventory** solution for Azure Automation is useful for identifying and tracking configuration changes to software, files, and Windows services. Having the ability to identify when changes occur is useful from both an operational and security perspective. To use the **Change Tracking** solution, a Log Analytics workspace must be connected to an Azure Automation account, such as the **Update Management** solution for managing updates on servers. The **Change Tracking** solution can be enabled through the automation account, on the virtual machine directly, or added as a solution using the Azure Marketplace. To open the solution once it's been enabled, open the Azure Automation account and choose **Change tracking** underneath **Configuration Management**. The following screenshot shows an overview of **Change tracking** showing software changes over the last 7 days:

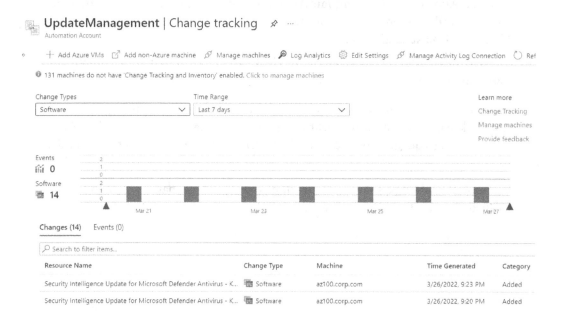

Figure 13.14 – The Change tracking solution

To view additional details about the change, click on **Resource Name** to view the flyout menu details. The solution can also be customized if you need to modify the changes that are tracked by default. To view the options, click on **Edit Settings** in the top toolbar from the solution overview. Here, you can enable and disable paths in the Windows registry, add Windows files, and adjust the collection frequency of which to capture Windows service changes.

More information about the **Change Tracking and Inventory** solution can be found at https://docs.microsoft.com/en-us/azure/automation/change-tracking.

Next, let's look at using **Service Map** to enable a visualized map of how applications communicate between services and networks.

Using Service Map

The **Service Map** solution helps you visually conceptualize your systems as a network of connected services, processes, and connections. It can help identify application and system dependencies for both inbound and outbound connections across all connected systems through a holistic view of their interconnectivity with each other. **Service Map** is a great troubleshooting tool for diagnosing communication problems that historically required many different monitoring solutions to gain these insights. Having the ability to identify active communications protocols allows security teams to lock down connectivity to servers, as well as only allow the necessary protocols for applications to function. The following screenshot of the **Service Map** solution shows an example of connections that have been established by the source system:

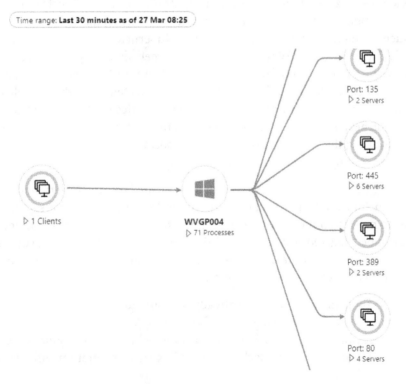

Figure 13.15 – Service Map solution

To enable the **Service Map** solution, Windows servers require **Dependency agent** to be installed on the system to collect the required data. After satisfying the prerequisite installation requirements, the **Service Map** solution can be viewed by clicking on **Insights** under **Monitoring** in the virtual machine overview or through Azure Monitor by clicking on **Virtual Machines** under **Insights** and finding the VM you wish to view.

More information about the **Service Map** solution can be found on the documentation page at `https://docs.microsoft.com/en-us/azure/azure-monitor/insights/service-map`.

Next, let's look at using Azure Monitor to analyze the telemetry data stored in Azure Monitor Logs.

Monitoring with Azure Monitor and activity logs

The Azure Monitor solution provides a centralized location for monitoring resources and services across many different workloads. Using telemetry data collected from a variety of sources, Azure Monitor provides insights into different aspects of the IT infrastructure, ranging from applications, networks, and virtual machines to other Azure resources such as storage accounts, SQL databases, Kubernetes containers, and more. The analysis tools that are used for Windows Server are largely driven by the log data that's collected in Azure Monitor Logs from AMA or MMA. Azure Monitor provides pre-built visualizations to help understand the data that's been collected through the built-in features, such as the solutions that are added to Log Analytics workspaces. Because the incoming data is near-real-time, alerts can be triggered in Azure Monitor based on data thresholds, including the configuration of notifications, which can be sent to interested parties using something called action groups.

An action group defines a series of events that are executed due to an alert being triggered. These actions include notifications in the form of sending emails, text messages, or push notifications to mobile phones or using custom webhooks. Using Azure Logic apps, automation can be triggered from alerts for advanced actions. For example, you can use thresholds to automatically scale up resources during periods of high demand and scale them back during low activity to reduce costs.

Azure Monitor's built-in dashboards are already preconfigured to transform log sources into rich visualizations. The following screenshot is from VM Insights in Azure Monitor showing the processor utilization of a virtual machine. VM Insights monitors the performance and health of virtual machines, including their integration with Service Map, all in one dashboard:

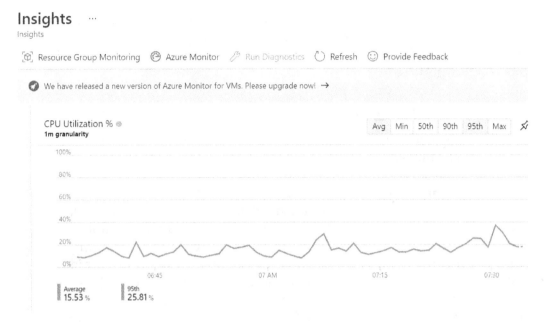

Figure 13.16 – VM Insights in Azure Monitor

The data in Azure Monitor can be tailored further using Azure Workbooks, by only displaying the relevant information for a specific workload and sharing it with the appropriate operational teams without exposing too much data. Next, let's look at using **role-based access control (RBAC)** to secure access to Azure Monitor.

Secure access to Azure Monitor

Monitoring data contains sensitive information about the IT infrastructure, systems, and processes. It's necessary to restrict access to only the appropriate teams to safeguard against its exposure. Azure Monitor allows for RBAC to the resource-level assignable scope using two built-in roles:

- **Monitoring Reader** can read all monitoring data in the scope of the assignment.

- **Monitoring Contributor** can read, create, or modify the monitoring settings or resources in the scope of the assignment.

Custom RBAC can also be configured through various monitoring permissions if the built-in roles are not fine-grained enough. For further information about the custom permissions and ways to secure monitoring data, visit https://docs.microsoft. com/en-us/azure/azure-monitor/roles-permissions-security.

For example, you can store telemetry data from different workloads in different resource groups, storage accounts, or workspaces, and then apply RBAC at those scopes to restrict what data can be viewed. Next, let's look at monitoring operations against resources in Azure using activity logs.

Monitoring Azure activity logs

Azure activity logs monitor operations that have been carried out on resources hosted in the Azure subscription. These logs are important for auditing and tracking from an operational and forensic security perspective, should an incident occur. Events stored in the activity log are broken down into different categories to help identify the types of events that occurred. Some examples include administrative events (create, update, and delete) that have been performed against resources, triggered alerts, security events, and changes to the policy. With the increase of cloud-hosted infrastructure, having an audit trail of activities is critical in any security investigation. Azure activity logs include detailed information about the who, what, and when of activities running in the cloud. For more information about the event schema, visit `https://docs.microsoft.com/en-us/azure/azure-monitor/essentials/activity-log-schema`.

By default, activity logs are stored for up to 90 days within your subscription for free. Long-term retention and archiving can be achieved by configuring diagnostics settings to send logs into Azure Monitor Logs, storage accounts, or to select Microsoft partners. Using Event Hubs or through API connections, activity logs can be forwarded to a SIEM solution outside of Azure for further analysis and automation. Power BI content packs are also available so that you can build business intelligence reporting dashboards that can be easily shared with IT and the business.

To access Azure activity logs, log into the Azure portal at `https://portal.azure.com` and search for **Activity Log**. These activity logs can also be accessed from Azure Monitor, under the **Overview** blade. The following screenshot shows an example of the operations that are stored in Azure activity logs. It includes details about the operation's name, status, time stamp, and who initiated the event:

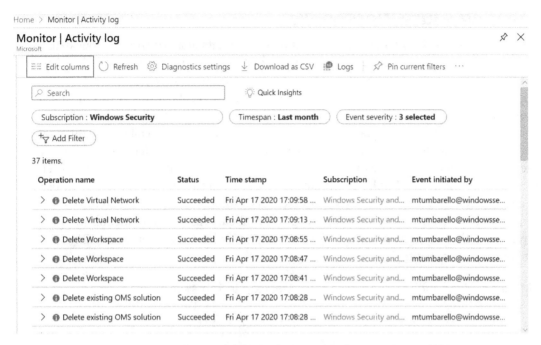

Figure 13.17 – An Azure activity log

Clicking an operational event will provide a detailed summary that includes a JSON formatting of the request and a change history that correlates changes on the same resource 30 minutes before and after the time of the operational event. Alerts can be created directly from the summary pane by clicking **New alert rule**. Activity logs can also be retrieved by downloading a CSV file from the portal, using the Azure Monitor PowerShell or CLI commands, or through REST APIs. Next, let's look at Azure Workbooks.

Creating Azure Workbooks

Azure Workbooks allow you to aggregate and perform data analysis from various data sources and data types, to create customized dashboards in a single integrated view. Some of the data sources that can be analyzed from within workbooks include Azure Monitor Logs, Azure Monitor data and metrics, Azure resource metadata, resource health, and more. This data can be collated and visualized using many of the available visualizations and graphs and saved as templates for reuse later. This customization allows for different perspectives on underlying datasets. To access Azure Workbooks, you can search for it in the Azure portal or click **Workbooks** under the **Overview** pane in Azure Monitor. There are many preconfigured **Public Templates** to choose from to help you get started or use as a reference when creating a blank workbook. An example of an available template can be seen in the Azure **Virtual Desktop Insights** dashboard, as shown in the following screenshot. The graphs show two perspectives regarding the volume of connection attempts by their result:

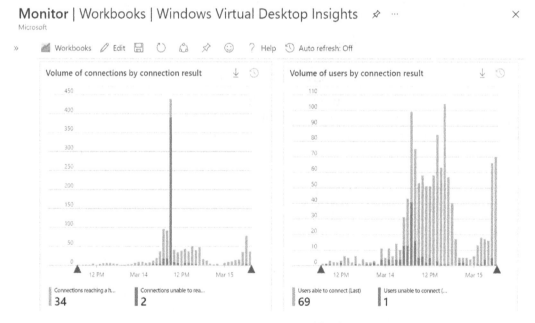

Figure 13.18 – Azure Workbooks

Workbooks can easily be shared between teams, but users will need to be granted read access to the underlying data directly to view the analysis.

For additional information about using Azure Workbooks, visit `https://docs.microsoft.com/en-us/azure/azure-monitor/visualize/workbooks-overview`.

Next, let's look at how to check for service issues and incidents in Azure using Service Health.

Azure Service Health

Hosting resources in any cloud environment makes them susceptible to service provider outages and any required maintenance of the underlying resources that support the cloud infrastructure. Most of the time, these events are out of your control, and can make planning for maintenance or troubleshooting incidents difficult. If your resources are hosted in Azure, Service Health can help you keep track of these events by reporting service issues, planned maintenance, and health and security advisories through the **Service Health** dashboard. Customizable dashboards can be created from the data for proactive monitoring of the applicable cloud resources in your environment. Service Health allows you to create configurable alerts, shareable reports, and view issue updates from Microsoft, including impact reports and downloadable **root cause analyses (RCAs)**.

To open Azure Service Health, search for `Service Health` in the Azure portal or click **Service Health** in the Azure Monitor overview pane. The following screenshot shows an example history of service issues over 1 month, with a downloadable RCA summary available:

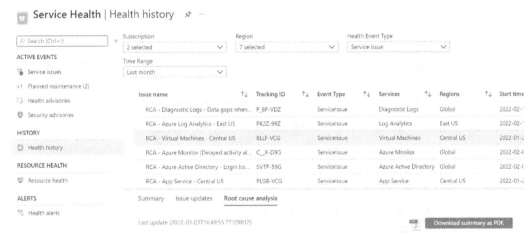

Figure 13.19 – Azure Service Health

In addition to using Service Health, a few useful resources for checking outages can be found at `https://downdetector.com/status/windows-azure/` and `https://status.azure.com/en-us/status`.

Next, let's look at vulnerability management for Azure resources using Microsoft Defender for Cloud.

Overview of Microsoft Defender for Cloud

Microsoft Defender for Cloud is a next-generation security management service that offers threat protection and vulnerability management for resources running on the cloud, on-premises, and in hybrid environments. Defender for Cloud leverages Microsoft's data ecosystem to process trillions of signals through machine learning algorithms to help detect, identify, and protect against many different types of threats across multiple workloads.

At the time of writing, all Windows and Linux operating systems are supported for versions that are compatible with the Log Analytics agent, as Defender for Cloud uses Azure Monitor Logs for analysis. Defender for Cloud includes two plans – one with enhanced security features off (free) and one with enhanced security features enabled for all plans. Enabling a Defender plan does come with an additional cost but can also be applied by resource type if you wish to selectively turn them off.. Leaving enhanced security off will provide continuous assessments, security recommendations, and access to the secure score. Defender for Cloud does support integration with third-party endpoint protection products from Symantec and McAfee. Additional information about the currently supported platforms can be found at `https://docs.microsoft.com/ en-us/azure/defender-for-cloud/os-coverage`.

Defender for Cloud provides the following capabilities:

- Protects resources using **Just-In-Time** (**JIT**) access control policies
- Hardens virtual machines with adaptive application controls and application safe lists
- Investigates security alerts and mitigates threats by following remediation actions or creating automated responses to alerts
- Evaluates resources against industry and regulatory compliance standards such as SOC, PCI, HIPAA, and ISO
- Creates and manages security policies to audit and remediate misconfigured controls – for example, requiring endpoint protection or a vulnerability assessment tool to be installed
- Tracks the overall health of a resource to ensure monitoring is enabled and reviews security recommendations and alerts

For a full list of features for virtual machines and servers, visit `https://docs. microsoft.com/en-us/azure/defender-for-cloud/supported- machines-endpoint-solutions-clouds-servers?tabs=features- windows`.

To access the Defender for Cloud console in Azure, follow these steps:

1. Log in to `https://portal.azure.com`.

2. Search for **Microsoft Defender for Cloud** and open it.

   The following screenshot shows the **Overview** pane of Microsoft Defender for Cloud:

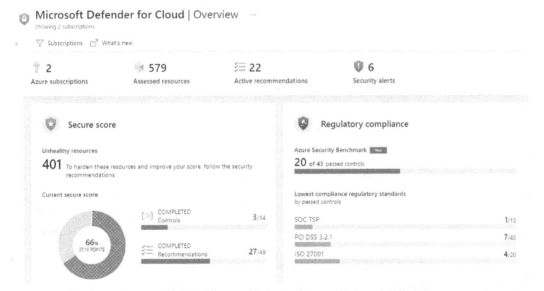

Figure 13.20 – The Defender for Cloud console

A significant benefit of Defender for Cloud is its ease of use. By default, it's enabled in your Azure subscription, and you can begin to receive many of its benefits immediately for free. Enabling the enhanced security option in Defender plans adds additional features that include JIT VM access, regulatory compliance dashboard and reports, adaptive application controls, threat protection and EDR for servers, and threat protection for PaaS services. To enable enhanced security within your subscription, follow these steps:

1. From the **Overview** pane of Microsoft Defender for Cloud, click on **Environment settings**.

2. Click on your subscription to open the settings for **Defender Plans**.

3. Click on **Enable all** and then click **Save** to enable the enhanced security features.

   The following screenshot shows all the features in the **On** state for enhanced security.

By default, all Defender for Cloud plans will be enabled. To toggle individual resource plans on or off, scroll down on the same page and change their statuses so that they fit your coverage needs:

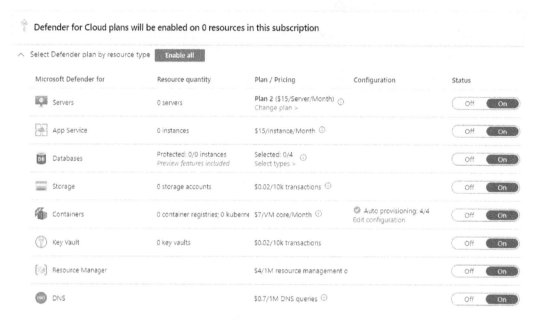

Figure 13.21 – Defender for Cloud plans

The pricing shown in the preceding screenshot may change, but the most updated pricing is available on the Azure pricing website at `https://azure.microsoft.com/en-us/pricing/details/defender-for-cloud/`.

Back on the **Settings** page, there are several other options for configuring Defender for Cloud to review:

- **Auto provisioning** allows you to automatically configure the Azure extensions needed to auto-onboard virtual machines into Log Analytics and enable EDR and threat protection. We mentioned this option earlier in the *Collecting telemetry with Azure Monitor Logs* section.

- **Email notifications** is where you can configure who should receive email notifications for alerts.

- **Integrations** allows Defender for Cloud to integrate with other Microsoft security services such as Defender for Cloud Apps and Defender for Endpoint.

- **Workflow automation** is where you can configure automation workflows based on event triggers. For example, you can customize notifications to specific users and schedule remediation actions.

- **Continuous export** allows data from Defender for Cloud to be exported or forwarded into Azure Event Hub or a Log Analytics workspace.

To onboard non-Azure Windows and Linux servers, you will need to install the Log Analytics agent or AMA through another method other than auto-provisioning. We covered different methods of onboarding servers in the *Onboarding Windows servers to Log Analytics* section of this chapter.

Once the server endpoints have been onboarded, the dashboards will begin to populate with data. To view the security recommendations and potential impact on the secure score, click on **Recommendations** on the **Microsoft Defender for Cloud** overview page under the **General** menu. Examples of some of the recommended controls can be seen in the following screenshot:

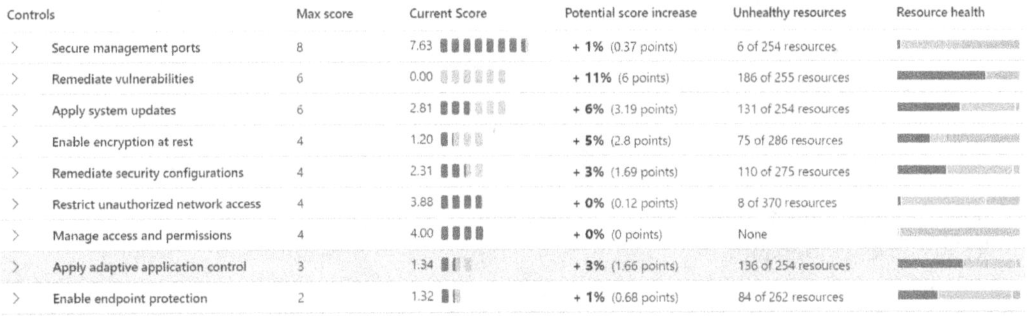

Controls		Max score	Current Score	Potential score increase	Unhealthy resources	Resource health
>	Secure management ports	8	7.63	+ 1% (0.37 points)	6 of 254 resources	
>	Remediate vulnerabilities	6	0.00	+ 11% (6 points)	186 of 255 resources	
>	Apply system updates	6	2.81	+ 6% (3.19 points)	131 of 254 resources	
>	Enable encryption at rest	4	1.20	+ 5% (2.8 points)	75 of 286 resources	
>	Remediate security configurations	4	2.31	+ 3% (1.69 points)	110 of 275 resources	
>	Restrict unauthorized network access	4	3.88	+ 0% (0.12 points)	8 of 370 resources	
>	Manage access and permissions	4	4.00	+ 0% (0 points)	None	
>	Apply adaptive application control	3	1.34	+ 3% (1.66 points)	136 of 254 resources	
>	Enable endpoint protection	2	1.32	+ 1% (0.68 points)	84 of 262 resources	

Figure 13.22 – Defender for Cloud recommendations

Any threats that Defender for Cloud detects will trigger an alert and assign a severity rating. To view any active alerts, click on **Security alerts** under the **General** menu:

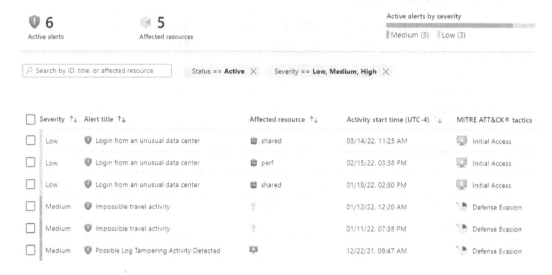

Figure 13.23 – Security alerts

To review a reference of the different types of security alerts, visit `https://docs.microsoft.com/en-us/azure/defender-for-cloud/alerts-reference`.

Microsoft Defender for Cloud categorizes security alerts by assigning them a MITRE ATT&CK tactic. Understanding these tactics can help you correlate events during a security incident when you're putting an attack timeline together.

Next, let's review the available report types in MEM for Windows systems managed with Intune or Configuration Manager co-management.

Reporting in MEM

MEM is the unified endpoint management solution that brings Intune and Configuration Manager together. Using the power of Microsoft cloud analytics, MEM provides unified reporting views to gain insights into device compliance, security updates, encryption status, and the overall health of managed endpoints. For security operations, having these views is a key component for monitoring endpoints and maintaining a good device security posture. With the increased adoption of remote work, admins also need to be able to collect logs from endpoints to help troubleshoot operational issues or for forensic security investigations.

In the next few sections, we will cover the different service-side reports available in MEM and where to collect client-side logs on managed devices. We will review how to enable device health telemetry collection, monitor software update deployments, and use Endpoint analytics to provide proactive reporting.

Security-focused reports in MEM

A great feature in MEM includes the built-in reporting that is available by default with the service. There is no immediate need to set up additional infrastructure to host data log collection, although retention past 90 days can be configured by changing the diagnostic settings to export platform logs into a Log Analytics workspace, storage account, or event hub. There are four main category report types in MEM, as follows:

- **Operational** reports include details about the status of individual devices. For example, you can check a **Device Configuration** profile assignment status, the encryption status report, or details about certificate issuance using the Intune certificate connector. These reports are most useful for reviewing or troubleshooting individual devices. To view these reports in MEM, choose **Devices** and click on **Monitor**.

- **Organizational** reports provide higher-level details about the ecosystem of endpoints and the overall status of the Intune service. Examples include device compliance reports and the status of endpoint security features such as Defender Antivirus and Firewall. To view these reports, choose **Reports** from the MEM home page.

- **Historical** reports allow you to view insights regarding past trends in your MEM environment, such as the trend of compliant devices over an amount of time. To collect and report on historical data, Microsoft Intune Data Warehouse can be used to connect MEM data to Power BI for reporting. To configure the Intune Data Warehouse, choose **Reports** and find **Data warehouse** in the navigation pane.

- **Specialist** reports include custom dashboards built by system admins and engineers using Azure workbooks from client data and MEM diagnostic data stored in Azure Monitor Logs. To configure diagnostic settings, choose **Reports** and click on **Diagnostic settings**.

Let's look at where to find the security-focused reports in MEM.

Device compliance reports

Device compliance reports help show if devices meet the criteria outlined in the conditions defined to access company resources. Device compliance is a key zero-trust component for protecting organization data by ensuring the devices accessing the resources meet both the safety and security standards that are outlined. Device compliance reports are available at both an operational and organizational level and can be found in the following locations in MEM:

- Operational reports can be found at **Devices | Monitor** under **Compliance.**

- Organizational reports can be found at **Reports | Device compliance** under **Device management.**

The following screenshot shows an example of an organizational report for **Device compliance**:

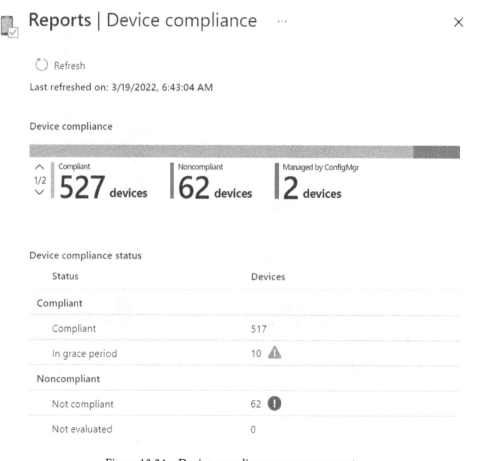

Figure 13.24 – Device compliance summary report

Within the summary overview page, choose **Reports** to view reports about the compliance status of individual devices and to view trends over the past 60 days.

Next, let's look at unhealthy endpoints and malware reports.

Unhealthy endpoints and active malware reports

The unhealthy endpoints report is an operational report that shows the status of the Defender Antivirus services running on endpoints that are reporting issues. You can use this report to find devices that may lack antivirus protection and have features such as the malware protection service disabled, real-time protection turned off, or are missing signature definition updates. Bulk remediation options are available directly in the reports and include running a quick and full antivirus scan and triggering a signature definition update on selected endpoints.

The active malware report is also an operational report that shows endpoints with unresolved threats detected from MDE. This includes the status of the malware and classifications, including its name, category, and severity. The active malware report also supports the same bulk remediation options as unhealthy endpoints.

To view these reports in MEM, go to **Endpoint Security | Antivirus** and choose **Unhealthy endpoints** or **Active malware**.

Antivirus agent status report

The antivirus agent status report is an organizational report that shows the status of Defender Antivirus features on the managed endpoints. Use this report to get the following details from clients:

- Device state
- Anti-malware version
- Tamper protection
- Real-time protection status (MDE Sense)
- MDE onboarding status
- Defender Engine Version

In addition to the antivirus agent status report is the detected malware report. You can use this report to view any detected malware on your devices, as reported from MDE. The report types can be seen in the following screenshot of the MEM console:

Figure 13.25 – Microsoft Defender Antivirus reports

To view the **Antivirus agent status** and **Detected malware** reports, go to **Reports**, choose **Microsoft Defender Antivirus** under **Endpoint Security**, and select **Reports**.

MDM Firewall status report

The **MDM Firewall status for Windows 10 and later** report shows details on the state of the Defender firewall on managed devices. The available statuses, as shown in the following screenshot, include **Enabled**, **Disabled**, **Limited**, **Temporarily disabled**, and **Not applicable**:

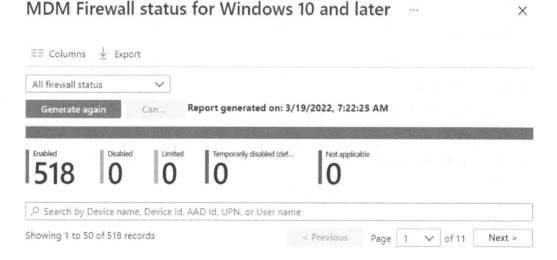

Figure 13.26 – MDM Firewall status

To view the MDM Firewall status report, go to **Endpoint Security | Firewall** and choose **MDM Firewall status for Windows 10 and later**.

Encryption report

The encryption report is another operational report that you can use to validate the status of encryption on your devices. The report includes details such as the device's name, operating system, OS version, TPM version, encryption readiness, and encryption status. Within the report, you can click on a device to view additional details, such as the status of an Intune BitLocker profile. The **Encryption readiness** column shows if the device supports enabling encryption using technology such as BitLocker or FileVault. To view the encryption report, choose **Devices | Monitor | Encryption report** under **Configuration**.

Windows Device Health Attestation report

The Windows **Device Health Attestation (DHA)** report provides insights into hardware-based security features that are enabled on your managed endpoints. The DHA feature in Windows can help ensure that the boot phase of a device remains secure and in a trusted state. The report includes a status for the following security features:

- BitLocker
- Code Integrity
- Early Launch AntiMalware
- Boot Debugging
- **Attestation Identity Key (AIK)**
- Secure Boot
- Data Execution Prevention Policy
- And more

To view the report in Microsoft Endpoint Manager, go to **Devices | Monitoring** and choose **Windows health attestation report**.

To configure devices to send data into the DHA report, deploy an Intune compliance policy that checks for BitLocker, Secure Boot, and Code Integrity. We covered how to deploy this policy in *Chapter 8*, *Keeping Your Windows Clients Secure*, in the *Configuring a Device Compliance Policy* section.

Next, let's look at enabling a profile in Intune to configure additional telemetry data collection for Windows updates and Endpoint analytics.

Enable Windows Health Monitoring

A Windows Health Monitoring profile can be assigned to managed endpoints using MEM, which allows Microsoft to configure the necessary data collectors to power Endpoint analytics reporting and enhanced telemetry analysis by the Windows Update Service. Once the profile has been created and assigned, Endpoint analytics will begin to be populated with insights and provide recommendations by calculating individual device scores. These scores are based on identified issues through slow startup, performance, and application reliability metrics. The Windows Health Monitoring profile also helps to power the Windows 11 readiness report in Endpoint analytics and enables proactive remediation functionality for deploying script packages to detect and fix common issues.

> **Tip**
> For co-managed devices, the client app's workload must be set to Intune to use proactive remediation scripts.

To deploy the health monitoring profile, use an Intune Device Configuration profile and select the **Windows health monitoring** profile type for **Windows 10 and later**. For the most benefit, select both the **Windows updates** and **Endpoint analytics** scopes, as shown in the following screenshot:

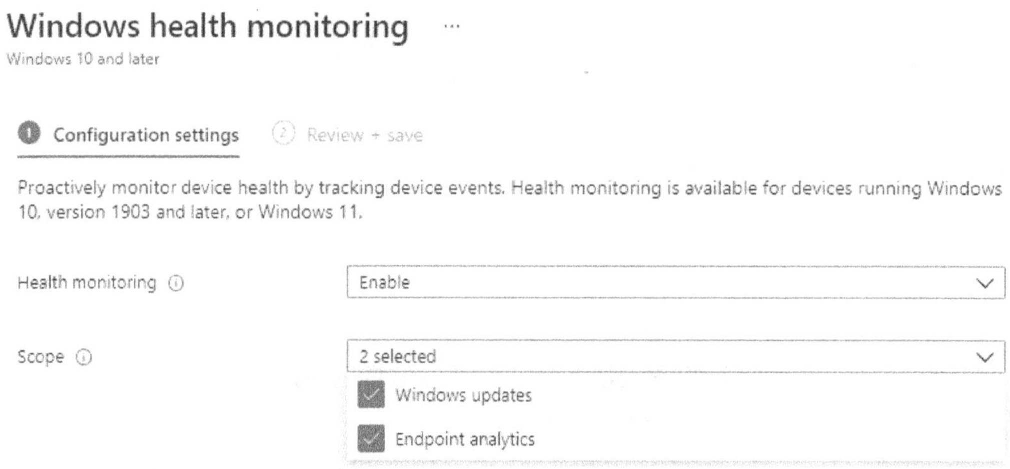

Windows health monitoring ...
Windows 10 and later

1 Configuration settings   (2) Review + save

Proactively monitor device health by tracking device events. Health monitoring is available for devices running Windows 10, version 1903 and later, or Windows 11.

Health monitoring (i)	Enable ⌄
Scope (i)	2 selected ⌄
	☑ Windows updates
	☑ Endpoint analytics

Figure 13.27 – Windows health monitoring profile

Once the profile has been assigned, Endpoint analytics data should begin to populate within 24-48 hours. Next, let's look at how to use Endpoint analytics reporting.

Using Endpoint analytics

Endpoint analytics allows you to create performance baselines and track how devices perform against the overall organizational score over time. Data collected by Endpoint analytics helps to identify issues with hardware, policy configurations, and applications that could be causing reduced performance for your end users. Some of the insights that can be gained by enabling Endpoint analytics are as follows:

- Startup score

- Model performance

- Device performance

- Startup processes

- Restart frequency

- Application reliability score

- App performance

- OS version performance

By tracking these performance metrics, IT staff can triage issues quicker with better insights and help reduce the investigation time, should a user open a support ticket. If trends are observed for multiple devices within the organization, proactive remediation scripts can be deployed to detect and fix common problems to reduce the burden on the support desk. For detailed information about using Endpoint analytics, visit https://docs.microsoft.com/en-us/mem/analytics/overview.

To use Endpoint analytics, devices can be onboarded from both Intune and Configuration Manager. To view the reports in the MEM admin center, click on **Reports** and choose **Endpoint analytics**. The following screenshot shows how the combined score of your devices stacks up against all organizations or the custom baseline for the **Startup performance**, **Application reliability**, and **Work from anywhere** metrics:

Figure 13.28 – Endpoint analytics

By clicking on **Device scores** next to **Overview**, you can dig deeper into the metrics for individual devices to troubleshoot issues. To find out more about the prerequisites to enable Intune on devices, visit `https://docs.microsoft.com/en-us/mem/analytics/enroll-intune`.

Now that we have reviewed service-side reporting, let's learn where to collect MDM diagnostic logs directly from clients.

Collecting client-side diagnostic logs

When the service-side logs don't provide enough detail for an individual device, sometimes, it's necessary to collect and review client-side logs. Devices managed with Intune create numerous client-side logs, depending on what you're trying to troubleshoot. Let's review what they are and how to gather them from clients.

Remotely collecting diagnostic logs

Windows device logs can be collected remotely by using the bulk device actions in MEM. After the collection action is complete, the diagnostic data is stored for 28 days in Intune until it's deleted by the service. The following screenshot shows the **Collect diagnostics** bulk device action on the **Device** overview page:

Figure 13.29 – Intune remote actions

The data that's collected includes information about the device's Azure AD status, system information, network configurations, Windows Event Viewer logs, and certificate store details. For detailed information on what data is collected, visit `https://docs.microsoft.com/en-us/mem/intune/remote-actions/collect-diagnostics`.

Next, let's look at collecting logs directly from a client.

Generating an MDM diagnostic report

An MDM diagnostic report can be generated directly on the client by going to **Settings | Access Work or School | Connected to Azure AD tenant | Info**. Then, choose **Create report** under **Advanced Diagnostic Report**. The exported report will be saved to the `C:\Users\Public\Public Documents\MDMDiagnostics` folder and contains an HTML-formatted file that shows general information about the device, connection information to MDM, managed certificates, and the profiles and policies that have been applied by Intune. To generate a full MDM diagnostic report, go back to **Access work or School** and choose **Export your management log files**. This will collect all possible MDM diagnostic logs, such as the remote diagnostic collection from Intune.

Intune Management Extension logs

The **Intune Management Extension (IME)** logs are used to review PowerShell scripts and Win32 app deployments that have been assigned to devices in Intune. To view the logs, navigate to `C:\ProgramData\Microsoft\IntuneManagementExtension\Logs`. These logs can help you troubleshoot both script and Win32 app deployments as you can review any output or error codes.

To view the registry parameter information for Win32 apps from the IME, navigate to `HKLM:\Software\Microsoft\IntuneManagementExtension\Win32Apps\<GUID>\<App GUID>`. To correlate the app in MEM, the app GUID will match the end of the URL in the browser when opening the overview pane of a Win32 app from the MEM portal. Some of the available information includes exit codes and the download start times of the app deployment.

Line-of-business or MSI app logs

To troubleshoot an MSI app deployment, you can review the following logs, depending on if the app was targeted at users or devices:

- **Device targeted** MSI deployment logs are written to `%windows% \temp\<MSIProductCode>.msi.log`.

- **User targeted** MSI deployment logs are written to `%temp%\<MSIProductCode>.msi.log`.

To view the registry information for MSI app deployments, including information about error codes, product version, and deployment status codes, visit the following locations:

- For **Device targeted** app deployments, go to `HKLM:\Software\Microsoft\EnterpriseDesktopAppManagement\<SID>\MSI\<MSIProductCode>`.

- For **User targeted** app deployments, go to `HKLM:\Software\Microsoft\EnterpriseDesktopAppManagement\<UserSID>\MSI\<MSIProductCode>`.

Next, let's look at where to check MDM logs in Event Viewer.

Event Viewer logs

The following location can be used to view information about MDM policy enforcement and device check-in details about the Intune service:

Event Viewer | Applications and Services Logs | Microsoft | Windows | Device Management-Enterprise-Diagnostics-Provider

Viewing applied policies

To view the settings that have been applied by the policy providers shown in the `DeviceManagement-Enterprise-Diagnostics-Provider` logs, go to `HKLM:\Software\Microsoft\PolicyManager\current\device` under **Registry Editor**. Here, you can view all the CSP-backed policies and the settings that have been applied to the client. To view the blocked group policies from the Group Policy engine because MDM has configured an equivalent policy, look for the **Blocked Group Policies** section in the `MDMDiagReport.html` file that was generated from the MDM diagnostic report.

Next, let's review the available reporting in MEM for Windows updates deployments to managed endpoints.

Monitoring update deployments

Monitoring the status of security updates is another key process to include in any security operations program. Ensuring updates are deployed timely is one of the best ways to keep devices secure against the latest threats. In *Chapter 8, Keeping Your Windows Client Secure*, we covered how to assign Windows Update for Business policies using update ring deployment profiles to managed devices. We can leverage those same deployment profiles to track the update statuses of clients that have been assigned to a deployment ring natively in MEM.

In MEM, choose **Devices | Update rings for Windows 10 and later** and select a deployment ring to view the overview. Under the **Monitoring** section, select **End User update status** to view details about the current device's **Quality Update Version**, **Feature Update Version**, **Last Scan Time**, and **Last Check-in Time**. While the report provides an overview of quality and feature updates, it doesn't provide much in-depth information about the overall compliance of your endpoints, nor any diagnostic details for failed or delayed update installations. For a better overview, we recommended using the **Update Compliance** solution in a Log Analytics workspace or building a custom Azure workbook. We discussed the Update Compliance solution earlier in the *Monitoring solutions and Azure workbooks* section of this chapter. The following screenshot shows the latest security update deployment status, sorted by OS build, as reported from the Update Compliance solution:

LATEST SECURITY UPDATE DEPLOYMENT STATUS

OS BUILD	VERSION	INSTALLED	IN PROGRESS OR DEFERRED	UPDATE ISSUES	STATUS UNKNOWN
22000.739	Win11-21...	149	7	1	9
19044.1766	21H2	289	22	4	49
19043.1766	21H1	2	0	0	0
19042.1766	20H2	1	0	0	0

Figure 13.30 – Update Compliance

If the reporting from Update Compliance doesn't meet your reporting needs, the log data can be used to create different views in an Azure workbook. The following screenshot shows an example of what this looks like:

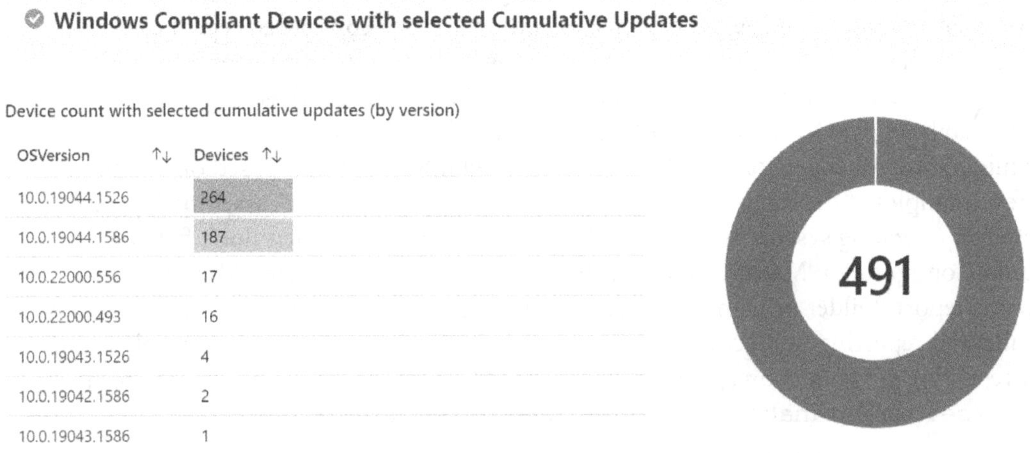

✅ **Windows Compliant Devices with selected Cumulative Updates**

Device count with selected cumulative updates (by version)

OSVersion ↑↓	Devices ↑↓
10.0.19044.1526	264
10.0.19044.1586	187
10.0.22000.556	17
10.0.22000.493	16
10.0.19043.1526	4
10.0.19042.1586	2
10.0.19043.1586	1

491

Figure 13.31 – Update Compliance report in an Azure workbook

Additional options for reporting the status of Windows updates are available in MEM if you're using the **Feature updates for Windows 10 and later** or **Quality updates for Windows 10 and later deployment** type. To view these reports in MEM, click on **Reports** and choose **Windows Updates** under **Device management** to view a summary or get a detailed report of an update deployment. Only devices that have been assigned either a feature update or expedited quality update policy will be available in the report.

Reports can also be created manually by exporting all Windows devices from MEM into a CSV file and comparing their current builds to the latest release builds from the Windows update release history website. Windows 11 update release history can be found at `https://support.microsoft.com/en-us/topic/windows-11-update-history-a19cd327-b57f-44b9-84e0-26ced7109ba9`.

Next, let's look at the reporting capabilities available in Microsoft Endpoint Configuration Manager.

Reporting in Microsoft Endpoint Configuration Manager

Configuration Manager has a wide variety of reports available from the data that's been collected from managed endpoints, including 500 pre-configured reports. Much of the data that populates these device reports has been collected from software and hardware inventory classes that are defined in the client settings. These reports leverage the capabilities of SQL Server Reporting Services and can be viewed directly in the Configuration Manager console. They are organized into subfolders based on category. A few examples of the categories that are available include hardware, asset intelligence, network, operating system, software, and users. If the built-in reports don't fit your needs, Configuration Manager supports building custom reports using Microsoft SQL Server Report Builder or with Power BI. If you use Configuration Manager and have not configured reporting, we recommend going to the following link to review the planning for reporting guide: `https://docs.microsoft.com/en-us/mem/configmgr/core/servers/manage/introduction-to-reporting`.

To access the reports in the management console, navigate to **Monitoring | Reporting | Reports**. The following are some of the security-related reports in Configuration Manager that can be reviewed against known vulnerabilities:

- **Lifecycle 05A – Product lifecycle dashboard** shows information about the Windows version distribution of your managed clients.

- **Details of firmware states on devices** shows the status of SecureBoot, UEFI, TPM version, and TPM state.

- **Hardware 01A – Summary of computers in a specific collection** shows details about managed device hardware such as the manufacturer, model, memory, processor, and disk space.

- **Summary compliance by configuration baseline** shows the overall compliance of deployed configuration baselines. You can drill into the report to get more details about the baselines, view the status of individual configuration items, and report on devices with non-compliant rules.

- **Count of all instances of software registered with Add or Remove Programs** provides a detailed software inventory of all managed devices.

- **Services – Services information for a specific computer** displays summary information about all the services running on a single computer.

This is only a very small sample of the reporting capabilities within Configuration Manager and many additional reports are available, depending on the enabled features and software and hardware inventory classes that have been set in the client settings within MECM. For a detailed list of the available reports, visit `https://docs.microsoft.com/en-us/mem/configmgr/core/servers/manage/list-of-reports`.

Next, let's use the Microsoft 365 Apps admin center to monitor the health and update status of Office apps.

Monitoring the health and update status of Office apps

The modern Microsoft 365 apps suite runs on a servicing channel that is independent of Windows and Windows Update. As a result, Office product updates are not delivered through the typical Windows updates servicing cycle. Office apps consistently remain a popular target for attackers who wish to exploit software bugs or security policy misconfigurations to launch malicious code, so it's important to have insight into the health and update status of Office apps. This helps ensure effective servicing and that you remain protected against the latest vulnerabilities. In *Chapter 9, Advanced Hardening for Windows Clients*, we covered deploying a security baseline for Office apps using the Microsoft 365 Apps admin center, also known as the Office Cloud Policy service. The admin center also includes dashboards that can monitor app health and security update status. To log in to the Office Cloud Policy service, go to `https://config.office.com`. Let's review the available dashboards in the Microsoft 365 Apps admin center.

Microsoft 365 Apps health dashboard

The Microsoft 365 Apps health dashboard provides insights into the overall health of the entire suite of Office apps in your organization by analyzing performance metrics across the different servicing channels. The metrics that are collected include details about crash rates, boot times, and measuring time delays in opening files to calculate a reliability metric. The **Overview** dashboard can help inform you of any deteriorations across app types, and then sorts them by app or impacted channel. The **Advisories** tab lists the latest performance advisories sorted by impacted application, channel, build, and severity. The following screenshot shows the app session crash rate for Outlook over the last 30 days by channel and build:

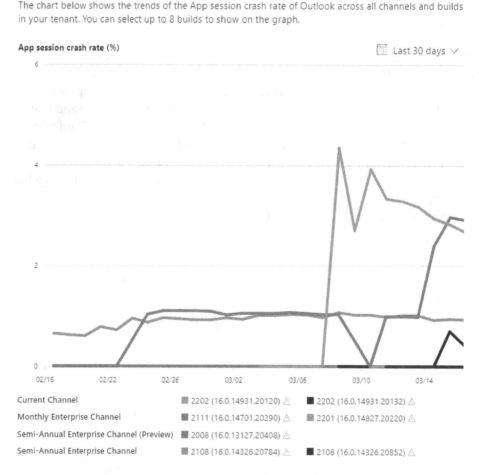

Figure 13.32 – App metrics for Outlook

Using the **App metrics** tab, you can view each app individually for a more fine-grained view of app health and sort by build or performance metrics such as the crash rate or application boot time.

Monitoring Security Update Status

The Security Update Status provides a dashboard of your security update compliance across all servicing channels by comparing them to the latest security update release date from Microsoft. To view the dashboard, go to **Health** and choose **Security Update Status** in the admin center. Here, you can easily track which devices are failing to update and build an action plan to get them fixed. In addition to reporting on security updates, you can use the **Goal** feature to set a target update goal and timeline to help track and report on update deployments. Creating a goal doesn't schedule update deployments or modify the policy and is only used for reporting purposes. The following screenshot shows the overall update status of all Office installations across all channels and devices compared to the latest release security updates:

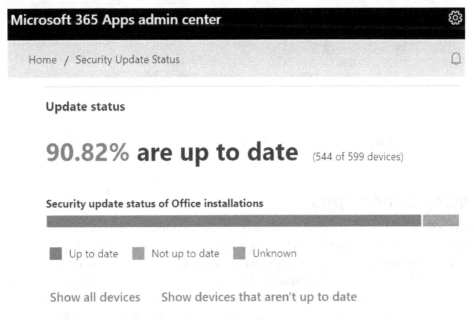

Figure 13.33 – Security Update Status

The dashboard also includes a link to a list of the latest security update releases from Microsoft.

Viewing the Office Inventory report

The Office Inventory dashboard provides reporting on the state of all servicing channels and builds across your environment. Here, you can easily view the build spread to understand the diversity across your environment and check if any devices include versions that are out of support. The **Data Insights** section reports the total number of devices checking into the Microsoft 365 Apps admin center and the device count by the architecture of the Office installation. The **Add-ins** section provides a report about the identified add-ins, the number of devices with the add-in installed, and the version numbers across all installations. The following screenshot shows the Office build spread from the Office Inventory dashboard.

Figure 13.34 – Office build spread

Devices are removed from the inventory if they have not sent a heartbeat within 30 days.

Servicing Office apps

The Microsoft 365 Apps admin center also allows you to configure automatic updates for Office apps using servicing profiles. Servicing profiles are used to automatically control and deliver Office updates to your clients and can be assigned to groups of users or devices. Assigning a servicing profile will automatically switch devices to the Monthly Enterprise servicing channel, uses the **Office Content Delivery Network (CDN)** to download updates, and takes precedence over other management tools for configuring automatic update settings, should there be a conflict. Windows Delivery Optimization policies can be used in conjunction with servicing profiles to help control bandwidth settings for update deployments, including configuring throttling based on network congestion and location. Should an issue arise during a rollout, updates can be paused or rolled back until a fix can be applied. To learn more about using servicing profiles, visit https://docs.microsoft.com/en-us/deployoffice/admincenter/servicing-profile.

Once a profile has been created, active devices will receive the profile within 24 hours.

Starting in June 2022, Microsoft recommends moving all Microsoft 365 app deployments to the Monthly Enterprise update channel to provide the best experience for end users.. If you're ready to switch channels, using a servicing profile is a great way to change the update channel for existing installations.

Summary

In this chapter, we covered different methods that can be used for monitoring and reporting Windows servers and clients. We covered MDE and reviewed the available dashboards that help track vulnerabilities, identify weaknesses, and report on machine health and software inventory. Then, we reviewed how to onboard Windows clients using the Microsoft Intune Connector and how to assign an EDR profile. Next, we learned about collecting telemetry data using Azure Monitor Logs and Log Analytics workspaces. We covered onboarding Windows clients and servers, as well as how solutions from the Azure Marketplace can help you configure data collectors and visualize the data being captured. We also provided an overview of Azure Monitor for viewing performance insights and reviewed the Azure activity logs for auditing resource operations.

In the next section, we discussed enabling Microsoft Defender for Cloud and showed you how to enable enhanced security and auto-provisioning to automatically onboard virtual machines. Then, we covered many of the reporting capabilities available in Microsoft Endpoint Manager, how to collect client-side diagnostic logs, and how to monitor security update deployments. Finally, we discussed the importance of monitoring Microsoft 365 Apps health and security updates using the Microsoft 365 Apps admin center.

In *Chapter 14*, *Security Operations*, we will provide an overview of the **Security Operations Center** (**SOC**) and its role within an organization. We will explain how organizations can use the different Microsoft 365 security tools to investigate potential security threats. Then, we will discuss the importance of planning for business continuity and disaster recovery using a cyber incident response plan.

14
Security Operations

In this chapter, we will cover security operations and how monitoring tools can be incorporated into their operational workflows. Like a technical operations team, it is just as important to have a security operations team or **Security Operations Center (SOC)** and program in place. This team's day-to-day responsibilities include 24/7 monitoring and response to any security-related incidents that affect your environment or end users. This is a critical component and a necessity of an overall security program. Recent trends have shown that attackers operate during off hours, weekends, and holidays to avoid detection and maximize damage.

In this chapter, we will focus on the Microsoft technologies available that can support your SOC and provide the insights needed to ensure that servers, end user devices, and users are safe. We will first cover an introduction to a SOC and provide an overview of the key concepts needed to implement a successful operation. We will define **eXtended Detection and Response (XDR)** and review the **Microsoft 365 (M365)** Defender portal, which provides a centralized place for monitoring Microsoft security solutions. Next, we will focus on **Microsoft Defender for Endpoint (MDE)** and review how to prevent, detect, investigate, and respond to threats.

In the following section, we will review Defender for Cloud alerts and incidents and provide an overview of **Microsoft Sentinel**, Microsoft's cloud **Security Information and Event Management (SIEM)** tool. Next, we will cover **Microsoft Defender for Cloud Apps (MDCA)** and how to configure and use **Microsoft Defender for Identity (MDI)**. To close the chapter, we will discuss data protection with M365 and finish with a brief overview of planning for business continuity.

To recap, this chapter will cover the following topics:

- Introducing the SOC
- Understanding XDR
- Using the M365 Defender portal
- Security operations with MDE
- Investigating threats with Defender for Cloud
- Enabling Azure-native SIEM with Microsoft Sentinel
- Protecting apps with MDCA
- Monitoring hybrid environments with MDI
- Data protection with M365
- Planning for business continuity

Technical requirements

To follow along with the overviews in this chapter and complete the how-to instructions, the following requirements are recommended:

- An Azure subscription with contributor rights
- MDCA (trial/MDCA and EMS E3 or E5)
- MDI
- Microsoft Defender for Cloud (free/enhanced)
- Microsoft Sentinel (per capacity or pay-as-you-go)
- MDE (Windows 10 E5/M365 E5)

Many of the services allow you to set up trials or include free versions to help you get started. We recommend working with a sales representative at Microsoft to understand the best options regarding licensing, as the requirements change frequently.

In addition to these products, we will also discuss an overview of Microsoft's data loss prevention technology using components of **Microsoft Purview Information Protection**, formerly **Microsoft Information Protection (MIP)**, and **Azure Information Protection (AIP)**. Throughout this chapter, the abbreviations may be used interdependent of each other.

Introducing the SOC

Operations within the technical world of a mature organization should be a standard and mature process. This function is core to the ongoing success of IT systems, users, and applications to ensure efficiency and availability for your business operations. If there is an outage or an issue, operations teams typically follow very strict **Service Level Agreements (SLAs)** to return a service to normal. This same concept is applicable to the security world. The concept of a SOC has grown exponentially over recent years, to the point where it is a necessity for maintaining normal business operations.

In short, a SOC manages and overlooks the day-to-day functions of your security operations. They should operate 24/7 to monitor and detect potential security risks and respond to alerts within an environment. If any alerts are detected, it is the SOC's responsibility to investigate, help remediate, and escalate to the appropriate team or personnel. A major part of this process also includes identifying the impact and potential damage your organization may face because of a security incident.

When looking at the security incident response life cycle, there are a few variations, but they all mostly overlap and follow a similar process. When referencing the *NIST Special Publication 800-61, Revision 2, Computer Security Incident Handling Guide*, a four-step process is used for the incident response life cycle:

1. Preparation
2. Detection and analysis
3. Containment, eradication, and recovery
4. Post-incident activity

To learn more about the NIST incident response life cycle, visit the *NIST Computer Security and Incident Handling Guide* at `https://nvlpubs.nist.gov/nistpubs/SpecialPublications/NIST.SP.800-61r2.pdf`.

One important document that needs to be clearly stated after any incident is the **Root Cause Analysis (RCA)**. Ensuring a root cause is identified validates that the vulnerability that caused the incident has been remediated. In addition, and sometimes dictated by the severity of the incident, the RCA may need to be provided to the leadership team for review. For this reason, the RCA must clearly define why there was an incident, what the impact or damage was, how the incident was remediated or mitigated, and an action plan to ensure any findings are resolved and don't happen again.

There are different variations of SOC models that can be adopted, and different factors will determine which model makes sense for your organization. These factors may include the company size, the type of industry of the business, regulatory reasons, and budget considerations. In the insourced model, a SOC is managed and operated by staff internal to the organization. This is a likely case for larger enterprises with bigger budgets who can afford to recruit top talent. Some organizations may opt for a fully outsourced model where you contract your SOC services to an external vendor who specializes in security services. This is referred to as a **Security Operations Center as a Service (SOCaaS)** or an outsourced SOC, which is typically offered through a **Managed Security Service Provider (MSSP)**. The SOCaaS model is an attractive service for medium to smaller-sized businesses that may not have the budget to implement a fully functioning insourced SOC. This hybrid model maintains some functions of the SOC internally and outsources a subset of specialty services to an MSSP.

An additional service that a mature SOC program should implement is a **Digital Forensic Incident Response (DFIR)** service. This is a very specialized service that provides a detailed analysis of artifacts during an investigation of a security incident. Due to the specialized skillset and tools used in digital forensics, the MSSP may not provide a DFIR as part of its standard service offerings. With the continuous increase of threats that organizations face today, having a DFIR service available to engage quickly is critical to providing a detailed forensic analysis of evidence in the attack timeline. If your business has purchased a cyber insurance policy (which is highly recommended), the cyber insurance company will have an approved list of vendors that you can engage with for any DFIR services.

A SOC's ability to be efficient will depend on the available tools to provide the best visibility into your environment and the activities that have taken place. Throughout this chapter, we will be reviewing many of the Microsoft tools that help make up a well-rounded and robust security operations program. Next, let's look at XDR and how relevant it has become for cyber defense.

Understanding XDR

XDR expands beyond the original detection and response capabilities of a single technology to include multiple technologies. You should now be familiar with **Endpoint Detection and Response (EDR)** for MDE and its comprehensive coverage across endpoints as a single technology. XDR not only covers endpoints but also expands into email, servers, cloud infrastructure, identity and access management, network, applications, and so on.

You are most likely monitoring, collecting data, running analysis, and responding to all the different technology areas mentioned previously as part of your security program. The challenge commonly faced is a lack of a unified view and context in the attack timeline within these technology areas, due to separate tools and solutions. XDR brings everything into a centralized view for greater efficiency, allowing for quicker response to incidents and alerts. XDR doesn't replace a SIEM or **security orchestration automation and response (SOAR)** deployment. It complements it by adding intelligence, automation, and advanced analytics to data to correlate events and allow focus on higher-priority incidents.

XDR is available from many vendors now, and it's highly recommended you review and understand XDR and the benefits it can provide to your organization. With the focus of this book on Microsoft technologies and the Microsoft security stack, the following solutions offer XDR capabilities:

- M365 Defender to protect and secure your end users with tools such as MDI, Defender for Office 365, MDE, and MDCA
- Microsoft Defender for Cloud to protect and secure your infrastructure with secure posture management and threat protection across Azure, hybrid, and other third-party cloud environments
- Microsoft Sentinel, a cloud-native SIEM and SOAR, for increased efficiency with your XDR capabilities

The following figure provides a high-level overview of Microsoft's XDR capabilities and what prevention, detection, and protection capabilities are provided:

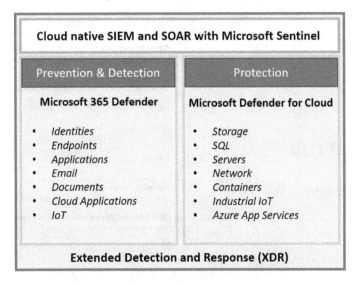

Figure 14.1 – Microsoft's XDR capabilities

Next, let's review the first component of XDR using the M365 Defender portal.

Using the M365 Defender portal

To help centralize the management and configuration of the Microsoft Defender products, the **M365 Defender** portal provides a *single pane of glass* experience. This helps simplify security operations and increase efficiencies by combining portals, allowing easier correlation of security threats across much of the Microsoft security suite to be viewed from one centralized place. The M365 Defender portal can be accessed by going to `https://security.microsoft.com`. The security team will need to be assigned a Global Administrator, Global Reader, Security Administrator, Security Operator, or Security Reader role to view data, with other fine-grained access controls available depending on the products they need to access. From the landing page, you can add and organize cards to quickly visualize activities across the products holistically. This can be seen in the following screenshot from the M365 Defender portal landing page:

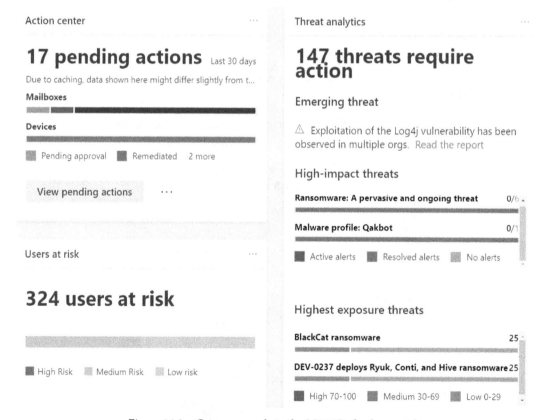

Figure 14.2 – Common cards in the M365 Defender portal

Within the M365 Defender portal, you can access the following products from within the menu options:

- MDE can be found under **Endpoints**. Here, you can manage the EDR capabilities of Windows Defender on clients and servers.

- **Microsoft Defender for Office (MDO)** can be found under **Email & Collaboration**. Security teams can manage threats targeting email and collaboration tools and configure MDO threat policies, such as anti-phishing and anti-spam.

- MDCA can be found under **Cloud Apps**. MDCA is a cloud app security broker that brings visibility and security controls to SaaS applications.

Access to reporting, unified audit logs, permissions and roles, and service health can all be found from the M365 Defender portal menu. Clicking **More resources** will provide links to the different management consoles that make up most of the security and compliance portfolio from Microsoft.

For more information about all the available Microsoft security portals, visit `https://docs.microsoft.com/en-us/microsoft-365/security/defender/portals?view=o365-worldwide`.

Next, let's look at using Microsoft Secure Score and the benefits it provides to help with your overall security posture.

Improving security posture with Microsoft Secure Score

Getting a reliable and consolidated overview of your organization's security posture can be challenging, especially if multiple security and vulnerability scanning solutions are being used across multiple vendors. Additionally, if your company goes through a yearly security audit exercise, it can be expensive and time-consuming, and the data can quickly become outdated. If you're heavily invested in the Microsoft security stack, one product that will help to provide an overview of your organization's security posture is Microsoft Secure Score. Microsoft Secure Score provides analysis across the identity, device, and application areas by comparing your current settings and configurations against Microsoft's recommended best practices. The output results in a percentage-based score across each category that is averaged to become your overall secure score, which includes recommendations and improvement actions. By completing these actions, you can increase your overall score. Each improvement action is weighted by priority and assigned points. As you complete each task, you will gain points that will be applied to the overall score.

To view your secure score, log in to the M365 Defender portal at `https://security.microsoft.com` and click on **Secure Score**. The **Overview** page will show your overall score, top improvement actions, and a comparison to other organizations of similar sizes. The landing page can be seen in the following screenshot:

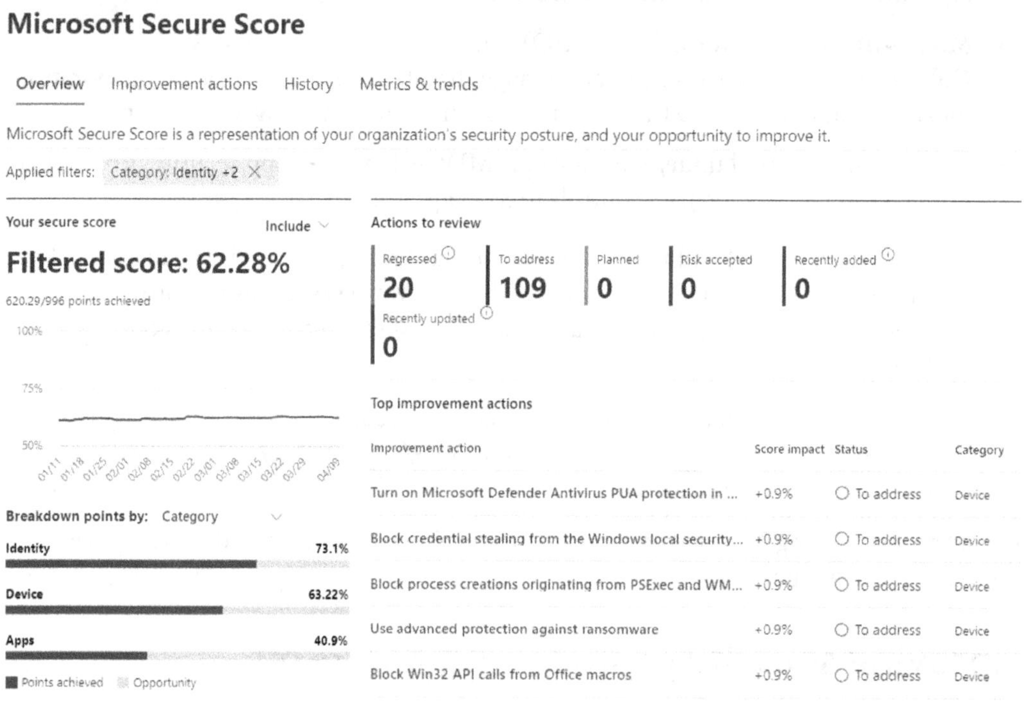

Microsoft Secure Score

Overview   Improvement actions   History   Metrics & trends

Microsoft Secure Score is a representation of your organization's security posture, and your opportunity to improve it.

Applied filters: Category: Identity +2 ✕

Your secure score   Include ⌄   Actions to review

Filtered score: 62.28%

620.29/996 points achieved

Regressed ⓘ	To address	Planned	Risk accepted	Recently added ⓘ
20	109	0	0	0

Recently updated ⓘ

0

Top improvement actions

Improvement action	Score impact	Status	Category
Turn on Microsoft Defender Antivirus PUA protection in ...	+0.9%	○ To address	Device
Block credential stealing from the Windows local security...	+0.9%	○ To address	Device
Block process creations originating from PSExec and WM...	+0.9%	○ To address	Device
Use advanced protection against ransomware	+0.9%	○ To address	Device
Block Win32 API calls from Office macros	+0.9%	○ To address	Device

Breakdown points by: Category ⌄

Identity	73.1%
Device	63.22%
Apps	40.9%

■ Points achieved   ▨ Opportunity

Figure 14.3 – Secure Score

Click the **Improvement actions** tab at the top of the page to view the list of available recommendations. Each improvement action includes more details about the recommendation, exposed entities, and an implementation plan that includes links to any correlated vulnerabilities identified by M365 Defender. After any recommended actions have been taken, your **Secure Score** dashboard will reflect the changes and add the points to your score. If the security recommendation doesn't apply, it can be exempt from reporting, and you will be credited for the score.

In the following example, Microsoft recommends enabling an **attack surface reduction (ASR)** rule to block Adobe Reader from creating child processes.

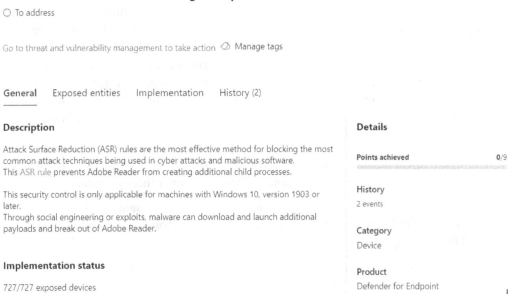

Block Adobe Reader from creating child processes

○ To address

Go to threat and vulnerability management to take action   ⟨⁄⟩ Manage tags

General   Exposed entities   Implementation   History (2)

Description

Attack Surface Reduction (ASR) rules are the most effective method for blocking the most common attack techniques being used in cyber attacks and malicious software.
This ASR rule prevents Adobe Reader from creating additional child processes.

This security control is only applicable for machines with Windows 10, version 1903 or later.
Through social engineering or exploits, malware can download and launch additional payloads and break out of Adobe Reader.

Implementation status

727/727 exposed devices

Details

Points achieved   0/9

History
2 events

Category
Device

Product
Defender for Endpoint

Figure 14.4 – Improvement actions

The current products included in the score weight include M365 (Exchange Online, OneDrive for Business, SharePoint Online, and Microsoft Purview Information Protection), Azure AD, MDE, MDI, MDCA, and Microsoft Teams.

> **Tip**
>
> To view the Microsoft Secure Score for Devices, click the **Dashboard** option under **Vulnerability management** from the M365 Defender portal. The score is broken into different controls, specific to security configurations on endpoints.

Next, let's look at how security teams can use the M365 Defender portal and MDE to investigate incidents and alerts on endpoints.

Security operations with MDE

The M365 Defender portal is a combined web-based interface that contains the various solutions in the Microsoft Defender security stack. This is where most security operations teams will spend their time investigating and monitoring threats that affect your endpoints across the application, identity, endpoint, and data domains. Using the available dashboards and built-in incident management system, the SOC can analyze threats, respond to security-related incidents, and perform real-time remediations. From the **Overview** page, they can quickly jump between domains, depending on where their investigations take them, making it easier to follow the threat should it perambulate across different domains. A few examples of the actions a SOC analyst can take in the M365 Defender portal include the following:

- Configure threat policies through **Email & collaboration** that control the actions to protect users from phishing, spam, and malware delivered through email.

- Configure rules for protecting the collaboration stack by configuring allow/block lists, filtering and quarantine policies, and configuring **DomainKeys Identified Mail (DKIM)** signatures.

- Use **Advanced Hunting** to build queries to find threats across all domains and create custom detection rules.

- Quarantine emails by deleting them or moving them to junk.

- View reports and service health.

- Review audit logs across all of Office 365.

Using **Permissions & roles**, located on the M365 Defender portal home page, security team members can be granted access to security-specific roles across Azure AD and MDE, and for tasks in the Security & Compliance center.

Next, let's look at how to configure role-based access in MDE using the portal.

Role-based access control in MDE

Access to MDE is locked down by design due to the sensitivity of the information that's available. Access can be managed using **Role-Based Access Control (RBAC)** and by assigning the appropriate roles in Azure AD. To provision basic access, the following Azure AD roles can be assigned directly to users:

- **Security Administrator** allows full access to the portal, including all system information, alerts, and administrative functions.

- **Security Reader** allows read-only access to log in and view all alerts and other information. You will not be able to perform administrative functions.

Fine-grained permissions can be assigned based on the SOC team member's job role, using the M365 Defender portal. For example, the security monitoring team may only need a Security Reader role for MDE data, while a level 2 team might need access to resolve actionable remediation recommendations. Neither role should have rights to change the security settings of your tenant or enable advanced features that the permissions of the Security Administrator Azure AD role grants.

To create a new role from the M365 Defender portal, go to **Permissions & roles** and choose **Roles** under **Endpoint roles & groups**. Click on **+ Add item** to create a new role. The following screenshot provides an example of the permissions a level 2 SOC analyst can be assigned.

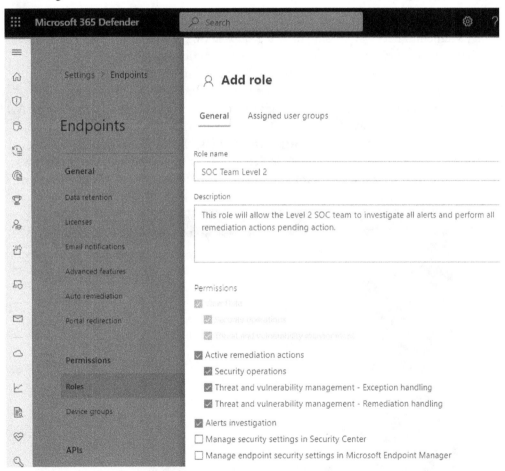

Figure 14.5 – Creating a role in MDE

This role will allow the analyst access to almost all features in MDE, except to manage the security settings in the Security & Compliance center and MEM.

Hovering your mouse over each of the permission settings will provide additional details about each setting. After the permissions have been selected, click on **Assigned user groups** to add an Azure AD security group for assignment. More information about managing portal access using RBAC can be found at `https://docs.microsoft.com/en-us/microsoft-365/security/defender-endpoint/rbac?view=o365-worldwide`.

Next, let's look at organizing devices using device groups and tags. Access can be further restricted to administrators based on these groups.

Creating device groups and tags

Creating device groups and tags can help organize endpoints and provide separation over their management if necessary. Using device groups allows you to configure different levels of automation for remediation actions from the Action center and automated investigations. For example, device groups can be created to separate servers from user workstations, or if the endpoints are in different geographic regions and managed by different security teams. User access to manage a device group is assigned using Azure AD security groups after they have been assigned to a role in MDE. To create a device group in the M365 Defender portal, go to **Permissions & roles**, choose **Device groups** under **Endpoint roles & groups**, and click **Add device group**. In the following screenshot, the **Device group name**, **Automation level**, and **Description** fields have been completed in the **General** tab.

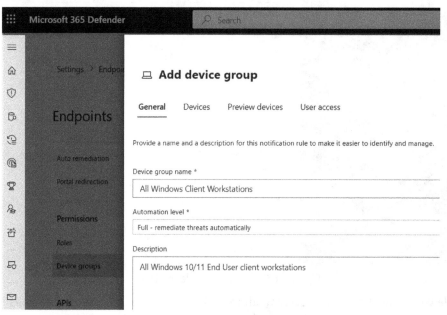

Figure 14.6 – Add device group

After filling out the required criteria, click on the **Devices** tab to specify the conditions to create a matching rule that determines which devices become members of the group. The following conditions are currently available for use:

- **Name**
- **Domain**
- **Tag**
- **OS**

In the example from the preceding screenshot, selecting Windows 11 and Windows 10 in the **OS** condition will select all Windows client workstations. You can see how to create a matching rule in the following screenshot.

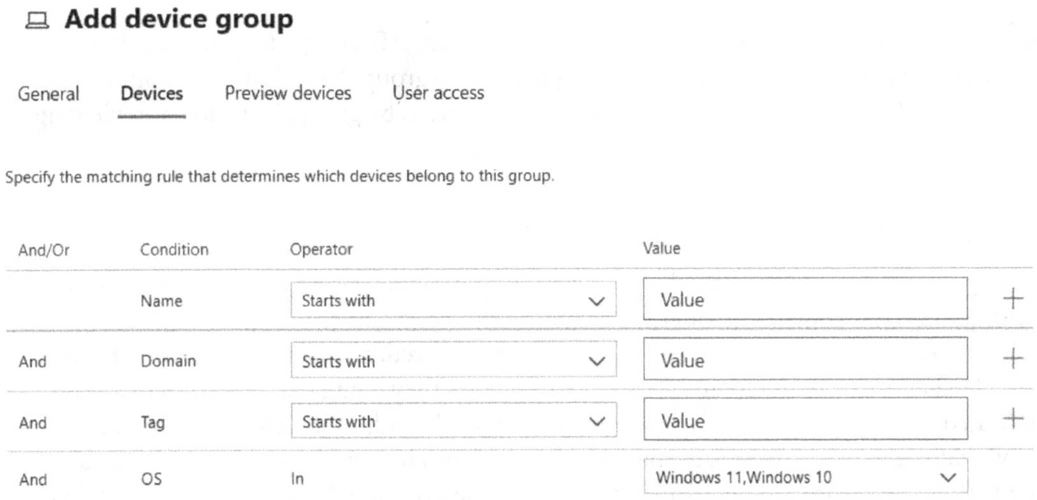

💻 **Add device group**

General   **Devices**   Preview devices   User access

Specify the matching rule that determines which devices belong to this group.

And/Or	Condition	Operator		Value	
	Name	Starts with	∨	Value	+
And	Domain	Starts with	∨	Value	+
And	Tag	Starts with	∨	Value	+
And	OS	In		Windows 11,Windows 10 ∨	

Figure 14.7 – Create a matching rule

Use the **Preview devices** feature to show up to 10 devices that match the conditions specified in the rule to validate that they are working as expected. If the available conditions don't provide enough flexibility, a tag can be used to help provide additional identification criteria. Device tags can be created using the portal or by setting a registry key value on your endpoints at the following location:

```
HKEY_LOCAL_MACHINE\SOFTWARE\Policies\Microsoft\Windows
Advanced Threat Protection\DeviceTagging
```

Then, set the following information to implement a custom device tag.

- Registry key value (REG_SZ): `Group`
- Registry data: `Name of tag`

> **Tip**
> There may be a delay before a tag shows up on a device in the portal after configuring the registry on the endpoint.

Registry keys can be deployed through Group Policy, MEM, or any other MDM method. More information about creating device groups and managing tags can be found at `https://docs.microsoft.com/en-us/microsoft-365/security/defender-endpoint/machine-groups?view=o365-worldwide`.

Finally, click on the **User access** tab and select the Azure AD groups to grant access to the endpoints, and click **Done** to finish creating the device group. Now that permissions have been assigned to the SOC team and devices are organized by group, let's look at reviewing incidents and alerts from the M365 Defender portal.

Reviewing incidents and alerts

Located on the home page of the M365 Defender portal navigation menu are links to the **Incidents & Alerts** queues. Incidents and alerts contain events from across the entire Microsoft security stack, where they can be viewed, filtered, and managed directly in the M365 Defender portal. Incidents differ from alerts, as they contain a collection of alerts correlated together across different domains or multiple endpoints. The alerts contain individual events of suspicious or malicious activity and help add context to the attack chain. They can be used to determine where and when the attack started, build a timeline of events, and add additional details about the processes used in an attack. Using the incidents and alerts queue, the SOC can build an incident response workflow that follows the example steps outlined as follows using the M365 Defender portal:

1. Review and manage incidents and alerts that are created in the M365 Defender portal.

2. Review the correlated alerts to analyze the attack and understand the impact, the total affected devices, the MITRE ATT&CK tactics used, and remediation recommendations.

3. Contain the attack using the mitigation features available from the M365 portal. For example, devices onboarded to MDE can have threats auto-remediated through behavioral blocking and containment features. Examples of these features include enabling ASR rules or enabling EDR in block mode. Additional actions can be taken directly on individual devices, such as isolating the device from the network, restricting app execution, running an antivirus scan, and collecting an investigation package to review the malicious artifacts, such as a file.

4. Remove all artifacts related to the attack across the various domains protected in the security stack. Examples would be to quarantine emails associated with a phishing attempt or delete malicious files stored locally on computers or uploaded to OneDrive and SharePoint.

5. Resolve the incident and close any relevant alerts. An RCA should be created for teams to review and identify security gaps through controls, processes, and misconfigurations. Any indicators of compromise should be added to blocklists, and custom detection rules can be created based on any artifact findings using advanced hunting.

Tip

Using the **Machine Risk Score** in a compliance policy can help prevent access to cloud apps connected to Azure AD during the investigation.

Let's review investigating an alert in more detail using the M365 Defender portal.

Reviewing an alert

The alerts queue lists events from the entire Microsoft Defender suite and are stored for up to 6 months. Using the filter, alerts can be sorted by severity, status, category, service source, tags, and policies. The following screenshot shows an example of the alerts queue in the M365 Defender portal, filtered to all sources from the MDE service source.

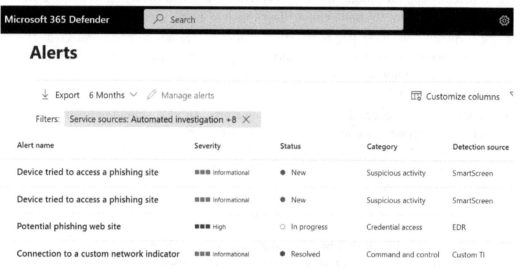

Figure 14.8 – The Alerts queue

The **Status** column indicates whether the alert is **New**, under investigation by an analyst (**In progress**), or **Resolved**. The **Category** column maps the detection to the **MITRE ATT&CK Framework** enterprise tactic definitions. The **Detection source** column shows which feature of MDE was responsible for the detection.

Clicking on the alert name will open the **Alert details** page and provide information about the alert story, attack timeline, impacted assets and users, and any linked incidents. The available actions include options to assign the alert to an analyst, review the alert timeline details, create a suppression rule for false positives, and consult a Microsoft threat expert if your company is licensed for the service.

Microsoft relies on the **MITRE** enterprise framework to categorize the attack tactics and will return the techniques used in the alert card if they match a known and documented technique. The following screenshot shows the alert details page about a potential **Ransomware** attack.

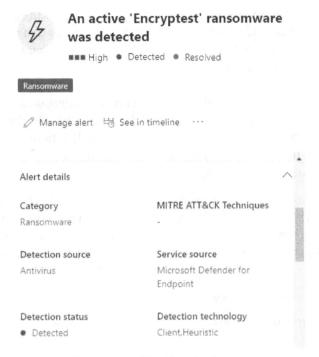

Figure 14.9 – The Alert details pane

From the alert page, the process tree is shown in the alert story and lists all the process IDs that can be corroborated, up to the execution of the malicious file and detection by the detection source, Defender Antivirus.

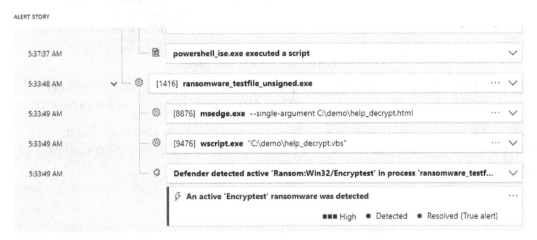

Figure 14.10 – The MDE alert story

In the preceding screenshot, the `ransomware_testfile_unsigned.exe` process was responsible for creating child processes (`msedge.exe` and `wscript.exe`) that executed the associated command-line arguments. Clicking the filename from the alert story will open the **File summary** pane that provides additional context about the executable for further research. This includes the process name, file path, command-line arguments, and a reference to a **VirusTotal** detection ratio. Mitigation actions can be taken on the file, such as adding the filename or hash value to the IOC list, downloading the file for detonation in a sandbox, **Stop and Quarantine File** actions using Defender Antivirus, submission to Microsoft for deep analysis, starting a consultation with a threat expert, and a link to **Advanced Hunting**.

Click **Open file page** to view the full-page detailed summary. In the **Overview** tab of the detailed summary, you can see the file prevalence within your organization to help detect and prevent the further spread of the malicious file. Luckily, in the following screenshot, the file has been contained, as it has not been detected in any email inboxes or other devices.

Figure 14.11 – The file summary page

To dig deeper into the alert and view the timeline of events leading up to the Defender Antivirus detection source, go back to the **Alert details** pane and click on **See in timeline**.

Next, let's review the incidents queue. Incidents correlate multiple alerts together to help add context to a potential threat and quickly identify whether multiple endpoints are involved.

Reviewing the incidents queue

The **Incidents** queue can be found under **Incidents & alerts** in the M365 Defender portal. Incidents help to correlate individual alerts from across multiple M365 services together to quickly gain insights about the entire scope of the attack chain. This helps build a timeline of events and answer questions such as where and how an attack originated and how many assets are impacted. Incidents include details such as severity, the number of associated active alerts, the service and detection source, the number of impacted assets, and an assigned category based on the MITRE ATT&CK enterprise tactic.

The SOC can incorporate the incidents queue into an **Information Technology Service Management (ITSM)**-style workflow by assigning incidents directly to an analyst, changing the status of the incident, applying a classification, and adding comments. The following screenshot shows the incident queue in the M365 Defender portal. It has been filtered for resolved incidents over the past 6 months with MDE as the service source.

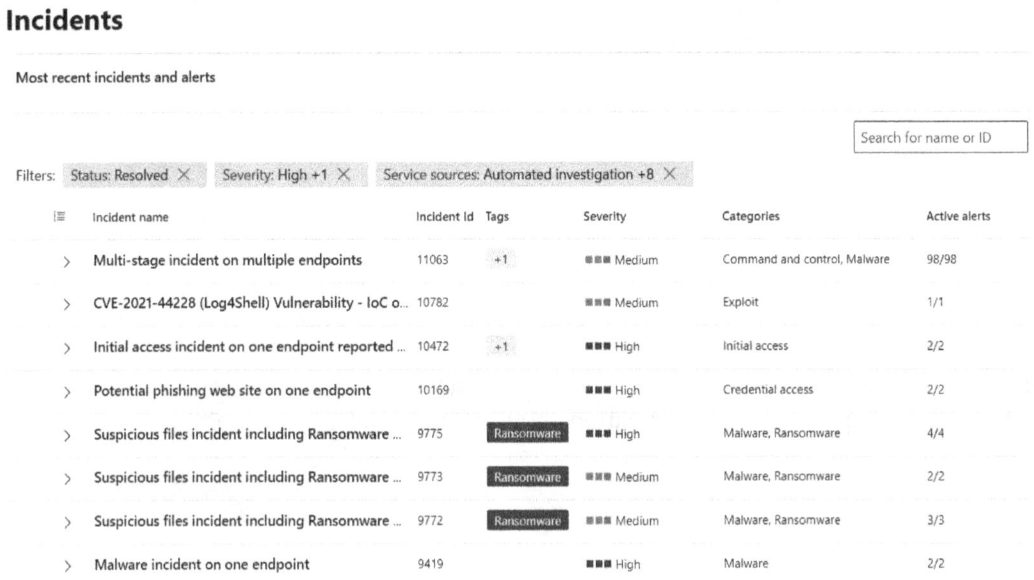

Incidents

Most recent incidents and alerts

Filters: Status: Resolved ✕   Severity: High +1 ✕   Service sources: Automated investigation +8 ✕

Search for name or ID

	Incident name	Incident Id	Tags	Severity	Categories	Active alerts
>	Multi-stage incident on multiple endpoints	11063	+1	▪▪▪ Medium	Command and control, Malware	98/98
>	CVE-2021-44228 (Log4Shell) Vulnerability - IoC o...	10782		▪▪▪ Medium	Exploit	1/1
>	Initial access incident on one endpoint reported ...	10472	+1	▪▪▪ High	Initial access	2/2
>	Potential phishing web site on one endpoint	10169		▪▪▪ High	Credential access	2/2
>	Suspicious files incident including Ransomware ...	9775	Ransomware	▪▪▪ High	Malware, Ransomware	4/4
>	Suspicious files incident including Ransomware ...	9773	Ransomware	▪▪▪ Medium	Malware, Ransomware	2/2
>	Suspicious files incident including Ransomware ...	9772	Ransomware	▪▪▪ Medium	Malware, Ransomware	3/3
>	Malware incident on one endpoint	9419		▪▪▪ High	Malware	2/2

Figure 14.12 – The Incidents queue in M365 Defender

Selecting an incident will open the incident **Summary** page to further analyze the artifacts that comprise each incident. Tabs include viewing the associated alerts, devices, users, mailboxes, apps, investigations, and evidence, and a graphical view to help understand the attack from a visual perspective. The following screenshot shows the detailed summary page of an incident. You can see there are three active alerts that are bundled into the incident.

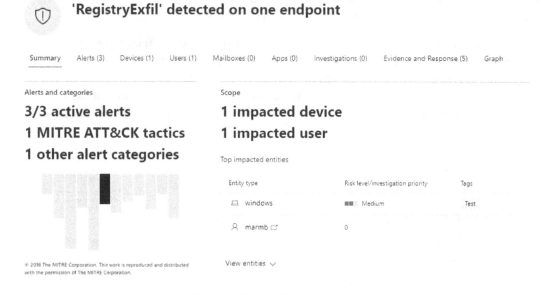

Figure 14.13 – Incident summary

Clicking the **Graph** tab will visually show how the attack played out. Clicking on each artifact will pull up additional information about the machine, file, or process identified. The graph view is seen in the following screenshot.

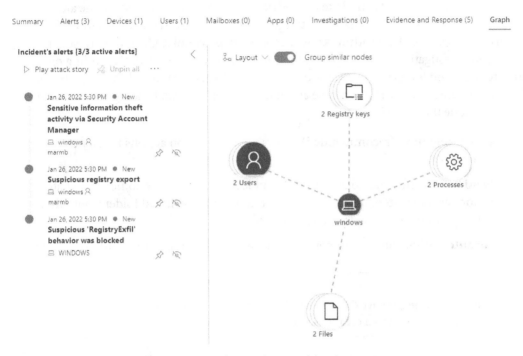

Figure 14.14 – The Graph view of the alert story

Additional information about managing incidents in the M365 Defender portal can be found at `https://docs.microsoft.com/en-us/microsoft-365/security/defender/manage-incidents?view=o365-worldwide`.

The **Incident & Alerts** queue can get noisy and make it difficult for the SOC to keep up with the entries. To help with this, Microsoft Defender has a feature known as **Automated Investigations** to automatically investigate and resolve known threats.

Automated investigations

Automated investigation and response (**AIR**) in MDE is a feature that can help automatically resolve alerts and take immediate action on threats to reduce the alert volume. An automated investigation is triggered by an initial alert where analysis then begins using a series of algorithms known as security playbooks. Once the investigation begins, an investigation ID is assigned and listed in the Action center. Depending on the auto-remediation level configured for device groups, items are tracked as completed or listed as pending review. The available automation types that can be assigned to device groups include the following:

- **Full automation** (recommended) allows for remediation actions to happen automatically.
- **Semi-automation** includes three different levels and can take some immediate action on core system directories or temporary and download folders, but other actions require approval by a team member.
- **No automation** (not recommended) means automated investigations do not run.

> Tip
>
> In the **Creating Device Groups and Tags** section, you can select the automation level when creating the device group.

The **Action Center** page lists all automated investigations across M365 security services and is not just limited to MDE. The analyst can see detailed information such as the action type, affected asset, entity type, action source, and status. In the following screenshot, the **1784** investigation ID was automatically remediated by automation.

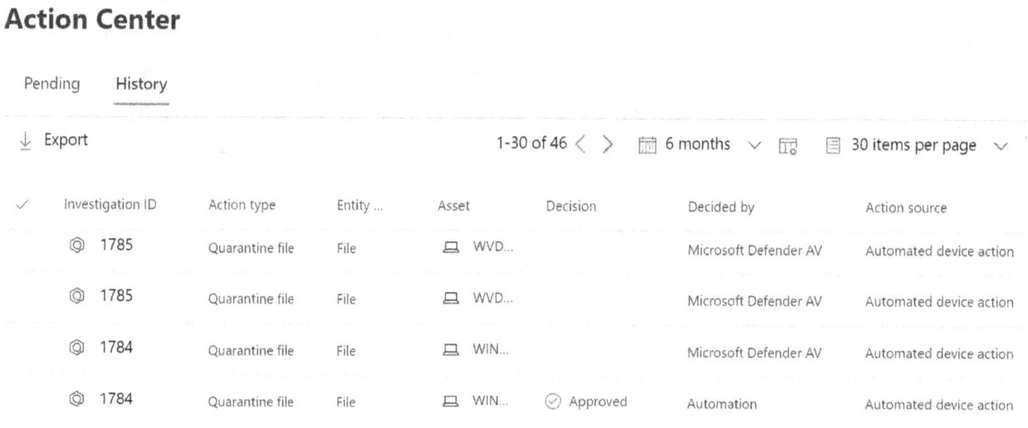
Action Center
Pending
History
Export
1-30 of 46
6 months
30 items per page
Investigation ID
Action type
Entity ...
Asset
Decision
Decided by
Action source
1785
Quarantine file
File
WVD...
Microsoft Defender AV
Automated device action
1785
Quarantine file
File
WVD...
Microsoft Defender AV
Automated device action
1784
Quarantine file
File
WIN...
Microsoft Defender AV
Automated device action
1784
Quarantine file
File
WIN...
Approved
Automation
Automated device action

Figure 14.15 – Action Center for automated investigations

During an open investigation in a pending status, any additional alerts generated from the compromised device or correlated across other workstations will automatically be grouped under the same investigation ID. Selecting the investigation ID will open a detailed investigation summary of the AIR process, as shown in the following screenshot:

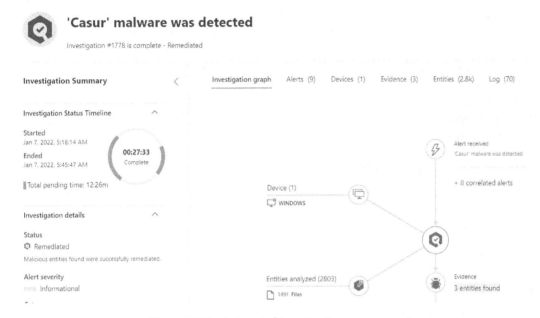

Figure 14.16 – Automated investigation summary

By setting the auto-remediation level to full, automated investigations can significantly cut down on alert fatigue and allow the SOC to focus on more valuable activities, such as complex investigations that require human intuition and deep analysis. To help assist with the investigation process and perform complex search queries, the SOC team can use the **Advanced Hunting** feature in M365 Defender. Let's review how advanced hunting can query for log events and build custom detection rules.

Using advanced hunting

Advanced Hunting in M365 Defender allows security teams to proactively hunt for threats across all the raw data that is collected by the connected M365 services. To search through the logs, the **Kusto Query Language (KQL)** is used to build queries and correlate data across the tables and columns. To access **Advanced Hunting**, choose **Hunting | Advanced Hunting** from the M365 Defender portal. The landing page includes a collapsible sidebar to view the schema, custom functions, saved queries, and detection rules. The query editor supports multi-tab editing and includes a list of basic queries to help get started using KQL. Clicking on the **Queries** tab in the schema sidebar, you can view predefined queries shared across the GitHub community and ones shared internally created by security teams. In the following screenshot, a KQL query was run to look for network logons that used local accounts. The summary results include **RemoteIP**, **LogonType**, **AccountName**, and **LogonAttempts** details.

Figure 14.17 – An Advanced Hunting query

Clicking a result from the **RemoteIP** column will open a side pane to inspect the record further, where you can add additional filters to fine-tune the results. Some records will also have links to a summary page such as the preceding example for the **RemoteIP** address. The summary can further analyze the artifact to correlate any known alerts and where it has been observed in the organization. To help search for additional artifacts from alerts, **Advanced Hunting** can also be opened with a preconfigured query by choosing the **Go hunt** button when investigating an alert story.

Advanced hunting is also a great way to filter for events that are not normally presented as an alert or incident but may indicate suspicious activity. Using custom queries, detection rules can be created that search for events on a schedule, can generate alerts, and take remediation actions on devices, adding great flexibility for the SOC to automate their workflows. Let's look at building a custom detection rule from advanced hunting.

Creating a custom detection rule

Custom detection rules in advanced hunting help proactively monitor an environment by looking for security events or flagging system misconfigurations and unhealthy endpoints. Detection rules are configured to run on a schedule, trigger alerts based on any findings, and take action on users, devices, or files. The alert details can be configured with a severity rating, attack category, description, recommended actions, and the ability to link to a threat analytics report or MITRE ATT&CK technique. Available actions include isolating a device, running an antivirus scan, restricting app execution, quarantining files, marking a user as high risk, collecting an investigation package for analysis, or opening an automated investigation. A custom detection rule is created directly from the query editor of **Advanced Hunting**. In the following screenshot, the **Create detection rule** button is shown, which opens the custom detection menu with available options for configuring the rule.

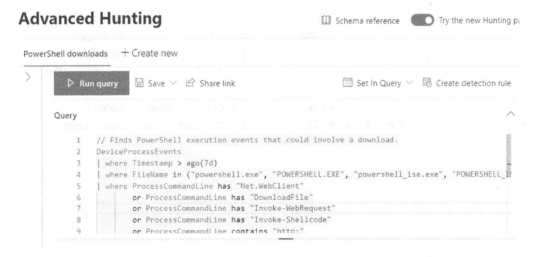

Figure 14.18 – Create detection rule

There are a mandatory set of columns that need to be returned from the query results; otherwise, the **Create detection rule** option will not be available. The list of available detection rules can be viewed from the M365 Defender portal by choosing **Hunting | Custom detection rules**. Select a rule to open the **Detection rules** page to review the run history, associated alerts, and trigger the rule to run immediately. An example of the available custom detection rules can be seen in the following screenshot of the M365 Defender portal.

Figure 14.19 – Custom detection rules

Now that we have reviewed how to use advanced hunting to search for events and build custom detection rules, let's look at how the SOC can track open vulnerabilities using a remediation request.

Tracking remediation requests

Remediation requests can be used to track the remediation efforts of identified vulnerabilities and security recommendations in MDE. Active requests are tracked by the remediation type, the related components (such as a vulnerable software title), and priority. The overall progress of the request is tracked based on the number of impacted assets and is completed once all devices are remediated. To view the active remediation requests from the M365 Defender portal, expand **Vulnerability management** under **Endpoints** and choose **Remediation**. The following screenshot shows a remediation request, tracking devices vulnerable to the Log4j exploit.

Remediation

Activities   Exceptions

Activity in progress
1

Activity past due
1

↓ Export

Activity

Attention required: Devices found with vulnerable Apache Log...

Attention required: Devices found with vulnerable Apache Log4j versions

⊘ Mark as completed   ▦ Export to CSV

Description

Uninstall Log4j to help lessen the security risk to your organization. Since this software has reach end-of-support, it is likely to have additionally vulnerabilities that are not published.

Device remediation status
In progress

Device remediation progress
0/9

Created on
1/18/2022

Created by
smun

Priority
Medium

Related component
Apache Log4j

Figure 14.20 – Remediation requests in M365 Defender

In the preceding screenshot, there are nine identified devices in the **Device remediation progress** field. The request is complete once all nine devices have been remediated.

New requests can be created by reviewing the available security recommendations in **Vulnerability management**. To do this, select a recommendation to open the details flyout pane and click **Request remediation**, as shown in the following screenshot.

Security recommendations

⚠ **5 discovered devices are not protected**
Onboard them now

↓ Export

Filters: Status: Active +1 ✕

Security recommendation

Update Tableau Reader

✔ Update Apache Log4j

Update Apache Log4j
○ Remediation required

⊙ Open software page   ⟳ Report inaccuracy

ⓘ The number of exposed onboarded devices may fluctuate due to changes to the inspection capabilit

General   Exposed devices   Vulnerable files   Associated CVEs   Activities

Description

Log4j 2 is a Java-based logging library that is widely used in business system development, included in various open-source libraries, and directly embedded in major software applications.

This recommendation provides a list of devices and paths where a vulnerable Apache Log4j 2 version was found.

To fix the vulnerability, either deploy a relevant software update issued by the vendor or upgrade the Apache Log4j library component to a newer version.

Associated CVEs

Request remediation   Exception options

Figure 14.21 – Request remediation

There is also an option to open **Exemption options** to remove the vulnerability from showing as active in the reporting. To use remediation requests, the Microsoft Intune connection must be enabled from MDE and MEM. We covered how to enable this in *Chapter 13, Security Monitoring and Reporting*, in the *Onboarding Windows clients to MDE* section.

Whether you're supplementing an existing IT service management solution, tracking vulnerabilities and weaknesses, or taking advantage of Microsoft's full XDR capabilities, the SOC will gain valuable insights by incorporating MDE into their workflows.

Next, let's look at using Microsoft Defender for Cloud to assess threats and track vulnerabilities for resources running in the cloud and on-premises.

Investigating threats with Defender for Cloud

In *Chapter 13, Security Monitoring and Reporting*, we enabled and configured the enhanced security features of Defender for Cloud to gain the benefits of all available advanced features. This included just-in-time VM access, regulatory compliance dashboard and reports, adaptive application controls, EDR for servers, and threat protection for PaaS services. With the enhanced security features enabled, you can investigate threats in Defender for Cloud by following these steps:

1. Sign into the Azure portal at `https://portal.azure.com`.
2. Search for `Microsoft Defender for Cloud` and open it.
3. Click on **Security Alerts** within the **General** section.

4. Here, you can view active alerts and the associated affected resources. Select one of the alerts to review additional information in the window that appears on the right, as shown in the following screenshot.

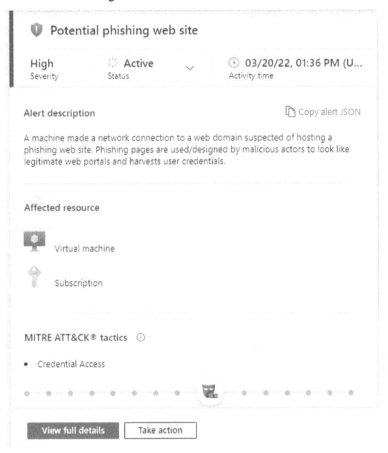

Figure 14.22 – Defender for Cloud high severity alert

5. To further investigate an alert, click on **View full details**. Clicking on the **Take action** tab will show any recommended mitigation steps, as shown in the following screenshot.

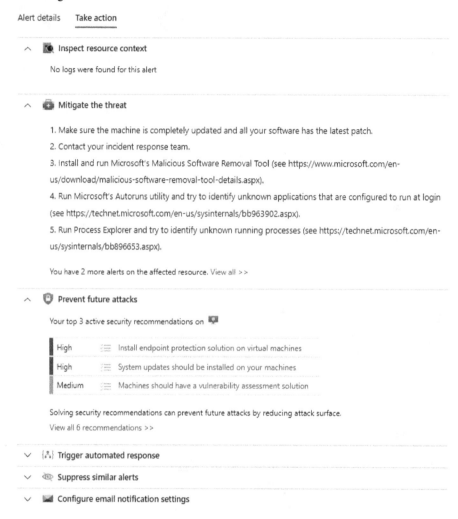

Figure 14.23 – Defender for Cloud mitigation actions

Defender for Cloud alerts can be sent to the SOC via email notifications by clicking on **Environment Settings** in the left-side navigation panel. Choose your subscription name (you may need to expand the menus), click on **Email notifications**, and then enter the email recipients to receive high-severity alerts. You can also use the **Workflow automation** section to configure automation using Logic Apps by setting conditions that are available through alerts. Finally, the **Continuous export** section allows you to export Defender for Cloud log data to external targets, including integration with a SIEM.

An added benefit that provides visual insight into your active alerts is the security alert map. To access the map, browse to **Security alerts**, and then click on **Security alerts map** in the top menu. The alerts map will look similar to the following screenshot for any open active alerts in your subscription.

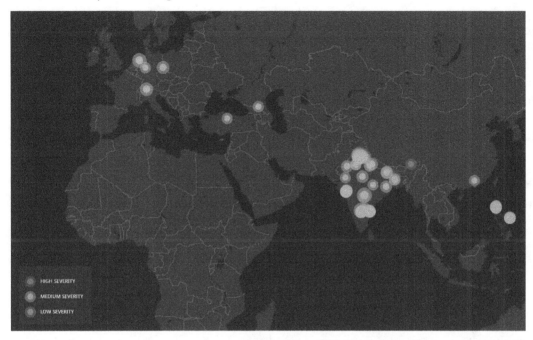

Figure 12.24 – The Microsoft Defender for Cloud security alert map

Within the map, you can click on the circles to view details about the alerts. In addition to managing alerts, Defender for Cloud is a great way to track vulnerabilities that have been identified on your cloud resources. The SOC can incorporate recommendations into their workflows and follow up with the appropriate app owners to close open security vulnerabilities and reduce the attack surface. Recommendations can be tracked directly in the Azure portal or exported to CSV.

In the next section, we will provide an overview of Microsoft's cloud-built SIEM service, Microsoft Sentinel.

Enabling Azure-native SIEM with Microsoft Sentinel

Microsoft Sentinel is a modernized **SIEM** and **SOAR** that is built on Microsoft Cloud technology. Microsoft Sentinel is a centralized SIEM solution that provides an intelligent robust life cycle to allow the collection of data, detection of threats, investigation of threats, and responses to incidents. Because Azure Sentinel is cloud-built in Azure, the ease of setup and integration makes the service extremely attractive to existing Microsoft customers that use Azure to host their cloud resources. The setup is simplified compared to a third-party SIEM, which typically requires additional infrastructure and storage to support log collection and analyze data. Let's look at how to connect data sources in Azure to Microsoft Sentinel.

Creating the connection

To set up Microsoft Sentinel within Azure, follow these steps:

1. Sign into the Azure portal at `https://portal.azure.com`.
2. Search for `Microsoft Sentinel` and open it.
3. Click on **Create Microsoft Sentinel** on the main screen.
4. Select a workspace to connect to, and then click on **Add**. (If no workspaces exist, click on **Create a new workspace** to build a new Log Analytics workspace.)
5. It will take a few minutes to create, and then you will be redirected to the Microsoft Sentinel console, which will look like the following screenshot.

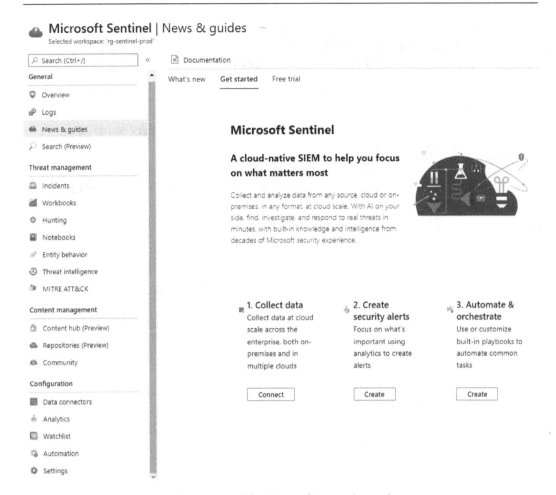

Figure 14.25 – The Microsoft Sentinel console

Note that the cloud-native SIEM was deployed within minutes. This is a great example of the power of cloud technology and the benefits it has for security. After the Microsoft Sentinel platform is set up and ready to use, you will first need to set up your data sources and begin collecting data. This can be completed by following these steps:

1. Ensure **Microsoft Sentinel** is open.

2. Click on **Data Connectors** within **Configuration**.

3. Within **Connector name**, select the connector you would like to connect, and then click on **Open connector page**. As an example, search for `Microsoft Defender for Cloud` and select the connector, and then click on **Open connector page**.

4. Follow the instructions to set up the connector. For Microsoft Defender, click on **Enable Microsoft Defender for all subscriptions** or select the individual subscriptions, click **Connect**, and then click **OK** to complete the connection.

5. Click **Enable** under **Create incidents** if you would like to allow the automatic creation of incidents.

6. Click on **Next steps** to review any recommended workbooks, query samples, or relevant analytic templates.

7. Once configured, you have just successfully set up a connector to collect logs within Microsoft Sentinel. The following screenshot shows Microsoft Defender for Cloud has been connected.

Figure 14.26 – The Microsoft Defender for Cloud connector

You can go back to set up any other relevant connectors for your environment. Additional connectors that provide insight specifically for users and Windows systems include Windows security events, Azure AD Identity Protection, Azure AD, Microsoft Defender for Office 365, and Office 365.

> **Tip:**
> You can view the latest pricing for Microsoft Sentinel at `https://azure.microsoft.com/en-us/pricing/details/microsoft-sentinel/`.

After your data sources have been connected, you can get a high-level overview of your environment by accessing the **Overview** menu item within the left-side navigation pane. The following screenshot shows the **Events and alerts over time** details, including a **Potential malicious events** graphical map.

Figure 14.27 – A Microsoft Sentinel overview

To get the most out of your Sentinel deployment, you will want to view your logs via **Workbooks**, create threat detection rules within **Analytics**, and orchestrate response to incidents by building automations. There is a lot to learn with Microsoft Sentinel, so whether you're designing the company's fist SIEM or supplementing existing services, visit the Microsoft Sentinel documentation library to learn more: `https://docs.microsoft.com/en-us/azure/sentinel/`.

In the next section, we will review MDCA to provide better visibility and protection for your SaaS applications.

Protecting apps with MDCA

MDCA is a **Cloud Access Security Broker** (**CASB**) that can identify, protect, and govern SaaS applications by providing visibility through identity-based analysis and traffic patterns from managed endpoints, and with API integration with SaaS providers. A CASB helps to provide visibility into shadow IT processes using its discovery mechanisms, which is traditionally a challenge for many organizations without a CASB solution in place. A benefit of using MDCA as your CASB is the native connectivity available for those heavily invested in Microsoft technologies. It also integrates easily with other cloud providers (AWS and Google), allowing for visibility into all combined cloud environments from a single console.

Planning for the implementation of MDCA can be divided into a few high-level steps. If your organization contains many apps, users, and devices, there can be a substantial amount of discovery data to review, and plan for a significant investment of time to fully configure all the advanced policies and governance controls the CASB offers. It is recommended to plan for a phased approach by following these steps:

1. Connect your application sources to identify and discover apps.

2. Review the discovery data.

3. Configure policies and alerts.

4. Set up governance and security controls.

To view the MDCA portal, go to `https://portal.cloudappsecurity.com`. First, let's review the methods for connecting applications to MDCA and the available connectors to solutions in the Microsoft security stack.

Connecting apps to MDCA

There are three primary methods to connect apps, users, and data to MDCA, as follows.

- Native integrations into M365 products that include MDE, MDI, Azure AD Identity Protection, Microsoft Purview Information Protection, and Microsoft Online Services (Office 365 and Azure AD).

- API connections to SaaS and cloud providers. Some examples include Amazon Web Services, Google Cloud Platform, Okta, Salesforce, ServiceNow, and Box.

- Azure AD Conditional Access App Control for apps using Azure AD or other non-Microsoft identity providers, using a reverse-proxy architecture.

Native integrations offer the most support regarding the available security features and protection capabilities for connected applications. As outlined previously, these are the native Microsoft solutions, and it's advantageous to adopt MDCA as a CASB if you're using multiple products in the Microsoft security stack.

API connections offer the next best option for connecting apps. This allows MDCA to be wired through exposed API endpoints from other third parties and enables enhanced governance controls for services not native to Microsoft. The features will vary by vendor, but you can visit the following link to learn about the capabilities available for the current supported connectors:

`https://docs.microsoft.com/en-us/defender-cloud-apps/enable-instant-visibility-protection-and-governance-actions-for-your-apps`

Azure AD Conditional Access typically provides the least amount of flexibility regarding protection capabilities but allows you to monitor and control user sessions by inspecting traffic and funneling session data through the MDCA reverse proxy. Some of the protection capabilities include controlling uploads and downloads of sensitive data, blocking user access to connected apps, and protection from malware through inspection. To learn more about the protection capabilities using Azure AD Conditional Access, visit `https://docs.microsoft.com/en-us/defender-cloud-apps/proxy-intro-aad`.

To add a connected app using native integration such as Office 365 or Microsoft Azure, follow these steps:

1. Sign in to `http://portal.cloudappsecurity.com/`.

2. Click the settings icon on the top-right toolbar and choose **App connectors** under **SOURCES**.

3. Choose + **Connect an app** to open the dropdown and select any of the native integrations available for MDCA. A preview of the available connectors can be seen in the following screenshot.

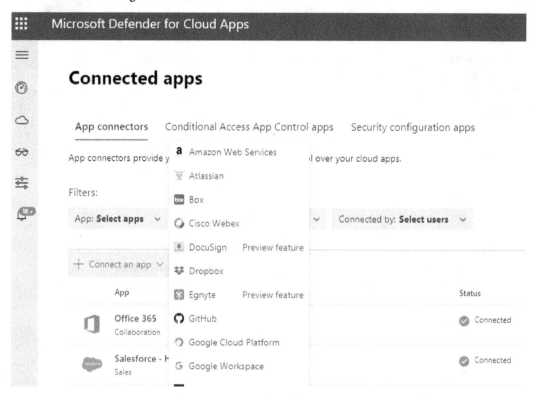

Figure 14.28 – Connecting apps in MDCA

To connect another cloud provider, click on the **Security configuration apps** tab, as shown in the preceding screenshot. For applications that are configured using Azure as the identity provider, you can easily add them by using session control settings in a conditional access policy, as shown in the following screenshot.

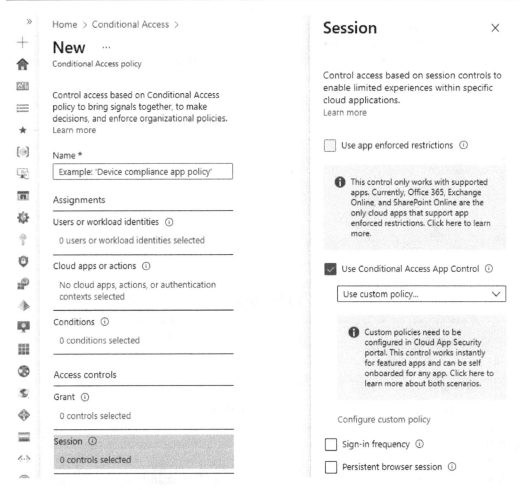

Figure 14.29 – Onboarding apps into MDCA using conditional access

For additional information about deploying conditional access app control, visit
`https://docs.microsoft.com/en-us/defender-cloud-apps/proxy-deployment-aad`.

Integration with MDE can be configured from the M365 Defender portal by going to `https://security.microsoft.com`, choosing **Settings**, selecting **Endpoints**, choosing **Advanced features**, and turning on **Microsoft Defender for Cloud Apps**.

Endpoints

Email notifications

Advanced features

Auto remediation

Portal redirection

Permissions

On   Microsoft Defender for Cloud Apps
Forwards Microsoft Defender for Endpoint signals to Defender for Cloud Apps, giving administrators deeper visibility into both sanctioned cloud apps and shadow IT. It also gives them the ability to block unauthorized applications when the custom network indicators setting is turned on. Forwarded data is stored and processed in the same location as your Cloud App Security data. This feature is available with an E5 license for Enterprise Mobility + Security on devices running Windows 10 version 1709 (OS Build 16299.1085 with KB4493441), Windows 10 version 1803 (OS Build 17134.704 with KB4493464), Windows 10 version 1809 (OS Build 17763.379 with KB4489899) or later Windows 10 versions.

Figure 14.30 – Enable MDCA connector

Next, let's look at how MDCA can discover apps and shadow IT by analyzing traffic logs from your connected sources.

Discovery

A main advantage of using a CASB solution is the ability to discover cloud application usage and shadow IT processes. MDCA primarily handles discovery using data from log snapshots, continuous reports from firewalls and endpoints, or through the Cloud Discovery API. An example of discovery sources that are used for continuous report evaluation include MDE, log collection using syslog or FTP, or using a secure web gateway through select vendors. Discovery data is analyzed up to four times a day from logs fed into the MDCA portal, where they are parsed and analyzed against a catalog of over 25,000 known cloud apps.

For additional information about cloud discovery in MDCA, including a list of the supported firewalls and proxies, visit `https://docs.microsoft.com/en-us/defender-cloud-apps/set-up-cloud-discovery`.

After cloud discovery sources have been connected, you can start to review the data in the **Cloud Discovery** dashboard. To view the dashboard, click the **Discover** dropdown from the main menu and choose **Cloud Discovery dashboard**. In the following screenshot, the discovery data is available through the connector to MDE.

Figure 14.31 – The Cloud Discovery dashboard

In the toolbar under **Cloud Discovery**, click on **Discovered apps** to view a list of all identified apps. Using the filtering options, you can sort apps by category, compliance risk, general risk, security risk, usage, and so on. Based on your filter results, custom queries can be created to quickly help you identify a subset of apps later. Additionally, the toolbar also includes links to dashboards to sort and review data by IP addresses, users, and devices.

Next, let's review the toolset available in MDCA that the security team can use to learn about the connected apps and data.

Investigate

MDCA lets security teams investigate apps and activities to help understand and identify risk in your environment. Using what you learn through the investigation tools is the first step before building out policies for alerts or configuring governance actions to protect users and data. To view the available investigation tools, expand **Investigate** from the main menu to reveal the following options:

- **Activity log**
- **Files**
- **Users and accounts**

- **Security configuration**
- **Oauth apps**
- **Connected apps**

The Activity log shows a timeline of events about a specific activity that occurred in a connected app. There are many different activity types that are logged, including administrative actions, downloads, failed logins, file and folder activities, risky security activity, and data sharing. By using the Activity log, if a risk is identified, security teams can quickly create new policies in near real time, based on the activities from an entry in the timeline.

Clicking on an activity will expand the activity drawer to provide additional insights, including user details, the connected app source, activity target, and geolocation. Selecting the **User** tab will show a quick view of the user profile analysis, which provides insights about the number of activities logged over the last 30 days, as shown in the chart in the following screenshot.

Figure 14.32 – The user activity log

The Activity log includes several pre-built queries to help identify known activities of interest, but custom queries can also be created by adding your own filter – for example, using the filters to view the activity of privileged users or keep a closer eye on users who have been flagged as a high risk for security incidents, based on previous history. The Activity log is a great way to begin understanding how users and devices are interacting with your connected apps.

The **Files** dashboard provides visibility into all files that were scanned from connected apps, and every time a file's content, metadata, or sharing is modified, MDCA will rescan the file to remain updated. Using the available filters, files can quickly be sorted by name, connected app, collaborators, access level, extension, file type, owner, sensitivity label, and so on. Just like the Activity log, applying filters allows you to quickly create and save queries for later use. For example, you can add a filter that shows all files from connected apps that are available directly over the public internet with or without a link, or all files shared externally to another domain. If you are using Microsoft Purview Information Protection labels, you can view the list of files by their applied sensitivity label. The following screenshot shows the **Files** view with a filter applied to show all files with a **Public, Public (Internet)** access level.

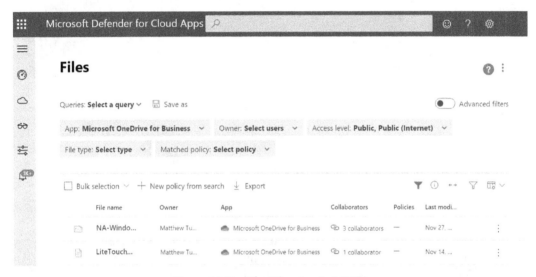

Figure 14.33 – The Files view in MDCA

Actions can be taken on files directly from the results by clicking the three vertical dots at the end of the log entry. Here, you can add additional conditions to your original file search query and apply governance actions, such as applying a sensitivity label, placing the item in trash, removing external collaborators, and making the file private if it is shared publicly.

The **Users and accounts** dashboard aggregates users by scanning connected apps and through the activity logs. Using the accounts page, filter options allow you to view accounts by affiliation (internal or external), application, and security group, and list accounts with administrative privileges inside the apps. In the following screenshot, the **Show admins only** filter was used to show all internal admins of the connected **Box** application.

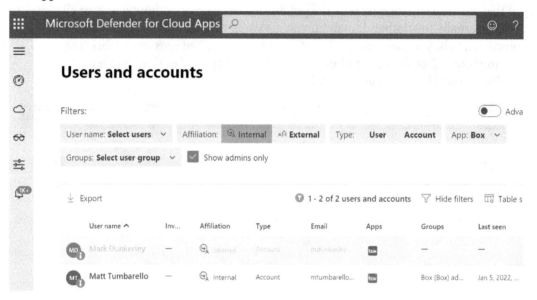

Figure 14.34 – Users and accounts

The **Users and accounts** view allows you to track the activity of a user, including which applications they access, and apply governance controls such as suspending a user, confirming a user is compromised, and requiring a user to sign in again.

> **Tip**
>
> Applying the **Confirm user compromised** action on an account will flag the Azure AD user risk as high. Any policy actions defined on the risk score will apply.

Click the username to view additional information about the user and to open the user profile page. Here, you can view any associated investigation score, recent activity, and an alert timeline, as shown in the following screenshot.

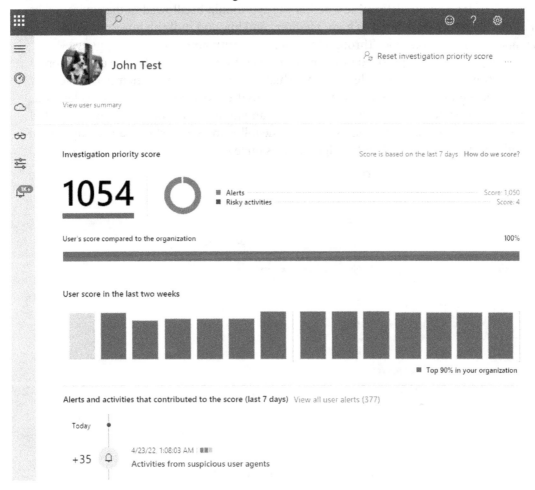

Figure 14.35 – Investigate a user

The **Security configuration** view provides recommendations for improving misconfigured security configurations that have been identified across any connected cloud platforms. The recommendations can be sorted by resource type, security benchmark, subscription, and severity. Some of the available benchmarks for Azure recommendations include the Azure Security Benchmark, ISO 27001, PCI DSS 3.2.1 and SOC TSP. For additional information about the available security recommendations for each cloud provider, visit `https://docs.microsoft.com/en-us/defender-cloud-apps/security-config`.

The **Manage Oauth apps** view shows any third-party apps that have permissions to view and interact with user data stored in Office 365, Google Workspace, or Salesforce and can sign in on the users behalf. Reviewing these app permissions helps security teams identify any risk of exposure to determine whether the app should be allowed in the environment. Permission levels are defined as high, medium, and low and indicate how much data an app may have access too. Through the **Manage Oauth apps** view, after apps are reviewed, they can be approved or banned (revoked for Google Workspace or Salesforce) right within the MDCA console. Any apps that are banned will have their permissions removed, and users will need to request consent to the app again. The filter view lets you find apps based on specific permissions – for example, access to your data anytime or full access to all mailboxes in your subscription. The following screenshot is an example of an Oauth app that was flagged with a high permission level.

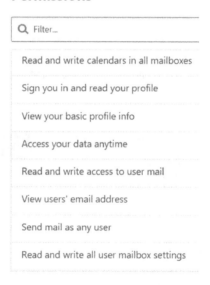

Figure 14.36 – Oauth permission level

As seen in the preceding screenshot, the level of permissions available to the app can be concerning and have a higher risk of exposing confidential data.

> **Tip**
> We strongly recommend configuring **Do not allow user consent** in the **User Consent** settings under **Enterprise Applications** in Azure AD.

Now that we have reviewed how to connect, discover, and investigate users and apps in MDCA, let's look at how to configure alert policies and controls.

Configuring policies and controls

Policies in MDCA are used to detect suspicious behavior, trigger alerts, and configure automatic governance actions and remediation workflows. Policies are configured based on nine different *policy types* across four categories to help correlate risk behavior. Each policy type will have different remediation options available. The main policy categories include threat detection, information protection, conditional access, and shadow IT. The different policy types are as follows:

- **Activity policies** are threat detection policies that track the activity of users throughout your cloud apps. Examples of activity types that can be created into a policy include multiple failed login attempts, logins from certain IP ranges, mass downloads or deletions, and administrative actions from non-specified IP ranges.

- **Anomaly detection policies** can be configured to alert on behaviors that are not commonly seen by users based on baselines built from entity behavior analytics and machine learning in MDCA. An anomaly detection policy can trigger an alert for impossible travel or an uncommon location, where a user's known to typically sign in from one geolocation but shortly after signs in from another location not commonly seen or physically impossible to travel to, based on the duration between sign-ins.

- **Oauth app policies** allows you to configure policies to alert when applications are approved in your organization. An example includes configuring a policy to alert when an Oauth app is approved with certain permissions or by a specific publisher. Governance actions can automatically revoke apps based on the configured filter conditions in the policy.

- **Malware detection policies** scan files in cloud repositories using Microsoft threat intelligence to identify known or potentially malicious files.

- **File policies** are used to identify specific file types based on their metadata. Examples of policies can include identifying files with sensitive information, shared externally or to personal email addresses. Governance actions on file policies include notifying users, removing external users, making a file private, applying sensitivity labels, or placing the file in the trash.

- **Access policies** monitor the access of cloud apps based on users, location, and devices. An example of access policies include blocking access to an application from a non-corporate IP range or triggering an alert if an app is accessed by a non-compliant Microsoft Intune device.

- **Session policies** allows you to control and monitor user session data. Session policies include actions such as blocking downloads based on certain conditions, applying sensitivity labels on upload or download, and blocking the upload of files containing malware.

- **App discovery policies** allow you to trigger alerts when a new app is discovered. This is a great way to monitor for shadow IT processes. Filters can be configured to alert based on the discovered application type – for example, whether the app is identified as a cloud storage repository, used for collaboration such as email or chat, or a **customer relationship management (CRM)** app.

- **Cloud discovery anomaly detection policies** are used to alert on anomalous behaviors by users within cloud apps. Examples include downloading data from an app that is not sanctioned or alerting when a user initiates a mass download or deletion event.

The governance actions, filters, and controls that are available will differ, depending on the policy type that you create. MDCA includes many preconfigured templates to help you get started, which are available in the MDCA portal under **Control | Templates**.

To review the governance actions that are available for each policy type, visit `https://docs.microsoft.com/en-us/defender-cloud-apps/governance-actions`.

Different actions will be available for connected apps and those found through cloud discovery or integrated with other solutions in the Microsoft security stack. Any governance actions that run automatically from being configured in a policy or manually through an administrators' actions can be reviewed in the governance log, found in the MDCA settings. The alert types are configured directly in the policies using email, text messages, or with playbooks in Power Automate.

With the increased adoption of SaaS applications for enterprise use, implementing a CASB can help to extend the protection of users and devices by governing the use and activity of these apps. Microsoft MDCA is a fully functional CASB that is easy to configure and a great addition to the list of tools available to the SOC to keep your Windows environments secure.

Next, we will review how to monitor hybrid environments and on-premises identities in hybrid AD environments using MDI.

Monitoring hybrid environments with MDI

MDI is a cloud security service that's used to analyze signals and events from AD domain environments to identify threats, compromised identities, and attacker activities inside the network. The solution provides incident tracking that the SOC can use to analyze timelines of security events, investigate alerts and user activities, and act on identified threats. The following list includes ways that MDI helps to identify suspicious activities inside your network:

- Monitors user activities and creates a behavioral profile about each user. Then, through intelligence, it creates alerts based on anomalies in the observed behavioral patterns. The heuristics used to build the user profile include details about permissions, group memberships, and account activities.

- Identifies attempts to enumerate information about a domain environment, such as queries for users, groups, and resources typically seen during reconnaissance activities.

- Alerts on attempts to move laterally through the network, using methods such as pass the hash, golden ticket, and silver ticket, or attacks over remote services such as **Server Message Block (SMB).**

- Identifies risks from compromised credentials by looking for brute-force attacks, numerous failed authentications, and changes in permissions or group memberships.

- Alerts on attempts for domain takeover, such as unsuspected domain replication, attempts to register or promote rogue domain controllers, or attacks using a compromised **Kerberos Ticket Granting Ticket Account (KRBTGT)** (Kerberos) account.

Telemetry data is collected from on-premises networks by installing the MDI sensor application on domain controllers or AD **Federation Services (FS)** servers. Information such as network traffic and Windows events are then forwarded to the MDI cloud service for investigation. The SOC team can then use the MDI or M365 Defender portal to monitor for suspicious activities and alerts. The MDI sensor will capture the following information from domain controllers or AD FS servers:

- Domain controller network traffic

- Windows events

- **Remote Authentication Dial-In User Service (RADIUS)** account information

- User and computer data from AD

For more detailed information about the MDI architecture, visit `https://docs.microsoft.com/en-us/defender-for-identity/architecture`.

Next, let's review the steps involved for planning an MDI deployment.

Planning for MDI

Microsoft docs have reference material that should be reviewed in detail before enabling MDI and installing the sensor on-premises. There are several prerequisites that need to be considered in the architecture and design, including licensing requirements. The official prerequisite documentation can be found at `https://docs.microsoft.com/en-us/defender-for-identity/prerequisites`.

Microsoft has a capacity planning guide and tool that can help you determine the appropriate resource requirements to run the MDI sensor. To download the MDI sizing tool, visit `https://docs.microsoft.com/en-us/defender-for-identity/capacity-planning`.

Activating your instance

After all prerequisites have been met, activating the MDI portal instance is straightforward. Visit the following link for step-by-step instructions on how to activate the instance, connect to AD, and install the MDI sensor:

`https://docs.microsoft.com/en-us/defender-for-identity/install-step1`

After the deployment is complete, data will automatically be collected and sent to the instance for analysis. Let's look at how MDI can recognize sequences of events to identify an attack chain.

Identifying attack techniques

MDI is designed to detect common attack methodologies from techniques defined in the **MITRE ATT&CK Matrix** to build a timeline of an attack kill chain. The kill chain is a sequence of events used by a malicious actor to gain knowledge of your environment and ultimately take over the domain. MDI recognizes these events in the following sequences:

- Reconnaissance
- Compromised credentials
- Lateral movements

- Domain dominance

- Exfiltration

Having the ability to correlate events to identify malicious activity during various stages of the kill chain can greatly increase the SOC's chances of stopping an attack. This is the real value of adding MDI monitoring as a security enhancement to your AD environment. The following table provides examples of the types of alerts that can be triggered from MDI during each stage of the kill chain:

Attack Phase	Alert
Reconnaissance	Account enumeration reconnaissance.
	Network mapping reconnaissance (DNS).
	Security principal reconnaissance (LDAP).
	User and IP address reconnaissance (SMB).
	User and group membership reconnaissance (SAMR).
Compromised Credential	Honeytoken activity.
	Suspected brute-force attack (Kerberos, NTLM, LDAP, and SMB).
	Suspected WannaCry ransomware attack.
	Suspected use of Metasploit hacking framework.
	Suspicious VPN connection.
Lateral Movements	Remote code execution over DNS.
	Suspected identity theft (pass-the-hash or pass-the-ticket).
	Suspected NTLM authentication tampering.
	Suspected NTLM relay attack.
	Suspected overpass-the-hash attack (encryption downgrade and Kerberos).
	Suspected SMB packet manipulation (CVSE-2020-0796 exploit).
Domain Dominance	Malicious request of the data protection API master key.
	Remote code execution attempt.
	Suspected DCShadow attack (domain controller promotion and replication requests).
	Suspected DCSync attack (replication of directory services).
	Suspected Golden Ticket usage (encryption downgrade, forged authorization, nonexistent account, ticket anomaly, and time anomaly).
	Suspected Skeleton Key attack (encryption downgrade).
	Suspicious additions to sensitive groups.
	Suspicious service creation.
Exfiltration	Suspicious communication of DNS.
	Data exfiltration over SMB.

Figure 14.37 – MDI alerts

Each alert that is identified as malicious activity will be presented on the **Timeline** page in MDI. The timeline is the default landing page after logging into the MDI portal.

MDI can easily be integrated with other solutions in the Microsoft security stack such as MDE and MDCA. Additionally, security and health alert events can be forwarded to a SIEM solution for analysis.

Let's look at using the attack timeline in the MDI portal to investigate alerts.

Looking at the attack timeline

To sign into the MDI portal, visit `https://portal.atp.azure.com`. The landing page provides a **Timeline** view of suspicious activities listed in chronological order, with the last-opened alert first. The filtering panel allows you to view the alert status by **Open**, **Closed**, and **Suppressed** options, as well as by **High**, **Medium**, and **Low** severity. An example of the timeline view can be seen in the following screenshot, showing both a **High** and a **Medium** alert.

Figure 14.38 – An attack timeline view

MDI uses entity tags to help identify sensitive groups, accounts, and assets. These tags are helpful for identifying suspicious activities related to privilege escalation, reconnaissance, and lateral movement. By default, many of the common account types, server roles, and groups are tagged sensitive – for example, domain admins, domain controllers, local administrators, and server operators. You can manually tag entries as sensitive too and create **Honeytoken** accounts to lure attackers.

To view **Entity tags** in the MDI portal, click on the **Configuration** menu and choose **Entity tags** under **Detection**. In the following screenshot, a **Honeytoken** account named `Break Glass [Admin]` is being configured.

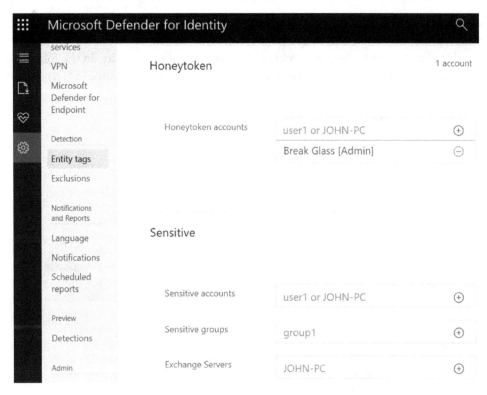

Figure 14.39 – Entity tags

In the **Sensitive** section, you can add additional accounts, groups, or servers that are considered sensitive for easy identification in the MDI portal. Any sensitive entity that shows up in an alert will have an **S** icon beside it, as shown in the following screenshot.

Figure 14.40 – A sensitive group with an entity tag

Heading back to the attack timeline, clicking on an alert will bring up additional information about the suspicious activity. Each alert tries to clearly define who and what is involved in the alert and when to quickly build a picture of the activity. In the following screenshot, an alert was created for a **User and IP address reconnaissance (SMB)** activity.

Figure 14.41 – A security alert in the MDI portal

Clicking + under **Evidence** will expand the details. In the preceding screenshot, you can see that the **Emily Young** account enumerated SMB session details on the MTLABDC01 domain controller from the WinSVR1 source system. The alert was labeled with a medium severity, but this activity can potentially indicate malicious intent and that the account has been compromised. Using this reconnaissance technique, a malicious actor could have successfully identified a list of network locations and potential accounts to target for further lateral movement. Based on the alert type, the SOC team can reach out to the user to reset the password or block all sign-ins until an investigation can be completed. Alerts can be cross-referenced using the M365 Defender portal and correlated with any relevant alerts from MDCA or MDE. Correlating alerts is an important step to identify any additional events and locate sources of compromise to thwart an attack.

Let's look at a few other alerts in the MDI portal. In the following screenshot, an attacker used the **overpass-the-hash** technique to acquire a Kerberos **Ticket-Granting Ticket (TGT)** using the BreakTheGlass account.

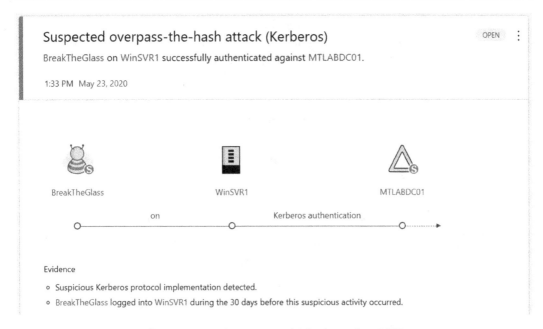

Figure 14.42 – An overpass-the-hash attack in MDI

If an attacker attempts to move laterally and execute code remotely throughout the network, using tools such as PSExec and mimikatz, MDI will trigger a **Remote code execution attempt** alert, as shown in the following screenshot.

Figure 14.43 – A remote code execution attempt alert

In *Chapter 10*, *Mitigating Common Attack Vectors*, we mentioned a credential dumping technique using domain controller replication, known as a DCSync attack. MDI can alert on any requested replication activities, as shown in the following screenshot.

Figure 12.44 – A suspected domain replication alert

If the replication is malicious in nature, additional privileged accounts can become compromised, or the KRBTGT password hash can be captured and exploited for use in a **golden ticket** attack. MDI will also alert on suspected golden ticket attacks, as shown in the following screenshot.

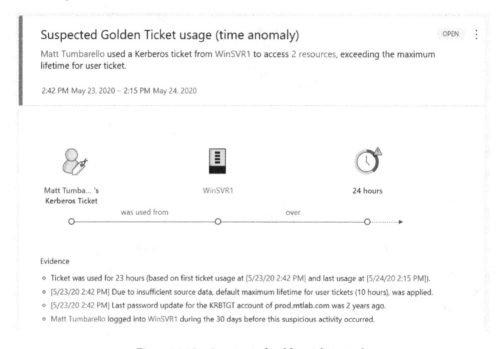

Figure 14.45 – A suspected golden ticket attack

Once this level of access has been obtained, the attacker now has control over resources that rely on Kerberos tickets for authentication and can timely plan a domain takeover. Having this visibility into the attack chain can help security teams take action and prevent these incidents from occurring.

Notifications and reports can be scheduled from the MDI portal by going to **Configuration** and choosing **Notifications** or **Scheduled reports** under **Notifications and Reports**. The available reports include the following:

- A summary of alerts and health issues
- Modifications to sensitive groups
- Passwords exposed in cleartext
- Lateral movement paths to sensitive accounts

For additional information about managing alerts in the MDI portal, visit `https://docs.microsoft.com/en-us/defender-for-identity/working-with-suspicious-activities`.

Next, we will review the solutions available from M365 to protect sensitive data.

Data protection with M365

Although not directly related with the security hardening your Windows devices or servers, it is critical that you are aware of the tools that will help protect the data your users are accessing from their Windows devices and any data stored on Windows servers. Like all technologies, there are options available from many vendors for data protection. For this publication, we will be focusing on the technology available with Microsoft. With Microsoft, a few of the technologies used to best protect your company data include the following:

- **Microsoft Purview Information Protection**
- **Data Loss Prevention (DLP)**
- **Windows Information Protection (WIP)**

In addition to the preceding data protection components is information governance. As you build and mature your data protection program, ensure your information governance program exists and matures. Some of the primary features Microsoft focuses on with information governance is data retention and deletion, record management, automation of retention labels, and the ability to prove compliance within your environment.

Purview Information Protection, DLP, and WIP can be used as separate products from Microsoft and may appear to provide some overlap, but they all have their unique usages and also complement each other to provide additional protection for your company's data. Let's look at each of these technologies in more detail.

Using Microsoft Purview Information Protection

In the first edition of this publication, we reviewed AIP. AIP now falls under the broader Microsoft Purview solution in which we will focus on. In short, Purview Information Protection is a technology that provides the ability to discover your data, classify your data, and provide visibility of your data by applying sensitivity labels. Microsoft has provided a simplified Purview Information Protection framework to follow. This includes the following three principles:

- Know your data by identifying the different types of data you need to protect.

- Protect your data by creating and applying sensitivity labels across all Microsoft services, cloud repositories, databases, and containers.

- Prevent data loss by building and deploying DLP policies within your environment.

To learn more about Purview Information Protection, visit the official documentation link at `https://docs.microsoft.com/en-us/microsoft-365/compliance/ information-protection?view=o365-worldwide`.

Before using Purview Information Protection, make sure you are familiar with the license requirements. Depending on the features in your license types, you may need to purchase add-ons to utilize advanced features. In the preceding link, scroll to the bottom of the page and click on the **PDF download** link within the **Licensing requirements** section to view more details about licensing.

To access Purview Information Protection, access the Office 365 Security and Compliance console by logging in to `https://compliance.microsoft.com/homepage` and then selecting **Information protection**. Here, you will view an overview and manage your labels, label policies, and auto-labeling. To review information on your sensitive info types, exact data matches, trainable classifiers, and so on, navigate to **Data classification** in the main menu on the left.

Next, we will provide an overview of DLP.

An overview of DLP

DLP tools help to identify, monitor, and prevent against unwanted sharing and exfiltration of sensitive company data from your environment. Using the Microsoft DLP solution, sensitive information can be prevented from leaking out of your environment by protecting the following services:

- Exchange Online, SharePoint Online, OneDrive for Business, and Microsoft Teams
- Word, Excel, and PowerPoint

- Windows 10, Windows 11, and macOS (Catalina 10.15+)
- Non-Microsoft cloud apps
- On-premises SharePoint and file shares

Within the DLP engine, you can create policies that allow you to select which type of data you would like to protect. With the data type, you can select from many pre-defined templates that scan for **Personal Identifiable Information** (**PII**), finance, medical and health, and privacy types of information. You can then define the locations that you would like to search within, and configure the policy settings by defining the conditions and protection actions against any identified sensitive information. You also can test the policy, which is highly recommended before enabling it for production.

To access Microsoft's DLP, log in to the M365 Compliance console at `https://compliance.microsoft.com/homepage` and select **Data loss prevention**. Click on **Policies** to set up and review your policies. You can also view **Alerts** and the **Activity explorer** within this section. As mentioned in the previous section, make sure you are familiar with the license requirements for DLP and review the license document referenced.

Next, we will look at the WIP technology.

WIP

The final technology we will review is WIP. This is the technology available to help prevent the accidental leakage or loss of data from your enterprise documents in Windows 10 version 1607 and later. With WIP, you can create policies that prevent data from being moved out of your environment, such as preventing data from being copied to USB drives. WIP also helps protect and separate corporate data from personal data if your company supports a *bring your own device* program for employees. In addition to running Windows 10 version 1607 or later, you will also need to be licensed for Microsoft Intune or manage devices with Microsoft Endpoint Configuration Manager to enable WIP policy. Third-party solutions can also apply WIP policies through the `EnterpriseDataProtection` **Configuration Service Provider** (**CSP**). Information about the CSP can be found at `https://docs.microsoft.com/en-us/windows/client-management/mdm/enterprisedataprotection-csp`.

To access Microsoft's WIP, log in to Microsoft Endpoint Manager at `https://endpoint.microsoft.com/` and click on **Apps**. Then, click on **App Protection Policies** under the **Policy** menu. Here, you can create and manage your WIP policies, including **Mobile Application Management (MAM)** policies for iOS and Android. For more information about creating a WIP policy using MDM, visit `https://docs.microsoft.com/en-us/windows/security/information-protection/windows-information-protection/create-wip-policy-using-intune-azure`.

We have just reviewed three different technologies for protecting data with M365. Protecting company data within your environment and endpoint devices is a journey requiring a partnership between IT, the business, and the data owners. It should be considered a priority in your overall hardening strategy and an integral component of a security program. Next, we will review business continuity and the importance of planning for disaster recovery and cyber incident response.

Planning for business continuity

To finish this chapter, we are going to cover business continuity planning and its importance as it relates to security, your infrastructure, a **Disaster Recovery Plan (DRP)** and a **Cybersecurity Incident Response Plan (CIRP)**. When we look at the **Business Continuity Plan (BCP)**, it is considered a business-wide plan that focuses on a business holistically to ensure the continued operation in the wake of major disruption. Within the BCP, multiple plans come together that comprise the overall plan, and there are many frameworks available to assist with the BCP program. The *SP 800-34 Rev. 1, Contingency Planning Guide for Federal Information Systems* NIST framework publication is a great resource for your BCP planning and can be found at `https://csrc.nist.gov/publications/detail/sp/800-34/rev-1/final`.

The following screenshot represents all the plans covered within the referenced NIST *SP 800-34 Rev.1* publication.

Figure 14.46 – The different plans that make up the BCP

The BCP is not a simple plan to create; it requires a lot of time and resources to make it successful. In addition to building a well-documented plan, it is just as important to ensure that everyone is familiar with the plan and that it has been coordinated and tested in some way. When it comes to executing the BCP in a real-world scenario, you don't want to be doing so for the first time without at least being familiar with the process and steps involved. To be better prepared, different exercises can be completed to become more familiar with the BCP, and one of the more common BCP exercises is in the form of a tabletop exercise. In a tabletop exercise, you bring together the key personnel in your organization and walk them through a crisis situation that may occur. When walking them through the situation, you request a response on how to handle the situation to identify any gaps in a real crisis.

Some examples that may trigger the execution of the BCP include a natural disaster, such as hurricanes, earthquakes, and floods. Depending on the severity, fires or power outages are other examples that may cause the BCP to execute. Today, more common threats that will trigger the BCP include cyberattacks, which can easily bring a business to a halt very quickly.

Learning DRP

The DRP is, for the most part, technical in nature and focuses on the recovery of IT infrastructure and systems. The DRP falls within a larger BCP plan for an entire organization, and the execution of a DRP plan will be unique, depending on the situation, and it may not impact the entire business. Events that trigger the BCP will typically have some form of impact on your IT systems and may require the execution of the DRP at the same time. Some examples include a natural disaster, such as a hurricane or fire destroying a data center.

As stated already, the entire DRP may not need to be fully executed in certain situations. For example, if a business-critical application becomes corrupted and is not highly available, you may only need to execute the DRP for that specific application to ensure ongoing operations. There are many different instances that can cause the DRP to be activated, so it's critical to account for each of these scenarios and make sure you are able to accommodate recovery for individual systems or entire data centers. A key component in DR planning is understanding the impact of a service or function and the **Maximum Tolerable Downtime** (**MTD**) that a service or function can withstand before your business is negatively impacted. Two important factors that also need to be understood are the **Recovery Time Objective** (**RTO**) and **Recovery Point Objective** (**RPO**). The RTO is the maximum amount of time a system can be unavailable for before negatively impacting a service or function within the business and preventing it from being able to operate normally. The RPO is the point in time at which the service or function can suffer data loss without being negatively impacted.

As you build your DRP and understand the expected restoration of each system based on the RTO and RPO, you are going to need to ensure that you have the process and infrastructure in place to meet those requirements. A few examples to consider include the following:

- Having **High-Availability** (**HA**) configurations for systems may help restoration after any local issues within a data center but will come at a cost.

- Understand your backup strategy as it relates to full backups, incremental backups, and differential backups, and also consider how often each backup needs to be taken and retained.

- Consider what type of failover is needed for a complete restoration of a data center, should you have a cold, warm, or hot site available.

All these considerations will depend on the business requirements and needs, along with the expense. Having a hot site on standby will come at a much greater cost than that of a warm or cold site. More importantly, as you implement your DRP and backup strategies, the role security plays in each should be well understood. Ensuring that your data is backed up securely, off-site storage of data is secure, and your standby data centers maintain the same level of security as your primary data centers should all be taken into consideration.

One last point to mention is testing. Like your BCP, it is critical that you test your DRP to ensure that everyone is familiar with the restoration procedures and to confirm that the DRP works. This is typically validated by executing the DRP to some extent, restoring part of or the entire infrastructure in a secondary location, and validating that services and applications are functioning. This should be an annual event at a minimum.

The importance of a CIRP

As we look at today's threats in the security space, there has continued to be an increase in advanced cyberattacks, and they are becoming more sophisticated each day. One of the more advanced and disruptive threats is ransomware, which is known to severely impact business operations. Many companies aren't prepared for these levels of attacks, which can take days, weeks, or even months to recover from and get back to normal operations. Depending on the size of the business and the severity of the breach, the impact could be as damaging as a complete business shutdown. With that, it is critical that your organization fully understands the impact of a potential breach and how to best be prepared to handle one. For optimal preparation, a CIRP can help organizations navigate through a breach when it occurs. Your CIRP should include elements such as identifying the roles and responsibilities of the incident response team, updated contact information of everyone involved including vendors, incident response procedures, communication procedures, and playbooks.

Once you have a production-ready CIRP, all parties involved in a cybersecurity incident should become familiar with the CIRP. You will also want to test the response of both your technical and executive teams with a tabletop exercise to address any gaps with your CIRP and better prepare the team for what decisions may need to be made in such a situation.

A thorough BCP, DRP, and CIRP will require a lot of thought and collaboration between the IT team, cybersecurity team, broader business stakeholders, and executive team. The success of these plans will require the support and involvement of leadership to ensure that the business is best prepared for when an incident does occur.

Summary

In this chapter, we covered security operations and reviewed the tools and technologies available from Microsoft that offer enterprise-class protection. We began the chapter with an introduction to the SOC and XDR and the importance of both in an enterprise. We then introduced the M365 Defender portal with an overview of the available features.

Next, we reviewed MDE before providing an overview of Microsoft Sentinel and MDCA, Microsoft's version of a CASB. Other tools and features reviewed in this chapter included MDI and the different data protection tools available from Microsoft (Purview Information Protection, DLP, and WIP). We then concluded the chapter with an overview of BCP, DRP, and CIRP.

In *Chapter 15, Testing and Auditing*, we will review validating controls to ensure that the security measures that have been agreed upon are in place. We will look at Microsoft's on-demand assessments and then review vulnerability scanning and testing to ensure that our controls are working correctly. The final sections will cover an overview of penetration testing and remediation, before finishing with security testing, awareness, and training.

15
Testing and Auditing

In this chapter, we will provide details about testing and auditing your environment, which will help validate controls and ensure due diligence has been executed within your security program. The challenge we face when deploying security controls, hardening configurations, and security baselines, is validating that they are in place and working as designed. The security department may have obligations to the leadership team, board stakeholders, shareholders, and regulators to prove that you have implemented the appropriate recommended controls depending on your business or industry. This is where testing and auditing comes into play by helping provide evidence that controls are in place. These activities can be performed internally or through a third-party company for added benefit. We test to ensure that our controls are doing what they are designed to do. Without testing, we fail to validate whether the controls work, and can't truly understand our risk level.

The first section will cover the validation of controls within your environment by reviewing the different types of audits and assessments. We will discuss vendor management and the importance of ensuring vendors meet the agreed-upon controls. This due diligence requires collecting and reviewing vendor questionnaires and security audits. We will also review the Microsoft Defender for Cloud regulatory compliance section to measure the status of your Microsoft environment against common regulations.

In the following section, we will review vulnerability scanning to confirm that your implemented controls and configurations are working correctly. We will look at the different scanning types, including scanning options within Azure. Next, we will provide an overview of penetration testing and review the different methodologies and steps taken as part of test execution. Finally, we will review security awareness, training and testing, and the importance of this program within your environment.

In this chapter, we will cover the following topics:

- Validating security controls
- Vulnerability scanning overview
- Planning for penetration testing
- An insight into security awareness, training, and testing

Technical requirements

To follow along with the overviews and instructions in this chapter, the following requirements are recommended:

- A Microsoft cloud services account to access the Service Trust Portal
- An Azure subscription with contributor rights
- Microsoft Defender for Cloud
- Microsoft Defender for Endpoint
- Microsoft 365 E5
- Microsoft Defender for Office 365 Plan 2

Let's start by looking at validating your environment controls.

Validating security controls

Validating that controls are in place is a significant task in the security program and one that should not be neglected. Building a validation program to ensure the documented controls are enforced will help provide additional certainty and peace of mind. Having a second set of eyes to review anything you implement in the IT and security fields is always a good idea. This doesn't necessarily mean an incident will never happen, but it does show that you are executing due diligence and doing what is right.

In addition, it's important to validate that the vendors you partner with also maintain the same level of detail in protecting their environments. The more we move data to vendor-managed cloud and SaaS services, the more due diligence is needed to audit access and validate controls in the vendor's environment. This is changing the dynamics of how we work in security compared to the standard model of hosting data internally on-premises. Having a detailed vendor onboarding process that includes collaboration with other business teams, such as the legal and procurement department, helps to ensure the due diligence process is completed thoroughly.

First, let's look at auditing and review the different types of audits and compliance reports.

Audit types

Auditing is the process that checks the intended controls are in place. At a high level, you will typically see two types of audits:

- An **internal audit** is one that is completed by an internal team employed by the organization to conduct audits within the business.

- An **external audit** is completed by a third-party company. The idea behind an external audit is to avoid any conflict of interest within the organization by using an independent party with no ties to the organization. Depending on your company's industry, this could be a legal requirement.

Commonly seen in an internal audit scenario, the auditing team reports directly to the CEO or the board of directors and not internally to the IT or security department. This helps ensure accountability, prevents any conflicts of interest, and provides a system of checks and balances.

When looking at audits, one of the more common practices and well-known reporting offerings is the **System and Organization Controls (SOC)**—not to be confused with the security operations center. The SOC services are part of the **American Institute of Certified Public Accountants (AICPA)**. It's a good idea to become more accustomed to SOC and other auditing standards within the industry, especially as you look to adopt more cloud services. As you meet with vendors and subscribe to services, it's important that these reports have been completed and the controls are in place for your vendors. Let's review SOC reports in more detail.

SOC reports

There are three different SOC reports to be familiar with:

- **SOC 1** reports focus on the financial aspect of an organization, ensuring the correct financial controls and reporting are in place. This report comes with two types of reports:

  - **Type 1** reports are based on the feedback and description of controls provided by management, completed at a specific time.

  - **Type 2** reports are more involved and look to test the effectiveness of the controls in place over a duration of time.

- **SOC 2** reports are focused more on the technology and security aspects of an organization to better protect users' data and reduce risk. The specific controls measured include security, availability, processing integrity, confidentiality, and privacy. Like SOC 1, there is a type 1 and type 2 report that follows the same format. SOC 2 reports are intended for internal use only and vendors will require a **non-disclosure agreement (NDA)** to be signed to view them.

- **SOC 3** reports take the output of SOC 2 reports and write them in a more readable format that is less technical. This allows them to be made available for anyone to access and view.

You can view more information about the AICPA SOC services at `https://www.aicpa.org/soc`.

> **Important Note**
>
> ISO/IEC 27000 is another common certification family to ensure best practices are implemented as part of your security program. The ISO/IEC certification is used more widely internationally than SOC. You can view more information about the ISO/IEC standards at `https://www.iso.org/isoiec-27001-information-security.html`.

Even if your business is not required to complete an audit, it may be a good idea to go through the process of a SOC engagement to ensure you are maintaining the highest level of standards for the organization. Next, let's look at managing the risk inherited by conducting business with third parties.

Vendor risk management

The job of managing vendors can be quite a challenge for organizations. Onboarding vendors can be a thorough and lengthy process involving different departments, such as legal and procurement, to help ensure contracts are written and executed correctly to reduce any liability. As more services shift to third-party cloud-based vendors that host our user and customer data, the onboarding process must be more rigorous than ever before. This becomes even more challenging as you need to deal with both current and new privacy requirements. As part of the onboarding process, the right personnel from both the technical and security teams should also be included for review. Nowadays, there are probably very few contracts that don't include some use of technology and I'd imagine that at some point, all contracts with vendors will involve technology that needs to be reviewed by the technical and security teams.

From a technical and security standpoint, one recommendation is to implement a security questionnaire or risk assessment that can help assess and better understand the vendor being onboarded. To assist with this, a couple of third-party questionnaires that are commonly used include the following:

- The **Standardized Information Gathering** (**SIG**) questionnaire, provided by Shared Assessments: `https://sharedassessments.org/sig/`

- The **Consensus Assessment Initiative Questionnaire** (**CAIQ**), which is made available by the **Cloud Security Alliance** (**CSA**): `https://cloudsecurityalliance.org/research/cloud-controls-matrix/`

In addition to collecting a questionnaire, you will want to request additional documents to ensure a thorough review and assessment can be completed. This includes the following items:

- Third-party audit reports—for example, a SOC 2 report or ISO 27001 certification
- Penetration test results or any security findings that have been flagged as critical/high-risk
- An information security management program policy or overview
- A business continuity plan , disaster recovery plan, and cybersecurity incident response plan
- Any other supporting audit/risk/security documentation that's available

There are also third-party services that help provide additional insight into your vendor's security posture by means of a security score or grading system. The score is built around scanning publicly available vendor domains and any associated assets reachable over the internet to identify any known vulnerabilities. A couple of examples are as follows:

- SecurityScorecard: `https://securityscorecard.com/product/security-ratings`
- BitSight: `https://www.bitsight.com/third-party-risk-management`

> **Important Note**
>
> Vendor risk management is not just a one-time exercise but one that needs a life cycle attached to it. At a minimum, annual reviews should occur as audits and certifications expire.

Next, let's review the Microsoft Service Trust Portal for access to Microsoft's compliance reports.

The Microsoft Service Trust Portal

Microsoft has a public portal dedicated to providing audit and compliance reports for all its cloud services, known as the Service Trust Portal. To visit the Service Trust Portal, go to `https://servicetrust.microsoft.com`. The landing page will look like the following screenshot:

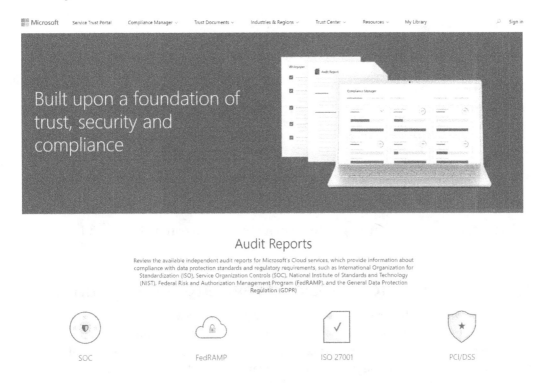

Figure 15.1 – The Microsoft Service Trust Portal

If you browse through the menu at the top of the web page, the audit-specific documents can be accessed by taking the following steps:

1. Ensure you are on the Service Trust Portal home page, found at `https://servicetrust.microsoft.com/`.

2. Click on **Trust Documents**.

3. Click on **Audit Reports**.

4. Scroll down to view the different reports and certificates available.

5. Click on the **SOC Reports** tab under **New and Archived Audit Reports** to view the SOC-related reports from Microsoft.

6. Select the report you would like to view, then sign in with your Microsoft account to access it:

Figure 15.2 – SOC reports for Microsoft

Next, let's look at accessing the **Regulatory compliance** dashboard in Microsoft Defender for Cloud.

Microsoft Defender for Cloud regulatory compliance

In addition to viewing the reports provided by Microsoft, you can also assess your own cloud environment from within Azure. As part of the Microsoft Defender for Cloud service, there is a **Regulatory compliance** section that will check your environment configurations against common compliance controls. Not all the controls will be automated, but the feature continues to be enhanced and it is a great tool to help with the assessment. There are many regulatory compliances available for reference, and they can be accessed by taking the following steps:

1. Log in to `https://portal.azure.com`.

2. Search for `Microsoft Defender for Cloud` and open it.

3. Click on **Regulatory compliance** in the left-side menu within the **Cloud Security** section:

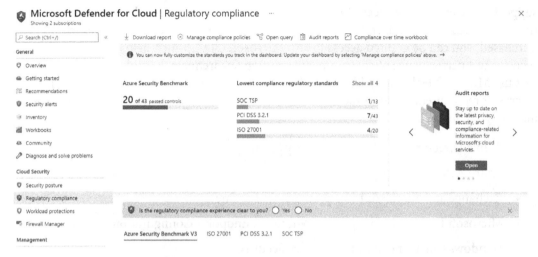

Figure 15.3 – Regulatory compliance dashboard

> **Important Note**
>
> To use the **Regulatory compliance** feature, you will need to upgrade your Microsoft Defender for Cloud plan from free to enhanced security features. The current costs can be found at `https://azure.microsoft.com/en-us/pricing/details/defender-for-cloud/`.

Within the dashboard, you can track individual controls based on assessments run by Defender for Cloud against your Azure resources and view them by control number, resource type, and the number of failed resources.

Next, let's look at using Microsoft **On-Demand Assessments** (**ODA**) to provide analysis of IT workloads running services with Microsoft.

Microsoft ODA

For Microsoft customers, as part of your support agreement, you gain access to the support hub called the Services Hub. The Services Hub can be accessed by going to `https://serviceshub.microsoft.com`. Depending on the level of support purchased, you can be granted access to run Microsoft ODA, which helps to analyze the security and performance of environments running on-premises, hybrid, or in the cloud that use Microsoft technology. The result is an output of your controls compared against Microsoft recommended best practices. There are currently over 49 assessments available, including the following:

- Active Directory and Active Directory Security
- Azure Active Directory
- Exchange and SharePoint (on-premises and online)
- Microsoft Endpoint Manager and Microsoft Endpoint Configuration Manager
- Windows Client and Windows Server Security
- SQL Server

To view the full list of available assessments, visit `https://docs.microsoft.com/en-us/services-hub/health/`.

There are several prerequisites needed to run ODA, and these will vary depending on the assessment you wish to run. Many of them require an active Azure Subscription, Log Analytics workspace, dedicated Windows server, and accounts with permissions to run remote scans against target systems. If you need help, as part of a Microsoft Unified Support agreement, engineers are available to assist you with the setup.

After an assessment is completed, the output includes an overview of all high/low priority recommendations and passed checks. The recommendations can be exported in Excel format for tracking, and an executive summary PowerPoint is available for presentation purposes.

Assessments

Workspace name   AzureAssessments

Total Assessments   7

Windows Server Assessment	Active Directory Assessment	Active Directory Security Assessment
Active Updated May 16, 2022	Active Updated May 16, 2022	Active Updated May 16, 2022
⚠ 10 High priority recommendations	⚠ 4 High priority recommendations	⚠ 13 High priority recommendations

Windows Server Assessment

High priority recommendations
10

Low priority recommendations
69

Resolved Recommendations
0

Passed checks
725

🖵 View all Recommendations

🗑 Remove Assessment

▣ Download Executive Summary

▣ Download All Recommendations

Create a Remediation Program

Figure 15.4 – ODA results

You can run the ODA as many times as needed, which will allow you to incorporate the results into your ongoing vulnerability management program to track identified items to resolution.

Next, let's review other options available for validating security and compliance controls.

Other validations

There are many other tools available that offer solutions to help validate controls within your environment. Not all have been covered, but some additional options we have discussed throughout the book include the following:

- CIS-CAT Lite, CIS-CAT Pro Assessor, and CIS CAT Pro Dashboard from the **Center for Internet Security** (**CIS**) are assessment tools that will review your systems against CIS benchmarks. With the assessment, you will be provided with a score along with remediation steps for missing controls. CIS-CAT Lite is the free version of the assessment tool versus CIS-CAT Pro, which provides additional benefits such as 80+ available benchmarks and evidence-based reports: `https://learn.cisecurity.org/cis-cat-lite`.

- The Microsoft Security Compliance Toolkit includes the Policy Analyzer tool for assessing Windows Security Baselines built from GPOs and local group policies compared to current systems settings.

- In *Chapter 2*, *Building a Baseline*, we discussed the NIST Cybersecurity Framework. If you implemented NIST, you will need to validate your responses using a third-party vendor to review and assess whether your stated responses are accurate. The NIST website also has some resources on additional tools to help with the implementation and tracking of this framework: `https://www.nist.gov/cyberframework/assessment-auditing-resources`.

In this section, we reviewed what is needed to validate controls, primarily focusing on auditing and the resources available to complete audits and assessments. In the next section, we will look at vulnerability scanning and testing tools to ensure the hardening of your Windows devices.

Vulnerability scanning overview

While auditing is important for validating that security controls are in place, it's strongly recommended to incorporate a vulnerability scanning program into your security operations. Vulnerability scanning will provide additional insight into your environment by identifying known exploits and weaknesses and raising awareness about how vulnerable your systems are against new and emerging threats. Scan reports can alert IT and security teams where immediate remediation is needed without relying on updated control frameworks. Let's look at what a vulnerability scan involves.

An introduction to vulnerability scanning

Vulnerability scans or assessments look for and identify known vulnerabilities within your environment or systems. For example, a scan might detect that an application, operating system version, or file seen on the network has a known vulnerability. After a scan completes, a report is generated that highlights these weaknesses and includes a list of improvement actions. Vulnerability scans are typically scheduled to run automatically after the initial assessment has been set up. The following is a list of common types of vulnerability assessments that are used:

- Network/wireless assessments
- Web application assessments
- Application assessments
- Database assessments
- Host-based assessments

There are many vulnerability assessment tools available at your disposal. Some of the more common tools that you may be familiar with include the following:

- Qualys: `https://www.qualys.com/`
- Tenable (Nessus): `https://www.tenable.com/`
- OpenVAS: `https://www.openvas.org/`
- Nikto: `https://cirt.net/nikto2`

Microsoft also has vulnerability scanning solutions available as part of their Defender security suite. Let's look at an example of vulnerability scanning with Microsoft Defender for Cloud and Microsoft Defender for Endpoint.

Vulnerability scanning with Microsoft Defender for Cloud

Microsoft Defender for Cloud includes a built-in vulnerability assessment for Windows and Linux servers running on-premises, in Azure, and connected with Azure Arc. This assessment is included with the enhanced security features and is powered by Qualys without the need for additional licensing. After the scanner is deployed to servers, data is collected and analyzed using Qualys analytics, and reports the findings back to the Defender for Cloud console in Azure. For a review of additional Defender for Cloud features, see *Chapter 13*, *Security Monitoring and Reporting*, and *Chapter 14*, *Security Operations*.

The following article provides additional information about the vulnerability scanner, including the supported operating systems:

`https://docs.microsoft.com/en-us/azure/defender-for-cloud/`
`deploy-vulnerability-assessment-vm`.

To enable the assessment on your virtual machines, follow these steps:

1. Log in to `https://portal.azure.com`.
2. Search for **Microsoft Defender for Cloud** and open it.
3. Click on **Recommendations** in the left-side menu under the **General** section.
4. Search for **Machines should have a vulnerability assessment solution** and click to open it.
5. Select the virtual machines you would like to deploy to, then click on **Fix**.

After clicking **Fix**, the Qualys extension will be installed on the virtual machine and will begin collecting data. After about 48 hours, you will be able to view any discovered vulnerabilities by following these steps:

1. From the Azure portal, browse to **Microsoft Defender for Cloud**.
2. Click on **Recommendations** from the left-side menu.
3. Within the main screen, look for **Remediate vulnerabilities** and expand the section.
4. Click on **Machines should have vulnerability findings resolved** to view all the findings, as in the following screenshot:

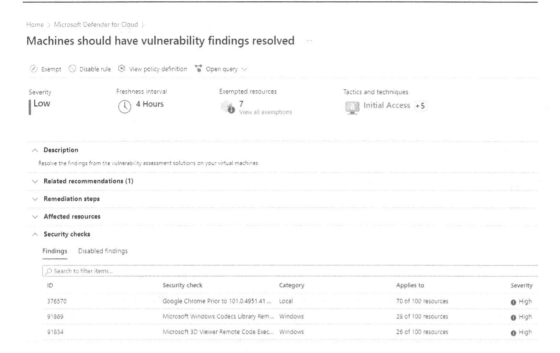

Figure 15.5 – Identified vulnerabilities

The following reference guide provides a list of the recommendations that you may see within the **Recommendations** section from the scans within Microsoft Defender for Cloud: `https://docs.microsoft.com/en-us/azure/defender-for-cloud/recommendations-reference`.

Next, let's look at reviewing weaknesses in the Microsoft 365 Defender portal.

The Microsoft 365 Defender portal

For devices onboarded to Microsoft Defender for Endpoint, you can access **Vulnerability management** from within the Microsoft 365 Defender portal to view identified vulnerabilities on your devices. To access this feature, take the following steps:

1. Log in to the Microsoft 365 Defender portal by going to `https://security.microsoft.com/`.

2. Click on the menu at the top left, then expand **Vulnerability management**.

3. Click on the **Weaknesses** option to view all the identified vulnerabilities within your environment.

4. Within the search field, you can refine the results by adding filters. For example, in the following screenshot, a filter was added to show vulnerabilities for Windows 10:

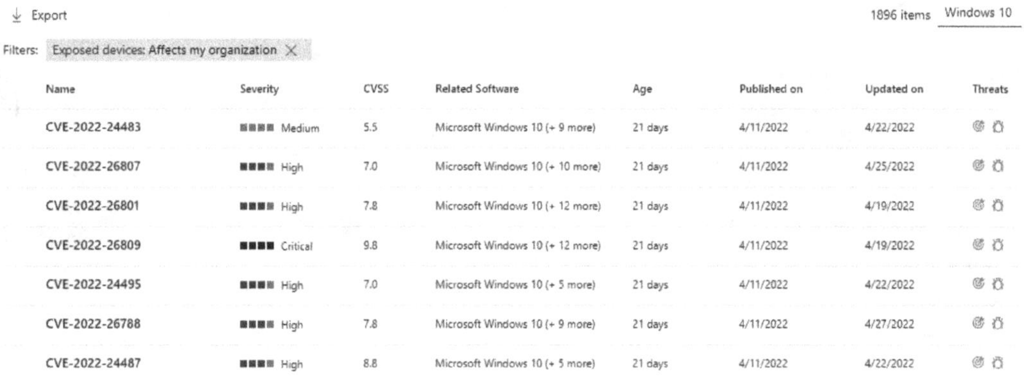
Export
1896 items
Windows 10
Filters: Exposed devices: Affects my organization
Name
Severity
CVSS
Related Software
Age
Published on
Updated on
Threats
CVE-2022-24483 Medium 5.5 Microsoft Windows 10 (+ 9 more) 21 days 4/11/2022 4/22/2022
CVE-2022-26807 High 7.0 Microsoft Windows 10 (+ 10 more) 21 days 4/11/2022 4/25/2022
CVE-2022-26801 High 7.8 Microsoft Windows 10 (+ 12 more) 21 days 4/11/2022 4/19/2022
CVE-2022-26809 Critical 9.8 Microsoft Windows 10 (+ 12 more) 21 days 4/11/2022 4/19/2022
CVE-2022-24495 High 7.0 Microsoft Windows 10 (+ 5 more) 21 days 4/11/2022 4/22/2022
CVE-2022-26788 High 7.8 Microsoft Windows 10 (+ 9 more) 21 days 4/11/2022 4/27/2022
CVE-2022-24487 High 8.8 Microsoft Windows 10 (+ 5 more) 21 days 4/11/2022 4/22/2022

Figure 15.6 – The Microsoft 365 Defender Weaknesses console

We also reviewed the **Weaknesses** dashboard in *Chapter 13, Security Monitoring and Reporting*.

Now that we have reviewed validating controls through audits and finding weaknesses with vulnerability scanning tools, let's look at planning for penetration testing. A penetration test is another great way to identify security gaps and exploits that affect your organization's systems.

Planning for penetration testing

Penetration testing, or pen testing, is another method for identifying security risks and an important function of a mature security program. Pen tests validate risk by performing specific activities against targets in your environment, such as system hosts, applications, users, and devices to exploit known vulnerabilities. Tests are executed by skilled security professionals, referred to as **ethical hackers**, to try and replicate the activities of a malicious actor. This practice is commonly referred to as **ethical hacking**.

Penetration tests can be executed externally to simulate an outside threat trying to break in, or internally, to simulate an insider threat that has breached your perimeter network. There are many different types of penetration tests, and commonly cover the following areas:

- Systems and servers including Active Directory
- Web, API, and mobile applications
- Databases

- Networks (internal/external/DMZ), including wireless
- Social engineering, such as phishing simulations
- Physical security tests against facility access and data center controls

Penetration testing also can cover software testing. This type of testing is most likely executed in conjunction with your developers and programmers. This includes static and dynamic testing, misuse case testing, fuzz testing or fuzzing, and interface testing against APIs and user interfaces. One excellent resource used in this type of testing is OWASP (`https://owasp.org/`), which was reviewed in *Chapter 1, Fundamentals of Windows Security*.

There are a few different testing types that can be used as part of penetration testing, known as **black box**, **gray box**, and **white box**:

- **Black box** testing, also referred to as no- or zero-knowledge testing, is where the tester is provided no information about your environment upfront. This best represents an actual hacker.
- **Gray box** testing, also referred to as some- or partial-knowledge testing, is where the ethical hacker is provided with limited information about the environment.
- **White box** testing is also referred to as complete- or full-knowledge testing. In this type, the ethical hacker is provided with complete knowledge of the environment.

Depending on the size of your organization and the maturity of the security program, it might make sense to outsource the testing service to a vendor. Most penetration testing vendors are equipped to handle small to large-sized environments and include experts who are very familiar with ethical hacking techniques and follow the rules of engagement to not inadvertently cause damage to critical systems. Many third-party companies offer penetration testing services, and a few examples include the following:

- Secureworks: `https://www.secureworks.com/services/adversarial-security-testing/penetration-testing`
- Mandiant: `https://www.mandiant.com/services/technical-assurance/penetration-testing`
- Rapid7: `https://www.rapid7.com/services/security-consulting/penetration-testing-services/`

Organizations that are well-staffed and have the in-house expertise to execute their own internal penetration test include a red team (offensive) and a blue team (defensive), which would likely be the SOC team. These teams work together to address the risks in an organization, and often complete purple team (red and blue team) exercises. For example, the red team will execute an attack on the organization and the blue team will attempt to validate whether they were able to detect the attack and respond to it. If the attack is detected, the blue team is successful with their role. If they don't identify the attack, they can address gaps for improvement.

> **Tip**
>
> Blue teams should also run threat-hunting exercises. A threat hunt uses IOCs as clues to *hunt* or run sweeps for potential active threats in the environment.

There are many tools available to help conduct your own tests within your environment. Some of the more common tools you may be familiar with include the following:

- Metaspoilt: `https://www.metasploit.com/`
- Wireshark: `https://www.wireshark.org/`
- NMAP: `https://nmap.org/`

Building a successful internal testing program will require a substantial investment of effort, experience, and skilled experts to operate efficiently. Next, let's review the steps included when executing a penetration test.

Executing a penetration test

Whether a penetration test is conducted internally or outsourced to a third-party company, you will need to ensure a rigid process is followed as part of its execution. Simply deploying some tools and trying to uncover vulnerabilities without a plan and proper approval can be extremely risky and potentially disruptive. The penetration test needs to be carefully planned out with the correct approval in place before executing it. There may be slight variations and different approaches, but the following diagram illustrates a standard process for executing a penetration test.

Figure 15.7 – The penetration testing process

Let's review each phase:

- The **Scoping and Planning** phase is where you will plan and define the scope of the test. This is where contracts will be signed, and the **Rules of Engagement (RoE)** document is agreed upon.

- **Reconnaissance** is the phase where information about the company or environment being targeted is gathered.

- In the **Vulnerability Assessment and Scanning** phase, you will scan the environment and search for and identify weaknesses and vulnerabilities.

- **Exploitation** is where you will attempt to exploit and breach the environment with any of the identified vulnerabilities.

- **Reporting and Analysis** is the final stage, where you will receive a report with an overview and detailed analysis of the test. This can also include recommended actions for remediation.

> **Tip**
> The *rules of engagement* document should be defined in clear language and approved and signed off by leadership before any work begins.

When you conduct a penetration test, you also need to be aware of other environments that could potentially be impacted if those services have rules of engagement defined. For example, Microsoft has a *rules of engagement* document that you will need to review if you're testing an environment hosted in Azure. You can find the Microsoft rules of engagement at https://www.microsoft.com/en-us/msrc/pentest-rules-of-engagement.

The following screenshot shows the Microsoft RoE introduction and scope:

Figure 15.8 – The Microsoft RoE

Make sure to review the section with the activities prohibited by Microsoft; the list is quite lengthy.

> **Tip**
> You should plan to execute a penetration test annually, at a minimum, within your environment.

After the test concludes, let's look at what the penetration test report findings may look like.

Reviewing the findings

The penetration test reports typically list any identified **Low**, **Medium**, **High,** and **Critical** vulnerabilities or concerns. It may also include an informational category that provides details on items of interest but may not be of high concern or applicable to your current systems. Once the test concludes and the report is finalized, the work doesn't stop. It is just as important that you take the report and build a plan around the remediation of those items.

After the remediation plan is developed, systems administrators and the security team should work together to mitigate any critical or high-risk items immediately and be held accountable by setting deadlines for completion. Once remediation and mitigation efforts have been completed, you will need to ensure you carry out re-testing to validate that the controls put in place work as designed so you can close the risk. In some instances, you might accept a risk if the cost to remediate outweighs the value of the data or the type of information at risk. In any case, building a remediation report that can be shared with leadership to show resolution to any identified vulnerabilities from a report is highly recommended. The following screenshot is an example of what could be used to document a vulnerability and accepted resolution:

Title of Vulnerability Identified

Findings	Resolution	Status: Risk Accepted or Remediated
Severity: Low or Medium or High or Critical		
Description: Description of vulnerability uncovered.	Here you will detail the steps taken to remediate the identified vulnerability or provide justification as to why the risk has been accepted.	
Systems/Applications Effected: List any systems or applications that were affected from this finding.		
Recommendation: Provide recommendation to remediate vulnerability.		

Figure 15.9 – Example of vulnerability remediation report

Any outstanding risks identified from a penetration test should also be included in a risk register, which is typically a document that tracks all known risks for your organization. If you don't have a risk register, it is highly recommended that you build one with visibility at the leadership level. Having a risk register is a great way for organizations to track known risks within their environment and document the risk owners and those who accepted the risk. Something as simple as an Excel spreadsheet could meet your needs, to begin with.

The output of a well-executed penetration test underscores the importance of its inclusion in a well-rounded security program. The results provide great value in identifying risks to your users, systems, and company data. Ensure you understand the technologies and concepts of penetration testing thoroughly to apply them efficiently within your environment.

> **Important Note**
> NIST has a very detailed document on security testing and assessment. You can view the *SP 800-115, Technical Guide to Information Security Testing and Assessment* publication at `https://csrc.nist.gov/publications/detail/sp/800-115/final`.

Next, we will review the importance of building a security awareness, training, and testing program.

An insight into security awareness, training, and testing

One area that shouldn't be overlooked is to ensure users participate in proper security awareness training and to test them using mock attacks. The human factor is often a weak link in the defenses, and providing awareness is critical to building a more resilient and secure organization.

There are three important components to consider in a robust employee security program. The first is **testing**, commonly executed through phishing campaigns, tabletop exercises (covered later in the chapter), or an attack simulator. The second is providing a robust and well-rounded **training** program, which includes keeping records of employees who have taken and passed the training. This shouldn't be a one-time event, but rather an ongoing event, as attack methodologies are constantly changing. Lastly, you need to provide consistent **awareness** to your users as deemed appropriate. Keeping security at the forefront by sending weekly communications with tips, tricks, and recommendations, or providing somewhere for users to learn more about security-related topics, such as the company intranet, will help strengthen your users' mindsets. Security should be considered a core company value.

> **Tip**
> Many security audits include showing proof of an active security awareness training program.

Just like with vulnerability and penetration testing, many vendors provide security awareness, training, and testing services and tools. A couple of the well-known vendors in the space include:

- KnowBe4: `https://www.knowbe4.com/`
- Proofpoint: `https://www.proofpoint.com/us/products/security-awareness-training`

Next, let's look at testing users by using the Microsoft 365 Defender attack simulation training tool.

Using attack simulation training with Microsoft 365 Defender

Microsoft has a tool available to test your users, named **Attack simulation training** as part of the Defender for Office licensing offering. With **Attack simulation training**, you can measure your user's awareness by choosing between different real-world scenarios commonly used in social engineering attacks, such as phishing emails. To use **Attack simulation training**, you will need to be a global administrator, security administrator, Attack simulation administrator, or an Attack Payload Author, and have a Microsoft 365 E5 or Microsoft Defender for Office 365 Plan 2 license. To access **Attack simulation training** and set up a simulation, take the following steps:

1. Log in to `https://security.microsoft.com/`.

2. From the left-side menu, click on **Attack simulation training**. The landing page will look like the following screenshot:

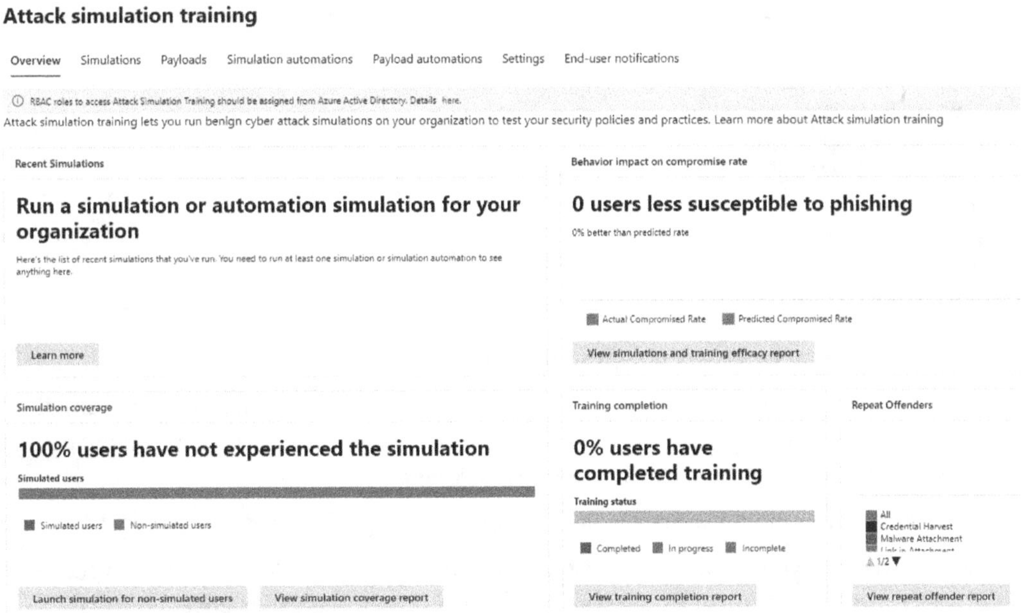

Attack simulation training

Overview   Simulations   Payloads   Simulation automations   Payload automations   Settings   End-user notifications

ⓘ RBAC roles to access Attack Simulation Training should be assigned from Azure Active Directory. Details here.
Attack simulation training lets you run benign cyber attack simulations on your organization to test your security policies and practices. Learn more about Attack simulation training

Recent Simulations

Run a simulation or automation simulation for your organization

Here's the list of recent simulations that you've run. You need to run at least one simulation or simulation automation to see anything here.

Behavior impact on compromise rate

0 users less susceptible to phishing

0% better than predicted rate

▪ Actual Compromised Rate   ▪ Predicted Compromised Rate

Learn more   View simulations and training efficacy report

Simulation coverage

100% users have not experienced the simulation

Simulated users

▪ Simulated users   ▪ Non-simulated users

Training completion

0% users have completed training

Training status

▪ Completed   ▪ In progress   ▪ Incomplete

Repeat Offenders

▪ All
▪ Credential Harvest
▪ Malware Attachment
▪ Link in Attachment
⚠ 1/2 ▼

Launch simulation for non-simulated users   View simulation coverage report   View training completion report   View repeat offender report

Figure 15.10 – Attack simulation training portal

3. Select **Simulations** on the top menu to view any current simulations.

4. To set up a new simulation, click on **+ Launch a simulation**, then select which attack technique you would like to use:

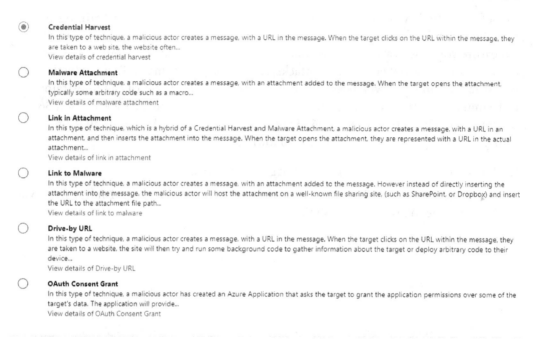

Figure 15.11 – Social engineering techniques

5. Click **Next**, then enter a simulation name and description.

6. Click **Next**, and you will then select the payload, which will be the malicious email sent to your users. Microsoft provides many different templates, or you can create your own. Once you select your payload, click **Next**.

7. Select **Target Users** by selecting all or by specifying specific users. Click **Next**.

8. Select whether you would like to assign training to the user. You can also redirect to a third-party platform or use Microsoft's native training experience. For this example, select **Microsoft training experience (Recommended)**, then select **Assign training for me (Recommended)**. Select a due date for training, then click **Next**.

9. Customize your landing page. Select your landing page preference and landing page layout, and edit your layout if desired. Click **Next**.

10. Next, select as required in **Select end user notification**. You will be required to complete additional steps depending on your selection. Click **Next**.

11. Configure **Launch Details** for when you would like to launch the simulation, then click **Next**.

12. Finally, review the simulation details and send a test before submitting to commit your selections. Click **Submit** when ready.

> **Tip**
> It is highly recommended to include all employee types, such as full-time employees, contractors, contingent workers, and interns, to complete the required training.

You can learn more about **Attack simulation training** by visiting the following link:

```
https://docs.microsoft.com/en-us/microsoft-365/security/
office-365-security/attack-simulation-training-get-
started?view=o365-worldwide.
```

Next, let's discuss the importance of executing a tabletop exercise in the event of a critical security incident.

Executing a tabletop exercise

In *Chapter 14, Security Operations*, we mentioned the importance of preparing your organization for when an incident occurs. It's best to maintain the mindset that in today's world, it's only a matter of *when*, and not *if* an incident occurs. To best prepare, you will need to assemble the personnel (as identified in your CIRP) that will be involved in an incident response through a tabletop exercise. A tabletop exercise walks your key stakeholders through a realistic incident scenario, asking questions that require the team to make timely decisions, troubleshoot, and communicate in a way that replicates an actual incident. A trained moderator will assess each response and provide feedback on strengths and areas for improvement. Tabletop exercises may include a wide range of personnel, which covers different real-world scenarios based on your business' industry, size of the organization, and IT footprint. There may be more specific target groups within your organization, but at the minimum, you should provide the following types of tabletops at least annually:

- A **technical tabletop exercise** tests your technical team's ability to respond to a cybersecurity incident. This exercise is mostly technical in nature and focused around identifying whether an incident has occurred, investigating, and the response needed from a technical perspective.

- An **executive tabletop exercise** brings your executive leadership team together to test how they will respond to an incident. This will focus more on the decision-making skills, how and whom to communicate with, the process and timing to engage cybersecurity insurance, and whether to pay a ransom.

Running through these exercises helps ensure your key stakeholders are familiar with the CIRP. This will increase the efficiency of an incident response should one occur, identify any gaps within the CIRP, educate stakeholders about their roles and responsibilities, and satisfy any regulatory requirements to execute them.

For the exercise scenario, you can build it around any relevant event that may occur within your environment. Some examples include a ransomware situation, a **business email compromise** (**BEC**), a disgruntled employee extracting top-secret information, a cyber extortion scheme, and disruptive **Distributed Denial-of-Service (DDoS)** attacks, to name a few. When executing a tabletop exercise, bringing in a third-party vendor can help provide insight from an outside perspective with real-world expertise. You could also run your own internal tabletop exercise by following readily available templates. CISA has a great repository of tabletop exercise packages at your disposal for use at `https://www.cisa.gov/cisa-tabletop-exercises-packages`.

It is important that you don't overlook your security awareness, training, and testing components. Preparing your users has never been more important, and ensuring your organization is prepared to respond to an incident will help facilitate a quicker recovery of normal business operations.

Summary

In this chapter, we covered validating controls within your environment, including internal and external auditing. We reviewed different types of SOC compliance and discussed the importance of vendor assessments as part of your vendor onboarding process. Next, we reviewed the Microsoft Service Trust Portal and how to view Microsoft's available audits and assessments. We then reviewed the Microsoft Defender for Cloud regulatory compliance offering, before finishing the section with Microsoft ODA.

In the next section, we covered what a vulnerability scan is and the different types of assessments that should be run. We reviewed how Microsoft Defender for Cloud can help run assessments against Windows and Linux hosts. We then discussed the importance of penetration testing and remediation. This included reviewing the different types of penetration tests, the test execution process, the importance of remediation, and an overview of the rules of engagement that Microsoft has published. In the final section, we covered security awareness, training, and testing, and reviewed Microsoft's attack simulation training, before finishing off with an overview of tabletop exercises.

In *Chapter 16, Top 10 Recommendations and the Future*, we will outline 10 key takeaways and to-dos to consider after reading through this book. Then, we will look at the future of device security and management, and an overview of what IT security means for our future.

16

Top 10 Recommendations and the Future

Welcome to the final chapter. We hope you've gained a better understanding of the tools and methodologies for securing Windows systems, and the value of building a well-rounded security program to protect your users and devices. We often hear of the challenges of keeping up to date with today's fast-growing technologies; therefore, the primary focus of the book is centered around solutions readily available in Microsoft's cloud. We hope to have provided you with the necessary knowledge to better understand these tools and the security solutions that can help support your transition to a more secure environment.

In this chapter, we will provide an overview of what we believe to be the 10 most important topics covered in this book. We hope these 10 recommendations will provide you with an actionable list of items to incorporate into your environment's security. Following these recommendations are additional items that we feel should be considered to strengthen your security program even further.

At the end of this chapter, we will provide our thoughts as they relate to the future of security and device management, and how the *anywhere-at-any-time* access model is forcing enterprises to modernize their access strategies using cloud technologies. We will discuss the role that security plays in the future and how everyday interactions need well-defined security models, and how a more autonomous world will require the right governance and security in place to stay protected.

In this chapter, we will cover the following topics:

- The 10 most important to-dos
- The future of device security and management
- Security and the future

The 10 most important to-do's

To finish the book, we wanted to highlight what we believe to be 10 of the most important areas covered within this book. These items are not listed in any priority order, but we feel they should be the focus of attention for your security program.

Implementing identity protection and privileged access

In a world that has shifted outside the walls of the office to an *anywhere-at-any-time* access model, identities have become a high target of attention and prone to weaknesses. They are a fundamental focus for attackers to gain access to your environment. Because of this, it is critical that your identities have multiple layers of protection and that preventative measures are in place.

Proper identity protection will require implementing account and access management tools and enforcing the principle of least privilege. A user must only be provided access to the specific data, applications, and systems that are necessary for their job role. Use **role-based access control** (**RBAC**) to streamline access, and enforce strong passwords or adopt passwordless technologies. Encourage users not to use the same password more than once, and provide an enterprise-grade password management tool to provide more efficient password management. Require **multi-factor authentication** (**MFA**) to access systems and implement conditional access controls that allow MFA to be bypassed from company-compliant devices for a better user experience. Enable biometric authentication when available and consider an end goal of working towards a passwordless-authentication world.

> **Tip**
>
> If you don't have MFA enabled for all users, ensure this is your highest priority. According to Microsoft, enabling MFA can prevent over 99.9% of account compromise attacks: `https://www.microsoft.com/security/blog/2019/08/20/one-simple-action-you-can-take-to-prevent-99-9-percent-of-account-attacks/`.

Information about access management and identity protection can be found by reading *Chapter 5*, *Identity and Access Management*. We also cover privileged access models in *Chapter 11*, *Server Infrastructure Management*. The important areas to focus on include adopting a tiered model for privileged access and following Microsoft's privileged access strategy, which is aligned with Zero Trust principles. Always enforce the principle of least privilege when assigning permissions to users. This includes Active Directory's built-in roles and Azure Active Directory roles. To manage access to resources in Azure, use Azure RBAC. Furthermore, enhance access security for your privileged users by deploying the following solutions:

- **Privileged access management (PAM)**
- **Just-in-time access (JIT)**
- **Privileged identity management (PIM)**

These solutions provide a well-rounded privileged access administration program for both your traditional on-premises environment and your cloud environment. If you don't have any privileged management tools available, create a secondary account for these purposes and ensure you educate your users not to use the same passwords between accounts.

Enact a Zero Trust access model

Ensure you adopt a Zero Trust access architecture for your systems, identities, applications, and infrastructure where applicable. In *Chapter 1*, *Fundamentals of Windows Security*, we covered Zero Trust access and its value in securing your environment. This is a model in which we trust no one until we can validate who they are, where they are coming from, and confirm their authorization. This approach will require an access model that consists of multiple layers and can evaluate several facets in the authentication and authorization chain, from the network and firewall to the physical devices, down to the user's identity. Implementing cloud-based security technologies will significantly help if you are looking to adopt a Zero Trust access model. You can read more about the Zero Trust access model at Microsoft here: `https://docs.microsoft.com/en-us/security/zero-trust/`.

Define a security framework

In *Chapter 2, Building a Baseline*, we covered the adoption of a security framework to serve as the foundation of your organization's security program. It should consist of recommendations from widely adopted frameworks, such as the following:

- **Control Objectives for Information and Related Technology (COBIT)**: `https://www.isaca.org/resources/cobit`

- **ISO (International Standards Organization)** 27000 standards: `https://www.iso.org/isoiec-27001-information-security.html`

- **NIST (National Institute of Standards and Technology) Framework for Improving Critical Infrastructure Cybersecurity**: `https://www.nist.gov/cyberframework`

We also covered the importance of well-documented policies, standards, procedures, and guidelines as part of your security program. The framework should consist of one or more security baselines that outline a minimum set of configurations for your devices. The security program should be sponsored by leadership and promoted throughout the organization to help educate users about the importance of security and the part they play.

Get current and stay current

Get current and stay current with the latest feature builds and security updates for your Windows clients and servers. In *Chapter 11, Server Infrastructure Management*, and *Chapter 8, Keeping Your Windows Client Secure*, we covered infrastructure and end user device management tools that assist with keeping your devices up to date. In *Chapter 6, Administration and Policy Management*, we reviewed how to administer your devices to ensure they remain current and compliant. For example, enforcing a compliance evaluation to ensure your devices meet a minimum operating system build version is helpful to flag non-updated devices that might be at risk. You can even enforce additional security controls, such as the requirement of MFA, based on this compliance evaluation using a conditional access policy. Configure **Windows Update for Business (WufB)** on Windows devices and **Windows Server Update Services (WSUS)** or **Azure Update Management** for Windows servers to keep your devices patched. This will help ensure that your devices are as secure as possible against ongoing threats. In addition to updating the Windows operating system, other business applications such as Google Chrome, Microsoft Office, and Adobe products need to be kept up to date as well. Plan to incorporate third-party applications into your update strategy.

Make use of modern management tools

Use modern management tools to enforce security configurations and administration of devices. Enterprise-grade solutions, such as Microsoft Endpoint Configuration Manager and Intune, can enforce security baselines, perform compliance evaluations, deploy applications, apply device configurations, and manage software updates. Use tools such as the **Microsoft Deployment Toolkit** (**MDT**) and Configuration Manager to build hardened images and deploy task sequences for in-place upgrades or migrations. Reduce the number of tools, if applicable, to avoid complexities in your environment. Simplicity with a reduced footprint helps to reduce the number of vulnerabilities. We primarily covered the management of your server infrastructure and end user devices in *Chapter 11, Server Infrastructure Management*, and *Chapter 8, Keeping Your Windows Client Secure.*

Certify your physical hardware devices

For end user physical devices and any physical servers within your environment, ensure that the hardware specifications pass a hardware certification program and can support virtualization-based security features. In addition to this, ensure that a process to securely update hardware and device firmware is built into your documented baseline procedures. In *Chapter 3, Hardware and Virtualization*, we covered hardware certification in more detail. As a reminder, make sure you review the **Windows Server Catalog** and **Windows Hardware Compatibility List** before procuring any hardware for your Windows operating systems, from the following links:

- `https://www.windowsservercatalog.com/`
- `https://partner.microsoft.com/en-us/dashboard/hardware/search/cpl`

Administer network security

In *Chapter 4, Networking Fundamentals for Hardening Windows*, we covered network security for your Windows environment. Even with recent trends focusing on securing the user devices and identity, network security still plays a pivotal role. The function of network security should focus on network appliances, physical offices, and data centers and include network-related configurations for end user devices and servers. These configurations should be included in your documented security baselines and could include settings for software-based firewalls and VPNs. For Windows devices, communications can be locked down by configuring Windows Defender Firewall with Group Policy, Intune, or Configuration Manager. Enhanced network protection that can block traffic to risky hosts or inappropriate websites can be added by deploying a proxy server or VPN or using advanced features of Microsoft Defender for Endpoint.

For servers running in Azure, apply a network security group to the subnet or network interface resource, and only allow the necessary communications to pass through. As your users become more decentralized, ensure that you implement a reliable and secure VPN service, such as Microsoft's *Always On* VPN, which we covered in *Chapter 4, Networking Fundamentals for Hardening Windows*.

Always encrypt your devices

Always require encryption for end user devices and servers. This should also include mobile devices that store company data or could be used to open company documents and email. For Windows clients, BitLocker disk encryption can easily be deployed using Intune and Azure Active Directory, Configuration Manager, or with Active Directory and Group Policy. For virtual machines in Azure, leverage Azure Disk Encryption and Key Vault for key storage. Additionally, ensure that backups for critical systems are configured. We covered encryption in detail for both end user devices and servers in *Chapter 8, Keeping Your Windows Client Secure*, and *Chapter 12, Keeping Your Windows Server Secure*.

Enable XDR protection beyond EDR

XDR is an extremely valuable strategic approach to implement in your cybersecurity program. XDR expands beyond the original EDR capabilities focused primarily on a single technology to join multiple technologies together. This allows security teams to help improve detection by adding context to alerts and seeing a holistic picture of a potential attack. An XDR strategy will combine capabilities that cover end user devices, email, servers, cloud infrastructure, identity and access management, network and applications, and data. This primary benefit of XDR helps consolidate everything into a centralized view. For EDR capabilities, which have replaced the traditional antivirus model, ensure you onboard your workstations and servers into Microsoft Defender for Endpoint, which provides next-generation endpoint protection with behavioral detection, native cloud-based analytics, and threat intelligence. We covered XDR in more detail within *Chapter 14, Security Operations*. In *Chapter 8, Keeping Your Windows Client Secure*, and *Chapter 12, Keeping Your Windows Server Secure*, we covered the Microsoft Defender for Endpoint onboarding process.

Deploy security monitoring solutions

Having the right security tools in place is a critical part of your security program, but if you don't have well-implemented operations and monitoring, the value of your security tools diminishes. Being a Microsoft customer means taking advantage of the XDR capabilities available, which allow instant reaction and automated remediation of any detected incidents within your environment. Take advantage of the enterprise-class security monitoring and reporting solutions readily available including Log Analytics, Microsoft Defender for Endpoint, Defender for Cloud Apps, Microsoft Defender for Cloud, and Microsoft Sentinel. Many of these solutions integrate with third-party SIEM tools should they be used in-house or from an outsourced security operations center. In *Chapter 13*, *Security Monitoring and Reporting*, and *Chapter 14*, *Security Operations*, we covered security operations, security monitoring, and reporting in detail.

Notable mentions

In addition to hardening devices and the top 10 list, we want to highlight other important items of the overall security program.

Stay educated

Stay current on the ever-evolving threat landscape in today's world. It is important as a security professional that you are aware of and understand the complexities of current threats to ensure you are applying the appropriate remediations. This is one way to ensure that you can help reduce the risk of a compromise. The following is a list of some of the resources referenced in *Chapter 1*, *Fundamentals of Windows Security*, and are great places to visit for up-to-date cybersecurity trends and new and emerging threats:

- *DarkReading*: https://www.darkreading.com/
- *The Hacker News*: https://thehackernews.com/
- *Risky Business podcast*: https://risky.biz/
- *Microsoft Patch Tuesday Dashboard*: https://patchtuesdaydashboard.com/
- *Common Vulnerabilities and Exposures (CVE) List*: https://cve.mitre.org/cgi-bin/cvekey.cgi?keyword=Windows

Validate controls

In *Chapter 15, Testing and Auditing*, we reviewed in detail the testing and auditing of your environment. It is critical to validate that your controls are in place by regularly scheduling audits. Additionally, at a minimum, schedule ongoing vulnerability assessments and annual penetration tests to help find and mitigate any new risks discovered in your environment. Don't exclude validating controls from your security program, as it could be a fatal mistake.

Application controls

Ensure you are only allowing access to applications you trust within your environment. Plan for and deploy **Windows Defender Application Control (WDAC)** adaptive application controls with Defender for Servers or using Microsoft Defender Vulnerability Management for full fine-grained control over what applications can run on your systems for your users. We covered WDAC in more detail in *Chapter 9, Advanced Hardening for Windows Clients*.

Security baselines and hardening

Ensure you harden your end user devices and servers by configuring Microsoft security baselines, the **Security Technical Implementation Guide (STIG)**, and CIS benchmarks. The hard work of building recommended controls has already been done by Microsoft and communities of other security professionals, so you don't have to. Security baselines can be enforced using Group Policy and modern management solutions such as Microsoft Endpoint Manager. Leverage reporting and auditing features with device compliance policies and configuration profiles in Intune or by deploying configuration baselines in Configuration Manager. Don't forget to include security baseline policies for other enterprise-based apps such as Microsoft 365 Apps, Zoom, Adobe, Google Chrome, Edge, and other web browsers. We covered the fundamentals of security baselines and hardening in *Chapter 2, Building a Baseline*. In *Chapter 8, Keeping Your Windows Client Secure*, and *Chapter 12, Keeping Your Windows Server Secure*, we reviewed how to deploy these baselines and hardening configurations to your workstations and servers.

Business continuity, disaster recovery, and cyber incident response

Having a well-defined **business continuity plan (BCP)**, **disaster recovery plan (DRP)**, and **cyber incident response plan (CIRP)** helps ensure that your organization is prepared for impactful events. Plans should not only include covering business operations but also the ongoing and evolving security threats that can have catastrophic ramifications for your company. Threats such as ransomware can prevent organizations from being able to operate normally, can lead to extortion and ransom demands, and have the potential for large amounts of data loss. We briefly covered business continuity, disaster recovery, and cyber incident response in *Chapter 14, Security Operations*.

Hopefully, this overview summarized some of the key takeaways in this book. Our goal was to provide you with insights into the critical components that need to be focused on to best protect your Windows workstations and servers. Next, we want to provide some personal insight into the future of security and device management.

The future of device security and management

As the technology we consume evolves and the access model becomes more internet-centric, the better our security posture and defense must be. We need to completely shift the way security has been implemented in the past beyond the four walls of the office. Users are far more dynamic today than ever and need access to company data from anywhere at any time. This challenge also extends beyond corporate devices to personally-owned mobile devices and even to **bring-your-own (BYO)** laptop/tablet models. Ensuring that your corporate data is protected and not exfiltrating the environment requires effective modern security tools and well-defined controls. At the same time, it's important to ensure we don't inhibit end user productivity; otherwise, they will look to circumvent the controls put in place and create a more vulnerable environment.

To help succeed with implementing your overall security strategy, it's recommended to simplify where you can. Having disjoined security tools can make maintenance unsustainable and as a result, make you more vulnerable due to their complexities. Because of this, plan a review of what you have implemented and set goals to consolidate your security footprint where applicable. Simplicity is key to a successful program, and Microsoft has done a great job in this regard, having evolved its security presence over the years.

During your consolidation efforts, we recommend reviewing next-generation security tools. Traditional security tools will no longer suffice in today's modernized world. Next-generation security tools can be enabled and deployed at scale using cloud technologies with limited or zero infrastructure and are, typically, always kept up to date by the vendor. These tools and services should support a level of automation, leverage artificial intelligence, analyze big data, and incorporate behavioral analytics. Without these features, organizations will miss out on valuable security insights that can help prevent attacks as opposed to reacting to a breach.

As already mentioned, your protection strategy needs to continue to incorporate an identity- and device-focused foundation. As next-generation security tools continue to improve and evolve, always assess new features, and enable them when applicable. They will be able to provide intelligent security insights using cloud-driven telemetry that analyze your users' and devices' behavior based on their location and alert on any atypical travel or anomalies based on user activity. Layering automation that automatically remediates incidents based on these anomalies is also a significant step to improving the effectiveness of security operations and results in a more secure organization. In addition, most modern technology now supports biometric-based authentication and can leverage fingerprint scans and face recognition. These technologies are pivotal in creating a path to a world without passwords. If you haven't already heard of **Fast Identity Online 2 (FIDO2)**, you should quickly become familiar with it, as this specification is currently driving the passwordless initiative.

It's commonly been said that *"Company data is the crown jewels of the organization."* To protect data in an available-anywhere-from-any-device model, data protection needs to be considered to prevent leakages. To do this, continue to grow and evolve your **data loss prevention (DLP)** program using cloud-based technologies. Enhance your protection with **information rights management (IRM)** and data classification tools, such as Microsoft Purview Information Protection. Your organization's data should be automatically labeled and classified based on industry-standard privacy regulations, along with custom rules used to identify sensitive data unique to your business. Based on the data classification, there should be automatic protections applied that include the ability to enforce encryption, require authentication, block data from leaving your devices, and restrict copy and paste to non-protected apps.

Beyond protecting the classic Windows and server operating system is the **Internet of Things (IoT)**. IoT has grown exponentially in recent years and continues to grow, as devices are now being built for everything imaginable. Microsoft also has a presence in this space with its Windows for IoT platform, which includes multiple versions available for building your IoT infrastructure including Windows IoT Enterprise, Windows Server IoT 2022, and Windows 10 IoT Core Services.

> **Tip**
>
> You can learn more about Windows IoT here: `https://azure.microsoft.com/en-us/services/windows-iot/`.

As IoT continues to grow, serious security concerns are beginning to surface as we become more dependent on IoT devices in today's world. We need to define a unified standard to govern, manage, and secure these devices. This includes unifying the management of all devices for centralized governance and a standard security approach. Currently, we are not aware of a true unified model. However, we are hopeful that this is something that may become available as the adoption of IoT continues to grow. The following is a diagram of the ideal future unified management model:

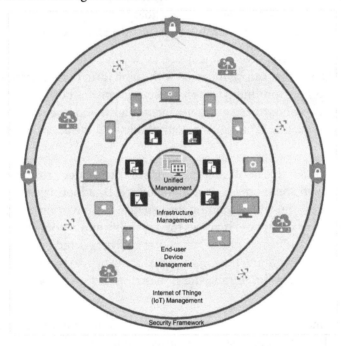

Figure 16.1 – The ideal unified management model in the future

Next, we will discuss security and the future.

Security and the future

We wanted to share some thoughts on the growth and future of security and the importance it will play in a world that becomes more connected every day. Technology continues to evolve at a significant pace. As this technology grows, security needs to be brought to the forefront, not only within the enterprise but also within the consumer space. Devices, gadgets, lightbulbs, health monitors, household appliances, entertainment, landscaping equipment, automobiles, accessories, and drones are all examples of the types of internet-connected *things* we can use today. Unfortunately, from a security perspective, usability is typically the primary design focus of these smart devices. Their internet connectedness, lack of security standards, and heavy usage in our daily lives expose a significant security risk. The following link provides some real-world examples of security weaknesses and challenges with IoT technology: `https://www.conosco.com/blog/iot-security-breaches-4-real-world-examples/`.

Hopefully, as we continue to evolve in this space, we will see the creation of a more universal standardization that can be followed with some form of certification showing whether a device meets the minimum security specifications for both enterprise and consumer usage. A few standardized examples include a PIN number for debit and credit cards, a fingerprint/face ID to unlock your phone, and the adoption of MFA across many services.

As highlighted throughout this book, shifting your security strategy to leverage cloud technologies will help you be more efficient in the future. By adopting the use of next-generation technologies, you will be gaining the benefit of an environment that has little to zero self-managed infrastructure. It allows for scalability and automation, makes use of artificial intelligence, and incorporates behavioral analytics using big data. This model is most suited for companies going through a digital transformation to cloud-based infrastructure and supports a decentralized user base that requires work access from anywhere at any time.

Moving forward, no matter the size of your organization or business, a security presence should be required. For a smaller business, security in the form of an outsourced model that leverages a **managed security service provider** (**MSSP**) might make more sense than hiring in-house. Having an MSSP available will give you the necessary resources to provide the expertise needed to handle security-related incidents. Larger organizations may opt for an in-house security team, but many MSSPs can cater to larger organizations as well. Data released by Frost & Sullivan predicts that the market for MSSPs will grow from $12.01 billion in 2020 to $18.81 billion by 2024: `https://www.msspalert.com/cybersecurity-research/managed-security-services-market-forecast/`.

Services considered as critical infrastructure, which includes **operational technology (OT)**, are another area that should be at the forefront of developing security-based technologies. Examples of critical infrastructure services include energy, emergency services, chemical and nuclear facilities, transportation, and the government sector. These are essential services that support our daily lives, and even a minor disruption could be catastrophic. We need to ensure these critical components used in everyday life are highly protected, as the stakes are too high.

Transportation services, from public to private to space travel, all involve internet-connected technology. These services are used by millions of people daily and any compromise could potentially diminish the safe operations of these vehicles and result in significant damage, injury, or loss of life. For example, in 2015, security researchers were able to remotely commandeer a Jeep Cherokee driving on the highway and disable the car's engine and brake systems: `https://www.wired.com/2015/07/hackers-remotely-kill-jeep-highway/`.

As our personal *digital* identity continues to evolve and be enhanced, where do we go beyond this? FIDO2 usage continues to grow and the benefits of adopting a passwordless approach have become obvious. Are we heading towards a unified identity where a single item we carry represents our *self*? We already see this, as smartphones can supplement car keys, wallets, digital identification, and more. The concept of Web3 introduces a decentralized internet model. Here, *you* control your digital identity and not the big tech companies that have profited tremendously by scraping your data and selling it to advertisers. Web3 projects a fundamental change in how we transact as our digital self in the future.

Recently, although considered controversial, the topic of microchip implants has become more than just a conversation. Neuralink, a company of which Elon Musk is a founder, has recently promoted beginning human trials on their brain chip implant by the end of 2022. Neuralink hopes to help people that suffer from neurological deficiencies by repairing certain cognitive and sensory-motor functions controlled by the brain. Will this ever become a reality or requirement for humans? Perhaps we will live in a world where a microchip could become mandated. Only time will tell.

Another interesting area with significant technological advances is that of robotics and autonomy. What does the future hold, and where do we draw the line in terms of how markedly intelligent robots can become. We all have most likely watched futuristic movies that entail robots becoming smarter than humans with the strength to overpower humanity. Could this ever become a reality? Could robots become programmed or compromised to do more harm than good? These are real conversations happening now and it's critical that we build a solid, core security model that includes protection against these threats as robotic technology continues to evolve. There should be no failure of security in this space.

From our discussion, the importance of security from a holistic approach should be clear, that is, one that does not overlook any area of the infrastructure, the physical device, or the underlying software down to the user identity. Security should be at the forefront when designing any solution and should be natively embedded into the product from the beginning. Failure to incorporate security shouldn't be a risk your company is willing to take.

Summary

In this chapter, we provided an overview of the 10 most important takeaways from the content of this book. We also covered additional items to keep in mind as you continue to harden and secure your Windows workstations and servers. Each of these items includes a reference back to the original chapter, where you can review the material in more detail to gain a better understanding.

We then provided our personal insights into the future of device security and management. Here, we covered a few essential areas related to securing devices, as well as the importance of security management in the IoT space. We finished the chapter with our thoughts on security and the future, especially as they relate to the ever-evolving innovation of new and futuristic technologies.

This chapter concludes the subjects in this book. We hope you enjoyed the content provided and were able to take away the necessary knowledge to help secure and strengthen your environment.

Index

Other Books You May Enjoy

If you enjoyed this book, you may be interested in these other books by Packt:

Cybersecurity Career Master Plan

Dr. Gerald Auger, Jaclyn "Jax" Scott, Jonathan Helmus, Kim Nguyen

ISBN: 9781801073561

- Gain an understanding of cybersecurity essentials, including the different frameworks and laws, and specialties

- Find out how to land your first job in the cybersecurity industry

- Understand the difference between college education and certificate courses

- Build goals and timelines to encourage a work/life balance while delivering value in your job

- Understand the different types of cybersecurity jobs available and what it means to be entry-level

- Build affordable, practical labs to develop your technical skills

- Discover how to set goals and maintain momentum after landing your first cybersecurity job

Mobile App Reverse Engineering

Abhinav Mishra

ISBN: 9781801073394

- Understand how to set up an environment to perform reverse engineering
- Discover how Android and iOS application packages are built
- Reverse engineer Android applications and understand their internals
- Reverse engineer iOS applications built using Objective C and Swift programming
- Understand real-world case studies of reverse engineering
- Automate reverse engineering to discover low-hanging vulnerabilities
- Understand reverse engineering and how its defense techniques are used in mobile applications

Packt is searching for authors like you

If you're interested in becoming an author for Packt, please visit `authors.packtpub.com` and apply today. We have worked with thousands of developers and tech professionals, just like you, to help them share their insight with the global tech community. You can make a general application, apply for a specific hot topic that we are recruiting an author for, or submit your own idea.

Share Your Thoughts

Now you've finished *Mastering Windows Security and Hardening*, we'd love to hear your thoughts! Scan the QR code below to go straight to the Amazon review page for this book and share your feedback or leave a review on the site that you purchased it from.

`https://packt.link/r/180323654X`

Your review is important to us and the tech community and will help us make sure we're delivering excellent quality content.